MW00760555

Foundations of Engineering Mechanics

Series editors

V.I. Babitsky, Loughborough, Leicestershire, UK
Jens Wittenburg, Karlsruhe, Germany

More information about this series at http://www.springer.com/series/3582

Leonid I. Manevitch · Agnessa Kovaleva
Valeri Smirnov · Yuli Starosvetsky

Nonstationary Resonant Dynamics of Oscillatory Chains and Nanostructures

 Springer

Leonid I. Manevitch
Institute of Chemical Physics
Russian Academy of Science
Moscow
Russia

Agnessa Kovaleva
Space Research Institute
Russian Academy of Science
Moscow
Russia

Valeri Smirnov
Institute of Chemical Physics
Russian Academy of Science
Moscow
Russia

Yuli Starosvetsky
Technion—Israel Institute of Technology
Faculty of Mechanical Engineering
Haifa
Israel

ISSN 1612-1384   ISSN 1860-6237   (electronic)
Foundations of Engineering Mechanics
ISBN 978-981-10-4665-0   ISBN 978-981-10-4666-7   (eBook)
DOI 10.1007/978-981-10-4666-7

Library of Congress Control Number: 2017943824

Printed on acid-free paper

This Springer imprint is published by Springer Nature
The registered company is Springer Nature Singapore Pte Ltd.
The registered company address is: 152 Beach Road, #21-01/04 Gateway East, Singapore 189721, Singapore

Acknowledgements

We are grateful to many colleagues who shared their views and discussed related problems with us. We are indebted to L.A. Bergman, O.V. Gendelman, I.P. Koroleva(Kikot), Yu.A. Kosevich, M.A. Kovaleva, C.-H. Lamarque, Yu.V. Mikhlin, A.I. Musienko, F. Pellicano, V.N. Pilipchuk, F. Romeo, A.V. Savin, D.S. Shepelev, M. Strozzi, and A.F. Vakakis for their collaboration. This book could not appear without their friendly help. The comments from A.I. Manevich, who read a part of the manuscript, are also highly appreciated. We would like to express our gratitude to E. Gusarova and E. Manevitch for their help in the manuscript preparation.

We also gratefully acknowledge financial support from the Russian Foundation for Basic Research (grants 16-02-00400 and 17-01-00582) and the Russian Science Foundation (grant 16-13-10302) the latter for the support of writing Chaps. 12, 14, 15 in the Part 3.

Book Homepage

We encourage all who wish to comment on the book to send e-mails to:
manevitchleonid3@gmail.com
agnessa_kovaleva@hotmail.com
smirnovvv@gmail.com
staryuli@technion.ac.il
All misprints and errors will be posted on the book homepage.

Contents

Abstract

We present a recently developed concept of limiting phase trajectories (LPTs) allowing a unified description of resonance highly non-stationary processes for a wide range of classical and quantum finite-dimensional dynamic systems with constant and varying parameters. This concept provides a far-going extension and adequate mathematical description of the well-known linear beating phenomenon in the systems of two weakly coupled oscillators to a diverse variety of nonlinear models. Contrary to stationary and non-stationary, non-resonance oscillations, described efficiently in the framework of nonlinear normal mode (NNMs) concept, the non-stationary resonance processes under consideration are characterized by strong modulation and intense energy exchange between different parts of the system. They include energy exchange in multi-particle systems, targeted energy transfer, non-stationary vibrations of carbon nanotubes, nonlinear quantum tunneling, autoresonance, and non-conventional synchronization. Besides the nonlinear beats, the LPT concept allows us to find the conditions of transition from intense energy exchange to strong energy localization. A special mathematical technique based on the non-smooth temporal transformations leads to the clear and simple description of strongly modulated regimes. The role of LPT as the fundamental non-stationary resonance process corresponding to the maximum possible energy exchange between different parts of the system turns out to be similar to that of NNMs in the stationary theory.

Keywords Nonlinear normal mode · Limiting phase trajectory · Weakly coupled nonlinear oscillators · Energy transfer and localization · Synchronization · Autoresonance · Quantum tunneling · Carbon nanotubes

Introduction

For a sufficiently long time, the main analytical tools for the study of both stationary and non-stationary nonlinear processes in the finite-dimensional systems, as well as the study of their instability and bifurcations, have been associated with the concept of nonlinear normal modes. Rosenberg (1960, 1962, 1966) defined a nonlinear normal mode (NNM) of an undamped discrete multi-particle system as synchronous periodic oscillation, where the displacements of all the material points of the system reach their extreme or zero values simultaneously. Then, the NNM concept was generalized on the autogenerators, and systems subjected to external force (see, e.g., (Manevitch et al. 1989; Vakakis et al. 1996) and references therein) to describe the forced stationary oscillations. In the autogenerators, a NNM can be considered as an attractor that corresponds to a synchronized motion of self-sustained non-conservative oscillators (Rand and Holmes 1980). The important features that distinguish NNMs from their linear counterparts are: (i) the superposition principle turns out to be invalid and (ii) there can be more NNMs than the number of the degrees of freedom the system has. In this case, the essentially nonlinear modes (they do not have an analogue in linear theory) are generated through the NNM bifurcations, breaking the symmetry of the dynamics and resulting in the nonlinear energy localization (motion confinement). This phenomenon is directly related to the strong asymmetry caused by the nonlinearity. The manifestation of spatially localized NNMs is extremely important in different fields of nonlinear physics. The excitations of this type determine elementary mechanisms of many physical processes and make a significant contribution into the thermal capacity. Such excitations exist, for instance, in molecules and polymer crystals (Ovchinnikov 1999).

Therefore, if we deal with stationary processes, a comprehensive theory for their analysis including analysis of their stability and bifurcations has been developed. Additionally, the combinations of non-resonant NNMs in quasi-linear systems can be used for the asymptotic description of non-stationary processes (Nayfeh and Mook 2004), the only difference being that the amplitudes in the nonlinear systems depend on the frequencies. However, when a nonlinear system is in resonance, intensive energy exchange and transport are characterized by a complicated

non-stationary behavior, which makes any analytical investigation very complicated. As a result, numerical studies are dominant in this field.

In this book, we discuss a new concept that allows us to derive a unified
description of such processes in finite-dimensional models. The main idea of the
concept is to introduce a fundamental non-stationary process that corresponds to the
maximum possible (under given conditions) energy exchange between weakly
coupled oscillators or different parts of the system (coherence domains). Such a
process is referred to as a limiting phase trajectory (LPT) (Manevitch 2005, 2007,
2009). In order to highlight the necessity for this new analytical background, we
discuss first the current state of the non-stationary resonance problems starting from
the beats in the system of two weakly coupled linear oscillators.

1. Beats between two weakly coupled classical linear oscillators represent the
   simplest example of energy exchange between the parts of an oscillator array. In
   quantum mechanics, the beats correspond to periodic oscillations of the probabilities for finding the system in one of two basis states (in the two-level
   model). Due to the superposition principle, the linear beats in the case of two
   degrees of freedom can be described as a combination of two normal modes
   corresponding to the stationary states. But this representation is not applicable
   for nonlinear oscillators wherein the superposition principle does not hold. The
   absence of an adequate analytical description of nonlinear strongly modulated
   processes prevents the understanding of their specifics and restricts the analysis
   by the numerical results only, which do not allow further theoretical general-
   izations. For instance, until recently there were no extensions of beat phenomena
   to multi-particle systems. In the framework of linear physics, energy transfer is
   usually associated with the propagation of the wave packets, that is, of the
   combinations of many normal modes with close frequencies. Typically, a wave
   packet gradually spreads due to dispersion (when the phase velocity of the
   normal modes depends on the frequency). The quantum harmonic oscillator, for
   which the linear Schrödinger equation has the solution in the form of a wave
   packet, provides a unique example of a non-spreading non-stationary process (in
   the presence of the force field). Such a non-spreading wave packet describes the
   non-stationary coherent state that is similar to the classical harmonic oscillator.
   The reason for this phenomenon is that the combination of the wave functions
   with close energies may result in their mutual compensation everywhere except
   for a relatively narrow area in which the "amplification" of the excitation occurs
   forming the localized profiles. This mechanism cannot be realized in the
   finite-dimensional nonlinear models due to dispersion. Therefore, the idea to
   present non-stationary energy transfer in such models as combinations of stationary processes turns out to be non-productive.

2. A special attention should be given to the problem of targeted energy transfer
   (TET), where energy of some form is directed from a source (donor) to a
   receiver (recipient) in a one-way irreversible fashion. Many processes in nature
   involve some type of energy transfer. It is not surprising that TET phenomena
   have received much attention in applications from diverse fields of applied

mathematics, applied physics, and engineering. Applications of energy localization and TET in diverse areas are given, e.g., in Dauxois et al. (2004) and Vazquez et al. (2003), including both theoretical and experimental results. Representative examples are given in earlier works on targeted energy transfer between nonlinear oscillators and/or discrete breathers (Aubry et al. 2001; Kopidakis et al. 2001), wherein the notion of TET was first introduced, and following works on the TET manifestation in physical systems (Maniadis et al. 2004; Maniadis and Aubry 2005; Morgante et al. 2002). The dynamic mechanisms considered in these works are based on imposing conditions of nonlinear resonance between interacting dynamic systems to achieve TET from one to another and then "breaking" this condition at the end of the energy transfer to make it irreversible. This principle was further employed in a wide variety of mechanical applications. A lightweight mass attachment referred to as a nonlinear energy sink (NES) was intensively studied over past two decades; a large number of examples can be found in monographs (Vakakis 2010; Vakakis et al. 2008), where the extensive bibliography is presented. The results of Vakakis et al. (2008) demonstrated that *the almost irreversible energy transfer in the systems coupled with NES could not be analytically described in the framework of the traditional approaches.* Recent researches (Kikot et al. 2015; Kovaleva and Manevitch 2010, 2013; Manevitch 2007, 2009, 2012; Manevitch and Gendelman 2011; Manevitch et al. 2007, 2010, 2011a, b; Manevitch and Kovaleva 2013; Manevitch and Musienko 2009; Manevich and Smirnov 2010a, b, c; Manevitch and Vakakis 2014; Smirnov and Manevitch 2011; Starosvetsky and Ben-Meir 2013; Starosvetsky and Manevitch 2011; Savadkoohi et al. 2011; Vaurigaud et al. 2011; Zhen et al. 2016) have proved that the explicit analytical results for impulsively loaded structures as well as for nonlinear oscillators with an external source of energy can be obtained with the help of the recently developed LPT methodology.

3. An analogy between classical and quantum energy transport is an interesting aspect for discussion. Examples of this analogy can be found in Hasan et al. (2013); Kosevich et al. (2010); Kovaleva and Manevitch (2012); Kovaleva et al. (2011); Manevitch et al. (2011c); Maniadis et al. (2004); Raghavan et al. (1999). In quantum mechanics, the transition between adjacent levels is realized by quantum tunneling, first analyzed contemporaneously but independently in (Landau 1932; Majorana 1932; Stückelberg 1932; Zener 1932). Generalization of the tunneling theory onto nonlinear quantum systems, such as superconducting Josephson junctions and Bose-Einstein condensates, is particularly interesting (Liu et al. 2002; Trimborn et al. 2010; Zelenyi et al. 2013; Zobay and Garraway 2000). However, existing analytical techniques that enable the calculation of the asymptotic values of transitional probabilities at large times are insufficient to describe tunneling at the finite times with arbitrary initial conditions.

4. Autoresonance (AR) in nonlinear oscillators, which is manifested as the growth of the resonance amplitude caused by periodic forcing with slowly varying frequency, has numerous applications in different fields of physics (see, e.g.,

Andronov et al. 1966; Barth and Friedland 2013; Ben-David et al. 2006; Bohm and Foldy 1947; Chacón 2005; Chapman 2012; Fajans and Friedland 2001; Kalyakin 2008; Marcus et al. 2004; McMillan 1945; Veksler 1944). After the first studies for the purposes of particle acceleration, AR has become a very active field of research. It was noticed in early works (Bohm and Foldy 1947; Livingston 1954; McMillan 1945; Veksler 1944) that the physical mechanism underlying AR can be interpreted as an adiabatic nonlinear phase-locking between the system and the driving signal. It is directly connected with the inherent property of nonlinear systems to remain in resonance when the driving frequency varies in time. However, an analytical quantitative analysis of the transition from bounded oscillations to AR has not been done yet. The only known critical threshold was obtained numerically, but its applicability to a large class of physical problems is questionable and can be a subject of a separate discussion. Moreover, this threshold is unusable in the case of the adiabatic AR (Kovaleva and Manevitch 2013a, b).

5. Synchronization is a fundamental problem discussed over the centuries, starting from Huygens' famous observations; an extensive bibliography and discussion of all known results can be found in, e.g., Pikovsky et al. (2001). Synchronization in the chains of coupled Van der Pol or Van der Pol-Duffing oscillators has numerous applications in different fields of physics and biophysics (Vazquez et al. 2003; Enjieu Kadjia and Yamapi 2006; Kovaleva M. 2013; Manevitch et al. 2013; Mendelowitz et al. 2009; Pankratova and Belykh 2012; Rand and Holmes 1980; Peles and Wiesenfeld 2003; Pikovsky et al. 2001; Rompala et al. 2007). Despite the long history, several questions are still open even for these benchmark models. In particular, the previous studies considered synchronization of these systems which is similar by spatial profile to nonlinear normal mode and did not discuss the possibility of the alternative types of synchronization.

6. Up to now our knowledge of the nonlinear dynamics of carbon nanotubes (CNTs) is very poor. The reasons are coupled with the limited possibilities of the numerical methods (including the *ab initio* quantum mechanical calculations and molecular dynamic simulations) as well as the strong complexity of the various continuous models of the CNTs. Such models allow us to estimate the linear spectrum of the CNT oscillations, but they seem as unsolvable for the analytical study of the nonlinear processes. Therefore, a few works in this area (Shi et al. 2009; Li and Shi 2008; Soltani et al. 2011, 2012) develop the numerical methods in the framework of the various approximations.

In this book, we present new concepts and methods of analysis, which provide the answers to the following open questions:

1. What is an adequate description of nonlinear beating?
2. Is it possible to generalize the concept of the beats on the linear and nonlinear multi-particle systems?

3. What are the conditions of the transition from intense energy transfer to energy localization in finite discrete systems?
4. What conditions provide irreversible energy transfer in time-dependent classical and quantum systems?
5. What are the conditions of transition from irreversible energy transfer (tunneling) to autoresonance in an oscillator with slowly varying frequency of the external excitation?
6. Can a non-conventional synchronization of weakly coupled autogenerators exist?
7. What are the basic non-stationary processes in the single-walled carbon nanotubes?

Recent studies have proved the answers to the questions above can be obtained in the framework of the LPT concept.

This book is organized as follows: in the first part, we introduce the LPT concept by analyzing finite-dimensional conservative models relating to the oscillators and oscillator arrays with constant parameters. These models describe the nonlinear beats in quasi-linear and strongly nonlinear systems of weakly coupled oscillators as well as in quasi-linear oscillatory chains.

The second part is devoted to the generalization of the LPT concept onto non-conservative systems with constant and variable parameters.

In the third part, we discuss the applications of the LPT concept to important mechanical and physical problems.

To capture a wider audience, the authors tried to set up the book in such a way that different chapters of the book could be read independently.

References

Andronov, A.A., Vitt, A.A., Khaikin, S.E.: Theory of Oscillators. Dover, New York (1966)

Aubry, S., Kopidakis, G., Morgante, A.M., Tsironis, G.P.: Analytic conditions for targeted energy transfer between nonlinear oscillators or discrete breathers. Physica B **296**, 222–236 (2001)

Barth, I., Friedland, L.: Two-photon ladder climbing and transition to autoresonance in a chirped oscillator. Phys. Rev. A **87**, 053420, 1–4 (2013)

Ben-David, O., Assaf, M., Fineberg, J., Meerson, B.: Experimental study of parametric autoresonance in Faraday waves. Phys. Rev. Lett. **96**, 154503, 1–4 (2006)

Bohm, D., Foldy, L.L.: Theory of the synchro-cyclotron. Phys. Rev. **72**, 649–661 (1947)

Dauxois, T., Litvak-Hinenzon A., MacKay R., Spanoudaki A. (eds.): Energy Localization and Transfer, World Scientific, Singapore (2004)

Enjieu Kadjia, H.G., Yamapi, R.: General synchronization dynamics of coupled Van der Pol–Duffing oscillators. Physica A **370**, 316–328 (2006)

Fajans, J., Friedland, L., Autoresonant nonstationary excitation of pendulums, Plutinos, plasmas, and other nonlinear oscillators. Am. J. Phys. **69**, 1096–1102 (2001)

Kalyakin, L.A.: Asymptotic analysis of autoresonance models. Russ. Math. Surv. **63**, 791–857 (2008)

Kopidakis, G., Aubry, S., Tsironis, G.P.: Targeted energy transfer through discrete breathers in nonlinear systems. Phys. Rev. Lett. **87**, 165501, 1–4 (2001)

Kovaleva, A.: Resonance energy transport in an oscillator chain. http://arxiv.org/submit/1151212 (2012)

Kovaleva, A., Manevitch, L.I.: Autoresonance energy transfer versus localization in weakly coupled oscillators. http://arxiv.org/abs/1410.6098

Kovaleva, A., Manevitch, L.I.: Classical analog of quasilinear Landau-Zener tunneling. Phys. Rev. E 85, 016202, 1–8 (2012)

Kovaleva, A., Manevitch, L.I.: Resonance energy transport and exchange in oscillator arrays. Phys. Rev. E 88, 022904, 1–10 (2013)

Kovaleva, A., Manevitch, L.I.: Emergence and stability of autoresonance in nonlinear oscillators. Cybernetics and Physics 2, 25–30 (2013)

Kovaleva, A., Manevitch, L.I.: Limiting phase trajectories and emergence of autoresonance in nonlinear oscillators. Phys. Rev. E 88, 024901, 1–6 (2013)

Kovaleva, A., Manevitch, L.I., Kosevich, Yu.A.: Fresnel integrals and irreversible energy transfer in an oscillatory system with time-dependent parameters. Phys. Rev. E 83, 026602, 1–12 (2011)

Koroleva, (Kikot) I.P., Manevitch, L.I., Vakakis, A.F.: Non-stationary resonance dynamics of a nonlinear sonic vacuum with grounding supports. J. Sound Vibr. 357, 349–364 (2015)

Landau, L.D.: ZurTheorie der Energieubertragung. II. Phys. Z. Sowjetunion. 2, 46–51 (1932)

Li, Q.M., Shi, M.X.: Intermittent transformation between radial breathing and flexural vibration modes in a single-walled carbon nanotube. Proc. R. Soc. A 464, 1941 (2008)

Liu, J., Fu, L., Ou, B.-Y., Chen, S.-G, Choi, D.-Il, Wu, B., Niu, Q.: Theory of nonlinear Landau-Zener tunneling. Phys. Rev. A 66, 023404, 1–7 (2002)

Livingston, M.S.: High-energy Particle Accelerators. Interscience, New York (1954)

Mahdavi, M., Jiang, L.Y., Sun, X.: Nonlinear vibration of a double-walled carbon nanotube embedded in a polymer matrix. Physica E 43, 1813–1819 (2011)

Majorana, E.: Atomiorientati in campo magneticovariabile, NuovoCimento 9, 43–50 (1932)

Manevitch, L.I., Gendelman, O.: Tractable Models of Solid Mechanics. Springer-Verlag, Berlin, Heidelberg (2011)

Manevitch, L.I., Gourdon, E., Lamarque, C. H.: Towards the design of an optimal energetic sink in a strongly inhomogeneous two-degree-of-freedom system. ASME J. Appl. Mech. 74, 1078–1086 (2007)

Manevitch, L.I., Kovaleva, A.: Nonlinear energy transfer in classical and quantum systems. Phys. Rev. E 87, 022904, 1–12 (2013)

Manevitch, L.I., Kovaleva, A., Manevitch, E.L., Shepelev, D.S.: Limiting phase trajectories and non-stationary resonance oscillations of the Duffing oscillator. Part 1. A non-dissipative oscillator. Commun. Nonlinear Sci. Numer. Simulat. 16, 1089–1097 (2011)

Manevitch, L.I., Kovaleva, A., Manevitch, E.L., Shepelev, D.S.: Limiting phase trajectories and non-stationary resonance oscillations of the Duffing oscillator. Part 2. A dissipative oscillator. Commun. Nonlinear Sci. Numer. Simulat. 16, 1098–1105 (2011)

Manevitch, L.I., Kovaleva, A., Shepelev, D.S.: Non-smooth approximations of the limiting phase trajectories for the Duffing oscillator near 1:1 resonance. Physica D 240, 1–12 (2011)

Manevitch, L.I., Kovaleva, M.A., Pilipchuk, V.N.: Non-conventional synchronization of weakly coupled active oscillators. Europh. Lett. 101, 50002, 1–5 (2013)

Manevitch, L.I., Kosevich, Y.A., Mane, M., Sigalov, G., Bergman, L.A., Vakakis, A.F.: Towards a new type of energy trap: classical analog of quantum Landau-Zener tunneling. Int. J. Nonlinear Mech. 46, 247–252 (2011)

Manevitch, L.I., Mikhlin, Yu.V., Pilipchuk, V.N.: The Normal Vibrations Method for Essentially Nonlinear Systems. Nauka Publ., Moscow (1989) (in Russian)

Manevitch, L.I., Musienko, A.I.: Limiting phase trajectories and energy exchange between anharmonic oscillator and external force. Nonlinear Dynam. 58, 633–642 (2009)

Manevitch, L.I., Smirnov, V.V.: Limiting phase trajectories and the origin of energy localization in nonlinear oscillatory chains. Phys. Rev. E 82, 036602, 1–9 (2010)

Manevitch, L.I., Smirnov, V.V.: Resonant energy exchange in nonlinear oscillatory chains and limiting phase trajectories: from small to large systems. In: Vakakis, A.F. (ed.) Advanced Nonlinear Strategies for Vibration Mitigation and System Identification. CISM Courses and Lectures **518**, pp. 207–258. Springer, Wien New York (2010)

Manevich, L.I., Smirnov, V.V.: Limiting phase trajectories and thermodynamics of molecular chains. Physics-Doklady. **55**, 324–328 (2010)

Maniadis, P., Kopidakis, G., Aubry, S.: Classical and quantum targeted energy transfer between nonlinear oscillators. Physica D **188**, 153–177 (2004)

Maniadis, P., Aubry, S.: Targeted energy transfer by Fermi resonance. Physica D **202**, 200–217 (2005)

Marcus, G., Friedland, L., Zigler A.: From quantum ladder climbing to classical autoresonance. Phys. Rev. A **69**, 013407, 1–5 (2004)

Masana R., Daqaq M. F.: Energy harvesting in the super-harmonic frequency region of a twin-well oscillator. J. Appl. Phys. **111**, 044501, 1–11 (2012)

McMillan, E.M.: The synchrotron—a proposed high energy particle accelerator. Phys. Rev. **68**, 143–144 (1945)

Mendelowitz, L., Verdugo, A., Rand, R.: Dynamics of three coupled limit cycle oscillators with application to artificial intelligence. Commun. Nonlinear Sci. Numer. Simul. **14**, 270–283 (2009)

Morgante, A.M., Johansson M., Aubry, S., Kopidakis, G.: Breather-phonon resonances in finite-size lattices: 'phantom breathers'?. J. Phys. A: Math. Gen. **35**, 4999–5021 (2002)

Nayfeh, A.H., Mook, D.T.: Nonlinear Oscillations, Wiley-VCH, Weinheim (2004)

Neishtadt, A.I.: Passage through a separatrix in a resonance problem with slowly varying parameter. J. Appl. Math. Mech. **39**, 594–605 (1975)

Ovchinnikov, A.A., Flach, S.: Discrete breathers in systems with homogeneous potentials: analytic solutions. Phys. Rev. Lett. **83**, 248–251 (1999)

Pankratova, E.V., Belykh, V.N.: Synchronization of self-sustained oscillators inertially coupled through common damped system. Phys. Lett. A **376**, 3076–3084 (2012)

Peles, S., Wiesenfeld, K.: Synchronization law for a Van der Pol array. Phys. Rev. E **68**, 026220, 1–8 (2003)

Pikovsky, A., Rosenblum, M., Kurths, J.: Synchronization: A Universal Concept in Nonlinear Sciences, Cambridge University Press (2001)

Rand, R.H., Holmes, P.J.: Bifurcation of periodic motions in two weakly coupled Van der Pol oscillators. Int. J. Nonlinear Mech. **15**, 387–399 (1980)

Rompala, K., Rand, R., Howland, H.: Dynamics of three coupled Van der Pol oscillators, with application to circadian rhythms. Commun. Nonlinear Sci. Numer. Simul. **12**, 794–803 (2007)

Savadkoohi, A.T., Manevitch, L.I., Lamarque, C.-H.: Analysis of the transient behavior in a two DOF nonlinear system. Chaos, Solitons & Fractals **44**, 450–463 (2011)

Smirnov, V.V., Gendelman, O.V., Manevitch, L.I.: Front propagation in a bistable system: how the energy is released. Phys. Rev. E **89**, 050901, 1–5 (2014)

Smirnov, V.V., Manevitch, L.I.: Limiting phase trajectories and dynamic transitions in nonlinear periodic systems. Acoust. Phys. **57**, 271–276 (2011)

Shi, M.X., Li, Q.M., Huang, Y.: Internal resonance of vibrational modes in single-walled carbon nanotubes. Proc. Royal Society. A **465**(2110), 3069–3082 (2009)

Soltani, P., Ganji, D.D., Mehdipour, I. Farshidianfar, A.: Nonlinear vibration and rippling instability for embedded carbon nanotubes. J. Mech. Sci. Tech. **26**, 985–992 (2012)

Soltani, P., Saberian, J., Bahramian, R., Farshidianfar, A.: Nonlinear free and forced vibration analysis of a single-walled carbon nanotube using shell mode. IJFPS. **1**, 47 (2011)

Starosvetsky, Y., Ben-Meir, Y.: Nonstationary regimes of homogeneous Hamiltonian systems in the state of sonic vacuum. Phys. Rev. E. **87**, 062919, 1–18 (2013)

Starosvetsky, Y., Manevitch, L.I.: Nonstationary regimes in a Duffing oscillator subject to biharmonic forcing near a primary resonance. Phys. Rev. E **83**, 046211, 1–14 (2011)

Stückelberg, E.C.G.: Theorie der unelastischenStössezwischenAtomen. Helv. Phys. Acta **5**, 369–422 (1932)

Trimborn, F., Witthaut, D., Kegel, V., Korsch, H. J.: Nonlinear Landau–Zener tunneling in quantum phase space. New J. Phys. **12**, 05310, 1–20 (2010)

Vakakis A.F. (ed.): Advanced Nonlinear Strategies for Vibration Mitigation and System Identification. CISM Courses and Lectures. **518**, Springer, Wien New York (2010)

Vakakis, A.F., Gendelman, O., Bergman, L.A., McFarland, D.M., Kerschen, G., Lee, Y.S.: Nonlinear Targeted Energy Transfer in Mechanical and Structural Systems. Springer, Berlin New York (2008)

Vakakis, A.F., Manevitch, L.I., Mikhlin, Yu.V., Pilipchuk, V.N., Zevin, A.A.: Normal Modes and Localization in Nonlinear Systems. Wiley, New York (1996)

Vaurigaud, B., Manevitch, L.I., Lamarque, C.-H.: Suppressing aeroelastic instability in a suspension bridge using a nonlinear absorber. In: Wiercigroch M., Rega G. (eds.) Proceedings of the IUTAM Symposium on Nonlinear Dynamics for Advanced Technologies and Engineering Design, pp. 263–277. Springer, Netherlands (2013)

Vazquez, L., MacKay, R., Zorzano, M.P. (eds.): Localization and Energy Transfer in Nonlinear Systems. World Scientific, Singapore (2003)

Veksler, V.I.: Some new methods of acceleration of relativistic particles. Comptes Rendus (Dokaldy) de l'Academie Sciences de l'URSS. **43**, 329–331 (1944)

Zelenyi, L.M., Neishtadt, A.I., Artemyev, A.V., Vainchtein, D.L., Malova, H.V.: Quasiadiabatic dynamics of charged particles in a space plasma. Physics-Uspekhi **56**, 347–394 (2013)

Zener, C.: Non-adiabatic crossing of energy levels. Proc. R. Soc. London A **137**, 696–702 (1932)

Zhen Zhang, I., Koroleva, I., Manevitch, L.I., Bergman, L.A., Vakakis, A.F.: Nonreciprocal acoustics and dynamics in the in-plane oscillations of a geometrically nonlinear lattice. Phys. Rev. E **94**(3), 032214 (2016)

Zobay, O., Garraway, B.M.: Time-dependent tunneling of Bose-Einstein condensates. Phys. Rev. A **61**, 033603, 1–7 (2000)

Part I
Conservative Systems

In this part, we introduce the LPT as fundamental notion in non-stationary resonance dynamics starting from the simplest conservative quasi-linear system of two weakly coupled identical oscillators (Sect. 1.1). The role of LPTs in the resonance energy exchange and transition to energy localization at initially excited oscillator is demonstrated. Then (Sect. 1.2), we define the LPTs in a more general conservative two-degrees-of-freedom (2DOF) system assuming a small frequency detuning between the oscillators. We suggest a thorough dynamical analysis, which helps us to highlight a similarity between classical and quantum systems and to find the sets of parameters corresponding to different types of dynamical behavior of the coupled systems. Besides, the suggested analysis underlies further study of coupled oscillators with slowly varying parameters in the Part II.

We show further (Sects. 2.1 and 2.2) that the LPT concept remains very productive even in the case when equations of motion cannot be linearized. Two such systems are considered which are similar to a weightless unstretched preliminarily string with two uniformly situated point-like masses. In the first model (Sect. 2.1), the motion of the particles occurs in the transversal (to the string) direction only (scalar case). In Sect. 2.2.4, this restriction is removed.

Section 3.1 is devoted to more complicated case of three weakly coupled nonlinear oscillators. We analyze a change in the types of fundamental non-stationary resonance processes described by LPTs with weakening the inter-particle coupling.

In the three following Sects. 4.1–4.3, it is shown how the LPT concept works in the case of finite multi-particle system. We consider subsequently Fermi–Pasta–Ulam (FPU), Klein–Gordon (KG), and dimer oscillatory chains and show that the LPT concept successfully works in multi-particle system if to introduce certain clusters of coupled particles (coherence domains) that allow, in particular, to generalize the beat notion on the systems with many degrees of freedom.

Finally, Sects. 5.1–5.3 are devoted to energy exchange between two oscillatory chains.

Chapter 1
Two Coupled Oscillators

1.1 Limiting Phase Trajectories of Two Weakly Coupled Identical Nonlinear Oscillators

1.1.1 The Model and Main Asymptotic Equations of Motion

Let us consider first a simple nonlinear problem of energy transfer in the system of two weakly coupled nonlinear oscillators with cubic restoring forces (Fig. 1.1). Its linearized version is a widely used example of beating phenomenon. In this limiting case, due to superposition principle, every vibrational process can be presented as a combination of two basic oscillations corresponding to in-phase and out-of-phase linear normal modes. If coupling between the oscillators is relatively weak and only one of the oscillators is initially excited, beat with a slow periodic inter-particle energy exchange is observed. So, from the mathematical viewpoint, in the linearized system only basic stationary vibrations, which are linear normal modes, are usually considered as fundamental solutions that allow us to describe also any non-stationary process. Analysis of the nonlinear model where the superposition principle is not valid, but internal (intermodal) resonance is present, shows clearly that resonant (and consequently strongly interacting) NNMs are not appropriate for an adequate description of such non-stationary process as beat. We will show that their adequate description is achieved in terms of fundamental non-stationary solutions (LPTs), without any modal representations.

The problem under consideration can be described by the following system of two nonlinear equations (in dimensionless form):

© Springer Nature Singapore Pte Ltd. 2018
L.I. Manevitch et al., *Nonstationary Resonant Dynamics of Oscillatory Chains and Nanostructures*, Foundations of Engineering Mechanics,
DOI 10.1007/978-981-10-4666-7_1

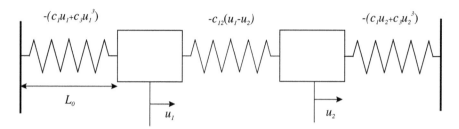

Fig. 1.1 Two linearly coupled nonlinear oscillators

$$\frac{d^2 U_1}{d\tau_1^2} + U_1 + 2\beta\varepsilon(U_1 - U_2) + 8\alpha\varepsilon U_1{}^3 = 0,$$
$$\frac{d^2 U_2}{d\tau_1^2} + U_2 + 2\beta\varepsilon(U_2 - U_1) + 8\alpha\varepsilon U_2{}^3 = 0, \tag{1.1}$$

Here

$$U_j = \frac{u_{j,0}}{L_0}, \quad \tau_1 = \sqrt{\frac{c_1}{m}}\,t, \quad 8\alpha\varepsilon = \frac{c_0 L_0}{c_1}, \quad 2\beta\varepsilon = \frac{c_{12}}{c_1}, \tag{1.2}$$

c_1 and c_3 are the linear stiffness and nonlinear stiffness of the first and second oscillators, respectively; c_{12} is the stiffness of the coupling spring; m is the mass of the particle; τ_1 is the dimensionless time; and $\varepsilon \ll 1$.

Introducing the complex variables:

$$\varphi_j = e^{-i\tau_1}\left(\frac{dU_j}{d\tau_1} + iU_j\right) \quad \varphi_j^* = e^{i\tau_1}\left(\frac{dU_j}{d\tau_1} - iU_j\right) \tag{1.3}$$

and slow time $\tau_2 = \varepsilon\tau_1$ (along with the fast time τ_1), one can use the following two-scale expansions for the functions:

$$\varphi_j(\tau_1, \varphi_2) = \sum_n \varepsilon^n \varphi_{j,n}(\tau_1, \tau_2), \quad j = 1, 2 \tag{1.4}$$

and for the differential operator $\frac{d}{d\tau_1} = \frac{\partial}{\partial\tau_1} + \varepsilon\frac{\partial}{\partial\tau_2}$.

Selecting the terms of different orders by parameter ε after substituting (1.4) into equations of motion and equating them to zero, one reveals that functions $\varphi_{j,n}$ do not depend on the fast time, and main asymptotic approach is described by the following equations (in slow time τ_2):

$$\frac{df_1}{d\tau_2} + i\beta f_2 - 3i\alpha|f_1|^2 f_1 = 0,$$
$$\frac{df_2}{d\tau_2} + i\beta f_1 - 3i\alpha|f_2|^2 f_2 = 0,$$
$$(1.5)$$

$$\varphi_j = e^{i\beta\tau_2} f_j, \quad j = 1, 2,$$
$$(1.6)$$

which describe, as well as the mechanical system discussed, others including two-level quantum systems (Kosevich and Kovalyov 1989) where functions f_j, do not depend on the fast time, τ_1. This system (Fig. 1.1) is fully integrable and has two integrals:

$$H = \beta(f_2 f_1^* + f_1 f_2^*) - \frac{3}{2}\alpha(|f_1|^4 + |f_2|^4),$$
$$(1.7)$$

$$N = |f_1|^2 + |f_2|^2.$$
$$(1.8)$$

(the second integral can be obtained by summation of Eq. (1.5) and conjugate equations after their multiplication on f_1^*, f_1, f_2^*, f_2, respectively).

The best way to address this is to use (1.7) and the coordinates θ and Δ, where

$$f_1 = \sqrt{N}\cos\theta\, e^{i\delta_1}, \quad f_2 = \sqrt{N}\sin\theta\, e^{i\delta_2}, \quad \Delta = \delta_1 - \delta_2.$$

Then the equations of motion have the form

$$\frac{d\theta}{d\tau_2} = \beta\sin\Delta, \quad \sin 2\theta\frac{d\Delta}{d\tau_2} = 2\beta\cos 2\theta\cos\Delta + \frac{3}{2}\alpha N\sin 4\theta,$$
$$(1.9)$$

and integral (1.7) can be written as follows:

$$H = (\cos\Delta + k\sin 2\theta)\sin 2\theta,$$
$$(1.10)$$

where $k = \frac{3\alpha N}{4\beta}$, $\alpha > 0$.

(we assume a hard nonlinearity here, but a soft nonlinearity can be taken into account similarly).

System (1.9) is strongly nonlinear even in the case of initially linear problem, but it is integrable. Before further analysis, let us present plots of the phase trajectories for different values of k in Fig. 1.2 (because of the phase plane periodicity, it is sufficient to consider two presented quadrants only).

Stationary points in Fig. 1.2a–f correspond to NNMs and can be found, as usual, from the conditions $d\theta/d\tau_1 = 0$; $d\Delta/d\tau_1 = 0$. These points are associated with the equilibrium states (in the slow time), and therefore, in the resonance case, the explicit definition of the stationary process can be found from Eq. (1.9).

The sets of trajectories encircling the stationary points correspond to non-stationary oscillations. Since system (1.9) is integrable, one can formally find its analytical solution for arbitrary initial conditions. However, it is not so if we

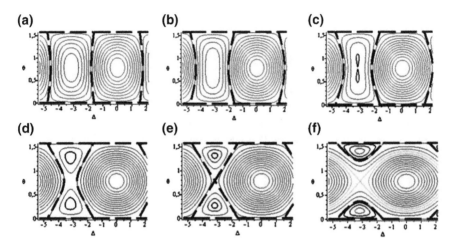

Fig. 1.2 Phase trajectories in the $\theta - \Delta$ plane for: **a** $k = 0.2$, **b** $k = 0.4$, **c** $k = 0.55$, **d** $k = 0.9$, e $k = 1$, **f** $k = 1.5$

consider more complicated models with many degrees of freedom, where the existence of two integrals is insufficient to find an analytical solution, or non-conservative systems having no integrals. In the mentioned models, the conventional approach can be frequently connected exactly with the NNM concept.

It is important to note that even for the considered simplest 2DOF model, the NNM concept may work if a chosen trajectory in the phase plane is at a relatively small distance from the stable stationary point. This approach fails when dealing with strongly non-stationary nonlinear resonance processes because of the *strong intermodal interaction*. Then, the NNMs are no more appropriate for analysis of non-stationary processes. In this case, we need fundamental solutions of other types which are *LPTs* describing maximum possible periodic inter-particle energy exchange (Manevitch 2005, 2007; Manevitch and Smirnov 2010). Since this particular solution represents an outer boundary for a set of trajectories encircling the basic stationary points (Fig. 1.2), we refer to it as the limiting phase trajectory. It is important to note that the LPT is defined by a certain set of *initial conditions*. The equality $\theta(0) = 0$ implies that $f_2(0) = 0$, and, in virtue of (1.2)–(1.4), $U_2 = 0$, $V_2 = 0$ at $\tau_0 = 0$. Therefore, the condition $\theta(0) = 0$ defines the LPT in the system with the initially excited first oscillator, while the second oscillator stays initially at rest. On the contrary, the condition $\theta(0) = \pi/2$ defines the same LPT in the system with the excited second oscillator, while the first oscillator is initially at rest. It is clear that we can interpret motion and energy transfer along LPT as strongly non-stationary process. Let us show that we can consider the LPT as another type of fundamental solution (alternative to NNMs). The LPT can then be used as a generating solution to construct close trajectories with strong energy transfer.

We can see that two dynamic transitions are clearly distinguished when the nonlinearity parameter k increases. The first transition consists in the appearance of two additional stationary points, corresponding to the NNMs, the number of which changes from 2 (if $k < 1/2$) to 4 (for $k > 1/2$), and of separatrix passing through unstable stationary point which is an image of out-of-phase NNM. This transition is distinctly shown in Fig. 1.2c. The first transition, which leads to the appearance of two new out-of-phase modes, does not qualitatively influence the behavior of the LPT, which is far from the stationary points.

The second transition, which occurs when $k = 1$, is connected with the behavior of the limiting phase trajectory (LPT) corresponding to complete energy exchange between the oscillators. Namely, this trajectory transforms into separatrix at this value of k (Fig. 1.2e) or one can say that the separatrix becomes the LPT. This means that the characteristic time for complete energy transfer turns out to be infinite. For $k > 1$, such a transfer becomes impossible (see Fig. 1.2f). Simultaneously, energy localization on the excited particle can be observed.

It is easy to check that the values $H = 1 + k^2$, $H = -1 + k^2$ correspond to the in-phase and out-of-phase cooperative NNMs, respectively. Regimes of this kind are not specific to systems with internal resonances only; they reflect synchronized motions that can be presented as lines in the configuration space of the initial variables (the existence of normal modes in strongly nonlinear systems was shown in Rosenberg (1960, 1962). Efficient techniques for their construction even in the case when they are not straight may be developed by applying the principle of least action in Jacobi's form and the corresponding equations for the trajectories in the configuration space. Such techniques allow to find the NNMs using power expansions by the independent variable (i.e., one of unknowns in this case) in the framework of nonlinear boundary problem (Manevitch et al. 1989; Vakakis et al. 1996).

For the discussed problem, there is no need for such the expansions because the NNMs are represented here by stationary points; this advantage is widely used in the papers devoted to NNMs and their bifurcations, as well as when searching for the close regimes in damped and forced weakly coupled systems (Vakakis et al. 1996; Manevitch and Gendelman 2011). The regimes close to stationary points in the $\theta - \Delta$ plane in the conservative system under consideration are beats with weak energy transfer between two oscillators. The equations of motions can be linearized in the vicinity of stationary points, and their solutions present small-amplitude oscillations of both θ and Δ around their values, corresponding to the NNMs. If one linearizes Eq. (1.9) after the transformation $\theta_1 = \theta - \pi/4$, $\Delta_1 = \Delta - \pi$ (the latter for the case $H = -1 + k^2$, only), one arrives at the equations of linear oscillators, which are valid for initial conditions close to those for the normal modes themselves:

$$\frac{d^2\theta_1}{d\tau_2^2} + \alpha_1^2\theta_1 = 0, \quad \frac{d^2\theta_1}{d\tau_2^2} + \alpha_2^2\theta_1 = 0, \tag{1.11}$$

where $\alpha_1^2 = 4\beta^2(1 + 2k)$, $\alpha_2^2 = 4\beta^2(1 - 2k)$.

These equations contain a contribution that depends on the nonlinear terms of the initial system. Moreover, they lead to the conclusion that instability is possible if $k > 1/2$, which corresponds to instability of the out-of-phase nonlinear normal mode (if $\alpha > 0$).

An alternative demonstration of the LPT evolution can be obtained by excluding the variable Δ from (1.9). Taking into account that (in the case when only the first oscillator is initially excited) $H = 0$ on the LPT, we can express $\cos \Delta$ as a function of θ and, finally, derive the following pendulum equation valid on the LPT (Manevitch 2009):

$$\frac{d^2\theta}{d\tau_2^2} + k^2 \sin 4\theta = 0, \quad 0 \le \theta \le \pi/2. \tag{1.12}$$

We come to an unexpected conclusion: LPTs corresponding to different values of parameter k satisfy the pendulum equation with a restriction on the amplitudes. Let us assume that the pendulum's energy is fixed but the value of parameter k can be chosen. We can interpret the process as pendulum oscillations as well as a one-dimensional motion of a ball or a point-like particle between two symmetrically located rigid walls. Figure 1.3 shows such systems ("ball between walls") with different curvature of the substrate (left panel). The point-like particle moves between two absolutely rigid walls along the substrate. This particle is reflected elastically at the points $\theta = 0$ and $\theta = \pi/2$, corresponding to the walls location. The central panel of the Fig. 1.3 shows the phase portrait with the trajectories, which correspond to the motion in the cases depicted on the left panel. The respective dependences $\theta(\tau)$ are shown on the right panel of the Fig. 1.3. The curvature of the substrate determines the value of the parameter k. When $k = 0$ (weakly coupled linear oscillators), this process is a uniform motion along a straight line with periodic reflections from the rigid walls and corresponding changes in the sign of velocity (Fig. 1.3, case a). It is clear that this motion can be described by a saw-tooth function of time (see Fig. 1.3, right panel). When $0 < k < 1$ (weakly coupled nonlinear oscillators), the motion occurs along a curved trajectory, but the periodic elastic collisions with both rigid walls still take place (Fig. 1.3b). Therefore, in this case motion differs from the linear case since the segments between the collisions are curved. Let us recall that in both cases motion of a particle corresponds to LPT with full energy exchange between the oscillators. When $k = 1$, LPT coincides with the separatrix of the pendulum, and the particle does not have a sufficient energy to overcome the energy barrier, hence the periodic collisions cannot be realized. This corresponds to a motion with infinite period (Fig. 1.3c).

Finally, if $k > 1$, the particle also does not have a sufficient energy to overcome the energy barrier, but a reverse motion to the left wall is possible (Fig. 1.3d). This corresponds to a transition from a "large" LPT with full energy exchange between the oscillators to a "small" LPT with partial (but maximum possible for the given nonlinearity parameter k) energy exchange (with predominant energy localization on the initially excited oscillator). To fully comprehend the fundamental role of LPTs in non-stationary resonance dynamics it is very important to understand that

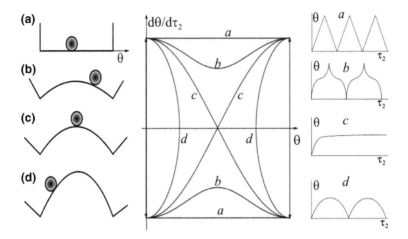

Fig. 1.3 Evolution of the Limiting Phase Trajactory as the motion of the ball between rigid walls. *Left panel*—the "ball between walls" systems at different values of the parameter k: **a** $k = 0$, **b** $k = 0.5$, **c** $k = 1.0$, **d** $k > 1.0$; θ is the "coordinate" of the ball. *Central panel*—the trajectories on the phase portrait corresponding to the systems on the *left panel*; the letter shows the respective system. *Right panel*—time evolution of the "coordinate" $\theta(\tau_2)$ for the systems on the *left panel*

NNMs do not take part in the process of intensive energy exchange between the oscillators. Therefore, we simply cannot analyze strongly non-stationary resonance processes in terms of interacting NNMs.

1.1.2 Analytical Solution for LPT

Because the value of H for the LPT in integral (1.10) is equal to zero. Therefore, the variables θ and Δ in this case are connected by the equation:

$$\cos \Delta = -k \sin 2\theta, \tag{1.13}$$

so that $\sin \Delta = \pm\sqrt{1 - k^2 \sin^2 2\theta}$. Then, the first of Eq. (1.8) can be written as follows:

$$\frac{d\theta}{d\tau_2} = \pm\beta\sqrt{1 - k^2 \sin^2 2\theta}, \tag{1.14}$$

The solution of Eq. (1.14) for the plus sign is the elliptic Jacobi's function: $\theta = (1/2)\, am(2\beta\tau_2, k)$. Because $0 \leq \theta \leq \pi/2$ by definition, one can use the negative sign for $(2n - 1)\pi < 2\beta\tau_2 < 2n\pi$, $n = 1, 2, 3\ldots$.

Thus, we arrive at the final solution:

$$\theta = \frac{1}{2}\,\text{mod}\left[|am(\beta\tau_2,\kappa)|,\frac{\pi}{2}\right],$$

$$\Delta = \pm\arccos\left(\kappa\sin\left(\frac{1}{2}\,\text{mod}\left[|am(\beta\tau_2,\kappa)|,\frac{\pi}{2}\right]\right)\right) \tag{1.15}$$

with period $K(k)$, i.e., the complete elliptic integral of the first kind (for the in-phase oscillations). The solution for the out-of-phase oscillations is:

$$\theta = \frac{1}{2}|am(2\beta\tau_2,k)|, \quad \Delta = \pi \pm \arcsin[k\,sn(2\beta\tau_2,k)], \tag{1.16}$$

The periodic functions (1.15) are not smooth; $\Delta(2\beta\tau_2)$ has breaks at the points $2\beta\tau_2 = (2n-1)\pi$, $n = 0,1,\ldots$, and $\theta(2\beta\tau_2)$ has discontinuities in the derivative at these points (in terms of distributions $d\theta/d\tau_2 = (2\beta/\pi)\,\Delta$). Plots of $\theta(2\beta\tau_2)$ for three values of parameter k are shown in Fig. 1.4.

The value $k = 0.5$ corresponds exactly to the first dynamic transition. However, the solution to the LPT (Fig. 1.4a) is still close to that of the linear case, except for a small change in period. Only for values of k that are close to unity, the deflections form an exact sawtooth profile and the change in period becomes noticeable (Fig. 1.4b).

The second dynamic transition occurs when $k = 1$. In this case, one can find a simple analytical solution corresponding to the LPT:

$$2\beta\tau_2 = \int\limits_{0}^{2\theta} \frac{d(2\theta)}{\cos 2\theta}, \quad \theta = \frac{1}{2}\arcsin\frac{1 - e^{-2\beta\tau_2}}{1 + e^{-2\beta\tau_2}} \tag{1.17}$$

It can be seen from (1.16) that the LPT actually becomes separatrix if $k = 1$: $\theta = \pi/4$.

Exact solutions for the LPTs in terms (θ, Δ) for $k = 0.5$, $k = 0.9$ and $k = 1.1$ are shown graphically in Fig. 1.4a–c, respectively. Two of them are obviously close to the sawtooth functions. The third one with smaller amplitude corresponds to energy localization after second transition.

It is convenient to introduce two non-smooth functions $\tau(\tau_2)$ and $e(\tau_2)$ (Fig. 1.5). Similar functions (but with alternating signs of dependent variables) were introduced first by Pilipchuk (2010). Here, we use similar notation.

We would like to show that the very natural area for application of these non-smooth basic functions is the description of beats (using the variables θ and Δ) and close trajectories with strong energy transfer. Actually in the case $k = 0$ (the linearized system), the solution of (1.9) can be rewritten in the form

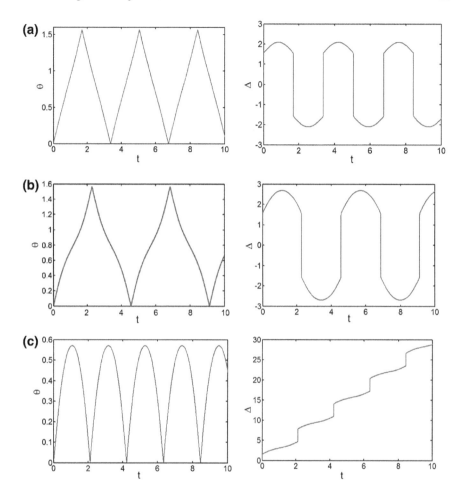

Fig. 1.4 Exact analytical solutions for LPTs for different values of parameter k: **a** $k = 0.5$, **b** $k = 0.9$; **c** $k = 1$

$\theta = (\pi/2)\tau$, $\Delta = (\pi/2)e$, $\tau = \tau(\tau_2/a)$, $e = e(\tau_2/a)$ where $a = \pi/2\beta$ (exactly as in a vibro-impact process with velocity $\Delta = \pi/2$). After introducing the basic functions $\tau(\tau_2/a)$, $e(\tau_2/a)$, we can present the solution as:

$$\Delta = X_2(\tau) + Y_2(\tau)\, e\left(\frac{\tau_2}{a}\right) \qquad \theta = X_1(\tau) + Y_1(\tau)\, e\left(\frac{\tau_2}{a}\right) \qquad (1.18)$$

Substituting expressions (1.18) into (1.9) we obtain that the smooth functions of non-smooth variable $X_i(\tau)$, $Y_i(\tau)$ satisfy Eq. (1.19):

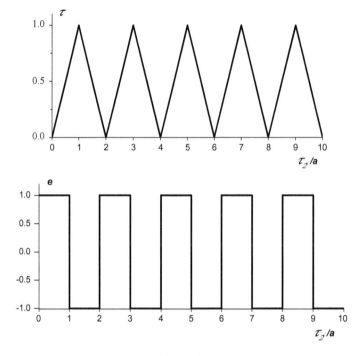

Fig. 1.5 Non-smooth basic functions $\tau(\tau_2/a)$, $e(\tau_2/a)$, where $2a$ is a halved period (in time τ_2)

$$\frac{\partial}{\partial \tau}\left\{\begin{array}{c} X_1 \\ Y_1 \end{array}\right\} = \frac{1}{2}a\beta[\sin(X_2 + Y_2) \mp \sin(X_2 - Y_2)]$$

$$\frac{\partial}{\partial \tau}\left\{\begin{array}{c} X_2 \\ Y_2 \end{array}\right\} = a\beta[ctg\,2(X_1 + Y_1)\cos(X_2 + Y_2) \mp ctg\,2(X_1 - Y_1)\cos(X_2 - Y_2)]$$

$$+ \frac{3a}{2}\alpha N[\cos\,2(X_1 + Y_1) \mp \cos\,2(X_1 - Y_1)].$$

$$(1.19)$$

Then, we can search for the solution of Eq. (1.19) in the form of power expansions in the independent variable τ:

$$X_i = \sum_{l=0}^{\infty} X_{i,\,l}\tau^l, \quad Y_i = \sum_{l=0}^{\infty} Y_{i,\,l}\tau^l, \quad j = 1, 2 \tag{1.20}$$

where a generating solution is the linear beat:

$$X_{1,0} = 0, \quad X_{1,1} = \frac{\pi}{2}, \quad Y_{1,0} = 0, \quad X_{2,0} = 0, \quad Y_{2,0} = \frac{\pi}{2}, \tag{1.21}$$

satisfying exactly the $\theta - \Delta$ equations for the case of the strongest beat. It can be proved that the presentation (1.18), taking into account (1.20), actually recovers the exact solution (in slow time) of the nonlinear problem for the most intensive energy transfer between the oscillators. As this takes place, the expansion (1.20) restores the exact local representation of the corresponding elliptic function (near $\tau = 0$), but the expression (1.18) allow predicting the exact global behavior of the system. It is important to note that, even for large enough values of k, the solution appears close to that of linear beats; the only change is the barely seen curvature of the lines that are straight for linear beats and a change in the period.

One can find corresponding corrections by considering the next order of approximations, namely $X_{1,0} = 0$, $X_{1,1} = \alpha \beta$, $X_{1,3} = -(2/3)(\alpha \beta)^3 k^2$, $Y_{2,0} = \pi/2$, $Y_{2,1} = 2\alpha \beta k$, which coincide with those in the expansions of the exact solution.

Contrary to previous applications of non-smooth transformations, in the considered case there is no need to formulate boundary problems to compensate for singularities. They arise due to the substitution of non-smooth functions into the second of the equations of motion (1.9) in order to derive these equations in terms of the smooth functions X_i and Y_i. However, these singularities are exactly compensated for LPT because $\sin 2\theta = 0$ at singular points.

The most important feature of the proposed technique is the unification of the local and global approaches. The local approach is invoked using power expansions, with unusually good results even in the zeroth approximation. For global characteristics such as the period of oscillations $T = 2a$, its expansion in the parameter k can be found, separately after construction of the analytical form of solution (with the period still unknown). The key point for the solution of this problem is preliminary knowledge of the amplitude values of the θ and Δ functions (in particular, $\theta(a) = \pi/2$). This is another important distinction from the previous applications of non-smooth transformations, in which the problem was solved step by step; in the zeroth approximation, $a = \pi/2\beta$.

It is worth discussing a principal question connected with the behavior of the arising power series. It was noted that the zeroth approximation turns out to be efficient even for large values of the nonlinearity parameter (i.e., going far from the first bifurcation point, corresponding to qualitative change in the phase plane). However, the convergence of these expansions is slow and practically does not depend on the magnitude of the nonlinearity parameter, that is, on the modulus of the elliptic integral of the first kind. This situation rather resembles the behavior of asymptotic series where only the first terms give a reliable representation. The solution may be found in a similar way to that for the case of asymptotic series: using the first terms of the expansion to construct a Pade approximation, which essentially allows the range of reliable representation of the solution to be extended. For example, the Pade approximation in the case $k = 0.62$, $\beta = 0.58$ gives for the

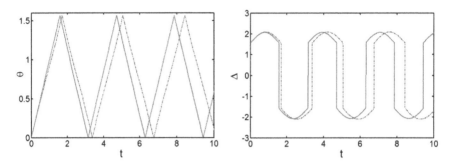

Fig. 1.6 Exact and approximate analytical solutions for LPTs for $k = 0.5$. *Dashed blue line* is the exact solution, *solid green line*—approximation that is described by expressions (1.22)

period the value $Tp = 6.18$, which is close to the numerical value $T = 6.14$. (when using the power expansion itself including nine terms, one finds $T = 4.28$).

The approximate solution for LPT can be found also with using the iteration procedure in which the starting approximation is accepted in the form of the basic non-smooth functions shown in Fig. 1.5. The first iteration gives solution (1.22) coordinated well with the exact solution (Fig. 1.6):

$$\theta = A\tau$$
$$\Delta = \left(\frac{\pi}{2} + \frac{3\alpha N}{4\beta}\sin(2A\tau)\right)e \tag{1.22}$$
$$A = \frac{\pi}{2}, \quad \alpha = \frac{\pi}{2\beta}$$

The proposed procedure may be applied in all cases where we are studying processes in systems with internal resonance, which are far from their stationary states and consequently close to a beat with complete energy transfer. We underline that LPT in systems that are linearized in terms of displacements but strongly nonlinear in terms of $\theta - \Delta$ is a good approximation for the LPT in nonlinear system. It is important that consideration of LPTs enables one to recognize the second dynamic transition that occurs in a system when the nonlinearity parameter N increases, caused by the transformation of the LPT into a separatrix. This means that complete energy transfer from the first mass to the second one becomes impossible, as mentioned above. When $k > 1$, the structure of the phase plane changes drastically that leads to the appearance of infinite trajectories. Simultaneously, the role of the two stable asymmetric normal modes, which appeared due to the bifurcation of the initial out-of-phase mode, becomes more important. These represent vibrations concentrated predominantly on one of the masses. The result of direct numerical integration of the initial system confirms that complete energy transfer (for $k < 1$) disappears when $k > 1$. When only the first

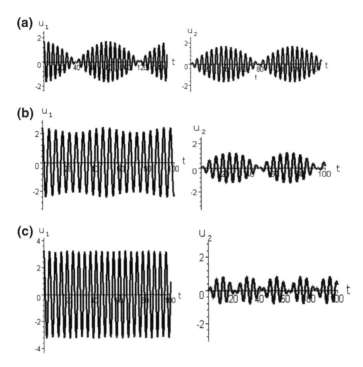

Fig. 1.7 Free oscillations with $\alpha = 0.125$, $\beta = 0.5$, $\varepsilon = 0.1$ for: **a** $k = 0.55$, $N = 2.933$, $u_1(0) = 1.7127$, $u_{1,t}(0) = u_2(0) = u_{2,t}(0) = 0$, **b** $k = 1.1$, $N = 5.867$, $u_1(0) = 2.422$, $u_{1,t}(0) = u_2(0) = u_{2,t}(0) = 0$, **c** $k = 2.0$, $N = 10.67$, $u_1(0) = 3.266$, $u_{1,t}(0) = u_2(0) = u_{2,t}(0) = 0$

particle is initially excited, one can see that for $k = 2$, the system oscillates in the attractive region of the localized nonlinear normal mode with energy concentrated predominantly on the first particle (Fig. 1.7).

Naturally, in the case of linear system ($k = 0$), one can find exact analytical solution using linear normal modes. However, the proposed description of the beating phenomena via non-smooth basic functions has the advantage of being physically adequate in both linear and nonlinear cases. It is clear that both linear and nonlinear beats close to LPT can be more adequately described in terms of the basic functions τ and e than in terms of smooth functions. We choose this rather simple system to illustrate the main ideas convincingly. Their application becomes necessary when dealing with non-conservative systems which are not integrable even in the slow time (see Parts 2 and 3).

Below, we present the solutions corresponding LPT on the initial (fast) timescale before the first transition, after the first and second transitions. It is clearly seen that LPT corresponds to envelope of the real beating process.

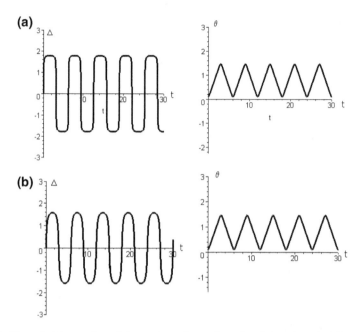

Fig. 1.8 Numerical integration of the modulated equations ($k = 0.4$, various initial conditions), **a** $N = 2.133$, $\theta(0) = 0.02$, $\Delta(0) = 0$, **b** $N = 2.133$, $\theta(0) = 0.01$, $\Delta(0) = 0$

1.1.3 Beating Close to LPTs

The LPT may be considered as the generating solution for the construction of close phase trajectories. If the trajectory is close to the LPT, the initial condition $\theta = 0$ for $\tau = 0$ has to be replaced by the condition $\theta = \theta_0$ for $\tau = 0$. Therefore, we can again use expansion (1.19), but $X_{1,0} = \theta_0$ is not equal to zero. The corresponding solution is shown in Fig. 1.8 for $k = 0.4$ and two different values of $\theta(0)$ (for $\Delta(0) = 0$ in both cases).

1.2 Effect of the Frequency Detuning Between the Oscillators

In Sect. 1.1 we considered a degenerate situation when both weakly coupled oscillators have the same dynamical characteristics. This allowed discussing the LPT concept in the most transparent case. In this section, we take into account a possible frequency detuning between the oscillators.

The LPT concept for asymmetric systems was introduced in (Manevitch and Kovaleva 2013). An explanation and a more detailed discussion of the LPT properties and their connection with the non-stationary processes in an asymmetric system ($g \neq 0$) is given below

1.2.1 Equations of Motion and Explicit Approximate Solutions

The equations of motion of a couple of oscillators are given by:

$$m\frac{d^2 u_1}{dt^2} + c_1 u_1 + c\, u_1^3 + c_{12}(u_1 - u_2) = 0,$$
$$m\frac{d^2 u_2}{dt^2} + c_2 u_2 + c\, u_2^3 + c_{12}(u_2 - u_1) = 0, \tag{1.23}$$

where m is the mass of each oscillator; c_1 and c_2 are the coefficients of linear stiffness of the corresponding oscillators; c is the coefficient of cubic nonlinearity; c_{12} is stiffness of linear coupling; u_1 and u_2 are the absolute displacements of the first and second oscillators, respectively. As in a single oscillator, the small parameter ε is defined through relative stiffness of weak coupling: $c_{12}/c_1 = 2\varepsilon \ll 1$. Assuming weak nonlinearity and taking into account resonance properties of the system, we redefine other parameters as follows:

$$c_1/m = \omega_0^2, \quad c_2/m = \omega_0^2(1 + 2eg), \quad \tau_0 = \omega_0 t, \quad c/c_1 = 8\varepsilon\alpha, \quad c_{12}/c_j = 2\varepsilon\lambda_j,$$
$$j = 1, 2, \tag{1.24}$$

where constant detuning parameter g is chosen to ensure the desired dynamics of the originally symmetric ($c_1 = c_2$) system. It follows from (1.24) that $\lambda_1 = 1$, $\lambda_2 = (1 + 2\varepsilon g)^{-1}$.

Substituting expressions (1.24) into (1.23), we obtain the following system:

$$\frac{d^2 u_1}{d\tau_0^2} + u_1 + 2\varepsilon(u_1 - u_2) + 8\varepsilon\alpha u_1^3 = 0,$$
$$\frac{d^2 u_2}{d\tau_0^2} + (1 + 2\varepsilon g)u_2 + 2\varepsilon(u_2 - u_1) + 8\varepsilon\alpha u_2^3 = 0 \tag{1.25}$$

It follows from (1.23), (1.25) that the only difference from degenerate case is a small difference in the linear frequencies of the oscillators. Selected initial conditions correspond to a unit impulse imposed to one of the oscillators in the system being initially at the equilibrium state, i.e., $u_1 = u_2 = 0$; $v_1 = du_1/d\tau_0 = 1$, $v_2 = du_2/d\tau_0 = 0$ at $\tau_0 = 0$ if the impulse applied to the first oscillators, or $v_1 = du_1/d\tau_0 = 0$, $v_2 = du_2/d\tau_0 = 1$ if the initial impulse $v_2 = 1$ is applied to the second oscillator. We take into account that for estimating the nonlinear contribution into the dynamics, it is sufficient to change corresponding coefficients in Eq. (1.25) without changing the intensity of excitation. Under given initial conditions, we determine the limiting phase trajectory (LPT) of system (1.25).

As in the previous section, an asymptotic solution of (1.25) for small ε is based on the complexification of the dynamics and the separation of the slow and fast constituents. To this end, we introduce the change of variables $v_j + iu_j = Y_j e^{i\tau_0}$, $j = 1, 2$, and then present the complex amplitude as $Y_j(\tau_0, \tau_1, \varepsilon) = \varphi_j^{(0)}(\tau_1) + \varepsilon\varphi_j^{(0)}(\tau_0, \tau_1) + O(\varepsilon^2)$, $\tau_1 = \varepsilon \tau_0$. Then, using the multiple-scale techniques and separating the fast and slow timescales (see Sect. 2.1), we derive the equations for the leading-order slow terms analogous to (2.6), from which we deduce that the main slow terms are presented in the form $\phi_1^{(0)}(\tau_1) = a(\tau_1)e^{i\tau_1}$, $\phi_2^{(0)}(\tau_1) = b(\tau_1)e^{i\tau_1}$ where the complex envelopes $a(\tau_1)$ and $b(\tau_1)$ are given by the equations similar to (1.5):

$$\frac{da}{d\tau_1} + ib - 3i\alpha|a|^2 a = 0,$$
$$\frac{db}{d\tau_1} + ia - 3i\alpha|b|^2 b - 2igb = 0. \tag{1.26}$$

Similar to the symmetric case, it is easy to prove that system (1.26) conserves the integral $|a|^2 + |b|^2 = 1$. We recall that the solutions of Eq. (1.26) with the above-mentioned initial conditions define not only the LPT of the averaged system but also the LPT of the initial system (1.25).

Equation (1.26) highlights similarity of the averaged equations of the classical oscillator to the equations of two-state atomic tunneling (Raghavan et al. 1999; Sievers and Takeno 1988). This similarity confirms a direct mathematical analogy between quantum and classical transitions, and the applicability of the results derived in this section to a wide class of physical problems. The polar representations $a = \cos \theta \, e^{i\delta_1}$, $b = \sin \theta e^{i\delta_2}$, $\Delta = \delta_1 - \delta_2$, yield the following real-valued equations:

$$\frac{d\theta}{d\tau_1} = \sin \Delta,$$
$$\sin 2\theta \frac{d\Delta}{d\tau_1} = 2(\cos \Delta + 2k \sin 2\theta) \cos 2\theta - 2g \sin 2\theta, \tag{1.27}$$

where $k = 3\alpha/4$. Initial conditions for system (1.27) are chosen as $\theta = 0$, $\Delta = \pi/2$ (the first oscillator is excited, but the second one is initially at rest) or $\theta = \pi/2$, $\Delta = \pi/2$ (the second oscillator is excited, but the first one is at rest). Both conditions correspond to the LPTs of system (1.27) in the plane (Δ, θ).

Note that system (1.27) conserves the integral of motion (preservation of energy):

$$H = (\cos \Delta + k \sin 2\theta) \sin 2\theta + g \cos 2\theta, \qquad (1.28)$$

whose properties are used in the dynamical analysis.

1.2.2   Stationary States and LPTs

The first step in the dynamical analysis is to define the stationary points of system (1.27) corresponding to NNMs. The first condition of stationarity $d\theta/d\tau_1 = 0$ yields $\sin \Delta = 0$; this implies that all steady states lie on the vertical axes $\Delta_1 = 0$ or $\Delta_2 = \pi$. The second condition $d\Delta/d\tau_1 = 0$ implies that the stationary values of θ are given by the equation:

$$F_{\pm}(\theta) = (\pm 1 + 2k \sin 2\theta) \cot 2\theta = g, \qquad (1.29)$$

where the signs "+" and "−" correspond to $\Delta = 0$ and $\Delta = \pi$, respectively.

Figures 1.9, 1.10, 1.11, and 1.12 depict phase portraits of system (1.27) in the plane (Δ, θ) for different values of the parameters k and g. In the quasi-linear case, when $k \leq 0.5$, there exists a unique solution of Eq. (1.29) on each of the axes $\Delta = 0$ and $\Delta = \pi$. It was recently shown (Kovaleva and Manevitch 2012) that in this case, the solution of the nonlinear system is close to that of the linear system. Phase portraits of system (1.27) for $k = 0.35$ and different g are shown in Fig. 1.9. It is easy to deduce from (1.27) that the change in the sign of the parameter $g \rightarrow -g$ entails the change in the solution: $\theta \rightarrow \pi - \theta$, $\Delta \rightarrow 2\pi - \Delta$. This allows the construction of the phase portraits only for $g \geq 0$ (Fig. 1.9). Here and below, stationary states correspond to nonlinear normal modes; bold lines depict LPTs. It is shown in Fig. 1.9 that the LPT represents an outer boundary for a set of closed trajectories encircling the stable center in the phase plane (Δ, θ).

Below, we study in detail the dynamics of the system with nonlinearity $k > 0.5$. It is shown in Figs. 1.10, 1.11, and 1.12 that if $k > 0.5$, there exists a certain value

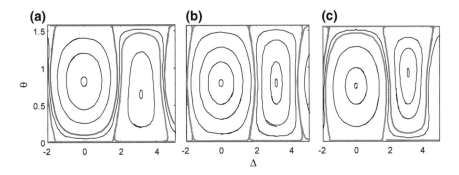

Fig. 1.9 Phase portraits of (1.27) for $k = 0.35$: **a** $g = -0.1$; **b** $g = 0$; **c** g 0.1

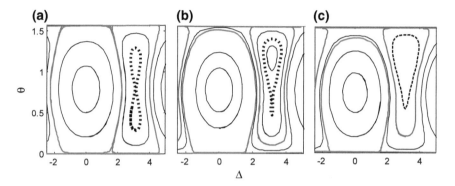

Fig. 1.10 Phase portraits of (1.27) for $k = 0.65$: **a** $g = 0$; **b** $g = 0.075$; **c** $g = 0.083$.

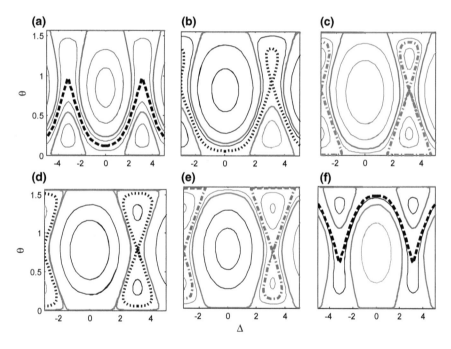

Fig. 1.11 Phase portraits of (1.27) for $k = 0.9$: **a** $g = -0.33$; **b** $g = -0.2$; **c** $g = -0.0945$; **d** $g = 0$; **e** $g = 0.0945$; **f** $g = 0.2$

g^* such that the system has 3 stationary points on the axis $\Delta = \pi$ for $|g| < |g^*|$ and a single point for $|g| \geq |g^*|$, and the transition occurs through the coalescence of two stationary states at a certain point θ_T such that $F_-(\theta_T) = g^*$. The condition $dF_-/d\theta = 0$ at a point of merging of two roots of Eq. (1.29) gives the following expressions for θ_T and g^*:

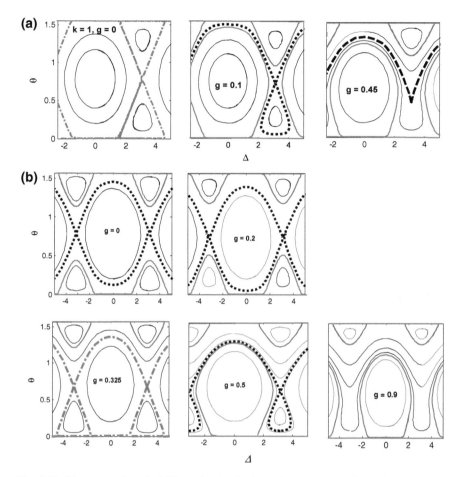

Fig. 1.12 Phase portraits of (1.27): **a** $k = 1$; (**b**) $k = 1.3$; the values of detuning $g > 0$ are indicated in the planes

$$\sin 2\theta_T = (2k)^{-1/3}, \quad g^* = \pm\left[(2k)^{2/3} - 1\right]^{3/2}. \quad (1.30)$$

It follows from (1.30) that four stationary states exist in the domain $k > 0.5$. Note that the parameter g^* coincides with the critical parameter presented without proof in Liu et al. (2002).

As shown below, the system with four stationary states exhibits two different types of dynamical behavior corresponding to moderately nonlinear and strongly nonlinear regimes. First, we consider the system with $k = 0.65$. It is easy to deduce from (1.27) that the change in the sign of the parameter $g \rightarrow -g$ entails the change in the solution: $\theta \rightarrow \pi - \theta$, $\Delta \rightarrow 2\pi - \Delta$. This allows us to construct the phase portraits only for $g \geq 0$ (Fig. 1.10). Bold lines in Fig. 1.10 depict the LPT of system (1.27) in the plane (θ, Δ); dotted lines correspond to the homoclinic separatrix. It is seen that with an increase in g, the lower homoclinic loop vanishes through the merging of the stable and unstable states, and the number of the stationary states changes from 4 to 2. The corresponding critical value $g^* = 0.083$ coincides with the theoretical value given by formula (1.30). It is easy to find that the number of the stationary points changes from 2 to 4 at $g^* = -0.083$.

Figure 1.10 illustrates the complete energy exchange between the symmetric oscillators ($g = 0$); i.e., the upper level $\theta = \pi/2$ is reached during the cycle of motion along the LPT starting at $\theta = 0$. Motion along the closed orbits within the domain encircled by the LPT obviously provides less extensive energy exchange than motion along the LPT. Vanishing of the homoclinic separatrix and the emergence of a new closed orbit of finite period characterize *moderately nonlinear* systems. The phase portraits for $g < 0$ can be constructed using symmetry. In these systems, the localization of energy near the lower center is replaced by the localization near the upper center. This phenomenon, responsible for the occurrence of nonlinear tunneling in the slowly time-dependent systems (Liu et al. 2002; Trimborn et al. 2010; Zobay and Garraway 2000), underlies the study of transient processes in Part II.

It is shown from Fig. 1.11 that the system with $k = 0.9$ exhibits a more complicated dynamical behavior. For clarity, the phase portraits for both $g < 0$ and $g > 0$ are shown; bold lines correspond to the LPTs; dash-dotted lines depict the homoclinic separatrix coinciding with the LPT; edges of the dashed "beaks" lie at the points of annihilation of the stable and unstable states.

The change from 2 to 4 fixed points at $g = -0.33$ leads to the emergence of a separatrix passing through the hyperbolic point and consisting of homoclinic and heteroclinic branches (by the heteroclinic separatrix, we mean a trajectory from the hyperbolic point on the axis $\Delta = \pi$ to the hyperbolic point on the axis $\Delta = -\pi$ that separates locked and unlocked orbits). At $g = -0.0945$, the heteroclinic loop coincides with the LPT at $\theta = 0$. Further increase in g leads to the occurrence of the homoclinic separatrix and then to the merging of the separatrix with the LPT at $\theta = \pi/2$ for $g = 0.0945$; finally, the merging of the lower stable center with the hyperbolic point results in the degeneration of the separatrix and the change from 4 to 2 fixed points. The numerically found value $|g^*| = 0.33$ coincides with the results of calculation by formula (1.30). Complete energy exchange is observed in the symmetric system ($g = 0$) moving along the LPT.

The phase portraits of system (1.27) with the parameters $k = 1$, $g > 0$, and $k = 1.3$, $g > 0$ are depicted in Fig. 1.12; the portraits for $g < 0$ may be constructed using symmetry. Figure 1.12b demonstrates in more detail the transformations of the separatrix and the LPT associated with the transition from energy localization to energy exchange.

The merging of the separatrix with the LPT and the transition from energy localization to energy exchange correspond to *strongly nonlinear regimes*. The merging of the stable and unstable states entails the emergence of a new unlocked orbit and an associated transition from weak to a strong energy exchange (Figs. 1.10, 1.11 and 1.12).

1.2.3 Critical Parameters

Transitions from energy localization near the stable state to energy exchange in the asymmetric system were investigated earlier (Tsironis 1993; Tsironis et al. 1993), but an analytical boundary between the corresponding domains of parameters was not derived. We present analytical conditions that ensure the transition from energy localization on the excited oscillator to strong energy exchange (Manevitch and Kovaleva 2013).

As mentioned previously, the dynamical behavior may be considered as strongly nonlinear if its separatrix coincides with the LPT. Using this condition, we find a set of parameters determining strongly nonlinear regimes. To this end, we consider integral of motion (1.28). It follows from the initial condition $\theta = 0$ that $H = g$ on the LPT and, therefore,

$$(\cos \Delta + k \sin 2\theta) \sin 2\theta - 2g \sin^2 \theta = 0 \tag{1.31}$$

on the LPT. Now, we obtain from Eqs. (1.27) and (1.31) that

$$d\theta/d\tau_1 = V = \sin \Delta; V = \pm[1 - (k \sin 2\theta - g \tan \theta)^2]^{1/2}. \tag{1.32}$$

$(k \sin 2\theta - g \tan \theta)^2 = 1$ at $V = 0$.

Using equality (1.31) to exclude Δ, we replace system (1.27) by the following second-order equation for LPTS

$$\frac{d^2\theta}{d\tau_1^2} + \frac{dU}{d\theta} = 0 \tag{1.33}$$

with initial conditions $\theta(0) = 0$, $V(0) = 1$. The potential $U(\theta)$ in (1.33) can be found from the energy conservation law $E = 1/2V^2 + U(\theta) = 1$, which gives

$$U(\theta) = 1 - 1/2V^2 = \frac{1}{2}\left[1 + (k \sin 2\theta - g \tan \theta)^2\right]. \tag{1.34}$$

The maximum value $U(\theta) = 1$ is attained at $V = 0$. Potential $U(\theta)$ and phase portraits for system (1.31) with different coefficients k and g are shown in Fig. 1.13.

The sought separatrix may exist if and only if $dU/d\theta = 0$ at $\theta_h \in (0, \pi/2)$, as in this case there exists a potential barrier corresponding to the local maximum $U(\theta_h)$ and attained at $V = 0$; the latter condition is equivalent to $\Delta = 0$. This implies that $(\theta_h, 0)$ is a hyperbolic point. It now follows from (1.34) that the equality $dU/d\theta^* = 0$ is equivalent to:

$$(k \sin 2\theta_h - g \tan \theta_h)(2k \cos 2\theta_h - g/\cos^2 \theta_h) = 0. \tag{1.35}$$

Combining (1.32) and (1.35), we find that Eq. (1.35) is reduced to a simple biquadratic equation:

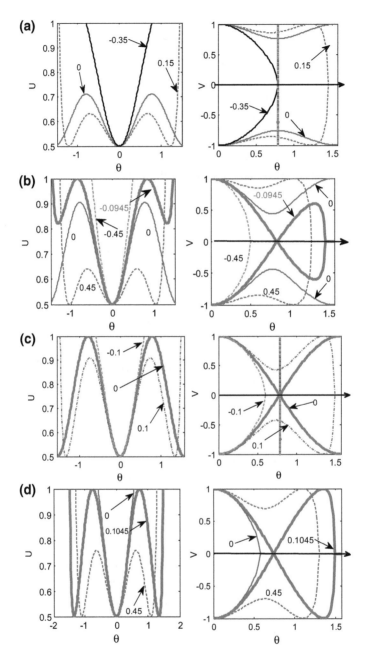

Fig. 1.13 Potential $U(\theta)$ (*left column*) and phase portraits in the plane (θ, v) (*right column*) of system (1.31) with $k = 0.65$ (**a**), $k = 0.9$ (**b**), $k = 1$ (**c**), $k = 1.1$ (**d**); detuning g is indicated on each *curve*; *bold lines* depict critical potentials and corresponding associated separatrices confluent with the LPT at $\theta = 0$

$$4k \cos^4 \theta_h - 2k \cos^2 \theta_h - g = 0, \quad \cos^2 \theta_h = \frac{1}{4} \pm \sqrt{\frac{1}{16} + \frac{g}{4k}}. \tag{1.36}$$

If $|g/k| \ll 1/4$, then $\cos^2\theta_h \approx 1/2 + g/2\,k$, $\theta_h \approx \pi/4 - g/2\,k$. Substituting these approximations into the last Eq. (1.32), we derive the simple condition providing the existence of the required separatrix:

$$|k - g_h| = 1. \tag{1.37}$$

It is easy to check that the theoretical threshold g_h closely agrees with the results of numerical calculations. In particular, $g_h = -0.1$ for $k = 0.9$, whereas the numerical threshold $g = -0.0945$, $g = g_h = 0.1$ for $k = 1.1$ (Fig. 1.13), $g_h = 0.3$ for $k = 1.3$, whereas the numerical threshold $g = 0.325$ (Fig. 1.12).

In a similar way, we can prove that, when the initial condition is taken at $\theta = \pi/2$, then the condition (1.37) is turned into the equality:

$$|k + g_h| = 1. \tag{1.38}$$

If $g < 0$, then solution (1.36) exists provided that $|g| \le k/4$; in the limiting case $g = -k/4$, we have $\cos \theta_h = 1/2$, $\theta_h = \pi/3$, and condition (1.36) becomes:

$$\frac{3\sqrt{3}}{4} k = 1, \quad k^* \approx 0.77. \tag{1.39}$$

Inequalities $k \ge k^*$ and $|g| \le k/4$ express *the necessary condition* for the existence of the separatrix coinciding with the LPT at $\theta = 0$. Additionally, we need to calculate the argument θ_h from (1.36).

References

Kosevich, A.M., Kovalyov, A.S.: Introduction to Nonlinear Physical Mechanics. Naukova Dumka, Kiev (1989) (in Russian)

Kovaleva, A., Manevitch, L.I.: Classical analog of quasilinear Landau-Zener tunneling. Phys. Rev. E **85**, 016202 (1–8) (2012)

Liu, J., Fu, L., Ou, B.-Y., Chen, S.-G., Choi, D.-I.L., Wu, B., Niu, Q.: Theory of nonlinear Landau-Zener tunneling. Phys. Rev. A **66**, 023404 (1–7) (2002)

Manevitch, L.I.: New approach to beating phenomenon in coupled nonlinear oscillatory chains. In: Awrejcewicz, J., Olejnik, P. (eds.) 8th Conference on Dynamical Systems—Theory and Applications, DSTA-2005, p. 289 (2005)

Manevitch, L.I.: New approach to beating phenomenon in coupling nonlinear oscillatory chains. Arch. Appl. Mech. **77**, 301–312 (2007)

Manevitch, L.I.: Vibro-impact models for smooth non-linear systems. In: Ibrahim, R. (ed.) Lecture Notes in Applied and Computational Mechanics. Vibro-Impact Dynamics of Ocean Systems and Related Problems, vol. 44, pp. 191–201. Springer, New York (2009)

Manevitch, L.I., Gendelman, O.: Tractable Models of Solid Mechanics. Springer, Berlin, Heidelberg (2011)

Manevitch, L.I., Kovaleva, A.: Nonlinear energy transfer in classical and quantum systems. Phys. Rev. E **87**, 022904 (1–12) (2013)

Manevitch, L.I., Smirnov, V.V.: Resonant energy exchange in nonlinear oscillatory chains and Limiting Phase Trajectories: from small to large systems. In: Vakakis, A.F. (ed.) Advanced Nonlinear Strategies for Vibration Mitigation and System Identification. CISM Courses and Lectures, vol. 518, pp. 207–258. Springer, Wien, New York (2010)

Manevitch, L.I., Mikhlin, Y.V., Pilipchuk, V.N.: The Normal Vibrations Method for Essentially Nonlinear Systems. Nauka Publ, Moscow (1989) (in Russian)

Pilipchuk, V.N.: Nonlinear Dynamics: Between Linear and Impact Limits. In: Lecture Notes in Applied and Computational Mechanics, vol. 52. Springer, Berlin, Heidelberg (2010)

Raghavan, S., Smerzi, A., Fantoni, S., Shenoy, S.R.: Coherent oscillations between two weakly coupled Bose-Einstein condensates: Josephson effects, π-oscillations, and macroscopic quantum self-trapping. Phys. Rev. A **59**, 620–633 (1999)

Rosenberg, R.M.: Normal modes of nonlinear dual-mode systems. J. Appl. Mech. **27**, 263–268 (1960)

Rosenberg, R.M.: The normal modes of nonlinear n-degree-of-freedom systems. J. Appl. Mech. **29**, 7–14 (1962)

Sievers, A.J., Takeno, S.: Intrinsic localized modes in anharmonic crystals. Phys. Rev. Lett. **61**, 970–973 (1988)

Trimborn, F., Witthaut, D., Kegel, V., Korsch, H.J.: Nonlinear Landau-Zener tunneling in quantum phase space. New J. Phys. **12**, 05310 (1–20) (2010)

Tsironis, G.P.: Dynamical domains of a nondegenerate nonlinear dimer. Phys. Lett. A **173**, 381–385 (1993)

Tsironis, G.P., Deering, W.D., Molina, M.I.: Applications of self-trapping in optically-coupled devices. Phys. D **68**, 135–137 (1993)

Vakakis, A.F., Manevitch, L.I., Mikhlin, Y.V., Pilipchuk, V.N., Zevin, A.A.: Normal Modes and Localization in Nonlinear Systems. Wiley, New York (1996)

Zobay, O., Garraway, B.M.: Time-dependent tunneling of Bose-Einstein condensates. Phys. Rev. A **61**, 033603 (1–7) (2000)

Chapter 2
Two-Particle Systems Under Conditions of Sonic Vacuum

In this section, we investigate resonance energy transport in a purely nonlinear system, wherein harmonic oscillations are prohibited by the properties of the system potential of degree higher than two. So, in contrast to the models considered in the previous sections, the potential of this system does not contain quadratic terms. Hence, the system does not exhibit the quasi-linear behavior; furthermore, there is no constant natural frequencies predetermining resonance properties of the system in the quasi-linear approximation. This implies that the study of resonance processes and resonance energy transport in a purely nonlinear system requires a special approach.

The basic model considered in this section comprises two weakly coupled purely nonlinear oscillators, wherein initial energy is imported to one of them. Numerical simulations reveal the existence of strong classical beat oscillations corresponding to complete recurrent resonant energy exchanges between the oscillators in the state of sonic vacuum, where no resonance frequencies can be defined. In this study, we show that both intense energy exchange and transition to energy localization are adequately described in the framework of the LPT concept. We show that the occurrence of the recurrent energy exchanges in this highly degenerate model strictly depends on the system parameters. For instance, choosing the parameter of coupling below a certain threshold leads to the significant energy localization on one of the oscillators; on the contrary, increasing the strength of coupling above the threshold brings to the formation of a strong beating response.

Analytical studies pursued in this section predict the occurrence of the strong beating phenomenon and provide necessary conditions for its emergence. Moreover, a careful analysis of the beating phenomenon reveals a qualitatively new global bifurcation of highly non-stationary regime.

Two systems considered in this chapter are similar to a weightless unstretched preliminarily string with two symmetrically located point-like masses. In the first model (Sect. 2.1), we restrict the motion by transversal (to the string) direction only (scalar case). In Sect. 2.2, this restriction is removed.

© Springer Nature Singapore Pte Ltd. 2018
L.I. Manevitch et al., *Nonstationary Resonant Dynamics of Oscillatory Chains and Nanostructures*, Foundations of Engineering Mechanics,
DOI 10.1007/978-981-10-4666-7_2

2.1 Weakly Coupled Oscillators Under Conditions of Local Sonic Vacuum

If the motion of the particles is transversal to the weightless string itself, the coupling between the particles is realized by the local strongly nonlinear interactions. If the preliminary stretching of the string is absent, they are cubic functions in the main approximation. However, we consider even more general case when the power may be not only three but an arbitrary odd integer.

2.1.1 Evidence of Energy Localization and Exchange in Coupled Oscillators in the State of Sonic Vacuum

We study energy localization and complete recurrent energy transport in a homogeneous system of two coupled anharmonic oscillators in the state of sonic vacuum. This model (in the case of cubic inter-particle interaction) is mathematically equivalent to weightless string with two symmetrically located identical particles if a preliminary stretching is absent and the motion occurs in the transversal direction only. The main goal is to describe the transition from initial energy localization on a single oscillator to complete recurrent energy exchanges (strong beating phenomenon) between the oscillators due to variations of the system parameters

The model under consideration comprises two identical anharmonic oscillators coupled with an anharmonic spring. The non-dimensional equations of motion are given by

$$
\begin{aligned}
\frac{d^2 x_1}{dt^2} + x_1^n &= \chi (x_2 - x_1)^n, \\
\frac{d^2 x_2}{dt^2} + x_2^n &= \chi (x_1 - x_2)^n.
\end{aligned}
\tag{2.1}
$$

here $n = 2k + 1$, $k = 1, 2, 3, \ldots$; the parameter χ denotes coupling stiffness. Initial conditions $x_1 = X_1$, $v_1 = dx_1/dt = V_1$, $x_2 = 0$, $v_2 = dx_2/dt = 0$ at $t = 0$ correspond to complete initial localization of the system energy on the first oscillator with the second oscillator being initially at rest.

It is important to emphasize that the system (2.1) is homogeneous, and therefore, its total energy E can be normalized to unity ($E = 1$) by choosing appropriate rescaling of the dependent and independent variables. This means that the global system dynamics is energy-independent and can be studied for an arbitrary value of the total energy level.

Below, we illustrate numerically the existence of two different regimes such as energy localization on the initially excited oscillator and complete energy exchanges between the oscillators. We also show that the first regime corresponds to the non-resonant behavior of the coupled oscillators while the second one is triggered by the formation of permanent 1:1 resonance capture resulting in the complete recurrent energy transport between the oscillators.

2.1.2 Energy Localization

Figure 2.1a depicts instantaneous energy $E_i(t) = \dot{x}_i^2/2 + (n+1)^{-1}x_i^{n+1}$ of each of the oscillators in the system with parameters $\chi = 0.12$, $n = 3$ and initial conditions $x_1 = 1$, $v_1 = 0$, $x_2 = 0$, $v_2 = 0$ at $t = 0$. Figure 2.1b, c illustrate the fast Fourier transform (FFT) transforms of $x_1(t)$ and $x_2(t)$, respectively.

In Fig. 2.1a, one can observe localization of the initial energy on the first oscillator. This result is not surprising, as the system under consideration is purely nonlinear (sonic vacuum), and strong spatial energy localization on certain fragments of the system is expected. Moreover, fast Fourier transform (FFT) diagrams presented in Fig. 2.1b, c show that this type of response exhibits a trivial non-resonant behavior, where the amplitude of a main frequency component of the first oscillator far exceeds the amplitude of the main components of the second oscillator. Also, the second oscillator possesses two comparable components with remote frequencies exhibiting a clear subharmonic motion. The detailed analysis of this type of response is performed in Sect. 2.1.4.

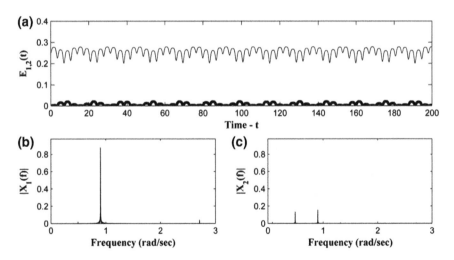

Fig. 2.1 Regime of energy localization. **a** Instantaneous energies recorded on the first and the second oscillators ($E_1(t)$ *thin solid line*, $E_2(t)$ *bold solid line*); **b** fast Fourier transform of the response of the first oscillator $x_1(t)$; **c** fast Fourier transform of the response of the second oscillator $x_2(t)$

2.1.3 Complete Energy Exchanges (Strong Beating Response)

Slightly increasing the coupling parameter χ, we observe a global change of the response. In particular, instead of energy localization on the first oscillator we observe the formation of a beating response characterized by complete energy exchange between the oscillators. Instantaneous energy $E_i(t) = \dot{x}_i^2/2 + (n+1)^{-1}x_i^{n+1}$ is plotted in Fig. 2.2a for each of the oscillators in the system with parameters $\chi = 0.18$, $n = 3$ and initial conditions $x_1 = 1$, $v_1 = 0$, $x_2 = 0$, $v_2 = 0$ at $t = 0$. Figure 2.2b, c illustrates the FFT transforms of $x_1(t)$ and $x_2(t)$, respectively.

Figure 2.2a depicts the beating response characterized by complete recurrent energy exchange between the oscillators. FFT diagrams in Fig. 2.2b, c reveal the resonant nature of the response ensuring the formation of 1:1 resonance between the oscillators, which, in turn, leads to the strong beating response.

It is important to note that beating oscillations are usually observed in linear or weakly nonlinear systems possessing at least one pair of close natural frequencies (Manevitch and Gendelman 2011; Manevitch and Smirnov 2010a, b, c). In these cases, the resonance frequency of the response is determined either by the natural frequency of the linear subsystem or by the frequency of a periodic excitation. However, in the system under consideration subjected to the state of acoustic vacuum the resonant frequency is determined by the energy level of initial excitation. Therefore, resonance frequencies observed in Fig. 2.2b, c are obviously amplitude-dependent. This means that an increase in the initial amplitude of the first oscillator results in an increase in the resonant frequency of oscillations.

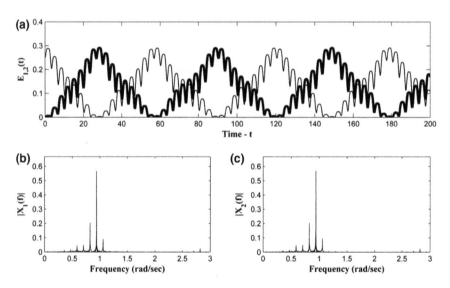

Fig. 2.2 Strong beating response. **a** Instantaneous energies recorded on the first and the second oscillators ($E_1(t)$—*thin dotted line*, $E_2(t)$—*bold solid line*); **b** FFT of $x_1(t)$; **c** FFT of $x_2(t)$

2.1.4 Asymptotic Analysis of Resonance Motion

This section suggests theoretical analysis of the near-resonant behavior of system (2.1). Special emphasis is given to an analytical description of the formation and the annihilation of a regular beating response, along with the local and global bifurcation analysis of the system dynamics.

Given the pure resonant nature of a strong beating regime, it is quite reasonable to anticipate its formation in the neighborhood of the 1:1 resonance manifold. It is well known that the phenomenon of 1:1 resonance capture provides maximum energy transfer between weakly coupled identical oscillators. Thus, in order to depict analytically the regime of complete energy transfer, as well as to find necessary conditions for its existence, it is convenient to consider the system dynamics in a neighborhood of the 1:1 resonance manifold.

Assuming resonance interactions, we rewrite (2.1) as follows:

$$\frac{d^2 x_k}{dt^2} + \Omega^2 x_k = \varepsilon\mu \left[\chi (x_{3-k} - x_k)^n - x_k^n + \Omega^2 x_k \right], \quad k = 1, 2, \qquad (2.2)$$

where Ω denotes the resonance frequency depending on the system energy, ε is a small parameter of the system, $\mu = 1/\varepsilon$. As in the previous sections, we assume that the sum in the square brackets is small but the expression $\mu \left[\chi (x_{3-k} - x_k)^n - x_k^n + \Omega^2 x_k \right]$ is of $O(1)$ in the vicinity of resonance.

We underline that the representation of equations of motion in the form (2.2) allows us to investigate the purely nonlinear system in the framework of the quasi-linear theory and to employ the earlier developed methods. In the first step, we introduce the new complex variables as follows:

$$u_k = \frac{1}{2i} \left(Y_k e^{i\Omega\tau_0} - Y_k^* e^{-i\Omega\tau_0} \right), \quad v_k = \frac{\Omega}{2} \left(Y_k e^{i\Omega\tau_0} + Y_k^* e^{-i\Omega\tau_0} \right), \qquad (2.3)$$

and then substitute (2.3) into (2.1) to obtain the equations for Y_k, Y_k^* with the right-hand sides of $O(\varepsilon)$ (see Starosvetsky and Ben-Meir 2013 for more details). In the next step, the complex amplitude $Y_k(t, \varepsilon)$ is sought in the form of the expansion $Y_k(t, \varepsilon) = \varphi_k^{(0)}(\tau_1) + \varepsilon \, \varphi_k^{(1)}(t, \tau_1) + \varepsilon^2 \ldots$, with the slow main term $\varphi_k^{(0)}(\tau_1)$, where $\tau_1 = \varepsilon t$ is the leading-order slow timescale. Then, applying the multiple scales methodology, we derive the following equation for the leading-order slow amplitudes $\varphi_k^{(0)}(\tau_1)$ (detailed arguments are provided in Starosvetsky and Ben-Meir 2013):

$$\frac{d\varphi_k^{(0)}}{d\tau_1} = i\mu \left(\frac{C_{(n-1)/2}^n}{(2\Omega)^n} \left| \varphi_k^{(0)} \right|^{n-1} \varphi_k^{(0)} - \frac{\Omega}{2} \varphi_k + \chi \frac{C_{(n-1)/2}^n}{(2\Omega)^n} \left| \varphi_{3-k}^{(0)} - \varphi_k^{(0)} \right|^{n-1} \left(\varphi_{3-k}^{(0)} - \varphi_k^{(0)} \right) \right)$$

$$k = 1, 2$$

$$(2.4)$$

where $C_{(n-1)/2}^n$ is the binomial coefficient. It is easy to conclude that system (2.4) possesses two integrals of motion

$$N^2 = \left|\varphi_1^{(0)}\right|^2 + \left|\varphi_2^{(0)}\right|^2$$

$$H = \left|\phi_1^{(0)}\right|^{(n+1)} + \left|\phi_2^{(0)}\right|^{(n+1)} + \mu^* \left|\phi_1^{(0)} - \phi_2^{(0)}\right|^{(n+1)}. \tag{2.5}$$

The first integral of motion allows for a convenient change of coordinates

$$\varphi_1^{(0)} = N \cos\theta e^{i\delta_1}, \qquad \varphi_2^{(0)} = N \sin\theta e^{i\delta_2}. \tag{2.6}$$

Substituting (2.6) into (2.4), considering the relative phase $\delta = \delta_1 - \delta_2$ as a new variable, and rescaling the independent variable by law $\tau = \left[C_{(n-1)/2}^n N^{n-1}/(2\Omega)^n\right]\tau_1$, we reduce (2.4) to the real-valued system

$$\frac{d\delta}{d\tau} = \mu\left[\cos^{n-1}\theta - \sin^{n-1}\theta + 2\chi(1 - \sin 2\theta \cos\delta)^{\frac{n-1}{2}}\cot 2\theta \cos\delta\right],$$

$$\frac{d\theta}{d\tau} = \mu\chi(1 - \sin 2\theta \cos\delta)^{\frac{n-1}{2}}\sin\delta. \tag{2.7}$$

It is important to note that system (2.7) does not involve the energy-dependent frequency Ω and governed only by the constant parameters χ and n, thus making the global dynamics of system (2.7) invariant to the slow amplitudes $\left|\varphi_1^{(0)}\right|, \left|\varphi_2^{(0)}\right|$. We will show that variations of the governing parameters χ and n lead to both local and global bifurcations.

2.1.5   *Fixed Points and NNMs in the Neighborhood of Resonance*

The further study of the system dynamics is concentrated on the analysis of system (2.7). First, we find fixed points of (2.7). By setting $d\theta/d\tau = d\delta/d\tau = 0$, we obtain the following algebraic equations defining the fixed points of (2.7):

$$\cos^{n-1}\theta - \sin^{n-1}\theta + 2\chi(1 - \sin 2\theta \cos\delta)^{\frac{n-1}{2}}\cot 2\theta \cos\delta = 0,$$

$$(1 - \sin 2\theta \cos\delta)^{\frac{n-1}{2}}\sin\delta = 0. \tag{2.8}$$

By setting $\cos 2\theta = 0$, $\sin\delta = 0$, we obtain the following set of fixed points:

$$\left(\delta_1^{(1)}, \theta_1^{(1)}\right) = (0, \pi/4); \quad \left(\delta_2^{(1)}, \theta_2^{(1)}\right) = (\pi, 3\pi/4);$$

$$(\delta_1^{(2)}, \theta_1^{(2)}) = (0, 3\pi/4); \; (\delta_2^{(2)}, \theta_2^{(2)}) = (\pi, \pi/4). \tag{2.9}$$

The first pair of fixed points corresponds to the in-phase nonlinear normal mode (NNM) of the original system (2.1), while the second one corresponds to the out-of-phase NNM.

Additional fixed points $(\delta_k^{(3)}, \theta_k^{(3)})$, $k = 1, 2$, are defined by the conditions cos $2\theta \neq 0$, sin $\delta = 0$, that is,

$$\left(\cos^{n-1}\theta - \sin^{n-1}\theta\right) \tan 2\theta - 2\chi(1 + \sin 2\theta)^{\frac{n-1}{2}} = 0, \quad \delta = \pi, \qquad (2.10)$$

$$\left(\cos^{n-1}\theta - \sin^{n-1}\theta\right) \tan 2\theta + 2\chi(1 - \sin 2\theta)^{\frac{n-1}{2}} = 0, \quad \delta = 0. \qquad (2.11)$$

As mentioned above, the global system dynamics is governed by the parameters χ and n. In Fig. 2.3, we plot the solutions of Eq. (2.10) corresponding to $\delta = \pi$ versus variations of the parameter χ for four different values of n. It is easy to see that the branches of solutions bifurcate from the point $(\pi, \pi/4)$. We will not illustrate solutions of (2.11) because they are similar to those of (2.10) but for the range of $\theta \in [\pi, 2\pi]$.

Figure 2.3 demonstrates a qualitative change of the solutions of (2.10) for $n > 5$. These topological changes represent the results of transition from a supercritical to a subcritical pitchfork bifurcation undergone by the fixed points described in (Zhupiev and Mikhlin 1981, 1984). Note that the appearance of the subcritical bifurcation of the fixed points has a significant effect on the occurrence of strong beats as well as on the shape of the response.

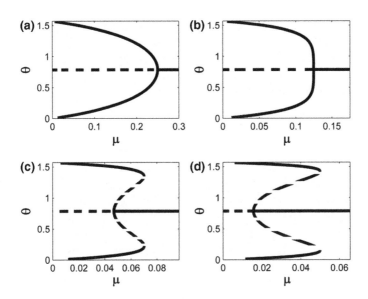

Fig. 2.3 Solutions of (2.10) versus variation of coupling strength χ: **a** $n = 3$; **b** $n = 5$; **c** $n = 5$; **d** $n = 9$. Stable branches of solution are denoted by *bold solid lines*, unstable branches of solutions are denoted by *dashed lines*. Horizontal axes $\theta = \pi/4$ correspond to the fixed point $\theta_2^{(2)}$

Before proceeding with the analysis of strong beating response, we analytically prove the occurrence of a subcritical pitchfork bifurcation at $n > 5$. In the first step, one can deduce that the branches of the solutions of (2.10) bifurcate from the fixed point $\theta_2^{(2)} = \pi/4$. To analyze motion near $\theta = \pi/4$, Eq. (2.10) is rewritten as

$$\chi = \frac{\left(\cos^{n-1}\theta - \sin^{n-1}\theta\right)\tan 2\theta}{2(1 + \sin 2\theta)^{\frac{n-1}{2}}}. \tag{2.12}$$

It follows from (2.12) that $\lim_{\theta \to \pi/4}(\partial\chi/\partial\theta) = 0$. In the next step, we insert the expansion $\theta = \pi/4 + \tilde{\theta}$, $(|\tilde{\theta}| \ll 1)$ into (2.12) to obtain

$$\chi = \frac{n-1}{2^n} + \frac{n(n-1)(n-2)}{3(2)^{n+1}}\tilde{\theta}^2 + O\left(\tilde{\theta}^3\right). \tag{2.13}$$

The first term in the right-hand side of (2.13) represents the first bifurcation value of coupling strength χ

$$\chi_{cr}^{(1)} = 2^{-n}(n-1), \tag{2.14}$$

at which the fixed point $(\delta_2^{(2)}, \theta_2^{(2)}) = (\pi, \pi/4)$ is transformed from the saddle point to the stable center for an arbitrary value of n. The transformation of the saddle point can be easily proved by performing a linear stability analysis (Starosvetsky and Ben-Meir 2013).

It follows from expansion (2.13) that the coefficient of the quadratic term is always positive for $n > 5$. In the special case of $n = 5$, this coefficient equals zero. In this case, one can easily show that the coefficient of the higher-order term in the expansion (i.e., the forth-order term) is negative. Hence, expansion (2.13) proves the formation of a subcritical pitchfork bifurcation for $n > 5$.

2.1.6 Limiting Phase Trajectories

Figures 2.3 and 2.4 depict phase portraits of system (2.7) for $n = 3$ and $n = 7$. The choice of these values of the parameter n is not arbitrary; it aims to show the global changes in the system dynamics caused by the transition from supercritical $(n > 5)$ to subcritical $(n > 5)$ pitchfork bifurcations.

In terms of the model (2.7), complete energy exchanges between the oscillators are associated with the *limiting phase trajectory* (LPT), which passes through zero point $\delta = \theta = 0$ and reaches the value $\theta = \pi/2$. We underline that, unlike the previous sections, there is no way to distinguish quasi-linear, moderately nonlinear, and strongly nonlinear regimes, because the system does not exhibit the quasi-linear behavior.

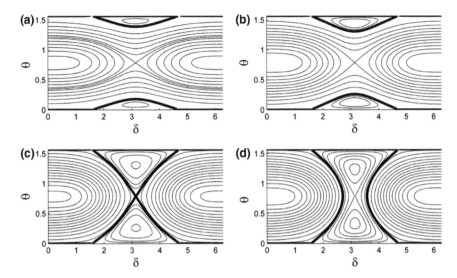

Fig. 2.4 Phase portrait of system (2.7) with $n = 3$ and different coupling strength: **a** $\chi = 0.075$, **b** $\chi = 0.1$; **c** $\chi = 0.1667$; **d** $\chi = 0.19$. Limiting phase trajectories (LPTs) are denoted by *bold lines* on each plot

Figure 2.4a, b demonstrate special orbits (bold lines) satisfying the initial condition $\delta = \theta = 0$. However, motion along these orbits cannot lead to complete energy exchange between the oscillators, as the trajectory does not reach the value $\theta = \pi/2$. Below, we refer to this kind of the phase trajectory as the LPT of the first kind.

Increasing the value of χ up to a certain critical value $\chi_{cr}^{(2)}$, one can observe the coalescence of the LPT of the first kind with the separatrix (Fig. 2.4c). This coalescence leads to the global bifurcation resulting in the formation of an LPT of the qualitatively different type (Fig. 2.3d), that is, of the LPT of the second kind.

Phase portraits in Fig. 1.18a are qualitatively similar to that ones in Fig. 2.4a. However, slightly increasing the strength of coupling χ (Fig. 2.5b), we observe the occurrence of an additional pair of unstable fixed points, and the transition of the fixed point $(\delta_2^{(2)}, \theta_2^{(2)})$ from the saddle point to the stable center. The latter observation is a result of the above described subcritical pitchfork bifurcation. Clearly, except the regular LPT starting at $\theta = 0$, there exists an additional branch of the LPT (we will refer to it as the "LPT bubble") encircling the center $(\delta_2^{(2)}, \theta_2^{(2)})$. This new type of LPT is illustrated in Fig. 2.5c. At a certain critical value $\chi = \chi_{cr}^{(2)}$, the regular LPT collides with the "bubble" LPT exactly at the saddle point, entailing the occurrence of the LPT of the second kind (Fig. 2.5d). The newborn LPT of the second kind has a topology different from that one in the case of $n = 3$ reported in previous sections.

It is seen in Fig. 2.5 that the LPT of the second kind in the case of $n > 5$ has a near rectangular shape instead of the triangle observed at $n = 3$ and $n = 5$.

In order to derive analytical conditions of the occurrence of the LPT of the second kind for any value of n, as well as to find a critical value $\chi_{cr}^{(2)}$, we consider the second integral of motion (2.5). Expressing H in terms of (θ, δ), one obtains

$$H(\delta, \theta, \chi) = \cos^{n+1}\theta + \sin^{n+1}\theta + \chi(1 - \sin 2\theta \, \cos \delta)^{(n+1)/2}. \qquad (2.15)$$

Equation (2.15) depicts phase trajectories of (2.7). We recall that the LPT of any kind passes through zero point $\delta = \theta = 0$. Substituting $\theta = 0$ into (2.15) yields the following exact value of $H(\delta, \theta, \chi)$ corresponding to the LPT:

$$H_{LPT}(\delta, \theta, \chi) = 1 + \chi. \qquad (2.16)$$

In the case of $n < 7$, an increase in the coupling parameter χ up to a critical value $\chi_{cr}^{(2)}$ corresponds to the transition from the LPT of the first kind to the LPT of the second kind. Since the LPT passes through the saddle point $(\delta_2^{(2)}, \theta_2^{(2)}) = (\pi, \pi/4)$, the value $\chi = \chi_{cr}^{(2)}$ can be found from the equality

$$H_{LPT}(\pi, \pi/4, \chi_{cr}^{(2)}) = 1 + \chi_{cr}^{(2)}. \qquad (2.17)$$

It follows from (2.15) and (2.17) that

$$\chi_{cr}^{(2)} = \left(1 - 2^{-\frac{(n-1)}{2}}\right)\left(2^{\frac{(n-1)}{2}} - 1\right). \qquad (2.18)$$

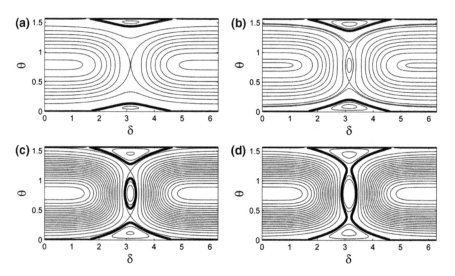

Fig. 2.5 Phase portrait of system (2.7) with $n = 7$ and different coupling strength: **a** $\chi = 0.040$; **b** $\chi = 0.055$; **c** $\chi = 0.061$; **d** $\chi = 0.063$. Limiting phase trajectories (LPTs) are denoted with the *bold line* on each plot

We recall that the critical value $\chi_{cr}^{(2)}$ is associated with the transition from the LPT of the first kind to the LPT of the second kind only for $n = 3$ and $n = 5$. If $n > 5$, then, requiring the fixed point $(\delta_2^{(2)}, \theta_2^{(2)})$ to lie on the LPT, we find a critical value $\chi_{cr}^{(3)}$ corresponding to the occurrence of the LPT "bubble." To this end, we require the branch of the LPT to cross a saddle point branching out from $(\delta_2^{(2)}, \theta_2^{(2)})$ through the pitchfork bifurcation. This branch of the solutions is given by Eq. (2.10). Thus, solving the following nonlinear system:

$$
\begin{aligned}
\left(\cos^{n-1}\theta - \sin^{n-1}\theta\right)\tan 2\theta - 2\chi_{cr}^{(3)}(1 + \sin 2\theta)^{\frac{n-1}{2}} &= 0, \\
\cos^{n+1}\theta + \sin^{n+1}\theta + \chi_{cr}^{(3)}(1 + \sin 2\theta)^{\frac{n+1}{2}} &= 1 + \chi_{cr}^{(3)},
\end{aligned}
\tag{2.19}
$$

one can derive a new criterion for the transition from localization to complete energy transfer for the case of $n > 5$. In Table 2.1, we summarize critical values of coupling strength χ corresponding to different types of dynamical transitions.

2.1.7 Numerical Analysis of the Fundamental Model

We perform numerical verifications of the theoretical model suggested in the previous section. We compare the response of the original system (2.1) with initial conditions $x_1(0) = 1$, $x_2(0) = 0$; $v_1(0) = v_2(0) = 0$ with the slow envelope (2.4) satisfying the corresponding initial conditions $\varphi_1(0) = i$, $\varphi_2(0) = 0$. To confirm the validity of the theoretical models for two different topologies of the LPT, we choose two representative values of n, namely $n = 3$ (Fig. 2.6) and $n = 7$ (Fig. 2.6).

From the results in Fig. 2.6a ($n = 3$), it is clear that the choice of coupling strength χ below the predicted threshold ($\chi < \chi_{cr}^{(2)} = 0.167$) leads to energy localization on the first oscillator. However, if the value of χ is increasing above the threshold ($\chi > \chi_{cr}^{(2)}$), we clearly observe the occurrence of a strong beating response, i.e., complete energy exchanges between the oscillators (Fig. 2.6b). Moreover, the slow flow envelop given by (2.4) is in a very good agreement with the full model (2.1). This means that the analytical model clearly recovers the mechanism of the transition from localization to recurrent energy transfer observed in the full model.

$$
\Sigma = \{x_1 = 0, \dot{x}_1 > 0\} \cap \{E(x_1, \dot{x}_1, x_2, \dot{x}_2) = 1\}
$$

Table 2.1 Critical values of coupling strength χ versus for different types of dynamical transitions

	Pitchfork bifurcation value	Formation of the LPT "bubble"	Transition from localization to a complete energy transfer
$n > 7$	$\chi_{cr}^{(1)} = 2^{-n}(n-1)$	–	$\chi_{cr}^{(2)} = \left(1 - 2^{-\frac{(n-1)}{2}}\right)\left(2^{\frac{n-1}{2}} - 1\right)$
$n > 7$	$\chi_{cr}^{(1)} = 2^{-n}(n-1)$	$\chi_{cr}^{(2)} = \left(1 - 2^{-\frac{(n-1)}{2}}\right)\left(2^{\frac{n-1}{2}} - 1\right)$	$\chi_{cr}^{(3)}$-solution of (25)

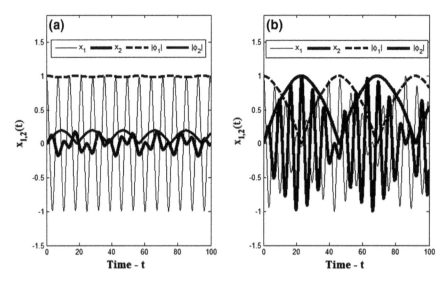

Fig. 2.6 Superposition of the slow envelope (4.4) on the full response (4.1) for $n = 3$ and different coupling strength: **a** $\chi = 0.08 < \chi_{cr}^{(2)}$; **b** $\chi = 0.19 > \chi_{cr}^{(2)}$. Initial conditions: $x_1(0) = 1$, $x_2(0) = 0$; $v_1(0) = v_2(0) = 0$ for the full system; $\varphi_1(0) = i$, $\varphi_2(0) = 0$ for the envelope

To better illustrate the effect of the transition from localization to a complete transport of energy, we construct the Poincaré maps corresponding to system chapter (4.1). To this end, we first restrict the system dynamics to an isoenergetic manifold. It is easy to show that, due to homogeneity of chapter (4.1), the total system energy can be normalized to unity ($E = 1$). Thus, fixing the total energy to a constant level, we restrict the dynamical flow of chapter (5.1) to the three-dimensional isoenergetic manifold $E(x_1, \dot{x}_1, x_2, \dot{x}_2) = 1$. By transversely intersecting the three-dimensional isoenergetic manifold by the two-dimensional cut plane $T : \{x_1 = 0\}$, one obtains the Poincaré map $P : Z \rightarrow Z$, where the Poincaré section is defined as.

Fundamental time-periodic solutions of a basic period T correspond to the period 1 equilibrium points in the Poincaré map. Additional subharmonic solutions of periods nT correspond to the period n equilibrium points of the Poincaré map, i.e., to the orbits that pierce the cut section n times before repeating themselves. Clearly, the construction of the Poincaré map ($P : Z \rightarrow Z$) effectively reduces the global system dynamics to the plane (x, v). The Poincaré section in Fig. 2.6a corresponds to the case of $\chi < \chi_{cr}^{(2)}$ (energy localization), while Fig. 2.6b corresponds to the case of $\chi > \chi_{cr}^{(2)}$ (strong beating response).

From the close observation of the results in Fig. 2.7, one can identify the formation of special orbits (marked with bold dots) passing through the origin. It is clear that this invariant set emanating from the origin constitutes a special orbit leading to a complete energy exchange between the oscillators (strong beating response). However, the results in Fig. 2.7a suggest that a similar orbit emanating from the origin do not lead to complete energy exchange. These special orbits of the Poincaré sections can be directly correlated to the LPT of the reduced model. This special orbit illustrated in Fig. 2.7a corresponds to energy localization (LPT of the first kind) while that one in Fig. 2.7b corresponds to complete energy exchange (LPT of the second kind).

Figures 2.8 and 2.9 present computational results for $n = 7$. Figure 2.8 depicts the time response for $n = 7$. Note that the response of the reduced-order model (2.4) agrees fairly well with the response of the full model (2.1) despite the relatively high power of nonlinearity. The theoretical prediction of the threshold value $\chi_{cr}^{(2)} = 0.0625$ is confirmed in Fig. 2.7.

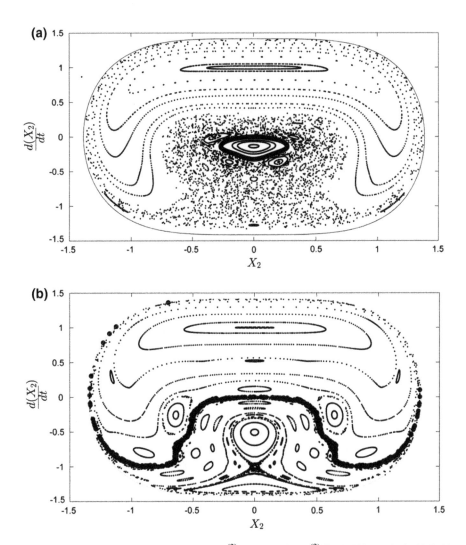

Fig. 2.7 Poincaré map for $n = 3$; $\chi = 0.08 < \chi_{cr}^{(2)}$ (**a**); $\chi = 0.2 > \chi_{cr}^{(2)}$ (**b**). LPT is marked with *bold dots*

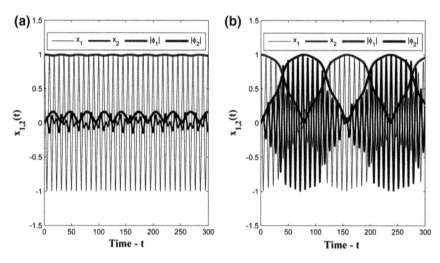

Fig. 2.8 Superposition of the slow envelope (2.4) on the full response (2.1) for $n = 7$ and different coupling strength: **a** $\chi = 0.05 < \chi_{cr}^{(2)}$; **b** $\chi = 0.05 < \chi_{cr}^{(2)}$. Initial conditions are indicated in Fig. 2.6

Figure 2.9 demonstrates that, despite the prevalence of the "chaotic sea" region, which covers almost the entire map, one can still observe the preservation of the special orbits corresponding to the LPTs of the first and the second types.

Numerical simulations show that, with an increase in the coupling parameter χ, the occurrence of the beating response is usually preceded by the regimes of a "mixed" type exhibiting a temporal energy localization followed by distinct irregular transitions into a beating-like response (Fig. 2.10) alternating with subsequent localizations.

In Tables 2.2 and 2.3, we compare the numerically obtained critical values of coupling χ with the theoretically derived values $\chi_{cr}^{(2)}$, $\chi_{cr}^{(3)}$ for different values of n. In numerical simulations, the total system energy is normalized to unity $E = 1$.

Tables 2.2 and 2.3 are constructed for two different initial excitations, namely $x_1(0) = 0$, $v_1(0) = \sqrt{2}$ (Table 2.2) and $x_1(0) = [(n + 1)(1 + \chi)^{-1}]^{1/(n+1)}$, $v_1(0) = 0$ (Table 2.3). We consider the following numerically found critical values of χ: χ_b, corresponding to the breakdown of localization followed by the occurrence of the response of a "mixed" type, and χ_{SB}, related to pure beatings (Fig. 2.11). This "mixed" type of response has been observed in the interval $\chi_b < \chi < \chi_{SB}$, or, in other words, in the interval between energy localization ($\chi < \chi_b$) and beating, associated with complete energy exchange ($\chi > \chi_{SB}$).

An insignificant difference in the critical values χ_b and χ_{SB} in Tables 2.2 and 2.3 can be explained by high sensitivity of system (2.1) to the change of the initial conditions. However, despite these deviations, one can note a good agreement between the numerical and analytical critical values in Tables 2.2 and 2.3.

It is important to note that the formation of regular beatings for $n \geq 9$ is problematic. However, even in the absence of a regular response, one can observe a transition from localization to complete energy exchanges at $\chi = \chi_{cr}^{(2)}$ provided the latter

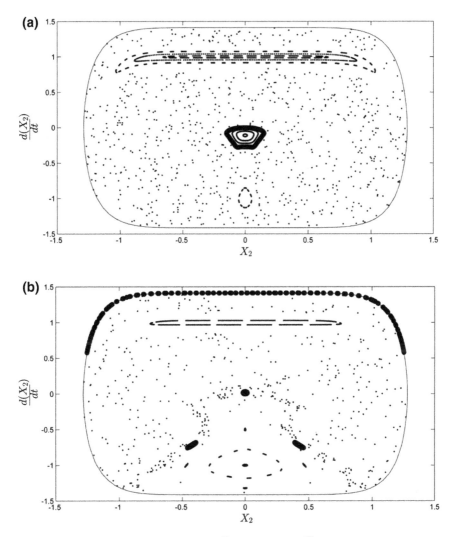

Fig. 2.9 Poincaré map for $n = 7$; $\chi = 0.05 < \chi_{cr}^{(2)}$ **(a)**; $\chi = 0.07 > \chi_{cr}^{(2)}$ **(b)**. LPT is marked with *bold dots*

exhibits highly irregular motion, i.e., the chaotic-like behavior. It follows from the results in Tables 2.2 amd 2.3 that, despite the absence of a regular beating response for $n \geq 9$, the analytical model predicts fairly well a critical value χ_b corresponding to the breakdown of localization followed by irregular energy transfer between the oscillators.

If the motion of the particles is transversal to the weightless string itself, the coupling between the particles is realized by the local strongly nonlinear interactions. If the preliminary stretching of the string is absent, they are cubic functions in the main approximation. However, we consider even more general case when the power may be not only three but an arbitrary odd integer.

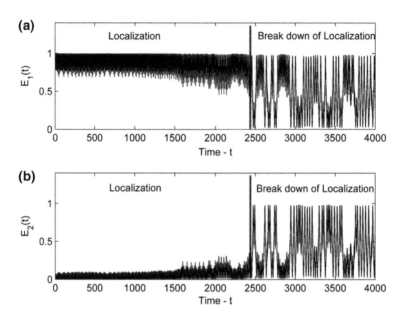

Fig. 2.10 Instantaneous energy corresponding to the mixed type of responses exhibiting temporal localization and sudden bursts of the beating-like behavior ($\chi_b = 0.09$, $n = 3$): **a** energy of the first oscillator; **b** energy of the second oscillator. Initial conditions: $x_1(0) = x_2(0) = 0$; $v_1(0) = \sqrt{2}$, $v_2(0) = 0$

Table 2.2 Comparison of the threshold values $\chi_{cr}^{(2)}$, $\chi_{cr}^{(3)}$ obtained from the theoretical analysis with the critical parameters χ_b, χ_{SB} obtained by numerical calculation of the response of system (2.1) subject to impulse excitation $x_1(0) = x_2(0) = 0$; $v_1(0) = \sqrt{2}$, $v_2(0) = 0$

	$n = 3$	$n = 5$	$n = 7$	$n = 9$	$n = 11$	$n = 13$
$\chi_{cr}^{(2)}$ ($n \leq 5$) $\chi_{cr}^{(3)}$ ($n \leq 7$)	0.1667	0.1071	0.0625	0.0445	0.0347	0.0285
χ_{SB}	0.17	0.112	0.069	–	–	–
χ_b	0.09	0.0875	0.054	0.038	0.03	0.027

Table 2.3 Comparison of the threshold values $\chi_{cr}^{(2)}$, $\chi_{cr}^{(3)}$ obtained from the theoretical analysis with the critical parameters χ_b, χ_{SB} obtained by numerical calculation of the response of system (2.1) subject to initial displacement $x_1(0) = [(n + 1)(1 + \chi)^{-1}]^{1/(n+1)}$, $x_2(0) = 0$; $v_1(0) = v_2(0) = 0$

	$n = 3$	$n = 5$	$n = 7$	$n = 9$	$n = 11$	$n = 13$
$\chi_{cr}^{(2)}$	0.1667	0.1071	0.0625	0.0445	0.0347	0.0285
χ_{SB}	0.173	0.11	0.0637	–	–	–
χ_b	0.165	0.108	0.0635	0.044	0.035	0.029

2.2 Non-local Sonic Vacuum

In this section, we remove the restrictions providing purely transversal motion of the particles accepted in the previous section. We formulate the problem for general multiparticle system assuming the presence of lateral springs without a preliminary stretching. However, we discuss here only the result obtained for two-particle model (Kikot et al. 2015), to compare them directly with those for the case of sonic vacuum.

2.2.1 The Model

Considered system is depicted in Fig. 2.11. It consists of n particles of identical mass m connected by linear inter-chain springs of elastic constant k_1; moreover, each particle is connected to the ground by two linear lateral springs of elastic constant k_2. It is assumed that all particles perform in-plane oscillations on the vertical plane (Oxy), and that all springs are unstretched at the equilibrium of the system corresponding to the line $y = z = 0$ (see Fig. 2.11). In addition, fixed-fixed boundary conditions are assumed for the particle chain, the unstretched length of the i-th inter-chain spring connecting particles $i - 1$ and i is being taken equal to l_i, for $i = 1, 2, \ldots, n$, and the unstretched lengths of the lateral springs are assumed to be equal to d. Considering the free in-plane oscillations of this system, the transverse and axial deformations of particle i are denoted by v_i and u_i, respectively, and the deformed length of the i-th inter-chain spring by l_i' and of the i-th lateral springs by d_i' (both lateral springs have equal stretched lengths due to symmetry). Without loss of generality gravity forces are disregarded, and it is assumed that no dissipation forces exist. Finally, without loss of generality, the normalization $\sum_{i=1}^{n+1} l_i = 1$ is imposed for the inter-chain springs. Then, the analysis follows the approach developed in (Manevitch and Vakakis 2014) for the corresponding system with no lateral grounding supports.

Applying Newton's law in the vertical and transverse directions, the equations of motion of i-th particle are expressed as,

$$
\begin{aligned}
m\ddot{u}_i + T_i \cos \phi_i - T_{i+1} \cos \phi_{i+1} + S_i \sin \theta_i \sin \psi_i = 0 \\
m\ddot{v}_i + T_i \sin \phi_i - T_{i+1} \sin \phi_{i+1} + S_i \sin \theta_i \cos \psi_i = 0
\end{aligned}
\tag{2.20}
$$

where $T_i = k_1 (l_i' - l_i)$ and $S_i = k_2 (d_i' - d)$ are the stretching forces (tensions) in the i-th inter-chain and lateral springs, respectively, ϕ_i is the angle between the i-th spring and the horizontal direction, θ_i is the angle between the deformed and undeformed positions of the i-th lateral springs, $\psi_i = \tan^{-1}(u_i/v_i)$, overdots denote differentiation with respect to the temporal variable t, and $i = 1, 2, \ldots, n$. At this

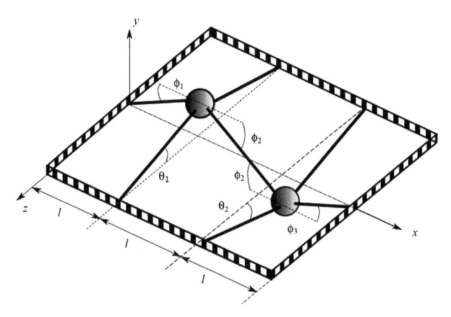

Fig. 2.11 Model of grounded nonlinear sonic vacuum consisting of 2 particles performing in-plane oscillations on the (*Oxy*) plane; the unstretched spring are depicted by *solid lines* and their undeformed states are shown by the *dashed lines*; *l* is the length of the springs; other notation—see text below

point, the elongation of the *i*-th inter-chain spring, $\varepsilon_{1i} = \left(l'_i - l_i\right)/l_i$, and the elongation $\varepsilon_{2i} = \left(d'_i - d\right)/d$ of the *i*-th lateral springs are introduced, and *the limiting case corresponding to predominantly low-energy transversal oscillations is considered*, taking into account only the corresponding leading-order geometrically nonlinear effects in the oscillating chain.

Accordingly, the trigonometric expressions in (2.20) are expressed in terms of the particle displacements v_i and u_i through the following geometric relations,

$$\cos \phi_i = \frac{l_i + u_i - u_{i-1}}{\left[\left(v_i - v_{i-1}\right)^2 + \left(l_i + u_i - u_{i-1}\right)^2\right]^{\frac{1}{2}}}$$

$$\sin \phi_i = \frac{v_i - v_{i-1}}{\left[\left(v_i - v_{i-1}\right)^2 + \left(l_{i-1} + u_i - u_{i-1}\right)^2\right]^{\frac{1}{2}}}$$

(2.21a)

$$\sin \theta_i = \frac{\left(u_i^2 + v_i^2\right)^{\frac{1}{2}}}{\left[u_i^2 + v_i^2 + d^2\right]^{\frac{1}{2}}}$$

(2.21b)

$$\cos \psi_i = \frac{v_i}{[u_i^2 + v_i^2]^{\frac{1}{2}}}$$

$$\sin \psi_i = \frac{u_i}{[u_i^2 + v_i^2]^{\frac{1}{2}}}$$

(2.21c)

and the spring deformations by,

$$l_i' - l_i = \left[(v_i - v_{i-1})^2 + (l_i + u_i - u_{i-1})^2 \right]^{1/2} - l_i \qquad (2.22a)$$

$$d_i' - d = \left[u_i^2 + v_i^2 + d^2 \right]^{1/2} - d \qquad (2.22b)$$

We will be interested in the limit of low-energy oscillations and sufficiently small angles ϕ_i, θ_i and ψ_i. Specifically, we will assume that the amplitudes of the transverse oscillations of the particles are much smaller than the axial distances (on the x-axis) between them; accordingly, we introduce the normalized displacements $\bar{u}_i = u_i/l$ and $\bar{v}_i = v_i/l$, and rescale the normalized displacements according to $\bar{u}_i \to \varepsilon^2 \bar{u}_i$ and $\bar{v}_i \to \varepsilon \bar{v}_i$. The small parameter $\varepsilon \ll 1$ is introduced to indicate the smallness of the transverse and axial normalized deformations and will be regarded as the small parameter in the perturbation analysis that follows. Moreover, we introduce the *slow timescale* $\tilde{t} = \varepsilon (k_1/m)^{1/2} t$ and assume for simplicity that $l_i = l = 1/(n+1)$, $i = 0, 1, \ldots, n+1$. Note that by the above rescalings, the axial displacements are assumed to be an order of magnitude smaller compared to the transverse ones.

Substituting these normalizations and rescalings into expressions (2.21a)–(2.21c) and (2.22a), (2.22b), omitting the overbars from the normalized displacements, and expanding in Taylor series with respect to the small parameter, we derive the following leading-order approximations that are valid in the low-energy limit:

$$\varepsilon_{1i} = \varepsilon^2 \left[(u_i - u_{i-1}) + \frac{1}{2} (v_i - v_{i-1})^2 \right] + \cdots, \qquad T_i = k_1 l \varepsilon_{1i}$$

$$\varepsilon_{2i} = \left(\frac{\varepsilon^2 l^2}{2d^2} \right) v_i^2 + \cdots, \qquad S_i = k_2 d \varepsilon_{2i}$$

$$\cos \phi_i = 1 + \cdots, \qquad \sin \phi_i = \varepsilon (v_i - v_{i-1}) + \cdots \qquad (2.23)$$

$$\sin \theta_i = \left(\frac{\varepsilon l}{d} \right) v_i + \cdots$$

$$\cos \psi_i = 1 + \cdots, \qquad \sin \psi_i = \frac{\varepsilon u_i}{v_i} + \cdots$$

In turn, substituting (2.23) into the equations of motion (2.20) governing the axial and transverse oscillations of the particles, we derive the following leading-order approximate equations valid in the low-energy limit:

$$\varepsilon^2 u_i'' + (2u_i - u_{i+1} - u_{i-1}) - \frac{1}{2}(v_{i+1} - v_i)^2 + \frac{1}{2}(v_i - v_{i-1})^2 + \mathcal{O}(\varepsilon^2) = 0$$

$$\text{(2.24a)}$$

$$v_i'' - (u_{i+1} - u_i)(v_{i+1} - v_i)$$
$$-\frac{1}{2}(v_{i+1} - v_i)^3 + (u_i - u_{i-1})(v_i - v_{i-1}) + \frac{1}{2}(v_i - v_{i-1})^3 + C v_i^3 + \mathcal{O}(\varepsilon^2)$$
$$= 0$$

$$\text{(2.24b)}$$

with $i = 1, \ldots, n$, $C = \left(\frac{k_2 l^2}{2k_1 d^2}\right)$, and prime denoting differentiation with respect to the slow timescale $\tilde{t}$.

We now express the Eqs. (2.24a) governing the $\mathcal{O}(\varepsilon^2)$ axial oscillations in terms of the axial tension developing in the chain,

$$\varepsilon^2 u_i'' + \bar{T}_i - \bar{T}_{i+1} + \cdots = 0 \tag{2.25}$$

where $\bar{T}_i$ denotes the rescaled tension in the i-th spring (after the previous normalizations and rescalings are imposed on the unscaled variable T_i), and with the understanding that $u_0 = u_{n+1} \equiv 0$ and $v_0 = v_{n+1} \equiv 0$ due to the fixed-fixed boundary conditions. Both (2.24a) and (2.25) are in singular form since in the limit $\varepsilon \to 0$ the derivative term vanishes, which enables the partition of the axial dynamics in terms of slow and fast components and the asymptotic treatment of the dynamics. Indeed, in the leading-order "slow" approximation with $\varepsilon = 0$, we may neglect the axial inertial effects in the dynamics to obtain $\bar{T}_1 = \bar{T}_2 = \cdots = \bar{T}_{n+1} \equiv \bar{T}$, which indicates that to leading-order, the rescaled tension in the springs is spatially uniform. In physical terms, this means that in slow timescale τ_1, the axial oscillations are relatively fast, so that an axial disturbance generated at a point of the chain propagates quickly to the remainder of the system. As a result, we can derive the following *non-local* representation of the average axial force in the low-energy limit:

$$\bar{T} = \frac{1}{n+1} \sum_{p=1}^{n+1} \bar{T}_p = \frac{1}{n+1} \sum_{q=0}^{n} \left[(u_{q+1} - u_q) + \frac{1}{2}(v_{q+1} - v_q)^2 \right]$$
$$= \frac{1}{2(n+1)} \sum_{q=0}^{n} (v_{q+1} - v_q)^2 \tag{2.26}$$

since $u_0 = u_{n+1} \equiv 0$. Note that the tension is uniform in the slow timescale, but when higher-order terms are taken into account, it becomes slowly varying (and not uniform) in space. This is equivalent to considering only the outer solutions in the axial equations of motion (2.24a) by neglecting the derivative term (or, setting $\varepsilon^2 = 0$) and deriving the following approximate (slow) expression,

$(2u_i - u_{i+1} - u_{i-1}) \approx \frac{1}{2}(v_{i+1} - v_i)^2 - \frac{1}{2}(v_i - v_{i-1})^2$, which combined with (2.26) brings the equations governing the transverse oscillations (2.24b) in the form:

$$\mu v_i''(\tilde{t}) + \mu \bar{T}[\underline{v}(\tilde{t})][2v_i(\tilde{t}) - v_{i+1}(\tilde{t}) - v_{i-1}(\tilde{t})] + v_i^3(\tilde{t}) + \cdots = 0, \quad i = 1, \ldots, n$$
$$v_0 = v_{n+1} \equiv 0$$

$$(2.27)$$

where $\bar{T}[\underline{v}(\tilde{t})] \equiv \frac{1}{2(n+1)}\left[\sum_{q=0}^{n}(v_{q+1}(\tilde{t}) - v_q(\tilde{t}))^2\right]$ is a quadratic term depending on the transverse displacement vector $\underline{v}(\tau) = [v_1 \quad \cdots \quad v_N]^T$, and $\mu \equiv C^{-1} = \left(\frac{2k_1 d^2}{k_2 l^2}\right)$.

The average longitudinal force may be regarded as a non-local force that is generated through the elastic extension of the chain by Hooke's law. The dynamical system (2.27) governs the leading-order slow transverse oscillations of the particles and is studied in detail in Sect. 2.3.

Considering the structure of the coupled oscillators (2.27), we note that there is complete absence of any linear stiffness component since all terms (except the inertia term) are of the third order. Accordingly, this system represents a nonlinear sonic vacuum since the linearized speed of sound in this medium is zero. Furthermore, similar to the nonlinear sonic vacuum derived in (Manevitch and Vakakis 2014), the system (2.27) governing the transverse oscillations of the particles contains strongly non-local terms even though starting system of Eqs. 2.20 involves only next-neighbor physical coupling between particles.

In the following analysis, we will consider system Eq. (2.27) in its simplest form corresponding to $n = 2$ in an attempt to highlight the highly complex dynamics that this system possesses. Following to the asymptotic treatment of the slow transverse oscillations of the chain, we will reconsider the axial oscillations governed by Eq. (2.24a) or Eq. (2.25) which will be shown to possess both slow and fast components.

2.2.2 Two-Particle System (n = 2): Slow Transverse Oscillations

For $n = 2$ system (2.27) reads,

$$\mu v_1'' + \frac{\mu}{6}\left[v_1^2 + (v_2 - v_1)^2 + v_2^2\right](2v_1 - v_2) + v_1^3 + \cdots = 0$$
$$\mu v_2'' + \frac{\mu}{6}\left[v_1^2 + (v_2 - v_1)^2 + v_2^2\right](2v_2 - v_1) + v_2^3 + \cdots = 0$$

$$(2.28)$$

and is in the form of two strongly nonlinear coupled oscillators. It can be also obtained directly from the variation of the following Lagrange function:

$$L = \frac{\dot{v}_1^2 + \dot{v}_2^2}{2} - \left[\frac{1}{4\mu} \left(v_1^4 + v_2^4 \right) - \frac{1}{6} \left(v_1^4 + v_2^4 - 2v_2^3 v_1 - 2v_1^3 v_2 + 3v_1^2 v_2^2 \right) \right]$$

Given that this system is homogeneous in its nonlinear terms, it admits similar nonlinear normal modes, i.e., time periodic oscillations corresponding to straight lines in the configuration plane (v_1, v_2). To compute them, we impose the linear relationship between the two coordinates, $v_2 = kv_1$, and substitute into (2.28) to find the modal constant k by the algebraic relation:

$$k \left\{ 1 + \frac{\mu}{6} \left[1 + (k-1)^2 + k^2 \right] (2-k) \right\} = k^3 + \frac{\mu}{6} \left[1 + (k-1)^2 + k^2 \right] (2k-1)$$

$$(2.29a)$$

There are always the solutions $k_{1,2} = \pm 1$ which correspond to in-phase and out-of-phase NNMs resulting due to the symmetry of the system. Two additional solutions are given by,

$$k_{3,4} = \frac{\mu - 3 \pm \sqrt{(\mu-3)^2 - 4\mu^2}}{2\mu} < 0$$

provided that $0 < \varepsilon < \mu < 1$ (note that if $\mu < \varepsilon$, we cannot separate transversal from longitudinal motions through the previous rescalings). This indicates that at $\mu = 1$, a bifurcation of the out-of-phase similar NNM takes place, generating two additional out-of-phase NNMs; this pitchfork bifurcation, together with other strongly nonlinear and non-stationary dynamical phenomena that can occur in system (2.28), will be studied in more detail in the following asymptotic analysis.

For each NNM, the leading-order approximations to the transverse displacements are computed by solving by quadratures the nonlinear equations for modal oscillators,

$$v_1'' + \left\{ \frac{1}{6} \left[1 + (k-1)^2 + k^2 \right] (2-k) + (1/\mu) \right\} v_1^3 + \cdots = 0,$$

$$v_2 = kv_1$$

$$(2.29b)$$

under the specified initial conditions of the problem.

2.2.3 Slow Flow Reduction of the Dynamics

We resort to a complexification/averaging methodology to analytically study its stationary and non-stationary resonance dynamics. To this end, we rewrite this system as,

$$\mu\left(v_1'' + \omega^2 v_1\right) + \varepsilon_1 \gamma \left\{ v_1^3 - \mu\omega^2 v_1 + \tfrac{\mu}{6}\left[v_1^2 + (v_2 - v_1)^2 + v_2^2\right](2v_1 - v_2)\right\} + \cdots = 0$$

$$\mu\left(v_2'' + \omega^2 v_2\right) + \varepsilon_1 \gamma \left\{ v_2^3 - \mu\omega^2 v_2 + \tfrac{\mu}{6}\left[v_1^2 + (v_2 - v_1)^2 + v_2^2\right](2v_2 - v_1)\right\} + \cdots = 0$$

$$(2.30)$$

where $\gamma = 1/\varepsilon_1$ and ε_1 are bookkeeping parameters used in the following asymptotic analysis, $\omega = (k_1/m)^{1/2}$, and identical terms were added and subtracted in the two equations. Then, we complexify the analysis by introducing the new complex variables,

$$\psi_1 = v_1' + j\omega v_1, \psi_2 = v_2' + j\omega v_2, j = (-1)^{1/2}$$

and the difference $\Phi = \psi_1 - \psi_2$. Expressing (2.11) in terms of the new complex variables, we obtain,

$$\mu\left(\psi_1' - j\omega\psi_1\right) = -\varepsilon_1 \gamma \left\{ \left(\frac{\psi_1 - \psi_1^*}{2j\omega}\right)^3 + \frac{\mu}{6}\left[\left(\frac{\psi_1 - \psi_1^*}{2j\omega}\right)^2 + \left(\frac{\phi - \phi^*}{2j\omega}\right)^2 + \left(\frac{\psi_2 - \psi_2^*}{2j\omega}\right)^2\right]\right.$$
$$\left. \times \left(2\frac{\psi_1 - \psi_1^*}{2j\omega} - \frac{\psi_2 - \psi_2^*}{2j\omega}\right) + \frac{j\mu\omega}{2}\left(\psi_1 - \psi_1^*\right)\right\}$$

$$\mu\left(\psi_2' - j\omega\psi_2\right) = -\varepsilon_1 \gamma \left\{ \left(\frac{\psi_2 - \psi_2^*}{2j\omega}\right)^3 + \frac{\mu}{6}\left[\left(\frac{\psi_1 - \psi_1^*}{2j\omega}\right)^2 + \left(\frac{\phi - \phi^*}{2j\omega}\right)^2 + \left(\frac{\psi_2 - \psi_2^*}{2j\omega}\right)^2\right]\right.$$
$$\left. \times \left(2\frac{\psi_2 - \psi_2^*}{2j\omega} - \frac{\psi_1 - \psi_1^*}{2j\omega}\right) + \frac{j\mu\omega}{2}\left(\psi_2 - \psi_2^*\right)\right\}$$

$$(2.31)$$

where $(\cdot)^*$ denotes complex conjugate. System (2.31) is now analyzed by applying the method of multiple scales by introducing the new timescales $\tilde{t}_0 = \tilde{t}, \tilde{t}_1 = \varepsilon_1 \tilde{t}$, $\tilde{t}_2 = \varepsilon_1^2 \tilde{t}, \ldots$ and considering the asymptotic expansions $\psi_i(\tilde{t}) = \psi_{i0}(\tilde{t}_0, \tilde{t}_1, \tilde{t}_2, \ldots)$ $+ \varepsilon_1 \psi_{i1}(\tilde{t}_0, \tilde{t}_1, \tilde{t}_2, \ldots) + \varepsilon_1^2 \psi_{i2}(\tilde{t}_0, \tilde{t}_1, \tilde{t}_2, \ldots) + \cdots, i = 1, 2$. Substituting in (2.31), expressing the time derivatives in terms of the new timescales and matching coefficients at different orders of ε_1, we obtain an hierarchy of subproblems governing the solutions of (2.31) at progressively higher of approximation. Since the following analysis will be restricted only up to terms of $\mathcal{O}(\varepsilon_1)$, only the two leading timescales will be considered.

The zeroth order approximation is obtained by matching coefficients of $\mathcal{O}(\varepsilon_1^0)$ and solving the following subproblem:

$$\frac{\partial \psi_{i0}(\tilde{t}_0, \tilde{t}_1)}{\partial \tilde{t}_0} - j\omega\psi_{i0}(\tilde{t}_0, \tilde{t}_1) = 0 \Rightarrow \psi_{i0}(\tilde{t}_0, \tilde{t}_1) = \varphi_{i0}(\tilde{t}_1)e^{j\omega\tilde{t}_0}, \quad i = 1, 2 \quad (2.32)$$

This provides the leading-order approximation to the solution and indicates that at the basic approximation, both particles perform slowly modulated transverse oscillations with common fast frequency ω. Hence, the following analysis is carried out under the assumption of 1:1 resonance in terms of the transverse oscillations, and all stationary and non-stationary dynamics discussed below satisfy this 1:1 resonance condition. The slowly varying complex amplitudes $\varphi_{i0}(\tilde{t}_1)$ are computed by substituting (2.32) into the problem governing the $\mathcal{O}(\varepsilon_1^1)$ approximation,

$$
\begin{aligned}
\mu \frac{\partial \psi_{11}}{\partial \tilde{t}_0} - j\mu\omega\psi_{11} = & -\mu \frac{\partial \varphi_{10}}{\partial \tilde{t}_1} - \gamma \Bigg\{ \left(\frac{\varphi_{10} e^{j\omega\tilde{t}_0} - \varphi_{10}^* e^{-j\omega\tilde{t}_0}}{2j\omega} \right)^3 \\
& + \frac{\mu}{6} \left[\left(\frac{\varphi_{10} e^{j\omega\tilde{t}_0} - \varphi_{10}^* e^{-j\omega\tilde{t}_0}}{2j\omega} \right)^2 + \left(\frac{\Phi_0 - \Phi_0^*}{2j\omega} \right)^2 \right. \\
& \left. + \left(\frac{\varphi_{20} e^{j\omega\tilde{t}_0} - \varphi_{20}^* e^{-j\omega\tilde{t}_0}}{2j\omega} \right)^2 \right] \\
& \times \left(\frac{\varphi_{10} e^{j\omega\tilde{t}_0} - \varphi_{10}^* e^{-j\omega\tilde{t}_0}}{j\omega} - \frac{\varphi_{20} e^{j\omega\tilde{t}_0} - \varphi_{20}^* e^{-j\omega\tilde{t}_0}}{2j\omega} \right) \\
& + \frac{j\mu\omega \left(\varphi_{10} e^{j\omega\tilde{t}_0} - \varphi_{10}^* e^{-j\omega\tilde{t}_0} \right)}{2} \Bigg\}
\end{aligned}
$$

$$
\begin{aligned}
\mu \frac{\partial \psi_{21}}{\partial \tilde{t}_0} - j\mu\omega\psi_{21} = & -\mu \frac{\partial \varphi_{20}}{\partial \tilde{t}_1} - \gamma \Bigg\{ \left(\frac{\varphi_{20} e^{j\omega\tilde{t}_0} - \varphi_{20}^* e^{-j\omega\tilde{t}_0}}{2j\omega} \right)^3 \\
& + \frac{\mu}{6} \left[\left(\frac{\varphi_{10} e^{j\omega\tilde{t}_0} - \varphi_{10}^* e^{-j\omega\tilde{t}_0}}{2j\omega} \right)^2 + \left(\frac{\Phi_0 - \Phi_0^*}{2j\omega} \right)^2 \right. \\
& \left. + \left(\frac{\varphi_{20} e^{j\omega\tilde{t}_0} - \varphi_{20}^* e^{-j\omega\tilde{t}_0}}{2j\omega} \right)^2 \right] \\
& \times \left(\frac{\varphi_{20} e^{j\omega\tilde{t}_0} - \varphi_{20}^* e^{-j\omega\tilde{t}_0}}{j\omega} - \frac{\varphi_{10} e^{j\omega\tilde{t}_0} - \varphi_{10}^* e^{-j\omega\tilde{t}_0}}{2j\omega} \right) \\
& + \frac{j\mu\omega \left(\varphi_{20} e^{j\omega\tilde{t}_0} - \varphi_{20}^* e^{-j(2.2.13)\omega\tilde{t}_0} \right)}{2} \Bigg\}
\end{aligned}
\tag{2.33}
$$

and eliminating secular terms proportional to $e^{j\omega\tilde{t}_0}$ from the right-hand sides which render the $\mathcal{O}(\varepsilon_1^1)$ approximations non-uniformly valid in the timescale $\tilde{t}_0$. This results in the following set of slowly modulated complex equations in the timescale $\tilde{t}_1$:

$$\mu\frac{\partial\varphi_{10}}{\partial\tilde{t}_1} = \gamma\left\{\frac{3j\varphi_{10}|\varphi_{10}|^2}{8\omega^3} - \frac{j\mu\omega\varphi_{10}}{2} + \frac{j\mu\left[\varphi_{10}^2 + (\varphi_{10} - \varphi_{20})^2 + \varphi_{20}^2\right]\left(2\varphi_{10}^* - \varphi_{20}^*\right)}{48\omega^3}\right.$$

$$\left. + \frac{j\mu\left[|\varphi_{10}|^2 + |\varphi_{10} - \varphi_{20}|^2 + |\varphi_{20}|^2\right]\left(2\varphi_{10} - \varphi_{20}\right)}{24\omega^3}\right\}$$

$$\mu\frac{\partial\varphi_{20}}{\partial\tilde{t}_1} = \gamma\left\{\frac{3j\varphi_{20}|\varphi_{20}|^2}{8\omega^3} - \frac{j\mu\omega\varphi_{20}}{2} + \frac{j\mu\left[\varphi_{10}^2 + (\varphi_{10} - \varphi_{20})^2 + \varphi_{20}^2\right]\left(2\varphi_{20}^* - \varphi_{10}^*\right)}{48\omega^3}\right.$$

$$\left. + \frac{j\mu\left[|\varphi_{10}|^2 + |\varphi_{10} - \varphi_{20}|^2 + |\varphi_{20}|^2\right]\left(2\varphi_{20} - \varphi_{10}\right)}{24\omega^3}\right\}$$

$$(2.34)$$

This represents the *complex slow flow* of the dynamics of system (2.28). As shown below, it is fully integrable and, hence, analytically solvable since it possesses two first integrals of motion. This indicates that the slow dynamics at the timescale $\tilde{t}_1$ can be exactly determined.

A first integral of the slow flow (2.34) is reflecting energy conservation and is given by

$$|\varphi_{10}(\tilde{t}_1)|^2 + |\varphi_{20}(\tilde{t}_1)|^2 = N \tag{2.35}$$

To compute a second independent first integral of motion, we express the slow flow (2.34) in terms of real variables by introducing the following polar transformations:

$$\varphi_{10} = \sqrt{N}\,\cos\theta\,e^{j\delta_1} \quad\text{and}\quad \varphi_{20} = \sqrt{N}\,\sin\theta\,e^{j\delta_2} \tag{2.36}$$

Substituting (2.36) into (2.34), setting separately the real and imaginary parts equal to zero, and manipulating the resulting equations, we obtain the following equations which represent the slow flow of the dynamics of system (2.28) in real coordinates, on the isoenergetic manifold $N = N_0$:

$$\frac{d\theta}{d\tilde{t}_1} = \frac{3\gamma N_0}{48\omega^3}(\sin 2\theta \ \sin 2\Delta - 2\sin\Delta) \tag{2.37a}$$

$$\mu\sin\theta\cos\theta\frac{d\Delta}{d\tilde{t}_1} = -\frac{3\gamma N_0}{32\omega^3}\sin 4\theta + \frac{\gamma\mu N_0}{24\omega^3}\left(\frac{3}{4}\sin 4\theta\,\cos 2\Delta 3\,\cos 2\theta\,\cos\Delta\right) \tag{2.37b}$$

where $\Delta = \delta_2 - \delta_1$. In particular, system (2.37a) and (2.37b) represents a reduction of the dynamics on the *isoenergetic slow flow two-torus* $(\theta, \Delta) \in [0, \pi/2] \times [0, \pi]$.

2.2.4 Stationary and Non-stationary Dynamics

The stationary (time periodic) solutions of system (2.28) correspond to $\frac{d\theta}{d\tilde{t}_1} = \frac{d\Delta}{d\tilde{t}_1} = 0$, since this leads to constant-amplitude fast oscillations of the leading-order approximations (2.32). It can be easily shown that $(\theta_e, \Delta_e) = (\pi/4, 0)$ is an equilibrium point on the torus corresponding to the in-phase similar NNM with modal constant $k_1 = 1$; the resulting solutions of system (2.28) are given by $v_1(\tilde{t}) = v_2(\tilde{t}) = A \sin(\omega\tilde{t} + \delta_1) + \mathcal{O}(\varepsilon)$, where the amplitude A and phase δ_1 are determined by the initial conditions. Similarly, the equilibrium point $(\theta_e, \Delta_e) = (\pi/4, \pi)$ corresponds to the out-of-phase similar NNM with $k_2 = -1$ with solutions $v_1(\tilde{t}) = -v_2(\tilde{t}) = A \sin(\omega\tilde{t} + \delta_1) + \mathcal{O}(\varepsilon)$. Additional equilibrium solutions (NNMs) correspond to $\sin 2\theta \sin 2\Delta - 2 \sin \Delta = 0$, which leads to two additional equilibrium positions $(\theta_e, \Delta_e) = \left(\theta^*_{e1,2}, \pi\right)$, where $\theta^*_{e1,2}$ are the two real roots of $\sin 2\theta = -2\mu/(\mu - 3)$ for $\mu < 1$; these correspond to the two bifurcating similar NNMs with modal constants $k_3, k_4 < 0$.

To study the non-stationary dynamics of system (2.28) and investigate changes in the global dynamics due to the previous bifurcation of NNMs, we reconsider the reduced slow flow on the isoenergetic two-torus (2.37a), (2.37b), and rescale the slow timescale as $\tilde{t}_1 = \frac{\gamma N_0}{32\omega^3}\tilde{t}_1$ to express it in the following simpler form:

$$\frac{d\theta}{d\tilde{t}_1} = 2(\sin 2\theta \sin 2\Delta - 2 \sin \Delta) \tag{2.38a}$$

$$\frac{d\Delta}{d\tilde{t}_1} = \frac{2}{\mu \sin 2\theta}[-3 \sin 4\theta + \mu (\sin 4\theta \cos 2\Delta - 4 \cos 2\theta \cos \Delta)] \tag{2.38b}$$

It can be shown that this system has the first integral of motion,

$$\mu\left(\frac{1}{2} \sin^2 2\theta \cos 2\Delta - 2 \sin 2\theta \cos \Delta\right) + \frac{3}{4} \cos 4\theta = H \tag{2.39}$$

which enables the exact analytic solution of system (2.38a), (2.38b). This represents a second independent first integral of motion of the slow flow (2.34), in addition to the energy-conservation first integral (2.35). Before proceeding to a more detailed analytical treatment of the dynamics on the two-torus (2.38a) and (2.38b) we discuss the bifurcations that occur in this system as the parameter μ changes.

There are two critical values for the parameter μ corresponding to two bifurcations affecting the stationary and non-stationary dynamics on the two-torus Eqs. (2.38a) and (2.38b). The first critical value, $\mu_{c1} = 1$, is the point of the pitchfork bifurcation of the out-of-phase NNM discussed in the introduction of this section, whereas the second critical value is concerned with a bifurcation of a *limiting phase trajectory* and the occurrence of strong energy exchanges between the coupled oscillators Eq. (2.28). As shown, a unified description of highly non-stationary resonance dynamics can be performed by introducing the concept of LPT. In fact, the LPT may be regarded as the orbit that is "maximally distant" from the stationary points (NNMs) on the two-torus (2.38a and 2.38b) and passes through the point $(\theta, \Delta) = (0, 0)$. In that context, the second critical value, $\mu = \mu_{c2}$, corresponds to the point where an LPT of the slow flow coincides with a separatrix, after which its topological structure changes drastically and from spatially extended becomes spatially localized. As shown in the numerical results presented below, critical value of orbits of the slow flow (2.38a and 2.38b) passing through the points $(\theta, \Delta) = (0, 0)$ and $(\pi/4, \pi)$ corresponds to the same value of the first integral Eq. (2.39). This estimates the second critical value as $\mu_{c2} = 0.6$.

The aforementioned transitions of the non-stationary dynamics of system Eq. (2.28) are depicted in Fig. 2.11 where the slow flow on the isoenergetic two-torus Eqs. (2.38a) and (2.38b) is depicted for varying values of μ. These results illustrate the transition of the dynamics of system (2.28) from complete energy exchanges between the two oscillators and energy localization in one of these oscillators. For $\mu > \mu_{c1}$, the torus possesses two stable similar NNMs, whereas after the NNM bifurcation at $\mu = \mu_{c1}$, there are four similar NNMs, with the out-of-phase NNM becoming unstable and the other NNMs being stable. The instability of the out-of-phase NNM for $\mu < \mu_{c1}$ affects drastically the global stationary dynamics on the two-torus since it generates two additional NNMs and two homoclinic loops (separatrices) that emanate from the unstable mode. However, the full energy exchange described by LPTs still remains possible. *Key to understanding non-stationary, resonant energy exchanges between the two coupled oscillators (Eq. (9)) is the topological changes of the LPTs* in Fig. 2.12 as μ varies. For $\mu > \mu_{c2}$, there occur strong energy exchanges along an LPT of the system, signified by strongly modulated transverse oscillations of the two particles of the system (or nonlinear beat phenomena). This is due to the fact that for $\mu > \mu_{c2}$, the LPT trajectory connects the two distinct localized states i.e. $\theta = 0$ (energy localization on the first oscillator) with $\theta = \pi/2$ (energy localization on the second oscillator) interval $(0, \pi/2)$; in turn, by the polar transformations Eq. (2.36) we deduce that intense energy exchanges between the two oscillators take place on the LPT (and on quasi-periodic orbits in its vicinity), with energy being continuously exchanged between the first oscillator (where the energy is localized for $\theta \approx 0$) and the second (for $\theta \approx \pi/2$). At $\mu = \mu_{c2}$ the LPT coincides with the speratrix emanating form the unstable out-of-phase NNM, signifying the end of strong energy exchanges in the dynamics. Indeed, for $\mu < \mu_{c2}$, the topological structure of the LPT changes from a single spatially extended orbit to two spatially localized orbits, with drastically decreased ranges in terms of θ for each orbit. This indicates a transition of the resonant non-stationary dynamics of the system, since for $\mu < \mu_{c2}$, the oscillations of system (2.28) become localized to either one of the two oscillators (within each of the two disjoint LPTs) and the intensity of energy exchanges between oscillators drastically decreases. Further decrease in μ enhances the intensity of energy

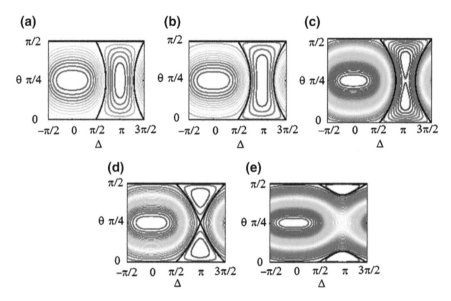

Fig. 2.12 Reduced slow flow on the isoenergetic two-torus (2.38a and 2.38b) for **a** $\mu = 1.4$, **b** $\mu = \mu_{c1} = 1.0$, **c** $\mu = 0.8$, **d** $\mu = \mu_{c2} = 0.6$, **e** $\mu = 0.3$; the LPTs are indicated by the bold lines. Different colors in contour lines correspond to values of the integral (2.39) and are used to highlight the topology of the dynamics on the torus

localization in either one of the two oscillators and energy exchanges between them increasingly diminish.

The change of the topological structure of the LPT for varying μ is numerically shown in Fig. 2.13, where the two LPTs on the two-torus (Eqs. 2.38a and 2.38b) for initial conditions $\theta(0) = 1 \times 10^{-3}$ and $\Delta(0) = 0$ (corresponding to energy initially localized to the first oscillator), and values of μ before and after the LPT bifurcation are depicted. We note that in the regime of strong energy exchanges ($\mu = 0.8 > \mu_{c2}$), the dependencies of the angles θ and Δ on the rescaled slow time $\bar{\tau}_1$ for the LPT are non-smooth, resembling sawtooth, and square-wave functions, respectively; whereas in the regime of localization ($\mu = 0.3 < \mu_{c2}$) the angles for the LPT have smoother waveforms. These observations will help us derive analytical approximations for these orbits representing non-stationary dynamics in the next section. In Fig. 2.14, we confirm the previous analytical predictions by direct numerical simulations of system (9) with initial conditions $v_1(0) = 1$ and $v_1'(0) = v_2(0) = v_2'(0) = 0$ and varying values of μ before and after the LPT bifurcation. The predicted strong resonant energy exchanges before the bifurcation and the localization of the motion after it are numerically verified from these results.

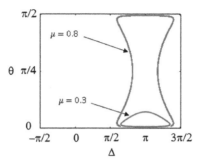

Fig. 2.13 Topological change of the LPT before and after the second bifurcation point $\mu = \mu_{c2} = 0.6$ and initial energy localized to the first oscillator: **a** $\mu = 0.8$ corresponding to strong energy exchanges between the two oscillators and spatially extended LPT on the two-torus, **b** $\mu = 0.3$ corresponding to localization in the first oscillator and spatially localized LPT

Fig. 2.14 Direct numerical simulations of system (2.28) for the case when initial energy is confined in the first oscillator ($v_1(0) = 1$ and $v_1'(0) = v_2(0) = v_2'(0) = 0$) and **a** $\mu = 0.8$, **b** $\mu = 0.3$

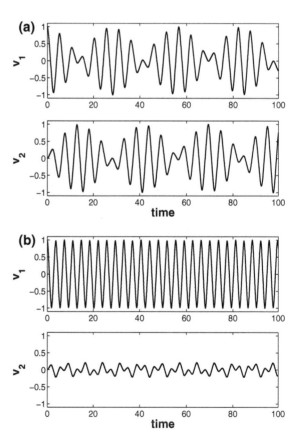

2.2.5 Analytical Approximations of the LPTs on the Two-Torus

The numerical results of the previous section highlight the importance of the LPT for understanding of resonant energy exchanges between the two coupled oscillators in system (2.28). Hence, in this section, we provide an analytical study of the LPTs in the regimes before and after their bifurcation with the separatrix on the two-torus. First, we start with the analysis of the LPT for $\mu > \mu_{c2}$, in the regime where intense energy exchanges between the two oscillators Eq. (2.28) take place. This analysis is motivated by the numerical results depicted in Fig. 2.13a which provide insight on the highly non-stationary dynamics on the two-torus Eqs. (2.38a) and (2.38b) corresponding to the LPT.

To this end, we reconsider the reduced slow flow Eqs. (2.38a) and (2.38b) in the regime before the second bifurcation and develop analytical approximations for the LPT by noting that on that orbit the dependencies of the angles θ and Δ with respect to the rescaled slow timescale $\bar{\tau}_1$ are periodic but discontinuous, resembling sawtooth and square-wave functions, respectively (see Fig. 2.13a). At this point, we need to emphasize that an LPT is a period orbit in the slow torus with *finite period*, contrary to a separatrix in the same torus which is an orbit of *infinite period*. Motivated by these observations, we resort to the method of non-smooth temporal transformations (Pilipchuk 2010) for analytically studying periodic orbits (and their bifurcations) of strongly nonlinear dynamical systems. We express the sought LPTs in terms of two new non-smooth variables, τ and e, as follows,

$$\begin{aligned} \theta(\bar{\tau}_1) &= X_1[\tau(\bar{\tau}_1/a)] + e(\bar{\tau}_1/a)Y_1[\tau(\bar{\tau}_1/a)] \\ \Delta(\bar{\tau}_1) &= X_2[\tau(\bar{\tau}_1/a)] + e(\bar{\tau}_1/a)Y_2[\tau(\bar{\tau}_1/a)] \end{aligned} \tag{2.40}$$

where $a = T/4$ represents the (yet unknown) quarter-period of the LPT in terms of the rescaled slow timescale $\bar{\tau}_1$. The non-smooth functions τ and e are used to replace the slow timescale $\bar{\tau}_1$ in the reduced system (2.38a) and (2.38b) and are defined by the expressions,

$$\tau(x) = \frac{1}{2}\left(\frac{2}{\pi}\sin^{-1}\left(\sin(\pi x - \frac{\pi}{2}\right) + 1\right), \quad e(x) = \frac{d\tau(x)}{dx} \tag{2.41}$$

The representations (2.4) provide decompositions of a periodic functions $\theta(\bar{\tau}_1)$ and $\Delta(\bar{\tau}_1)$ in terms of symmetric (the X-components) and anti-symmetric (the Y-components) parts. Moreover, considering the numerical time series of these angles for the LPT of Fig. 2.13a, we need to set $Y_1[\tau(\tilde{\tau}_1/a)] = X_2[\tau(\tilde{\tau}_1/a)] = 0$ in Eq. (2.40) so that the analytical representations for the angles $\theta(\bar{\tau}_1)$ and $\Delta(\bar{\tau}_1)$ resemble sawtooth and a square-wave waveforms, respectively, and retain the symmetric and anti-symmetric features, respectively, of the numerical results. In the following analysis, we consider only the leading-order approximations for the functions X_1 and Y_2, and of the quarter-period a; for a regular perturbation scheme for estimating higher-order approximations, we refer to (Manevitch and Smirnov 2010c: 3).

Considering the first integral (2.39) of the reduced slow flow, and imposing the condition that the LPT passes through the point $(\theta, \Delta) = (0, 0)$, we compute $H = 3/4$, so the equation describing the LPT on the two-torus is given by,

$$\mu\left(\frac{1}{2}\sin^2 2\theta \cos 2\Delta - 2\sin 2\theta \cos \Delta\right) + \frac{3}{4}\cos 4\theta = \frac{3}{4} \qquad (2.42a)$$

leading to the following functional relationship between the angles θ and Δ:

$$\cos \Delta = \frac{1}{\sin 2\theta} \pm \sqrt{\frac{1}{\sin^2 2\theta} + \frac{\mu + 3}{2\mu}} \qquad (2.42b)$$

In the following analysis, we consider only the $(-)$ sign in this expression so the result holds for $0 < \bar{t}_1 < a$, i.e., over a quarter-period of the LPT [in fact, this result estimates the component $Y_2[\tau(\bar{t}_1/a)]$ in (2.40)]. Over the same quarter-period, the angle θ varies linearly in $\bar{t}_1$ between the two limiting values 0 and $\pi/2$, corresponding to energy confined in the first and the second oscillator, respectively, (this estimates the component $X_1[\tau(\bar{t}_1/a)]$ in Eq. (2.40). To extend these analytical results over the entire time domain and obtain a symmetric periodic representation for the angle $\theta(\bar{t}_1)$, and an anti-symmetric representation for the angle $\Delta(\bar{t}_1)$, we take into account Eq. (2.40) and express the leading-order analytical approximation for the LPT as:

$$\theta(\bar{t}_1) = \frac{\pi}{2}\tau\left(\frac{\bar{t}_1}{a}\right)$$
$$\Delta(\bar{t}_1) = -e\left(\frac{\bar{t}_1}{a}\right)\cos^{-1}\left[\frac{-\frac{\mu+3}{2\mu}\sin\left(\pi\tau\left(\frac{\bar{t}_1}{a}\right)\right)}{1 + \sqrt{1 + \frac{\mu+3}{2\mu}\sin^2\left(\pi\tau\left(\frac{\bar{t}_1}{a}\right)\right)}}\right] \qquad (2.43)$$

Note that by the definitions of the non-smooth variables Eq. (2.41), we get a symmetric (odd) sawtooth-like periodic extension for $\theta(\bar{t}_1)$, and an anti-symmetric (even) square-wave-like periodic extension for $\Delta(\bar{t}_1)$. These results are in full agreement with the numerical waveforms depicted in Fig. 2.14.

Now, it remains to analytically estimate the period $T = 2a$ of the LPT. This can be performed in terms of a definite integral, since Eq. (2.42b) yields the dependence $\Delta = \Delta(\theta)$. In addition, from Eq. (2.38a), we can write that

$$d\bar{t}_1 = \frac{d\theta}{2(\sin 2\theta \sin 2\Delta - 2\sin \Delta)} \Rightarrow \int_0^a d\bar{t}_1 = \int_0^{\pi/2} \frac{d\theta}{2(\sin 2\theta \sin 2\Delta - 2\sin \Delta)}$$

$$(2.44a)$$

or

$$a = \int_0^{\pi/2} \frac{d\theta}{2(\sin 2\theta \, \sin 2\Delta(\theta) - 2 \, \sin \Delta(\theta))} \tag{2.44b}$$

This is the exact expression for the half-period of the LPT in slow time and yields excellent agreement with the results of direct numerical integration of system (2.20) (cf. Fig. 2.15). This last computation completes the leading-order analytical approximation for the LPT before the second bifurcation (i.e., in the regime of intense energy exchanges between the oscillators), given by Eq. (2.43) and (2.44b). The derived analytical approximations can be used for analytically studying the strongly nonlinear, non-stationary dynamics of system Eq. (2.28) involving highly intense energy exchanges between the two oscillators.

Considering the original system Eq. (2.28) governing the transverse oscillations of the two particles, assuming initial conditions $v_1(0) = 1$ and $v_1'(0) = v_2(0) = v_2'(0) = 0$, and that $\delta_1 = 0, \delta_2 = \Delta$, the leading-order approximation for the LPT in the highly intense energy exchange of the dynamics is expressed as $v_1(\tilde{t}) \approx \cos(\theta(\tilde{t}/32\omega)) \, \sin(\omega\tilde{t})$ and $v_2(\tilde{t}) \approx \sin(\theta(\tilde{t}/32\omega)) \, \sin(\omega\tilde{t} + \Delta(\tilde{t}/32\omega))$. In Fig. 2.15, we compare the analytical approximations with the numerical solutions for the LPT corresponding to $\mu = 0.8$ (cf. Fig. 2.13a) and the aforementioned initial conditions, and note satisfactory agreement.

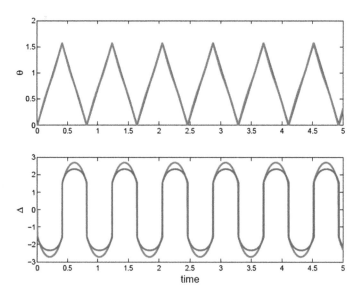

Fig. 2.15 Comparison between the analytical solution for the LPT derived by Eqs. (2.43) and (2.44b) (*blue* (*solid*) *line*), and the LPT computed by direct numerical simulation of Eq. (2.28) (*red* (*solid*) *line*) for $\mu = 0.8$ (regime of intense energy exchanges between the oscillators)

We now analyze the LPT for $\mu < \mu_{c2}$, i.e., in the regime where weak energy exchanges between the two oscillators Eq. (2.28) take place, and the oscillations are localized in either one of them. This analysis is motivated by the numerical results depicted in Fig. 2.13b which provide insight on the near-stationary dynamics on the two-torus (2.38a) and (2.38b) corresponding to the LPT in that regime of the dynamics. We note that although we employ to term "localization," to describe the transverse oscillations of system (2.28) and the absence of *complete* energy exchange, the two particles' *partial* energy exchange is still possible. In this case, the LPT corresponds to the orbit where maximum possible (partial) energy exchange between the two oscillators (2.28) occurs. Moreover, in contrast to the previous regime of intense resonant energy exchange, for $\mu < \mu_{c2}$, there are two disjoint LPTs corresponding to localized resonant oscillators to either one of the two oscillators.

Similar to the previous case of full energy exchange, the LPT is defined as the trajectory which passes through the point $(\theta, \Delta) = (0, 0)$. Hence, all previous calculations still hold and the functional relation between the angle variables on the LPT can be expressed as Eq. (2.42b). In turn, the temporal dependencies of the two angle variables are expressed as

$$\theta(\tilde{t}_1) = A\,\tau\!\left(\frac{\tilde{t}_1}{a}\right)$$

$$\Delta(\tilde{t}_1) = e\!\left(\frac{\tilde{t}_1}{a}\right)\left\{\pi - \cos^{-1}\left[\frac{1}{\sin\left(2A\,\tau\left(\frac{\tilde{t}_1}{a}\right)\right)} \pm \sqrt{\frac{1}{\sin^2\left(2A\,\tau\left(\frac{\tilde{t}_1}{a}\right)\right)} + \frac{\mu+3}{2\mu}}\right]\right\} \qquad (2.45a)$$

where the amplitude A is determined from the energy-conservation first integral (2.39) as

$$A = \frac{1}{2}\sin^{-1}\left(\frac{4-\mu}{3\mu}\right) \qquad (2.45b)$$

and the period $T = 2a$ by employing the methodology outlined for the LPT in the regime of intense energy exchanges:

$$a = \int_0^A \frac{d\theta}{2\left(\sin 2\theta\,\sin 2\Delta(\theta) - 2\,\sin\Delta(\theta)\right)} \qquad (2.45c)$$

As shown in Fig. 2.16, the comparison between the analytical solutions (2.45a), (2.45b) and direct numerical simulations of the original system (2.28) is satisfactory.

This completes the analytical and numerical study of the transverse oscillations of system (2.28), including both the stationary and non-stationary regimes and bifurcations of these motions. In the next section, we consider the axial motions of the system of two particles governed by the singularly perturbed Eqs. (2.24a)

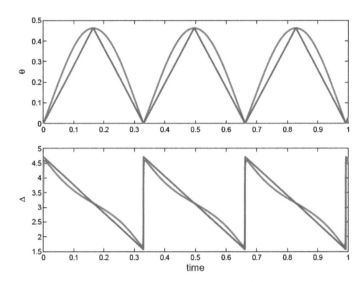

Fig. 2.16 Comparison between the analytical solution for the LPT derived by (2.45a)–(2.45c) (*blue* (*solid*) *line*), and the LPT computed by direct numerical simulation of (2.28) (*red* (*solid*) *line*) for $\mu = 0.3$ (regime of weak energy exchanges between the oscillators)

2.2.6 Mixed Slow/Fast Axial Oscillations for n = 2

To obtain the leading-order analytical approximations for the axial oscillations of the particles, we reconsider system (2.24a) which for $n = 2$ is expressed as:

$$
\begin{aligned}
\varepsilon^2 u_1'' + (2u_1 - u_2) &= -\tfrac{1}{2}\left[v_1^2 - (v_2 - v_1)^2\right] + \cdots \\
\varepsilon^2 u_2'' + (2u_2 - u_1) &= -\tfrac{1}{2}\left[(v_2 - v_1)^2 - v_2^2\right] + \cdots
\end{aligned}
\tag{2.46}
$$

This problem can be solved by noting that having already computed the analytical approximations for the transverse displacements v_i, the dynamical system (2.46) can be regarded as a system of two coupled *linear* oscillators forced by "pseudo-forcing" terms on the right-hand sides, and so can be studied by standard linear modal analysis. Then, the leading-order approximations of the amplitudes of the axial oscillations contain both *slow* (represented by the particular solutions) and *fast* (resulting from the homogeneous solutions) components. We have seen that the slow dynamics is critical for realizing the nonlinear acoustic vacuum in terms of transverse oscillations (providing the approximately uniform, in the slow scale, axial tension in the system), and play the main role in constructing the leading-order approximations. Regarding the axial oscillations of the particles, however, in addition to the slow dynamics, the fast dynamics need also to be explicitly taken into account in the leading-order approximation in order to satisfy the initial conditions for the axial oscillations of the problem (this is demonstrated in the following example for in-phase transverse oscillations). The combined effect of this mixed slow/fast dynamics is a fast redistribution of axial forces from an arbitrary initial state to a nearly uniform state in slow time which is necessary for the realization of the nonlinear sonic vacuum.

We demonstrate this computation by considering the in-phase mode for transverse oscillations, but a similar procedure can be followed for the other NNMs or for non-stationary resonant motions such as the LPTs. Accordingly, we set $v_1 = v_2 \equiv v$ and introduce the modal coordinates $\xi_1 = u_1 + u_2$ and $\xi_2 = u_2 - u_2$ to express Eq. (2.46) as:

$$\varepsilon^2 \xi_1'' + \xi_1 = 0 + \cdots$$
$$\varepsilon^2 \xi_2'' + 3\xi_2 = -v^2 + \cdots \tag{2.47}$$

where the transverse displacement $v(\tilde{t})$ is computed by solving for the nonlinear modal oscillator (2.29b) with $k = 1$

$$v'' + [(1/3) + (1/\mu)]v^3 = 0 \Rightarrow v(\tilde{t}) = V \operatorname{cn}\left(\sqrt{L}\tilde{t}, 1/\sqrt{2}\right) \tag{2.48}$$

where $L = (1/3) + (1/\mu)$, $\operatorname{cn}(\cdot, \cdot)$ is the Jacobi elliptic cosine function, and the initial conditions for the transverse oscillations are assumed in the form $v(0) = V, v'(0) = 0$. Substituting (2.48) into (2.47), solving the resulting nonhomogeneous system, and transforming back to original coordinates, we derive the following leading-order approximations for the axial displacements

$$u_1(\tilde{t}) = \underbrace{C_1 \, \cos(\tilde{t}/\varepsilon) + C_2 \, \sin(\tilde{t}/\varepsilon) + C_3 \, \cos\left(\tilde{t}\sqrt{3}/\varepsilon\right) + C_4 \, \sin\left(\tilde{t}\sqrt{3}/\varepsilon\right)}_{\text{Fast oscillations}} + \underbrace{v^2/6}_{\text{Slow oscillations}} + \cdots$$

$$u_2(\tilde{t}) = \underbrace{C_1 \, \cos(\tilde{t}/\varepsilon) + C_2 \, \sin(\tilde{t}/\varepsilon) - C_3 \, \cos\left(\tilde{t}\sqrt{3}/\varepsilon\right) - C_4 \, \sin\left(\tilde{t}\sqrt{3}/\varepsilon\right)}_{\text{Fast oscillations}} - \underbrace{v^2/6}_{\text{Slow oscillations}} + \cdots$$

$$\tag{2.49}$$

where the coefficients of the fast oscillations are determined by imposing the initial conditions for the axial oscillations, and we recall the rescaling of the time variable introduced in Sect. 2.2, $\tilde{t} = \varepsilon(k_1/m)^{1/2}t$, in terms of the physical time t. Note that apart from the mixed slow/fast dynamics, the axial dynamics also mix the in-phase and out-of-phase axial modes at the fast timescale of the problem.

As discussed above and confirmed by (2.49), the leading-order approximations for the axial oscillations consist of two parts, namely of fast oscillations with two different fast frequencies, and of slow oscillations which have the same frequency as the leading-order (slow) approximations of the transversal oscillations. These fast axial oscillations play an important role in the dynamics since the generate the non-local effects in the transverse dynamics by providing the fast redistribution of axial forces from an arbitrary initial distribution to a nearly uniform state in the slow timescale which ultimately generates the nonlinear sonic vacuum.

To demonstrate the validity of the asymptotic analysis, in Fig. 2.17, we depict the direct numerical integration of the equations of motion of the original system (2.24a), (2.24b) for $n = 2$, for the case of the LPT in the regime of intense energy exchange (after the first but before the second bifurcation), and compare them with the leading-order analytical results derived by the slow/fast decompositions. Satisfactory agreement between the asymptotic and numerical results is noted, confirming the validity of our theoretical predictions.

2.2.7 Global Dynamics

Since contrary to the reduced slow flow on the two-torus Eqs. (2.38a), (2.38b), system (2.28) is not integrable (since it possesses only one first integral of motion corresponding to energy conservation), we performed an additional study to clarify its global dynamics and confirm the theoretical predictions of the previous section which were based on asymptotic analysis of the slow dynamics. This analysis was based on the construction of numerical Poincaré maps and clarified the role of the resonance dynamics and of NNM and LPT bifurcations on the non-stationary and chaotic dynamics of the strongly nonlinear system (2.28).

To this end, we consider the four-dimensional phase space $(v_1, v_2, v'_1, v'_2) \in R^4$ of system (2.28) and reduce the dynamics to a three-dimensional isoenergetic manifold by fixing the total energy (hamiltonian) of this system to a constant value $F(v_1, v_2, v'_1, v'_2) = f \Rightarrow v'_2 = F^{-1}(v_1, v_2, v'_1; f)$. This renders $\dot{v}_2$ a dependent variable. Then, we introduce the Poincaré cut section $\Sigma = \{(v_1, v_2, v'_1, F^{-1}(v_1, v_2, v'_1; f)) \in R^3 / v_2 = 0, v'_2 > 0\} \cap F(v_1, v_2, v'_1, v'_2) = f$ and define the Poincaré map,

$$P : \Sigma \to \Sigma : \{(v_1(t_k), v'_1(t_k)) \to (v_1(t_{k+1}), v'_1(t_{k+1}))\}$$

where t_k and t_{k+1} denote consecutive time instances of crossings of orbits on the isoenergetic manifold with the cut section Σ. By this construction, the Poincaré is orientation preserving and captures all the global dynamics on the cut section (since it can be proven that the isoenergetic flow is always transversal to Σ, with the only exception being at the boundary of the map where $v'_2 = 0$).

In the following numerical results, we fix the hamiltonian of system (2.28) to $F = 10$, and consider the global dynamics for varying values of μ; moreover, we denote the LPTs on the Poincaré maps by bold lines. In Fig. 2.18, we depict the Poincaré maps of (2.28) for $\mu = 1.6, 0.8$ and 0.3, and the global numerical analysis fully confirms and validates our asymptotic results. The NNMs of the system are fixed points of the map on the horizontal axis $v'_1 = 0$ and the bifurcation of these stationary solutions at the first bifurcation point $\mu_{c1} = 1$ can be clearly observed in

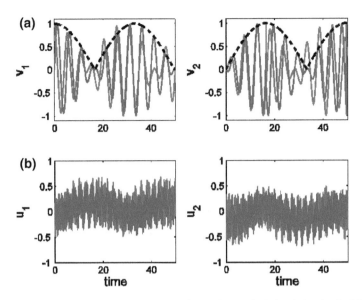

Fig. 2.17 Comparison between theoretical prediction and numerical simulation for the LPT of the system with $n = 2$, $\mu = 0.8, \varepsilon = 0.05$ and initial conditions $v_1(0) = 1$, $v_1'(0) = v_2(0) = v_2'(0) = 0$, and $u_1(0) = u_1'(0) = u_2(0) = u_2'(0) = 0$: **a** Transverse displacement $v_1(t)$, and **b** axial displacement $u_1(t)$; the asymptotic predictions (2.43, 2.44a, 2.44b and 2.46 are denoted by *blue solid*) lines, while the numerical simulations of Eqs. (2.24a), (2.24b) by *red* (*solid*) *lines*; the *bold dashed line* in (**a**) shows the analytical prediction of the (slow) envelope

the numerical results. Moreover, the stability of the two NNMs before the bifurcation (with the NNMs are depicted as centers—cf. Fig. 2.18a) and the instability of the out-of-phase NNM after the bifurcation (with the mode depicted as a saddle point—cf. Fig. 2.18b) are clearly inferred. In addition, the change of the topological structure of the LPT from spatially extended to spatially localized after the second bifurcation point $\mu_{c2} = 0.6$ is also confirmed. In particular, in Fig. 2.17c, we note the result of the LPT bifurcation, which is signified by the existence of two disjoint LPTs, with each of them being localized in the neighborhood of either one of the stable bifurcated NNMs.

Additionally, we note that the region of chaotic motions depends strongly on the value of parameter μ. When μ is large enough (far from both topological transformations resulting from the two bifurcations of the global dynamics), the regions of chaos are nearly negligible (cf. Fig. 2.18a), and the entire phase space of the system seems to be foliated by isoenergetic invariant tori where the global dynamics are topologically similar to the reduced slow flow two-torus (2.38a and 2.38b); given, however, that the dynamical system (2.28) is non-integrable, we infer that even in this case chaotic layers still exist by they are spatially confined. Hence, in this case, the asymptotic analysis based on slow flow reduction fully reflects the behavior of the non-integrable system (2.28).

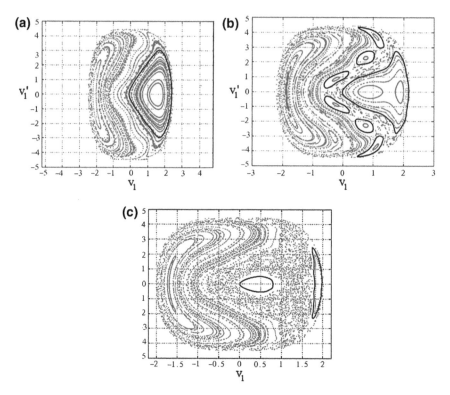

Fig. 2.18 Poincaré maps depicting the global dynamics of system (2.28) for **a** $\mu = 1.6$, **b** $\mu = 0.8$, and **c** $\mu = 0.5$; LPTs are denoted by *bold lines*

After the first topological transition (cf. Fig. 2.18b), a weak chaotic layer appears, together with a set of secondary stationary points. These points correspond to periodic motions satisfying conditions of 1:3 internal resonance so they are not captured by our previous asymptotic analysis which is based on the assumption of 1:1 resonance. This higher-order resonance dynamics can be also treated analytically, e.g., with the methodologies developed in (2.32 and 2.33). The 1:3 resonant stationary points are surrounded by secondary LPTs which correspond to less intense energy exchanges between the two oscillators in Eq. (2.28). Finally, after the second transition takes place (cf. Fig. 2.18c), the chaotic region expands, and the regime of regular dynamics is confined only in the neighborhoods of the bifurcated NNMs where the two bifurcated localized LPTs can be seen. In this case, the two LPTs provide the boundaries between the regimes of chaotic and regular dynamics.

Thus, the presence of strongly nonlinear elastic supports changes qualitatively both the stationary and non-stationary resonance dynamics of the nonlinear sonic vacuum. Considering the simplest case of the system with only two particles, we showed that, in contrast to the ungrounded sonic vacuum studied, there occur two main bifurcations (dynamical transitions) in the dynamics under condition of 1:1

resonance. First, a bifurcation of NNMs generates a pair of localized bifurcating NNMs and an unstable out-of-phase NNM; such NNMs bifurcations are not possible in the ungrounded sonic vacuum whose NNMs are identical to those of the corresponding linear spring-mass chain. Second, a bifurcation of the LPT which changes drastically the intensity of energy exchanges occurring between the two oscillators of the system; again this type of LPT bifurcation is not possible in the ungrounded sonic vacuum where, under certain conditions, only a single LPT can occur corresponding to intense energy exchanges between different NNMs in 1:1 resonance.

The analytical and numerical results presented here show that, alongside with well-known bifurcation and instability of NNMs (Manevitch and Smirnov 2010c: 3), full energy exchanges between different parts of the sonic vacuum become possible in the regime of spatially extended LPTs. For different values of the system parameters and depending on the energy level of the oscillation, the LPTs can become localized in neighborhoods of bifurcated NNMs, in which case the intensity of energy exchanges in the resonant non-stationary dynamics of the sonic vacuum drastically decreases. This type of strongly non-stationary phenomena can be adequately described by LPTs whose analytical presentation was obtained in terms of non-smooth basic functions. An asymptotic analysis performed under condition of 1:1 resonance and based on slow/fast decomposition of the dynamics fully predicted these previous global dynamical changes. Moreover, it was shown that chaotic effects due to the non-integrable nature of the sonic vacuum could not affect the bifurcations and the corresponding topological changes in the phase space predicted by the slow flow asymptotic analysis. Another interesting finding was that the LPTs appeared to separate the regions of regular dynamics in-phase space from the surrounding local or global chaotic layers.

The results presented in this section can be extended to sonic vacua with larger number of particles, but the analysis of these higher-dimensional systems is much more complex (we present such analysis in Part 3). In addition, a new family of sonic vacua composed of a number of parallel coupled chains of particles similar to the one shown in Fig. 2.11 can be constructed and its stationary and non-stationary resonant dynamics examined. Then, it would be of interest to examine the NNMs, LPTs, and their bifurcations of this class of essentially nonlinear "discrete membranes." Finally, from a practical viewpoint, it is significant that the number of resonating modes in the discrete model of untensioned grounded string may exceed that of the corresponding ungrounded system and that the non-stationary dynamics and nature of resonant energy exchanges and LPTs in these two systems are completely different.

Let us note that the obtained results for the two-particle modes are similar to those presented in the previous section in the case of cubic nonlinearity. Despite the parameters determining a qualitative change of non-stationary and non-stationary dynamics are different, the types of dynamical behavior turn out to be similar (excluding the longitudinal motion which is absent in the case of local sonic vacuum).

References

Manevitch, L.I.: Complex representation of dynamics of coupled nonlinear oscillators, In: Uvarova, L., Arinstein, A.E., Latyshev, A.V. (eds.), Mathematical Models of Non-Linear Excitations. Transfer, Dynamics, and Control in Condensed Systems and Other Media, pp. 269–300. Springer, New York (1999)

Manevitch, L.I.: The description of localized normal modes in a chain of nonlinear coupled oscillators using complex variables. Nonlinear Dyn. **25**, 95 (2001)

Manevitch, L.I.: New approach to beating phenomenon in coupled nonlinear oscillatory chains. In: Awrejcewicz, J., Olejnik, P. (ed.) 8th Conference on Dynamical Systems-Theory and Applications (DSTA-2005), p. 289 (2005)

Manevitch, L.I.: New approach to beating phenomenon in coupled nonlinear oscillatory chains. Arch. Appl. Mech. **301** (2007)

Manevitch, L.I., Gendelman, O.V.: Tractable Models of Solid Mechanics. Springer, Berlin (2011)

Manevitch, L.I., Musienko, A.I.: Limiting phase trajectories and energy exchange between an anharmonic oscillator and external force. Nonlinear Dyn. **58**, 633–642 (2009)

Manevitch, L.I., Smirnov, V.V.: Localized nonlinear excitations and interchain energy exchange in the case of weak coupling. In: Awrejcewicz, Jan (ed.) Modeling, Simulation and Control of Nonlinear Engineering Dynamical System, pp. 37–47. Springer, Netherlands (2009)

Manevich, L.I., Smirnov, V.V.: Limiting phase trajectories and thermodynamics of molecular chains. Phys. Doklady **55**, 324–328 (2010)

Manevitch, L.I., Smirnov, V.V.: Limiting phase trajectories and the origin of energy localization in nonlinear oscillatory chains. Phys. Rev. E **82** (2010b)

Manevitch, L.I., Smirnov, V.V.: Resonant energy exchange in nonlinear oscillatory chains and limiting phase trajectories: from small to large systems. In: Vakakis, A.F. (ed.) Advanced Nonlinear Strategies for Vibration Mitigation and System Identification, CISM Courses and Lectures, 518, pp. 207–258. Springer, New York (2010c)

Manevitch, L.I., Vakakis, A.F.: Nonlinear oscillatory acoustic vacuum. SIAM J. Appl. Math. **74** (6), 1742–1762 (2014)

Pilipchuk, V.N.: Nonlinear Dynamics: Between Linear and Impact Limits. Springer, Berlin (2010)

Starosvetsky, Y., Ben-Meir, Y.: Nonstationary regimes of homogeneous Hamiltonian systems in the state of sonic vacuum. Phys. Rev. E **87**, 062919 (1–18) (2013)

Vakakis, A.F., Manevitch, L.I., Mikhlin, YuV, Pilipchuk, V.N., Zevin, A.A.: Normal Modes and Localization in Nonlinear Systems. Wiley, New York (1996)

Zhupiev, A.L., Mikhlin, Y.V.: Stability and branching of normal oscillation forms of non-linear systems. J. Appl. Math. Mech. **45**, 328–331 (1981)

Zhupiev, A.L., Mikhlin, Y.V.: Conditions for finiteness of the number of instability zones in the problem of normal vibrations of non-linear systems. J. Appl. Math. Mech. **486–489**, 48 (1984)

Chapter 3
Emergence and Bifurcations of LPTs in the Chain of Three Coupled Oscillators

The natural question is: How the LPT concept may be extended to the systems with more than two degrees of freedom? To answer this question, we consider first a simplest extension to the case of three weakly coupled identical oscillators.

3.1 "Hard" Nonlinearity

3.1.1 Bifurcations of Limiting Phase Trajectories and Routes to Chaos in the Anharmonic Chain of the Three Coupled Particles

The corresponding non-dimensional equations of motion read

$$\frac{d^2 x_n}{dt^2} + x_n + \varepsilon x_n^3 = \mu\varepsilon\{(1 - \delta_{1n})(x_{n-1} - x_n) - (1 - \delta_{3n})(x_n - x_{n+1})\} \tag{3.1}$$
$$1 \leq n \leq 3$$

Here ε is small parameter $(0 < \varepsilon \ll 1)$ which reflects both weak coupling and weak nonlinearity. Presence of this parameter provides existence of the slow timescale, exactly as in the case of 2DOF system.

© Springer Nature Singapore Pte Ltd. 2018
L.I. Manevitch et al., *Nonstationary Resonant Dynamics of Oscillatory Chains and Nanostructures*, Foundations of Engineering Mechanics,
DOI 10.1007/978-981-10-4666-7_3

3.1.1.1   Multi-scale Analysis and Derivation of the Slow Flow Model

Analyzing the dynamics of the system in (3.1) is based on development of the asymptotic series by multiple scales method. We employ a complex multi-scale expansion procedure introduced previously $\psi_n = \dot{x}_n + i x_n$ and adopt the following multi-scale expansion

$$\psi_n = \psi_{n0}(\tau_0, \tau_1) + \varepsilon \psi_{n1}(\tau_0, \tau_1) + O(\varepsilon^2), \quad \frac{\partial(\bullet)}{\partial t} = \frac{\partial(\bullet)}{\partial \tau_0} + \varepsilon \frac{\partial(\bullet)}{\partial \tau_1} + O(\varepsilon^2) \quad (3.2)$$

Introducing (3.2) in (3.1) and proceeding systematically up to the order of $O(\varepsilon)$ and removing the secular terms, leads to the following slow flow model,

$$\frac{d\varphi_n}{d\tau_1} = i\frac{\gamma}{2}|\varphi_n|^2\varphi_n + \frac{i\mu}{2}\{(1 - \delta_{1n})(\varphi_n - \varphi_{n-1}) + (1 - \delta_{Nn})(\varphi_n - \varphi_{n+1})\},$$
$$n = 1, 2, 3$$

$$(3.3)$$

where $\psi_n = \varphi_n(\tau_1)\exp(i\tau_0)$. Performing additional rescale $\tau_1 = \gamma/2^*\tau_1$ and designating the rescaled coupling $\beta = \frac{\mu}{\gamma}$ yields

$$\frac{d\varphi_n}{d\tau_1} = i|\varphi_n|^2\varphi_n + i\beta\{(1 - \delta_{1n})(\varphi_n - \varphi_{n-1}) + (1 - \delta_{Nn})(\varphi_n - \varphi_{n+1})\},$$
$$\varphi_1(0) = iN, \quad \varphi_k(0) = 0, \quad k \neq 1, \quad n = 1, 2, 3 \quad (3.4)$$

It is easy to show that system (3.4) possesses two integrals of motion,

$$N^2 = |\varphi_1|^2 + |\varphi_2|^2 + |\varphi_3|^2$$
$$H = \gamma\frac{i}{2}\sum_{k=1,2,3}|\varphi_k|^4 + i\mu|\varphi_1 - \varphi_2|^2 + i\mu|\varphi_2 - \varphi_3|^2 \quad (3.5)$$

In fact, the dimensionality of the system under consideration is six. Interestingly enough, the global system dynamics can be reduced to the three-dimensional manifold. To demonstrate that let us start from introduction of spherical coordinates:

$$\begin{cases} \varphi_1 = N\cos(\theta)\cos(\varphi)e^{i\delta_1} \\ \varphi_2 = N\sin(\theta)e^{i\delta_2} \\ \varphi_3 = N\cos(\theta)\sin(\varphi)e^{i\delta_3} \end{cases} \quad (3.6)$$

Substitution of (3.6) into (3.4) and using the coordinates of relative phases, i.e., $\Delta_{12} = \delta_1 - \delta_2$, $\Delta_{23} = \delta_2 - \delta_3$ allows for the global flow reduction

$$\frac{d\theta}{d\tau_1} = \beta \sin(\varphi) \sin(\Delta_{12}) - \beta \cos(\varphi) \sin(\Delta_{23})$$

$$\frac{d\varphi}{d\tau_1} = -\beta \tan(\theta)(\sin(\Delta_{12}) \cos(\varphi) + \sin(\Delta_{23}) \sin(\varphi))$$

$$\frac{d\Delta_{12}}{d\tau_1} = \begin{bmatrix} \cos^2(\theta) \sin^2(\varphi) - \sin^2(\theta) - \beta + \beta \cot(\theta) \cos(\varphi) \cos(\Delta_{23}) \\ -\beta \dfrac{2}{\sin(2\theta) \sin(\varphi)} \left(\sin^2(\theta) - \cos^2(\theta) \sin^2(\varphi) \right) \cos(\Delta_{12}) \end{bmatrix}$$

$$\frac{d\Delta_{23}}{d\tau_1} = \begin{bmatrix} \sin^2(\theta) - \cos^2(\theta) \cos^2(\varphi) + \beta - \beta \cos(\theta) \cos(\varphi) \cos(\Delta_{12}) \\ -\dfrac{2}{\sin(2\theta) \cos(\varphi)} \left(\cos^2(\theta) \cos^2(\varphi) - \sin^2(\theta) \right) \cos(\Delta_{23}) \end{bmatrix}$$

(3.7)

The second integral of motion can be expressed in terms of the new coordinates $(\theta, \phi, \Delta_{12}, \Delta_{23})$:

$$H_{4D} = \begin{bmatrix} \frac{i}{2} \left(\sin^4(\theta) + \cos^4(\theta) \left(\cos^4(\varphi) + \sin^4(\varphi) \right) \right) - \\ -i\beta \sin(2\theta) \sin(\varphi) \cos(\Delta_{12}) - i\beta \sin(2\theta) \cos(\varphi) \cos(\Delta_{23}) \\ -i\frac{\beta}{2} \cos(2\theta); \end{bmatrix}$$

(3.8)

In contrast to the 2DOF counterpart, the reduced slow flow model (3.7) of the 3DOF given in its angular form is not integrable, thus obviously enough the response of (3.7) can also be chaotic. However, as it will become clear from the results brought below there exist certain domains in the space of system parameters where the regular response regimes coexist with the non-regular ones.

3.1.2 Nonlinear Normal Modes (NNMs)

Similarly to the 2DOF system, NNMs turn out to be stationary points in the phase space of 3DOF system in the slow timescale. The rest of solutions correspond to ordered non-stationary processes, among which one can distinguish LPT with maximum possible energy exchange, and chaotic motions.

Let us start the present discussion with the bifurcation analysis of nonlinear normal modes. Thus, seeking for the stationary solutions of (3.7), we set the derivatives of (3.7) to zero $\left(\frac{d\theta}{d\tau_1} = \frac{d\varphi}{d\tau_1} = \frac{d\Delta_{12}}{d\tau_1} = \frac{d\Delta_{23}}{d\tau_1} = 0 \right)$, yielding the algebraic system of nonlinear equations. In Figs. 3.1 and 3.2, we plot the amplitudes $(|\varphi_1|, |\varphi_2|, |\varphi_3|)$ of each one of the oscillators corresponding to the stationary solutions of the out-of-phase modes $(\Delta_{12} = \Delta_{23} = \pi)$ (see Fig. 1.1) and the mixed mode $(\Delta_{12} = 0, \Delta_{23} = \pi)$. In-phase mode is not presented (i.e. $\Delta_{12} = \Delta_{23} = \pi$) as it yields the rather trivial result of $\left(|\varphi_1| = |\varphi_2| = |\varphi_2| = \frac{1}{\sqrt{3}}\right)$.

It is worthwhile noting that in the present diagrams (Fig. 3.1), we have shown solely the nonlinear normal modes with the spatial energy localization spanned over the first two oscillators (i.e., energy localization on either first or the second oscillators or alternatively on both first and the second oscillators). Interestingly

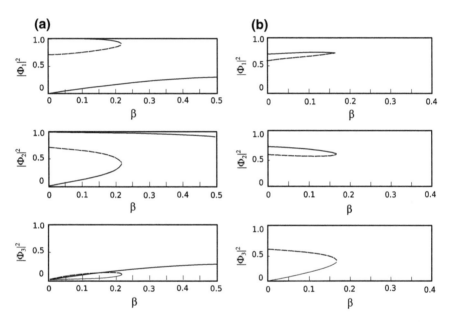

Fig. 3.1 Bifurcation diagram of NNMs: **a** $\Delta_{12} = \Delta_{23} = \pi$, **b** $(\Delta_{12} = 0, \Delta_{23} = \pi)$. Unstable modes are denoted with the *dashed lines*

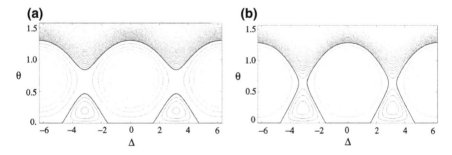

Fig. 3.2 Phase portraits of the reduced system with $\chi = 0.1667$ (**a**) and $\chi = 0.18$ (**b**)

enough that the stable and the unstable localized modes bifurcate through the saddle-center bifurcation, while the additional mode localized on the central oscillator is stable and persists for all the values of coupling—β.

3.1.3 Emergence and Bifurcations of Limiting Phase Trajectories (LPTs) in the System of Three Coupled Oscillators

As it was shown in the previous section, the concept of the limiting phase trajectories plays a crucial role in the analysis of bifurcations of the highly non-stationary regimes manifested by the massive energy transport between the coupled anharmonic conservative systems, generators, externally driven systems as well as the oscillatory chains and this under various conditions of the system parameters (i.e., constant parameters and time-varying parameters). However, in all these works, the usage of the ideology of limiting phase trajectories has been mostly limited to the analysis of the two coupled effective particles or alternatively to the externally driven, single particle.

In the present section, we would like to show the emergence of a somewhat similar phenomenon of limiting phase trajectories in the system of three coupled effective particles. Moreover, we show that limiting phase trajectories can also play a major role in predicting the formation and bifurcations of a special class of non-stationary regimes which can be characterized by the weak and strong beating established between the first and the second oscillators having insignificant energy leakage to the third oscillator. In the same section, we show that breakdown of these non-stationary regimes yields the significant energy transport to the third oscillator ending up in chaotic-like motion manifested by the energy equipartition between all the three particles.

Before proceeding with the further analysis, let us start with the definitions of the new type of non-stationary regimes directly related to the limiting phase trajectories discussed above.

First type of non-stationary regimes manifests itself by the significant energy localization on the first oscillator with the moderate energy exchanges between the first oscillator and the rest of the chain (i.e., second and third oscillators). This type of the response is analogous to the weakly modulated regime obtained in the system of two oscillators (e.g., two coupled pendula) subject to the impulsive excitation of the first oscillator. As is clear from the above discussion in the framework of the two coupled oscillators, this type of the response can be attributed to the type-I limiting phase trajectory.

Second type of non-stationary regimes manifests itself by the significant energy localization on the first and the second oscillators in the form of the high-energy exchanges between the first and the second oscillators with the negligibly small energy leakage to the third coupled oscillator. This type of the response is analogous to the strongly modulated regime obtained in the system of two oscillators (e.g., two coupled pendula) subject to the impulsive excitation of the first oscillator. As is clear from the above discussion in the framework of the two coupled oscillators, this type of the response can be attributed to the type-II limiting phase trajectory.

The last type of the response discussed in the present section corresponds to the breakdown of the type-II limiting phase trajectory bringing about the irregular energy wandering between all the three oscillators of the chain, obviously enough the latter type of the response has no analogy in the system of two coupled oscillators.

For the sake of convenience, we refer to this special type of regularly pulsating regimes as the quasi-periodic LPT of type I and type II accordingly due to their modulated nature. It is worthwhile noting that these special LPT orbits emerging in the system of three coupled effective particles highly differ from their two DOF counterparts due to their aperiodicity. Clearly enough for some particular choice of initial conditions (fairly close to the impulsive ones), one could also find the perfectly periodic LPT similarly to the 2DOF systems. The periodic LPTs will be illustrated in the following subsections.

To better understand the phenomenon of the consecutive transitions between the different types of LPTs as well as predicting the formation of the fully delocalized response, we resort to the construction of the reduced-order model based on "master and slave" decomposition.

Thus, system (3.4) can be further reduced by assuming that the third element is hardly excited in comparison to the 1st and the 2nd oscillators. Based on that assumption, we reduce the slow flow system (3.4) as follows,

$$
\begin{aligned}
\frac{d\varphi_1}{d\tau_1} &= i|\varphi_1|^2\varphi_1 + i\chi(\varphi_1 - \varphi_2), \\
\frac{d\varphi_2}{d\tau_1} &= i|\varphi_2|^2\varphi_2 + i\chi(2\varphi_1 - \varphi_1), \\
\frac{d\varphi_3}{d\tau_1} &= i\chi\varphi_2 = i\chi \sin\theta\, e^{i\delta_2}.
\end{aligned}
\tag{3.9}
$$

The system of the first two equations in (3.9) possesses two integrals of motion

$$
\begin{aligned}
N &= \sum_{k=1\,to\,2} (|\phi_k|^2), \\
H &= \sum_{k=1\,to\,2} (|\phi_k|^4) + \chi * \left[|\phi_1 - \phi_2|^2 + |\phi_2|^2 \right]
\end{aligned}
\tag{3.10}
$$

A proper change of variables $\varphi_1 = e^{i\delta_1}\cos\theta$, $\varphi_2 = e^{i\delta_2}\sin\theta$ reduces the first two equations in (3.9) to a planar system with two angular variables θ and $\Delta = \delta_1 - \delta_2$

$$\frac{d\theta}{d\tau_1} = \chi \sin \Delta,$$

$$\frac{d\Delta}{d\tau_1} = \cos^2 \theta - \sin^2 \theta + \chi[2 \cot 2\theta \cos \Delta - 1]. \tag{3.11}$$

It is easy to prove that system (7.11) is Hamiltonian, with the Hamiltonian

$$H = 1/2(\cos^4 \theta + \sin^4 \theta) - c(\sin 2\theta \cos \Delta - \sin^2 \Delta).$$

Phase portraits of system (3.11) for different values of the parameter χ are given in Fig. 3.2a, b. One can observe localization of energy on the first oscillator at $\chi = 0.1667$ (Fig. 3.2a), and formation of intense beating at the higher value $\chi = 0.18$ (Fig. 3.2b). The LPTs are indicated by the bold solid lines.

It was shown in the previous sections that the transition from the LPT of the first type (moderate energy exchange) to the LPT of the second type (strong energy exchange) corresponds to the passage of the LPT of the first type through the separatrix. Using the same arguments as in Sects. 2 and 3 (see e.g. the analysis brought in Sect. 2, Eqs. 2.15– 2.19 for the sonic vacuum case), we derive the critical parameter corresponding to the passage through separatrix as $\chi = \chi_1 = 0.1746$.

We now consider the occurrence and the disappearance of the LPTs of the second type. We show that, at certain coupling strength, the LPT vanishes, yielding significant energy transfer to the third oscillator and generating chaotic energy transport in the entire system.

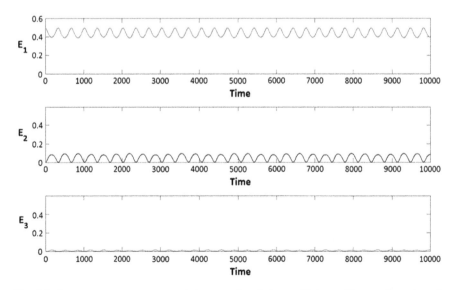

Fig. 3.3 Instantaneous energy recorded on each of the oscillators with coupling strength $\chi = 0.11 < \chi_1 = 0.175$

Table 3.1 Dependence of the frequency of modulations Ω and angular frequency ω on coupling strength χ

χ	Ω	ω
0.3	0.67	0.83
0.365	0.835	0.835

We restrict our consideration by the truncated system (3.9). Taking into account the resonant interaction between the LPT of the first pair of oscillators and the third oscillator, we formulate a robust analytical criterion for the annihilation of strong beating between the first and second oscillators.

It follows from Eq. (3.9) that the unbounded growth of the response φ_3 can be attained under the condition of resonance, i.e., when the spectrum of $\cos \theta$ coalesces with that of $\exp(i\delta_2)$, thus yielding the secular growth of φ_3. First, we note that the phases $\delta_1(\tau)$ and $\delta_2(\tau)$ represent relatively small oscillations near a slowly increasing drift, namely,

$$\delta_j(\tau) = \omega\tau + \theta_j(\tau), \quad j = 1, 2, \tag{3.12}$$

where ω is the average rate of rotation corresponding to the LPT, $\vartheta_j(\tau)$ is an oscillating function. Next, $\theta(\tau)$ is assumed to be a periodic function with the period $T_{\text{mod}} = 2\pi/\Omega$ corresponding to the period of modulation of strong resonant beating between the first and second oscillators. It now follows that the slow envelope φ_3 exhibits the secular growth if the angular frequency ω coincides with the modulation frequency Ω, that is, $\omega = \Omega$. This condition allows one to determine the critical value χ_2 from the equality

$$\omega(\chi_2) = \Omega(\chi_2). \tag{3.13}$$

The derivation of analytical expressions for ω and Ω and the analytical calculation of χ_2 are beyond the scope of this report. To illustrate the criterion for the second transition phase, we used Fast Fourier Transform to calculate both the frequency of modulation Ω and the angular frequency ω for the two distinct values of coupling $\chi = 0.3 < \chi_2$ and $\chi = 0.365 = \chi_2$. The results are summarized in Table 3.1.

3.1.4 Numerical Results

Numerical results of this section illustrate the occurrence of the aforementioned quasi-periodic LPTs due to gradual increase of the coupling parameter χ. In addition to quasi-periodic LPTs, we demonstrate complete spontaneous energy delocalization for the high values of coupling. The numerical results have been obtained for system (3.1) with initial conditions (3.2).

1. *Quasi-periodic LPTs of the first type.* Figure 3.3 depicts instantaneous energy of each oscillator for weak coupling $\chi < \chi_1$. One can observe significant energy localization on the first oscillator together with weak energy exchange between the oscillators (weak modulation). In analogy to the 2DOF system, the response with significant energy localization on the first oscillator is further referred to as the quasi-periodic LPT of the first type.

2. *Quasi-periodic LPTs of the second type.* Given that the coupling strength $\chi > \chi_1$, we reveal a global change of the response (Fig. 3.4). Instead of energy localization on the first oscillator or alternative energy delocalization (i.e., energy exchange through the whole chain), the formation of an intermediate state characterized by regular strongly modulated beating between the first and the second oscillators is observed.

It is noted in Fig. 3.4 that an insignificant amount of energy is transported to the third oscillator. This type of highly modulated response with significant energy localization on the first and the second oscillators is referred to as a quasi-periodic LPT of the second type.

3. *Complete energy delocalization.* Further increase of the coupling parameter entails an additional global change of the system response. In this case, regimes of regular local pulsations (quasi-periodic LPTs) are completely destroyed, yielding irregular chaotic behavior in the chain (Fig. 3.5).

The diagrams in Fig. 3.6 demonstrate consecutive transitions between the locally pulsating regimes as well as the emergence of highly delocalized ones. At $\chi =$

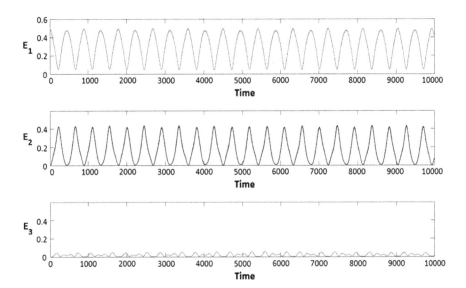

Fig. 3.4 Instantaneous energy recorded on each oscillator provided the coupling strength $\chi = 0.18$ ($\chi_1 < \chi < \chi_2$) is above the first threshold $\chi_1 = 0.175$ and below the second threshold $\chi_2 = 0.36$

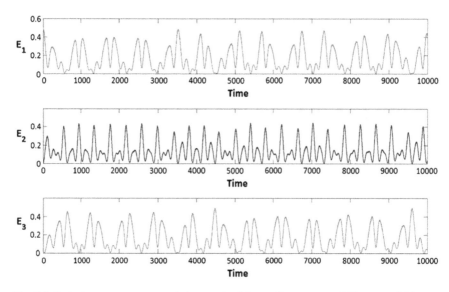

Fig. 3.5 Instantaneous energy recorded on each of the oscillators at $\chi = 0.33 < \chi_2 = 0.36$

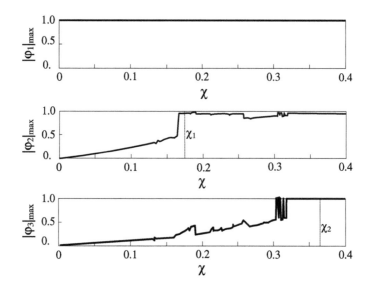

Fig. 3.6 Transition diagrams

$_1 = 0.175$, we observe the first transition from energy localization on the first oscillator to the intermediate state of energy localization on the first two oscillators with intense energy exchange between them. The second transition at $\chi \approx \chi_2 = 0.365$ is from the moderately localized state (strongly pulsating regimes) to complete energy delocalization, i.e., maximal amplitudes of the response recorded on all the three oscillators are nearly equal.

It is obvious that the first transition is well approximated in the framework of the LPT concept for the simplified 2DOF model, wherein the effect of the third weakly excited oscillator introduces asymmetry similar to that considered in the 2DOF oscillators (Sect. 3). As a result, the numerical critical value for the 3DOF system $\chi_{1num} = 0.165$ is slightly lesser than the critical value $\chi_{1LPT} = 0.175$ defined for the 2DOF system (the difference is of the order of 6%). The point of the second transition from the localized to entirely delocalized states is also in a fairly good agreement with the transition diagram, namely, the numerical critical point $\chi_{2num} = 0.31$ is close to the theoretically found value $\chi_{2th} = 0.36$ (with the difference $\sim 14\%$).

3.1.5 Spatially Localized Pulsating Regimes

To assess the global dynamics of regular regimes as well as transition to chaos, we construct the Poincaré sections for the slow flow model (3.7) at fixed energy levels.

As shown above, the global system dynamics can be reduced to the three-dimensional (3D) manifold. To this end, we fix the value of the Hamiltonian H to a constant h, thus restricting the flow of (3.7) to the 3D isoenergetic manifold.

$$H_{4D}(\theta, \phi, \Delta_{12}, \Delta_{23}) = h. \tag{3.14}$$

Transversely intersecting the 3D manifold by a two-dimensional (2D) cut plane $S: \{\theta = \theta_0\}$, we construct the 2D Poincaré map $P : \Sigma \to \Sigma$, where the Poincaré section is defined as

$$\Sigma = \{\theta = \theta_0, d\theta/d\tau_1 > 0\} \cap \{H_{4D}(\theta, \phi, \Delta_{12}, \Delta_{23}) = h\}, \tag{3.15}$$

The restriction $d\theta/d\tau_1 > 0$ is imposed to indicate the orientation of the Poincaré map.

Despite apparent limitations, the constructed Poincaré maps faithfully reveal periodic orbits. The fundamental time-periodic solutions of a basic period T correspond to the period-1 equilibrium points of the Poincaré map, i.e., to orbits of system (3.7), which recurrently pierce the cut section at the same point. Additional subharmonic solutions of periods nT may exist corresponding to the period-n equilibrium points of the Poincaré map, i.e., to orbits that pierce the cut section n times before repeating themselves. In the present study, we construct the Poincaré map $P : \Sigma \to \Sigma$ such that the global system dynamics is mapped onto the (Δ_{12}, ϕ) plane.

To study consequent transitions from the highly localized to delocalized states, we set the value of the integral of motion corresponding to the initial state of the system strictly localized on the first oscillator. This restriction corresponds to the following choice of the initial conditions for the reduced slow flow model:

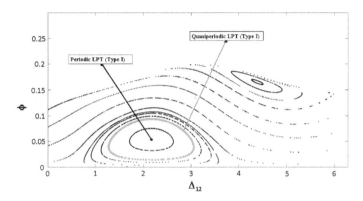

Fig. 3.7 Poincaré section: system parameters: $\theta_0 = 1.4$, $\chi = 0.15$

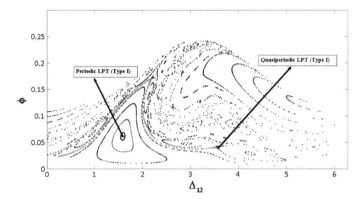

Fig. 3.8 Poincaré section: system parameters: $\theta_0 = 1.23$, $\chi = 0.1632$

$\theta(0) = \pi/2$, $\phi(0) = \Delta_{12}(0) = \Delta_{23}(0) = 0$. This choice of initial conditions yields the following value of the second integral of motion

$$H_{4D}(\pi/2, 0, 0, 0) = H_\chi = \frac{i}{2}\left(1 - \frac{\chi}{2}\right). \tag{3.16}$$

The Poincaré sections illustrated in Figs. 3.7, 3.8, 3.9, 3.10 and 3.11 correspond to different values of χ and H_χ. On each Poincaré diagram, the points of the quasi-periodic LPT are denoted with red color.

Starting from coupling $\chi < \chi_1$, we obtain the orbits encircling the center (Fig. 3.7). The center corresponds to the periodic LPT characterized by weak energy pulsations and mostly localized on the first oscillator. This response, referred to as the periodic LPT of the first type, allows us to determine the transition from intensive energy exchange to energy localization in the reduced 2DOF model. The

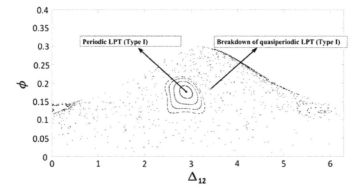

Fig. 3.9 Poincaré section: system parameters: $\theta_0 = 1.23$, $\chi = 0.18$

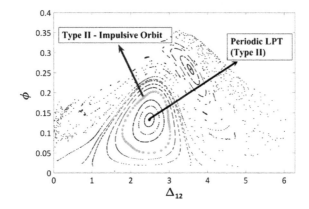

Fig. 3.10 Poincaré section: system parameters: $\theta_0 = 1.23$, $\chi = 0.1991$

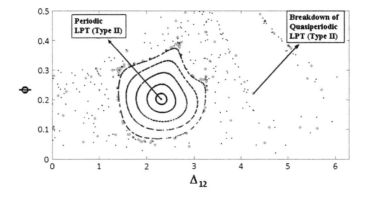

Fig. 3.11 Poincaré section: system parameters: $\theta_0 = 1.23$, $\chi = 0.31$

boundary of the resonant island corresponding to this type of the LPTs is characterized by stronger modulation, which, however, does not describe the maximal possible energy transport between the oscillators (quasi-periodic LPT of the first type). Note that this conclusion is in the perfect agreement with the results obtained for the 2DOF systems.

An increase of coupling strength χ entails the growth of the quasi-periodic LPT (Fig. 3.8), which then destructed while passing the chaotic region (Fig. 1.10). The annihilation of this regular orbit corresponds to the first transition from energy localization on the first oscillator to recurrent strong energy exchange between the first and second oscillators. It is worth noting the intermittent nature of highly pulsating regime emerging right after the breakdown of the quasi-periodic LPT. This intermittency is emphasized by its chaotic-like behavior. Naturally, this peculiarity cannot be observed in the framework of the reduced system (Fig. 3.11).

Further increase of the strength of χ results in the termination of the intermittent response with the following formation of the quasi-periodic LPT of the second type (Figs. 3.10 and 3.11) encircling the new center. This stationary point corresponds to the perfectly synchronous periodic LPT with strong energy exchange between the first two oscillators. As discussed above, the second transition (i.e., breakdown of local energy pulsations spanned over the first two oscillators) is characterized by chaotic energy transport between all oscillators. This spontaneous transition of the second kind is manifested by a sudden blow up of the regular quasi-periodic LPT of the second type.

3.2 "Soft" Nonlinearity

Let us consider now the case of the "soft" nonlinearity. In the case of two oscillators, the transition from "hard" to "soft" nonlinearity leads to the exchange of the roles of NNMs, namely which mode becomes unstable. For the "hard" nonlinearity, this is the out-of-phase NNM, when for the "soft" nonlinearity, this is the in-phase NNM. Now we are going to show that in the case of three oscillators, the behavior of the "soft" system can differ significantly from the behavior of the "hard" one. The basic model under consideration is the same chain of three weakly coupled oscillators as in the previous paragraph, but the type of nonlinearity is different ("soft" instead of "hard"). The dimensionless equations of motion are similar to (3.1):

$$\frac{d^2x_1}{dt^2} + x_1 - 8\varepsilon\alpha x_1^3 = 2\beta\varepsilon(x_2 - x_1)$$

$$\frac{d^2x_2}{dt^2} + x_2 - 8\varepsilon\alpha x_2^3 = 2\beta\varepsilon(x_1 - 2x_2 + x_3) \qquad (3.17)$$

$$\frac{d^2x_3}{dt^2} + x_3 - 8\varepsilon\alpha x_3^3 = 2\beta\varepsilon(x_2 - x_3)$$

where x_n is the displacement of the n-th oscillator, β is the coupling parameter, $\alpha > 0$ characterizes "soft" nonlinearity, and ε is a formal small system parameter. Below we present a brief analysis of this system without detailed calculations.

Again, we introduce the complex coordinates ψ_k:

$$\psi_n = \dot{x}_n + ix_n, \quad n = 1, 2, 3 \tag{3.18}$$

Substituting (3.18) into (3.17) and using the multi-scale procedure

$$\psi_n = \psi_{n0}(t_0, t_1) + \varepsilon\psi_{n1}(t_0, t_1) + O(\varepsilon^2), \frac{\partial(\bullet)}{\partial t} = \frac{\partial(\bullet)}{\partial t_0} + \varepsilon\frac{\partial(\bullet)}{\partial t_1} + O(\varepsilon^2),$$

The main asymptotic approximation yields:

$$\psi_{n0} = \varphi_n(t_1)\exp(it_0), \quad n = 1, 2, 3;$$

$$\frac{d\varphi_1}{dt_1} = -3\alpha i|\varphi_1|^2\varphi_1 + i\beta(\varphi_1 - \varphi_2)$$

$$\frac{d\varphi_2}{dt_1} = -3\alpha i|\varphi_2|^2\varphi_2 + i\beta(\varphi_2 - \varphi_1) + i\beta(\varphi_2 - \varphi_3) \tag{3.19}$$

$$\frac{d\varphi_3}{dt_1} = -3\alpha i|\varphi_3|^2\varphi_3 + i\beta(\varphi_3 - \varphi_2)$$

System (3.19) has two conserved quantities

$$N^2 = \sum_{k=1}^{3}|\varphi_k|^2$$

$$\tag{3.20}$$

$$H = -\frac{3\alpha i}{2}\sum_{k=1}^{3}|\varphi_k|^4 + i\beta\left(|\varphi_1 - \varphi_2|^2 + |\varphi_2 - \varphi_3|^2\right)$$

Let us recall that the first conserved quantity is usually referred to as an occupation number, whereas the second one is the Hamilton function (energy) of the slow flow (3.19). It is important to emphasize here that without any loss of generality, we can let $N = 1$, for example, by rescaling the amplitudes of the response. Obviously, the presence of the occupation number as one of the integrals of motion in slow timescale enables us to reduce the global six-dimensional phase space to a four-dimensional subspace, similarly to the case of "hard" nonlinearity.

To this end, we introduce the spherical coordinates (the same way as above):

$$\begin{cases} \varphi_1 = \cos(\theta)\cos(\varphi)e^{i\delta_1} \\ \varphi_2 = \sin(\theta)e^{i\delta_2} \\ \varphi_3 = \cos(\theta)\sin(\varphi)e^{i\delta_3} \end{cases} \tag{3.21}$$

Substituting (3.21) into (3.19) and introducing the relative phases $\Delta_{12} = \delta_1 - \delta_2, \Delta_{23} = \delta_2 - \delta_3$, we can describe the slow flow by four angular coordinates. The reduced system takes the form:

$$\frac{d\theta}{d\tau_1} = \beta \cos(\varphi) \sin(\Delta_{12}) - \beta \sin(\varphi) \sin(\Delta_{23})$$
$$\frac{d\varphi}{d\tau_1} = \beta \tan(\theta)(\sin(\Delta_{12}) \sin(\varphi) + \sin(\Delta_{23}) \cos(\varphi))$$
$$\frac{d\Delta_{12}}{d\tau_1} = \begin{bmatrix} -3\alpha\left(\cos^2(\theta) \cos^2(\varphi) - \sin^2(\theta)\right) - \beta + \beta \cot(\theta) \sin(\varphi) \cos(\Delta_{23}) \\ -\beta \frac{2}{\sin(2\theta)\cos(\varphi)} \left(\sin^2(\theta) - \cos^2(\theta) \cos^2(\varphi)\right) \cos(\Delta_{12}) \end{bmatrix}$$
$$\frac{d\Delta_{23}}{d\tau_1} = \begin{bmatrix} -3\alpha\left(sin^2(\theta) - \cos^2(\theta) \sin^2(\varphi)\right) + \beta - \beta \cot(\theta) \sin(\varphi) \cos(\Delta_{12}) \\ -\frac{2}{\sin(2\theta)\sin(\varphi)} \left(\cos^2(\theta) \sin^2(\varphi) - \sin^2(\theta)\right) \cos(\Delta_{23}) \end{bmatrix}$$

$$(3.22)$$

The Hamilton function given in (3.20) can be rewritten in terms of the new coordinates $(\theta, \phi, \Delta_{12}, \Delta_{23})$ as follows

$$H_{4D} = \begin{bmatrix} \frac{3i\alpha}{2}\left(|\sin(\theta)|^4 + |\cos(\theta)|^4\left(|\cos(\varphi)|^4 + |\sin(\varphi)|^4\right)\right) - i\beta \sin(2\theta) \cos(\varphi) \cos(\Delta_{12}) \\ -i\beta \sin(2\theta) \sin(\varphi) \cos(\Delta_{23}) - i\frac{\beta}{2}\cos(2\theta) \end{bmatrix}$$

$$(3.23)$$

Note that (3.22) is not integrable and may exhibit chaotic regimes. However, as it will become clear from the results shown below, there exist certain domains in the parametric space where the regular response regimes coexist with the non-regular ones. The intensive energy exchange regime (Fig. 3.12) displays the phase trajectory corresponding to the initial conditions for system (3.22)

$$\theta(0) = \varphi(0) = \Delta_{12}(0) = \Delta_{23}(0) = 0 \qquad (3.24)$$

We can observe an almost complete energy exchange between the oscillators 1 and 3 along with energy localization on the initially excited oscillator. The

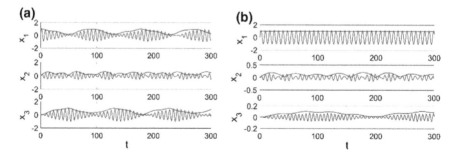

Fig. 3.12 Intensive energy exchange for the initial excitation of the first element, parameter: $\varepsilon = 0.1$, **a** $\beta = 0.75$; **b** $\beta = 0.15$. *Blue lines* depict the evolution of the initial system, *red lines*—the envelopes obtained from the solution of system (3.23)

qualitative distinction from the "hard" nonlinearity can be explained as follows. In both cases, the oscillator 2 interacts with two neighboring oscillators, while the oscillators 1 and 3 interact only with one neighboring oscillator each. The "hard" nonlinearity cannot compensate this asymmetry, only the "soft" one can do this.

Thus, we demonstrate the intensive energy transport from one side of the short chain to another one. The middle element manifests the energy pulsations with the doubled frequency of the side element pulsations. The considered energy exchange involves two characteristic timescales, namely the scale of fast oscillations (t_0) and the scale of slow energy exchange (t_1) between the coupled oscillators.

We will conclude this section with some observations and remarks concerning the relation of the quasi-periodic LPTs in the 3DOF system to the general theory of LPTs. In the case of two coupled anharmonic oscillators, emergence of the LPTs (of both types) is determined by an initial excitation applied to the first oscillator. The role of the coupling parameter in the 2DOF and 3DOF systems is similar. Choosing the coupling strength between the two oscillators below a certain threshold leads to the formation of the LPT of the first type with energy localization on the first oscillator and weak energy exchange between the oscillators. However, increasing the strength of coupling above the threshold leads to global bifurcations of the LPT of the first type manifested by the formation of the LPT of the second type. In the case of hard nonlinearity, the latter falls under category of pure beating with complete energy exchange between the oscillators.

However, the dynamics of the 3DOF system with "hard" nonlinearity is more complicated than the behavior of its 2DOF counterpart. In particular, the mechanism of transition from a locally pulsating quasi-periodic LPT to a strong energy exchange can be quite different. The Poincaré sections demonstrate that the first transition occurs because the quasi-periodic LPT enters a chaotic region. This leads to an intermittent response corresponding to an almost irreversible energy transfer from the first to the second oscillator. As the coupling strength increases, we observe the emergence of a new quasi-periodic LPT (of the second type) characterized by the recurrent energy exchange between the first and second oscillators. The breakdown of a strong local energy exchange occurs due to the penetration of the orbit of the second type LPT into the chaotic region. It is clear from the consideration of the Poincaré sections that the birth and death of the regular quasi-periodic LPTs of the first and second type goes not through their reconnection (as is in the case of the periodic LPTs) but through the sudden formation of a stable strongly pulsating regime. As for the case of "soft" nonlinearity, almost complete energy exchange is possible between the first (initially excited) and third oscillators.

Thus, the possibility of the generalization of the LPT concept on multi-particle systems as well as the mechanism of its breaking is clarified. In the case of "hard" nonlinearity, the reduced 2DOF model plays a key role here, as it does not depend on the number of weakly interacting oscillators. However, as the number of particles increases, another significant possibility of the extension of the LPT concept appears for both types of nonlinearities. This possibility is strongly connected with the notion of *coherence domains*. Detailed presentation of this approach is given below.

Chapter 4
Quasi-One-Dimensional Nonlinear Lattices

In this section, it is shown how the LPT concept can be extended to finite-dimensional oscillatory chains. The systems under consideration are finite-dimensional analogues of several classical infinite models which were initially used for analysis of such significant physical phenomena as recurrent energy transfer and localization.

Over the last several decades, these phenomena remain the subjects of primary theoretical and experimental interest in various aspects of modern theoretical and applied physics such as physics of plasma, nonlinear optics, soft matter, fluid mechanics, and granular matter (Scott 2003). We consider several such problems in Part 3.

The problem of vibration energy localization in the infinite Fermi–Pasta–Ulam (FPU) chain can be asymptotically reduced to finding a localized solution (soliton or breather) of the continuum Korteweg–de Vries (KdV) equation (this long wavelength limit is not significant for our consideration) or nonlinear Schrodinger equation (NLSE), which are completely integrable systems (Scott 2003). After such reducing, the mathematical origin of the phenomenon becomes clear. However, it can be shown that not an infinite but a finite (periodic) β-FPU chain has to be chosen to clarify the physical nature of vibration energy localization (Manevitch and Smirnov 2010). This model has been intensively studied (Poggi and Ruffo 1997; Chechin et al. 2002; Rink and Verhulst 2000; Rink 2001; Henrici and Kappeler 2008a, 2008b; Flach et al. 2005) in the framework of nonlinear normal mode (NNM) concept (Manevitch et al. 1989; Vakakis et al. 1996). On the contrary to NLSE, finite discrete systems are not completely integrable (except the three particle β-FPU periodic model (Feng 2006). Therefore, the knowledge of NNMs leads to a unique possibility for constructing a wide class of solutions using a perturbation technique (Scott 2003). However, an increase of the particle number in the symmetric (β-FPU) chain leads inevitably to the resonant interaction between the high-frequency NNMs because of linear spectrum crowding. We consider the

© Springer Nature Singapore Pte Ltd. 2018
L.I. Manevitch et al., *Nonstationary Resonant Dynamics of Oscillatory Chains and Nanostructures*, Foundations of Engineering Mechanics,
DOI 10.1007/978-981-10-4666-7_4

concepts of "coherence domains" and limiting phase trajectories (LPTs) to understand and describe both complete energy exchange between different parts of the chain and transition to energy localization.

This concept has been extended to finite Klein–Gordon (KG) (Smirnov and Manevitch 2011) and dimer chain (Starosvetsky and Manevitch 2013).

4.1 Finite Fermi–Pasta–Ulam Oscillatory Chain

The main problem discussed in the present section is whether we can extend the LPT concept onto the description of energy exchange in the symmetric β-FPU chain or similar processes in asymmetric $\alpha-\beta$-system. We analyze the influence of the asymmetry of the potential on the resonant interaction of the NNMs and the processes of energy localization. We use an analytical description of energy exchange between "coherent domains" in the $(\alpha-\beta)$-FPU chain.

One should note that the effective renormalization of the quartic coupling constant was introduced for modulation instability analysis of normal modes close to the zone-boundary mode in the rotating-wave approximation in (Sandusky and Page 1994; Burlakov et al. 1996). The current section considers a more wide class of the problems, and the key result is in an analytical description of intensive energy exchange between the "coherent domains" and energy localization on the "coherent domain" in terms of the non-smooth basic functions.

4.1.1   The Model

We consider the $(\alpha-\beta)$-FPU system defined by Hamilton function

$$H_0 = \sum_{j=1}^{N}\left(\frac{1}{2}p_j^2 + \frac{1}{2}\left(q_{j+1} - q_j\right)^2 + \frac{\alpha}{3}\left(q_{j+1} - q_j\right)^3 + \frac{\beta}{4}\left(q_{j+1} - q_j\right)^4\right) \quad (4.1)$$

and periodic boundary conditions ($q_{N+1} = q_1$, $p_{N+1} = p_1$), where q_j and p_j are the coordinates and the conjugate momenta, respectively, and N is the number of particles.

The transition to normal coordinates is given by the linear canonical transformation

$$q_j = \sum_{k=0}^{N-1}\sigma_{j,k}\xi_k \quad (4.2)$$

where

$$\sigma_{j,k} = \frac{1}{\sqrt{N}}\left(\sin\frac{2\pi kj}{N} + \cos\frac{2\pi kj}{N}\right)$$

The transformation (4.2) allows us to present the quadratic part of Hamilton function as the total energy of the independent oscillators:

$$H_2 = \sum_{k=1}^{N-1} \frac{1}{2}\left(\eta_k^2 + \omega_k^2 \xi_k^2\right) \tag{4.3}$$

Here, ξ_k and η_k are the amplitudes and the momenta of NNMs; the coordinate ξ_0 associated with motion of the center of mass is removed from (4.3). The eigenvalues ω_k have the form

$$\omega_k = \omega_{N/2}\sin\frac{\pi k}{N}, \quad \omega_{N/2} = 2, \ k = 0,\ldots,N-1 \tag{4.4}$$

In what follows, we consider an even number of particles N. In this case, the frequencies ω_k are bounded by the highest value $\omega_{N/2} = 2$. All eigenvalues are twice degenerated. The zero eigenvalue $\omega_0 = 0$ corresponds to motion of the chain as a rigid body. If the number of particles increases, the frequency gap between the highest frequency mode and nearby modes quickly decreases.

The anharmonic part of the Hamilton function (4.1) for the periodic $(\alpha-\beta)$-FPU chain has the following form:

$$H_n = \frac{\alpha}{6\sqrt{N}}\sum_{i,j,k=1}^{N-1}\omega_i\omega_j\omega_k D_{ijk}\xi_i\xi_j\xi_k + \frac{\beta}{8N}\sum_{k,l,m,n=1}^{N-1}\omega_k\omega_l\omega_m\omega_n C_{klmn}\xi_k\xi_l\xi_m\xi_n \tag{4.5}$$

where

$$D_{n,m,k} = -\Delta_{n+m+k} + \Delta_{n+m-k} + \Delta_{n-m+k} + \Delta_{n-m-k}$$
$$C_{k,l,m,n} = -\Delta_{k+l+m+n} + \Delta_{k+l-m-n} + \Delta_{k-l+m-n} + \Delta_{k-l-m+n}$$

and

$$\Delta_r = \begin{cases} (-1)^r & \text{if } r = mN, \ m \in Z \\ 0, & \text{otherwise} \end{cases}$$

So the equations of motion can be written as:

$$
\begin{aligned}
\frac{d^2\xi_k}{dt^2} + \omega_k^2\xi_k + \frac{\alpha}{2\sqrt{N}}\omega_k \sum_{l,m=1}^{N-1} \omega_l\omega_m D_{lmk}\xi_l\xi_m \\
+ \frac{\beta}{2N}\omega_k \sum_{l,m,n=1}^{N-1} \omega_l\omega_m\omega_n C_{klmn}\xi_l\xi_m\xi_n = 0
\end{aligned}
\tag{4.6}
$$

To analyze the dynamics of the chain, we introduce the complex variables corresponding to the combination of displacement and velocity in the new basis

$$
\Psi_k = \frac{1}{\sqrt{2}}\left(\frac{1}{\sqrt{\omega_k}}\eta_k + i\sqrt{\omega_k}\xi_k\right)
$$

The inverse transformation allows us to define the coordinates and momenta as follows:

$$
\xi_k = \frac{-i}{\sqrt{2\omega_k}}(\Psi_k - \Psi_k^*); \quad \eta_k = \sqrt{\frac{\omega_k}{2}}(\Psi_k + \Psi_k^*)
$$

In terms of the complex amplitudes, Eq. (4.6) can be rewritten as follows:

$$
\begin{aligned}
i\frac{d\Psi_k}{dt} + \omega_k\Psi_k - i\frac{\alpha}{4\sqrt{2N}}\sum_{l,m=1}^{N-1} \sqrt{\omega_k\omega_l\omega_m}D_{lmk}(\Psi_l - \Psi_l^*)(\Psi_m - \Psi_m^*) \\
- \frac{\beta}{2N}\sum_{l,m,n=1}^{N-1} \sqrt{\omega_k\omega_l\omega_m\omega_n}C_{klmn}(\Psi_l - \Psi_l^*)(\Psi_m - \Psi_m^*)(\Psi_n - \Psi_n^*) = 0
\end{aligned}
\tag{4.7}
$$

Taking into account the dependence of the chain properties on the number of the particles, one can consider $1/N$ as a small parameter ε.

4.1.2 Basic Asymptotic

We employ the multiple scales procedure in order to consider the processes in the "slow" time with the specific scale greatly exceeding $2\pi/\omega_k$.

Following this procedure, we introduce the timescales:

$$
\tau_0 = t, \tau_1 = \epsilon\tau_0, \tau_2 = \epsilon^2\tau_0, \ldots
$$

where the "fast" time τ_0 corresponds to the initial timescale, while the slow times τ_1, τ_2 correspond to the slowly varying envelopes. The envelope functions φ_k are defined by the relation

$$\Psi_k = \varphi_k e^{i\omega_k \tau_0}$$

We construct the asymptotic representation of φ_k in the form

$$\varphi_k = \left(\chi_{k,0} + \epsilon \chi_{k,1} + \epsilon^2 \chi_{k,2} + \cdots\right)$$

The multiple scales expansion based on these relations will be used to derive the equations of the leading-order approximation.

Terms with ε^0:

$$\frac{\partial \chi_{k,0}}{\partial \tau_0} = 0$$

Terms with ε^1:

$$\left(\frac{\partial \chi_{k,0}}{\partial \tau_1} + \frac{\partial \chi_{k,1}}{\partial \tau_0}\right) e^{i\omega_k \tau_0}$$

$$-\frac{\alpha\sqrt{2}}{8} \sum_{l,m=1}^{N-1} \sqrt{\omega_k \omega_l \omega_m} D_{lmk} \left(\chi_{l,0} e^{i\omega_l \tau_0} - \chi_l^* e^{-i\omega_l \tau_0}\right)\left(\chi_{m,0} e^{i\omega_m \tau_0} - \chi_m^* e^{-i\omega_m \tau_0}\right)$$

$$= 0$$

There is one secular term in this equation:

$$\frac{\partial \chi_{k,0}}{\partial \tau_1} = 0$$

So, the function $\chi_{k,0}$ does not depend on the time τ_1.

Then, we can get the expression for $\chi_{k,1}$ and use it at the further steps of approximation:

$$\chi_{k,1} = -i\frac{\alpha\sqrt{2}}{8} \sum_{l,m=1}^{N-1} \sqrt{\omega_k \omega_l \omega_m} D_{lmk}$$

$$\left(\chi_{l,0}\chi_{m,0} \frac{e^{i(\omega_l + \omega_m - \omega_k)\tau_0}}{(\omega_l + \omega_m - \omega_k)} - 2\chi_{m,0}\chi_{l,0}^* \frac{e^{i(\omega_l - \omega_m - \omega_k)\tau_0}}{(\omega_l - \omega_m - \omega_k)} + \chi_{m,0}^*\chi_{l,0}^* \frac{e^{i(\omega_l + \omega_m + \omega_k)\tau_0}}{(\omega_l + \omega_m + \omega_k)}\right)$$

So, one can see that the part of this equation corresponding to the symmetric nonlinear component of the Hamilton function depends only on $\{\chi_{\kappa,0}\}$, and the part corresponding to the asymmetric component can be expressed in $\{\chi_{\kappa,0}\}$ too. Then, we integrate this equation with respect to τ_0 and obtain equation with the renormalized constants.

After some calculations, one can get the equations for the pair of modes with the same eigenvalue (we denote $\chi_{k,0} = \chi_k$ and suppose that the other modes are not excited):

$$i\frac{\partial \chi_k}{\partial \tau_2} + c_{1,k}|\chi_k|^2\chi_k + c_{2,k}|\chi_{N-k}|^2\chi_k + c_{3,k}\chi_{N-k}^2\chi_k^* = 0$$

$$i\frac{\partial \chi_{N-k}}{\partial \tau_2} + c_{1,k}|\chi_{N-k}|^2\chi_{N-k} + c_{2,k}|\chi_k|^2\chi_{N-k} + c_{3,k}\chi_{N-k}^*\chi_k^2 = 0$$

(4.8)

where

$$\begin{cases} c_{1,k} = \left(\frac{9}{8}\beta - \frac{3}{4}\alpha^2\right)\omega_k^2 + \alpha^2 \\ c_{2,k} = \left(\frac{3}{4}\beta + \frac{1}{2}\alpha^2\right)\omega_k^2 - 2\alpha^2 \\ c_{3,k} = \left(\frac{3}{8}\beta - \frac{3}{4}\alpha^2\right)\omega_k^2 + 3\alpha^2 \end{cases}$$

This system possesses the integrals

$$X = |\chi_k|^2 + |\chi_{N-k}|^2$$

and

$$G_- = i(\chi_k\chi_{N-k}^* - \chi_k^*\chi_{N-k}).$$

So, Eqs. (4.8) can be rewritten as follows:

$$i\frac{\partial \chi_k}{\partial \tau_2} + c_{1,k}X\chi_k + ic_{3,k}G_-\chi_{N-k} = 0$$

$$i\frac{\partial \chi_{N-k}}{\partial \tau_2} + c_{1,k}X\chi_{N-k} - ic_{3,k}G_-\chi_k = 0$$

One should note that the effective linearization of equations (4.8) is possible only due to the existence of the integral G_- on the manifold which contains two degenerate modes only. On the other side, such integral manifold exists at the specific initial conditions which prevent an excitation of other nonlinear normal modes.

The intensity of the energy exchange between conjugate modes is governed by the value of the parameter G_-. While this parameter is small, the energy of the mode varies slowly, but the complete amount of it is transferred to the conjugate mode. In the coordinate space, this process corresponds to the slow rearrangement of the oscillation energy along the system. The rate of energy exchange increases as the value of G_- grows, but amount of the energy participating in the exchange

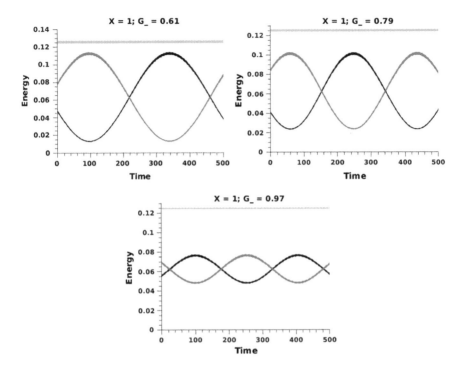

Fig. 4.1 The change of the energy exchange rate with the variation of parameter G_-. The later grows from the left to the right, while the "occupation number" X is constant. The *green line* corresponds to the total energy of the chain, and the *black* and *red curves* correspond to the energies of resonant modes. The amount of energy transferred from one mode to another one decreases, while the parameter G_- grows

decreases (Fig. 4.1). The dynamics of the normal modes in the essentially discrete system (small value of N) should be considered in the assumption that the gap between π-mode and nearby conjugate ones is large enough. However, if the number of particles grows, then a new opportunity arises. Really, while the parameter ε decreases, the gap between the eigenvalues corresponding to the π-mode and the nearby ones becomes smaller:

$$\omega_{\frac{N}{2}} - \omega_{\frac{N}{2}-1} \sim \frac{\pi^2}{2N^2}\,\omega_{\frac{N}{2}}$$

Therefore, the resonant interaction between the highest frequency modes becomes possible. Let us choose the top frequency $\Omega = \omega_{N/2}$ as the basic one. First of all, because of the closeness of frequencies, additional linear terms appear in the equations for $\chi_{N/2-1}$ and $\chi_{N/2+1}$. Moreover, we have to take into account the resonant interactions in the nonlinear part of equations. The problem turns out to be more complicated than that for the β-FPU chain. After some manipulations, we have got the equations

$$i\frac{\partial \chi_1}{\partial \tau_2} - \frac{\pi^2}{2}\Omega^2 \chi_1 + C\Omega^2 \left(|\chi_1|^2 \chi_1 + \chi_1^* \chi_2^2 + 2\chi_0^2 \chi_1^* + 2|\chi_0|^2 \chi_1\right) = 0$$

$$i\frac{\partial \chi_2}{\partial \tau_2} - \frac{\pi^2}{2}\Omega^2 \chi_2 + C\Omega^2 \left(|\chi_2|^1 \chi_2 + \chi_2^* \chi_1^2 + 2\chi_0^2 \chi_2^* + 2|\chi_0|^2 \chi_2\right) = 0 \qquad (4.9)$$

$$i\frac{\partial \chi_0}{\partial \tau_2} + 2C\Omega^2 \left(\left(|\chi_1|^2 + |\chi_2|^2\right)\chi_0 + \left(\chi_1^2 + \chi_2^2\right)\chi_0^*\right) = 0$$

where $\chi_0 = \chi_{N/2}$, $\chi_1 = \chi_{N/2-1}$, $\chi_2 = \chi_{N/2+1}$ and

$$C = \frac{3}{4}\beta - \alpha^2 = \frac{3}{4}\beta_1$$

The obtained equations coincide with those for the β-FPU chain if the constant of the nonlinear interaction is renormalized ($\beta \rightarrow \beta_1$). In particular, (4.9) possess an integral of motion—the total "occupation number":

$$X = |\chi_0|^2 + |\chi_1|^2 + |\chi_2|^2$$

We would like to note that the presence of asymmetric term in the potential function leads to decreasing of the effective nonlinearity up to linearization of (4.9). It is a good agreement with the numerical simulation results (see Fig. 4.2, where the processes of energy exchange between different parts of the really linear chain ($\alpha = 0$, $\beta = 0$) and effectively linear chain ($\beta_1 = 0$) are compared). Because the "renormalized" equations for the $\alpha\beta$-FPU chain coincide with the equations described the β-FPU chain, all effects discussed in (Manevitch and Smirnov 2010) occur in $\alpha\beta$-FPU chain also. Therefore, in the framework of this asymptotic approximation, there is no difference between $\alpha\beta$- and β-FPU chains. It is very interesting that the dynamics of α-FPU system has to correspond to the dynamics of the β-FPU chain with $\beta < 0$. In order to complete the description of the energy exchange processes, we perform the analytical solution of (4.9) for LPT.

4.1.3 From "Waves" to "Particles"

Since we consider of the energy migration in the nonlinear chain, the description of system's dynamics in the terms of NNMs has to be replaced by that in terms of the "coherent domains" and LPTs. This allows us describing the intensive energy

Fig. 4.2 Distribution of the
energy among the particles of
the chain; initial condition
corresponds to LPT (excited
one half of the chain—
coherent domain): **a** really
linear system ($\alpha = \beta = 0$) and
b "effectively linear" system
($\alpha = 0.5$, $\beta = 0.333$)

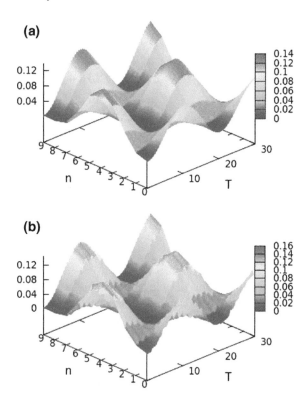

exchange between different parts of the chain, which can be identified as "coherent
domains" with "coordinates" Ψ_1 and Ψ_2:

$$\psi_1 = \frac{1}{\sqrt{2}}\left(\chi_0 - \left(\sqrt{1-c^2}\chi_1 + c\chi_2\right)\right)$$

$$\psi_2 = \frac{1}{\sqrt{2}}\left(\chi_0 + \left(\sqrt{1-c^2}\chi_1 + c\chi_2\right)\right)$$

$$\varphi = \left(c\chi_1 - \sqrt{1-c^2}\chi_2\right)$$

where c is a constant defined by initial conditions ($0 \le c \le 1$). Such transfor-
mation preserves the total occupation number in the form $X = |\Psi_1|^2 + |\Psi_2|^2 + |\varphi|^2$.
For the value $c = 1/\sqrt{2}$ which corresponds to equal initial conditions of modes, we
have the equations in the coordinates of "coherent domains"

$$i\frac{\partial \psi_1}{\partial \tau_2} + \frac{\pi^2}{4}\Omega^2(\psi_2 - \psi_1) + C\Omega^2(|\psi_1|^2\left(\frac{9}{4}\psi_1 - \frac{1}{2}\psi_2\right) - |\psi_2|^2\left(\frac{3}{2}\psi_1 + \frac{1}{4}\psi_2\right)$$

$$+ \frac{1}{4}(\psi_1{}^2\psi_2{}^* - \psi_2{}^2\psi_1{}^*) + |\varphi|^2(\psi_1 + \psi_2) + \varphi^2\left(\frac{3}{2}\psi_1{}^* + \frac{1}{4}\psi_2{}^*\right) = 0$$

$$i\frac{\partial \psi_2}{\partial \tau_2} + \frac{\pi^2}{4}\Omega^2(\psi_1 - \psi_2) + C\Omega^2(|\psi_2|^2\left(\frac{9}{4}\psi_2 - \frac{1}{2}\psi_1\right) - |\psi_1|^2\left(\frac{3}{2}\psi_2 + \frac{1}{4}\psi_1\right)$$

$$+ \frac{1}{4}(\psi_2{}^2\psi_1{}^* - \psi_1{}^2\psi_2{}^*) + |\varphi|^2(\psi_1 + \psi_2) + \varphi^2\left(\frac{3}{2}\psi_1{}^* + \frac{1}{4}\psi_2{}^*\right) = 0$$

$$i\frac{\partial \varphi}{\partial \tau_2} - \frac{\pi^2}{2}\Omega^2\varphi + C\Omega^2(\varphi(|\psi_1|^2 + |\psi_2|^2 + |\varphi|^2) + (\psi_1\psi_2{}^* + \psi_1{}^*\psi_2)$$

$$+ \varphi^*\left(\frac{3}{2}(\psi_1{}^2 + \psi_2{}^2) + \psi_1\psi_2\right)) = 0$$

$$(4.10)$$

Since the notion of coherent domains is of great importance, we discuss it in more details. When dealing with the mechanisms of intensive energy exchange, one needs to reveal the elementary agents which exchange energy and become the domains of its localization (after exceeding some critical level of excitation). In gaseous media, they are the interacting particles (atoms or molecules) themselves, which participate in almost free motion.

On the contrary, the particles in oscillatory chains as well as in all crystalline solids undergo a strong mutual interaction. Therefore, the elementary agent here is all oscillatory chain performing oscillations corresponding to one of non-interacting normal modes. However, an increase of the particle number leads to the resonance relations between several normal modes and then between multiple modes with close frequencies. If the initial conditions are strongly asymmetric (they are far from those corresponding to every resonating normal modes), it leads to the strong intermodal interaction.

As a result, the resonating normal modes are no longer the appropriate elementary agents. In the system of two weakly coupled oscillators, their role is played by the particles themselves changing slowly by the energy in the beating process (their displacements can be presented as a sum and a difference of the modal variables).

As the number of particles increases, the appearance of the resonating modes does not mean that we have to come back to the real particles. Now, the "coherent domains" play a role of elementary agents. Their displacements can be constructed as combinations of resonating modal variables only. These combinations manifest a beating process described by limiting phase trajectory. Thereby, the beating notion is extended to multi-dimensional systems. Besides, introducing the LPT allows us to describe adequately the transition from intensive energy exchange to energy localization on one of the coherent domain due to an increase of intensity of excitation. By this manner, the connection with continuum systems, having the breathers as localized solutions, is clarified. At last, a most simple analytic presentation of intensive energy exchange can be attained exactly in terms of the "coherent domains" and limiting phase trajectories.

Let us consider a particular solution corresponding to $\varphi = 0, \chi_1 = \chi_2$. In this case, Hamilton function is

$$
H = \Omega^2 \left(-\frac{\pi^2}{4} |\psi_1 - \psi_2|^2 + C \left(9 \left(|\psi_1|^2 + |\psi_2|^2 \right)^2 + 6 |\psi_1|^2 |\psi_2|^2 \right. \right.
$$
$$
\left. \left. -2 \left(|\psi_1|^2 + |\psi_2|^2 \right) \left(\psi_1 \psi_2^* + \psi_1^* \psi_2 \right) + \psi_1^2 \psi_2^{*2} + \psi_1^{*2} \psi_2^2 \right) \right)
\tag{4.11}
$$

Since occupation number $X = |\Psi_1|^2 + |\Psi_2|^2$ is the integral of motion, the variables Ψ_1 and Ψ_2 admit the polar representation as follows:

$$
\psi_1 = \sqrt{X} \cos \theta e^{i\delta_1}; \quad \psi_2 = \sqrt{X} \sin \theta e^{i\delta_2}
\tag{4.12}
$$

Substitution of (4.12) with $\phi = 0$ into (4.10) yields the following equations

$$
\frac{\partial \theta}{\partial \tau_2} + K_1 \sin \Delta + K_2 \sin 2\theta \sin 2\Delta = 0
\tag{4.13}
$$

$$
\sin 2\theta \frac{\partial \Delta}{\partial \tau_2} + 2K_1 \cos \Delta \cos 2\theta - 2K_2 \sin 4\theta \left(8 - \cos^2 \Delta \right) = 0
$$

where $\Delta = \delta_1 - \delta_2$ and

$$
K_1 = \Omega^2 \left(\frac{\pi^2}{4} - \frac{3}{32} \beta_1 X \right); \quad K_2 = \frac{3}{64} \beta_1 X \Omega^2
\tag{4.14}
$$

Equations (4.13) correspond to Hamilton function

$$
H(\theta, \Delta) = X \left(\frac{27 \beta_1 X - 16\pi^2}{64} \Omega^2 + K_1 \sin 2\theta \cos \Delta - K_2 \left(8 - \cos^2 \Delta \right) \sin^2 2\theta \right)
\tag{4.15}
$$

At fixed value of X, it is convenient to analyze the solutions of Eqs. (4.13) by the phase portrait method. Figure 4.3 shows the phase portrait at a small value of occupation number X.

There are two stationary points corresponding to NNMs:

$$
(a) \left\{ \theta = \frac{\pi}{4}, \ \Delta = 0 \right\}; \quad (b) \left\{ \theta = \frac{\pi}{4}, \ \Delta = \pi \right\}
\tag{4.16a, b}
$$

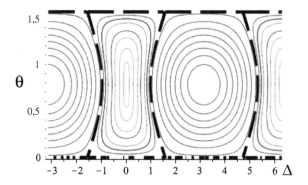

Fig. 4.3 Phase portrait of (4.16) with $X < X_c$ (LPT—*dashed line*)

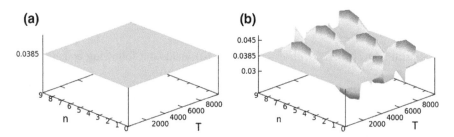

Fig. 4.4 Distribution of the energy among the particles of the chain (numerical solution); initial condition corresponds to π-mode: **a** before the threshold of instability X_c and **b** after it

The points (a) and (b) correspond to $\Psi_1 = \Psi_2$; by definition, this leads either to the pure π-mode (a) or to the pure ($\chi_{N/2-1} + \chi_{N/2+1}$)-mode (b).

When the amplitude of excitation grows, the in-phase stationary point (4.16a) becomes unstable (Fig. 4.4). This instability leads to the appearance of two new stationary points. Namely, if the occupation number X exceeds the value X_c, then there exist three stationary solutions of (4.13) with the phase shift $\Delta = 0$:

$$
\begin{align}
&\text{(a)} \quad \Delta = 0, \theta = \pi/4 \\
&\text{(b)} \quad \Delta = 0, \theta = \frac{1}{2}\arcsin\frac{K_1}{14K_2} \\
&\text{(c)} \quad \Delta = 0, \theta = \frac{\pi}{2} - \frac{1}{2}\arcsin\frac{K_1}{14K_2}
\end{align}
\tag{4.17a–c}
$$

The instability threshold of the π-mode is equal to $X = \pi^2/3\beta_1$, that is in a good accordance with the estimation obtained in the framework of the "narrow packet" approximation (Lichtenberg et al. 2008). Any trajectories in the neighborhood of stationary points (4.17b, c) correspond to weakly localized solutions, for which the energy of one half of the chain—"coherent domains"—only slightly exceeds the energy of the second one. At the same time, the trajectories starting beyond this vicinity and passing through the points Ψ_1 ($\theta = 0$) or Ψ_2 ($\theta = \pi/2$) correspond to the combinations of modes with approximately equal energies. It means that an initial excitation of one "coherence domain" entails the complete energy exchange between both ones, i.e., the transition from the state Ψ_1 into the state Ψ_2 and inversely. This implies that a possibility of complete energy exchange between different parts of the chain exists for the excitation level exceeding the instability threshold X_c (Fig. 4.5).

A growth of the amplitude X entails an enlargement of the domain encircled by the separatrix passing through an unstable stationary point (out-of-phase NNM); at last, the separatrix coincides with the LPT. At this point, the topology of the phase plane changes drastically (Fig. 4.5). Namely, any energy exchange between the mixed states Ψ_1 and Ψ_2 disappears; this implies that a trajectory starting at a point corresponding to $\theta < \pi/4$ (or $\theta > \pi/4$) and for any Δ cannot reach a point corresponding to $\theta > \pi/4$ (or $\theta < \pi/4$) (excepting the trajectories surrounding in-phase stationary points $\Delta = \pi$ and bounded by the separatrix going through the unstable point $\theta < \pi/4$, $\Delta = 0$). Therefore, the energy initially concentrated near the states Ψ_1 or Ψ_2 remains confined in the excited effective oscillator Fig. 4.6.

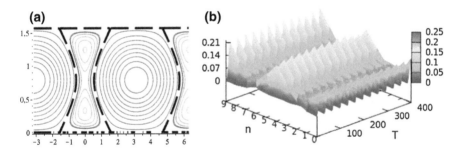

Fig. 4.5 **a** Phase portrait of (4.15) (LPT—*dashed line*) and **b** distribution of energy among the particles in the chain along LPT (numerical solution), $X < X_c < X_{loc}$

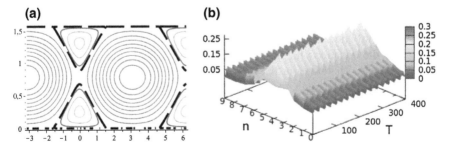

Fig. 4.6 **a** Phase portrait of (4.15) (LPT—*dashed line*) and **b** distribution of energy among the particles in the chain along LPT (numerical solution), $X > X_{loc}$

Epy energy threshold associated with the above-mentioned localization can be found from the condition of equality of the energy corresponding to the LPT and the energy at the unstable π-mode. It is shown in Figs. 4.3, 4.5 and 4.6 that the LPT goes through the points $\theta = \pi/2$ and $\theta = 0$. This means that

$$H(\theta, \Delta)_{|\text{LPT}} = \frac{\Omega^2 X}{64} \left(27\beta_1 X - 16\pi^2 \right)$$

and

$$\sin 2\theta \cos^2 \Delta + \frac{K_1}{K_2} \cos \Delta - 8 \sin 2\theta = 0$$

Since the energy of the π-mode is equal to zero, it is easy to calculate the respective occupation number $X_{\text{loc}} = 16\pi^2/27\beta_1$ and the energy of the chain $E_{\text{loc}} = 16\pi^2/27\beta_1 N$. It now follows that above the excitation level E_{loc} we can observe the localized vibration excitation (a breather).

If the initial energy is concentrated at the state ψ_1 or ψ_2, the representing point in the phase plane moves along a trajectory encircling the respective stationary point. Then, the temporal evolution of the breather corresponds to the regular variation of its profile (the "breathing" mode of the localized excitation). The period of breathing can be calculated as the integral taking along the LPT.

$$T = \oint dt = \frac{1}{\epsilon^2} \oint \frac{d\Delta}{\partial \Delta / \partial \tau_2}$$

Contrary to "breathing" breathers corresponding to the motion along the LPT, the stationary points (4.18) determine new normal modes with invariable non-homogeneous energy distribution, or the breather-like excitations. In spite of these mode existence at any $X > X_c$, the true threshold of localization is equal to X_{loc} because the possibility of the complete energy exchange is preserved for motion along the LPT till $X = X_{\text{loc}}$.

Thus, we get the solution (4.17) corresponding to immobile breather. Finally, we want to clarify the nature of traveling breathers. We recall that the above results have been obtained under assumption $\varphi = 0$ that corresponds to equal amplitudes of the conjugate modes. Now, we assume that the amplitudes of the conjugate modes are slightly different, so we consider small enough $\varphi \neq 0$. One can show that in this case the equations for functions Ψ_1 and Ψ_2 include only quadratic terms depending on φ. Therefore, in the framework of the linear approximation, if the value $\varphi \neq 0$ is small enough, there is no qualitative change in the phase portrait in

Figs. 4.3, 4.5a and 4.6a. As for the behavior of φ in the vicinity of any stationary points (4.17), it is described as follows:

$$i\frac{\partial\varphi}{\partial\tau_2} - \frac{\pi^2}{2}\varphi + \frac{3}{8}\beta_1 X\left(2\sin 2\theta\varphi + \left(3 + e^{2i\delta_1}\sin 2\theta\right)\varphi^*\right) = 0$$

The respective eigenvalue

$$\lambda = \pm\frac{1}{28}\sqrt{891\beta_1^2 X^2 + 60\pi^2\beta_1 X - 96\pi^4}$$

is imaginary if $X < X_{\text{loc}}$, i.e., if the amplitude of excitation is less than the threshold corresponding to coincidence of LPT and separatrix.

It is easy to demonstrate that small nonzero φ in the equations of motion leads to small oscillations of the breather center when the eigenvalue λ is imaginary. From the other hand, appearance of a real part in the eigenvalue λ at $X > X_{\text{loc}}$ leads to the directional motion of the breather (Fig. 4.7).

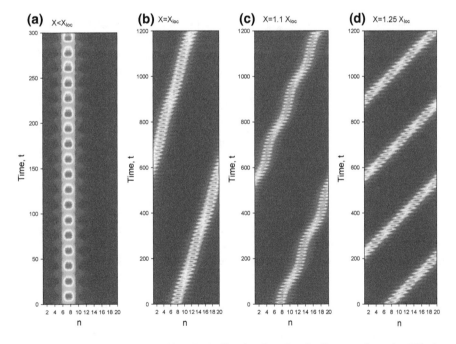

Fig. 4.7 Standing breather (**a**) and breather's directional motion (**b–d**) at growing value X in the β-FPU chain with 20 particles. Dimensionless time is measured in the period of π-mode; n is the number of particle

4.1.4 Analytical Solution for the LPTs

Using the energy value corrsponding to the LPT, one can get the equation for variable Δ, which is solvable in terms of elliptical integrals:

$$\frac{\partial \Delta}{\partial \tau_2} - 2\sqrt{K_2^2(8 - \cos^2 \Delta)^2 - K_1^2 \cos^2 \Delta} = 0$$

or

$$\tau_2 = \int\limits_{-\pi/2}^{\Delta} \frac{d\Delta}{2\sqrt{K_2^2(8 - \cos^2 \Delta)^2 - K_1^2 \cos^2 \Delta}}$$

We get the formal solution of (4.15) to LPT, but since its use is prohibitively difficult, we analyze the system in angle variables (4.15) with LPT equation with $X < X_{\text{loc}}$, $X > X_{\text{loc}}$ and derive an asymptotic formula to a periodic solution of these equations.

It is shown in Fig. 4.8 that $\theta(\tau_2)$ ($X < X_{\text{loc}}$) and $\Delta(\tau_2)$ ($X > X_{\text{loc}}$) are close to the straight line, with an almost instant reverse at $\pi/2$ in the case of θ and instant step at $-\pi/2$ in the case of Δ. In order to describe the solution, we introduce new variable $\tau(\phi)$, $\phi = \Omega_* \tau_2$, Ω_*—constant, which will be yielded from the first-order approximation.

Function $\tau(\phi)$ and its derivative $\varepsilon(\phi) = d\tau(\phi)/d\tau_2$ have the form

$$\tau(\phi) = \frac{2}{\pi} \left| \arcsin\left(\sin \frac{\pi \phi}{2} \right) \right|$$

$$e(\phi) = \begin{cases} 1, & \text{if } 0 < \phi < 1 \\ -1, & \text{if } 1 < \phi < 2 \end{cases} \tag{4.18}$$

We construct an approximate solution as a function of τ; the inverse transformation $\tau_2 = \tau_2(\tau)$ automatically yields a periodic solution in τ_2. Taking into account the discontinuity of $\Delta(\tau_2)$, we construct the solution of (4.9) in the form

$$\theta(\tau_2) = \Theta(\tau); \quad \Delta(\tau_2) = e(\phi)Y(\tau); \quad \frac{\partial}{\partial \tau_2} = \Omega_* e \frac{\partial}{\partial \tau} \tag{4.19}$$

To derive the equations for $\Theta(\tau)$, $Y(\tau)$, we insert (4.18) into (4.14). This yields the set of equations

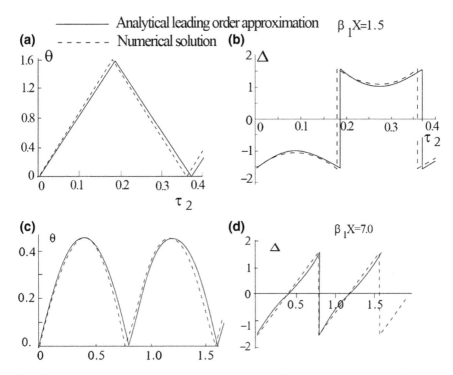

Fig. 4.8 Comparing of the numerical solution of (4.15) with the analytical first-order approximations in the case of beating between the "coherent domains" (**a, b**) and in the case of strongly localized solution (**c, d**)

$$\Omega_* \frac{\partial \Theta}{\partial \tau} + K_1 \sin Y + K_2 \sin 2\Theta \sin 2Y = 0$$

$$\Omega_* \frac{\partial Y}{\partial \tau} \sin 2\Theta + 2K_1 \cos Y \cos 2\Theta - 2K_2 \sin 4\Theta \left(8 - \cos^2 Y\right) = 0 \qquad (4.20)$$

$$\Theta(0) = 0; \;\; Y(0) = -\frac{\pi}{2}$$

Keeping in mind the relation between Y and θ corresponding to the LPT

$$\sin 2\Theta \, \cos^2 Y + \frac{K_1}{K_2} \cos Y - 8 \sin 2\Theta = 0$$

one can get the leading-order approximation of the solution of system (4.20) in the case of beating between "coherent domains" far from localization:

$$\Theta_0(\tau) = A_0\tau$$

$$Y_0 = -\frac{\pi}{2} + \arctan\left(2\sqrt{\frac{2}{7}}\tan\left(\frac{\sqrt{14}K_1}{A_0\Omega_{*0}}\sin(2A_0\tau)\right)\right) \qquad (4.21)$$

$$A_0 = \frac{\pi}{2}; \; A_0\Omega_{*0} = K_1$$

and in the case of strongly localized solution:

$$\Theta = A_* \sin \pi\tau$$

$$Y = \frac{\pi}{2}\left(\tau - \frac{1}{2}\right)$$

$$A_* = \arcsin\frac{K_1}{7K_2}; \; A_*\Omega_{*0} = K_1$$

Figure 4.8 demonstrates that the numerically constructed solution of (4.18) is in a good agreement with the yielded approximations.

4.2   Klein–Gordon Lattice

4.2.1   The Model

This section is devoted to the generalization of results obtained in the previous one to another important class of nonlinear chains, Klein–Gordon models. Unlike Fermi–Pasta–Ulam chains, the Hamiltonian for these models, along with the gradient component depending on deformations (in this case, quadratic), contains a component depending on displacements.

Let us consider the periodic system of weakly coupled particles in the field of the local (on site) potential. It was noted in the introduction that the best known examples of such systems are the Frenkel–Kontorova and Klein–Gordon models, in particular, the model ϕ^4. In the linear approximation, these models result in equations of the same type, linear discrete Klein–Gordon equations, the properties in the continual limit of which are well studied. However, in relatively small systems, the discrete character plays an important role, and in this case, the traditional approach consists in the application of the technique of normal oscillations, linear normal modes. In the quasi-linear case, at first glance, the idea of nonlinear normal modes (Manevitch et al. 1989; Vakakis et al. 1996) should also be used. The stability of nonlinear normal modes is usually studied in the linear approximation. However, this analysis for finite systems turns out to be quite cumbersome and does not give information on the process realized as a result of loss of stability. The only factor obtained from linear analysis is the existence of the maximum growth mode.

Let us represent the Hamiltonian of the considered system in the form

$$H = \sum_{j=1}^{N} \left(\frac{p_j^2}{2} + \frac{c}{2}(q_{j+1} - q_j)^2 + V(q_j) \right) \tag{4.22}$$

where

$$V(q) = \omega_0^2 \left(1 - \cos\left(\frac{2\pi}{d}q\right) \right) \tag{4.23a}$$

for the Frenkel–Kontorova model and

$$V(q) = \frac{\omega_0^2}{2}q^2 + \frac{\beta}{4}q^4 \tag{4.23b}$$

for the Klein–Gordon model.

Further, the quasi-linear expansion in form (4.23b) will be used for the Klein–Gordon as well as Frenkel–Kontorova model (with the respective renormalization of the parameters). The behavior of the system essentially depends on the sign of the nonlinearity parameter β. In the Klein–Gordon model, this parameter is assumed to be positive (hard nonlinearity), and in the Frenkel–Kontorova model, it is assumed to be negative (soft nonlinearity). The FK model describing many important physical processes will be considered in the Part 3.

Taking into account the generalization to the case of infinite system below, we use periodic boundary conditions, $q_{N+1} = q_1$, $p_{N+1} = p_1$. Then, normal modes are introduced according to the same rules as in the Fermi–Pasta–Ulam lattice (Poggi and Ruffo 1997):

$$q_j = \sum_{k=0}^{N-1} \sigma_{j,k} \xi_k$$

where $\sigma_{j,k} = \frac{1}{\sqrt{N}}\left(\sin\left(\frac{2\pi kj}{N}\right) + \cos\left(\frac{2\pi kj}{N}\right)\right)$, and ξ_k is the amplitude of the kth mode. The eigen frequency spectrum is determined by the dispersion relation

$$\omega_k^2 = \omega_0^2 + 4c^2 \sin^2 \frac{\pi k}{N}.$$

The main difference from the spectrum of the Fermi–Pasta–Ulam lattice is on the right-hand side, i.e., it is manifested for small values of the "wave number" k (since the spectrum is symmetric with respect to the point $k = N/2$, the value $k = 0$ is eliminated). Figure 4.9 shows the comparison of eigenfrequency spectra for periodic Klein–Gordon (solid points) and Fermi–Pasta–Ulam (open points) lattices with 20 particles (the value of the "wave vector" k is increased by unity in order to shift end frequencies from the ordinate axis).

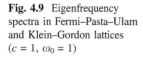

Fig. 4.9 Eigenfrequency
spectra in Fermi–Pasta–Ulam
and Klein–Gordon lattices
($c = 1$, $\omega_0 = 1$)

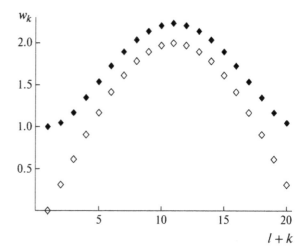

Unlike the Fermi–Pasta–Ulam lattice, the spectrum of the Klein–Gordon model has ω_0 as the left boundary; this value corresponds to homogeneous particle oscillations in the local potential of the substrate. Thus, the spectrum of the Klein–Gordon lattice includes two boundary frequencies corresponding to the wave numbers $k = 0$ and $k = N/2$. The following fact should be pointed out: In the Klein–Gordon lattice, $k = 0$ and $k = N/2$ are the points of extremum of the function $\omega(k)$. Therefore, there are no linear terms in the neighborhood of these points in the frequency expansion and eigen frequencies are separated by smaller intervals than in the intermediate part of the spectrum. The largest values of eigenfrequencies correspond to the sawtooth (or π) mode realizing out-of-phase shifts of neighboring particles for the Fermi–Pasta–Ulam and Klein–Gordon lattices. Therefore, it is natural to assume that the behavior of this mode in both models is similar (to corresponding renormalization). This study is connected with the analysis of stability of the lower boundary mode in the Klein–Gordon lattice. Similar to the case of the π mode, the density of eigenvalues near the end grows quickly with increasing the number of the particles. Therefore, for systems containing more than 10–15 particles, the resonance conditions are satisfied near the left end of the spectrum.

In order to elucidate the consequences of the resonance action, let us consider the initial conditions corresponding to the excitation of two modes with close frequencies ("wave numbers" $k = 0$ and $k = 1$). Strictly speaking, three interacting modes should be taken into account ($k = 0$, $k = 1$, $k = N - 1$), since the two latter modes have the same eigenfrequency, i.e., are degenerate. However, the generalization to the case of three interacting modes does not present any fundamental difficulties (Manevitch and Smirnov 2010). Therefore, in order to avoid cumbersome calculations, we consider only two modes, preserving fundamentally important specific features of the studied process. Then, particle displacements in the lattice are expressed as follows:

$$q_j = \frac{1}{\sqrt{N}}\left[\xi_0 + \left(\sin\frac{2\pi j}{N} + \cos\frac{2\pi j}{N}\right)\xi_1\right]$$

and the potential energy of the system takes the form

$$U(\xi_0, \xi_1) = \frac{1}{2}\left(\omega_0^2\xi_0^2 + \omega_1^2\xi_1^2\right) + \frac{\beta}{N}\left(\frac{1}{2}\xi_0^4 + 3\xi_0^2\xi_1^2 + \frac{3}{4}\xi_1^4\right)$$

$$\omega_1^2 = \omega_0^2 + 4c^2\sin^2\frac{\pi}{N}$$

With increasing number of particles, the gap between the frequencies decreases according to the dependence

$$\omega_1 - \omega_0 \cong 2\frac{c^2\pi^2}{\omega_0 N^2}. \tag{4.24}$$

In considering systems with the number of particles $N \gg 1$, we choose $1/N$ as the small parameter ε and rewrite relation (4.24) in the form

$$\delta\omega = v\varepsilon^2, \quad v = 2\frac{c^2\pi^2}{\omega_0}$$

Then, the motion equations can be written as follows:

$$\frac{d^2\xi_0}{dt^2} + \omega_0^2\xi_0 + \varepsilon\beta\left(\xi_0^3 + 3\xi_0\xi_1^2\right) = 0$$

$$\frac{d^2\xi_1}{dt^2} + \omega_1^2\xi_1 + \varepsilon\beta\left(\frac{3}{2}\xi_1^3 + 3\xi_1\xi_0^2\right) = 0 \tag{4.25}$$

For analysis of the obtained equations, it is reasonable to introduce the complex variables. This procedure, as well as subsequent asymptotic analysis based on the multi-scale method, was described in detail in Manevitch and Smirnov (2010). Therefore, here, we briefly mention the corresponding calculations.

4.2.2 Asymptotic Analysis

The complex variables are determined as follows:

$$\Psi_k = \frac{1}{\sqrt{2}}\left(\frac{d\xi_k}{dt} + i\omega_k\xi_k\right)$$

Since the evolution of the envelope of the lower boundary mode ξ_0 is of greatest interest, it is reasonable to separate the carrier frequency and introduce the time hierarchy. Moreover, it is necessary to expand the amplitudes with respect to the small parameter. Thus, we have

$$\Psi_k = \Phi_k e^{i\omega_k t} = \sqrt{\varepsilon}\big(\chi_k + \varepsilon\chi_{k,1} + \varepsilon^2\chi_{k,2} + \cdots\big)e^{i\omega_k t} \tag{4.26}$$

$$\tau_0 = t, \ \tau_1 = \varepsilon\tau_0, \ \tau_2 = \varepsilon^2\tau_0$$

Substituting expansions (4.26) into Eq. (4.25), collecting terms at different powers of the small parameter, and equating them to zero, we obtain the following system of equations:

$$\varepsilon^{1/2} : i\frac{\partial\chi_k}{\partial\tau_0} = 0,$$

$$\varepsilon^{3/2} : i\frac{\partial\chi_k}{\partial\tau_1} = 0.$$

These equations show that the amplitudes of the principal approximation for both modes are independent on the "fast" times τ_0 and τ_1. In the next order, with respect to the small parameter, we obtain the equations for amplitudes in terms of the slow time τ_2 that correspond to the main asymptotic approximation

$$\varepsilon^{5/2} : i\frac{\partial\chi_0}{\partial\tau_2} + \frac{3\beta}{4\omega_0^3}\Big[|\chi_0|^2\chi_0 + 2|\chi_1|^2\chi_0 + \chi_1^2\chi_0^*\Big] = 0$$

$$i\frac{\partial\chi_1}{\partial\tau_2} + v\chi_1 + \frac{3\beta}{8\omega_0^3}\Big[3|\chi_1|^2\chi_1 + 4|\chi_0|^2\chi_1 + 2\chi_0^2\chi_1^*\Big] = 0 \tag{4.27}$$

It should be noted that Eq. (4.27) describes nonlinear interaction of modes with the frequency difference v. In this case, the cross-terms containing squared absolute value of the amplitude are included in the equations in the absence of resonance interaction; however, their presence does not influence the mode dynamics and in reality results in frequency shift only. On the contrary, terms containing complex conjugate quantities occur only in resonance conditions that results in nonlinear mode interaction. Equation (4.27) corresponds to Hamilton function

$$H_\chi = v|\chi_1|^2 + \frac{3}{16}\frac{\beta}{\omega_0^3}\Big[2|\chi_0|^4 + 3|\chi_1|^4 + 8|\chi_0|^2|\chi_1|^2 + 2\big(\chi_0^2\chi_1^{*2} + \chi_0^{*2}\chi_1^2\big)\Big]. \tag{4.28}$$

Equations (4.27), unlike the original system, admit a second integral, the occupation number,

$$X = |\chi_0|^2 + |\chi_1|^2 \tag{4.29}$$

and therefore, system (4.27) is integrable. Hamilton function (4.28) describes the system dynamics in the modal, i.e., "wave," representation. However, it is natural to analyze the processes resulting in energy localization in the "particle" representation, which preserves the simplicity of description inherent in Eq. (4.27). For this purpose, we introduce the new variables

$$\varphi_0 = \frac{1}{\sqrt{2}}(\chi_0 + \chi_1); \quad \varphi_1 = \frac{1}{\sqrt{2}}(\chi_0 - \chi_1)$$
$$|\varphi_0|^2 + |\varphi_1|^2 = X$$

which due to relation (4.28) provide representation of the system dynamics in terms of polar coordinates—the amplitude and phase (Kosevitch and Kovalev 1989).

$$\varphi_0 = \sqrt{X}\cos\theta e^{i\delta_0}; \quad \varphi_1 = \sqrt{X}\sin\theta e^{i\delta_1}$$

Then, the change of the system behavior with increasing the excitation intensity can be clearly traced on the phase plane $(\theta, \Delta = \delta_0 - \delta_1)$. Omitting details of calculations, we obtain the equations in terms of angular variables

$$\sin 2\theta \frac{\partial \Delta}{\partial \tau_2} - \frac{1}{16\omega_0^3}\cos 2\theta\left[(3\beta X + 16v\omega_0^3)\cos\Delta - 3\beta X(\cos^2\Delta + 9)\sin 2\theta\right] = 0$$
$$\sin 2\theta\left[\frac{\partial \theta}{\partial \tau_2} - \frac{1}{32\omega_0^3}\left((3\beta X + 16v\omega_0^3 - 3\beta X\cos\Delta\sin 2\theta)\sin\Delta\right)\right] = 0$$

$$\tag{4.30}$$

For low-excitation levels X, Eq. (4.30) has two stationary points corresponding to NNMs. The point $(\Delta = 0, \theta = \pi/4)$ corresponds to the lower boundary mode χ_0 and the point $(\Delta = \pi, \theta = \pi/4)$ to the closest mode χ_1. States with $\theta = 0$ and $\theta = \pi/2$ describe the mixture of these modes; these points are intersected by trajectories maximally remote from the stationary points. The LPTs, describe the most intense energy exchange between parts of the system. In the case of weak excitation, the topology of the phase plane corresponds to Fig. 4.10.

The growth of excitation level results in a qualitative change of the phase portrait of the system. Indeed, along with the stationary points mentioned above

Fig. 4.10 Topology of phase
plane of Eq. (4.30) for
low-excitation levels in the
Frenkel–Kontorova model.
Dashed line corresponds to
the limiting phase trajectory

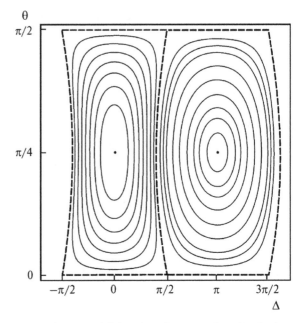

corresponding to the NNMs, there exist two additional stationary points, beginning
from some excitation level.

$$(a)\begin{cases} \sin 2\theta = -\dfrac{3\beta X + 16v\omega_0^3}{21\beta X} \\ \cos \Delta = 1 \end{cases}$$

$$(b)\begin{cases} \sin 2\theta = \dfrac{3\beta X + 16v\omega_0^3}{21\beta X} \\ \cos \Delta = -1 \end{cases}$$

(The cases (a) and (b) relate to the FK and KG chains, respectively.)

Similarly to the cases of two weakly coupled oscillators and finite FPU chain,
these stationary points occur as a result of the loss of stability of the boundary mode
if the excitation level X_{cr} is achieved. Its value can be estimated by (4.31a) for the
Frenkel–Kontorova model or by (4.31b) for the Klein–Gordon lattice

$$X_{\text{cr}} = -\frac{2v\omega_0^3}{3\beta} \quad (\beta < 0) \tag{4.31a}$$

$$X_{\text{cr}} = \frac{8v\omega_0^2}{9\beta} \quad (\beta > 0) \tag{4.31b}$$

Fig. 4.11 Topology of phase plane of Eq. (4.30) for excitation levels exceeding the stability threshold in the Frenkel–Kontorova model. *Short-dashed* and *long-dashed lines* correspond to the limiting phase trajectory and to the separatrix, respectively

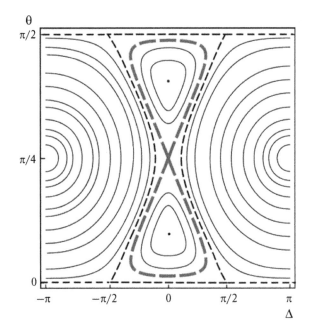

The consequence of instability of these normal modes is the formation of the separatrix which transforms with increasing excitation level (Fig. 4.11). However, in this case, the possibility of complete energy exchange between parts of the system via motion along limiting phase trajectories is retained.

One should note that for the Klein–Gordon lattices, the topology of the phase plane is similar to that shown in Figs. 4.10 and 4.11 with the phase shift by π. However, this conclusion is related to the mode closest to the boundary mode, rather than to the boundary mode itself. Further growth of excitation level results in the second dynamic transition due to the merging of the limiting phase trajectory and the separatrix (Fig. 4.12). In this case, complete energy exchange becomes impossible, since any trajectories beginning at $\theta < \pi/4$ ($\theta > \pi/4$) cannot achieve the domain $\theta > \pi/4$ ($\theta < \pi/4$). The latter means that the energy initially localized in some part of the system (coherent domain) is preserved in the excited region in the course of motion. In order to determine the critical value of excitation level corresponding to the second dynamic transition, i.e., merging of the limiting phase trajectory and the separatrix, let us equate the energies corresponding to these trajectories

$$E\left(\theta = \frac{\pi}{4}, \Delta = 0\right) = E\left(\theta = 0, \Delta = \frac{\pi}{2}\right)$$

Fig. 4.12 Phase portrait of
the Frenkel–Kontorova
system above the second
dynamic transition

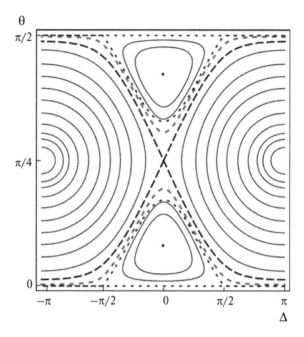

The energy is calculated based on Hamilton function representation (4.28) in terms of polar variables

$$E(\theta, \Delta) = \frac{X}{256\omega_0^3} \left(159X\beta + 128v\omega_0^3 + 45X\beta \cos 4\theta + 6X\beta \cos 2\Delta \sin^2 2\theta \right. \\ \left. - 8 \cos \Delta \sin 2\theta (3X\beta + 16v\omega_0^3) \right)$$

As a result, we obtain the following value of the critical excitation level for the Frenkel–Kontorova model:

$$X = -\frac{32v\omega_0^3}{27\beta} \tag{4.32a}$$

and for the Klein–Gordon model:

$$X = -\frac{32v\omega_0^3}{27\beta} \tag{4.32b}$$

It should be noted that while before the second dynamic transition the phenomena similar to pulsations in the system of two coupled oscillators are observed both in the Frenkel–Kontorova and in Klein–Gordon models, above the second critical excitation level, the behavior of these systems becomes different. The matter is that the energy localization is connected with perturbation of homogeneous, i.e., lower boundary mode, and in Klein–Gordon lattices, this mode turns out to be stable. The

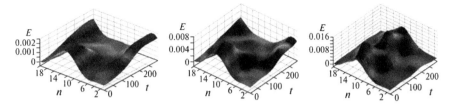

Fig. 4.13 Total energy surface for the system of 20 particles in the Frenkel–Kontorova model

loss of stability and further energy "blocking" take place for the phase $\Delta = \pi$. At the same time, it is the homogeneous mode that is modulationally unstable in the Frenkel–Kontorova model, which results in the energy localization in the originally excited part of the system (coherent domain). Thus, at the second critical value of the nonlinearity parameter in the Frenkel–Kontorova model, the transition "complete energy exchange–energy localization" is observed, but in the Klein–Gordon model, there is no such transition (near the upper boundary of the spectrum, the situation is the opposite). Thus, it becomes clear why localized oscillatory excitations (breathers) are absent near the left end of the spectrum in Klein–Gordon type systems. The data of numerical simulation prove the results of analytical study.

Figure 4.13 shows the results of simulation for the periodic system with 20 particles in the Frenkel–Kontorova model. The surface of the total particle energy is shown, and the particle numbers and time in units corresponding to the period of self-oscillations of the lower boundary mode are shown along the axes. Excitation levels go from left to right as follows: to the threshold of loss of stability of the boundary mode, after the loss of stability below the localization threshold, and above the localization threshold. It can be clearly seen that before the threshold of true localization, the energy originally concentrated in the central part of the chain (maximum at approximately 12th particle) is transmitted to particles with small numbers, while above the localization threshold the energy is preserved in the central part of the chain. The simulation performed for the Klein–Gordon model in this spectral region showed the absence of such localization processes.

4.3 Intense Energy Exchange and Localization in Periodic FPU Dimer Chains

We considered till now the oscillatory chains of homogeneous structure. System under consideration in the present section is FPU dimer chain given to periodic boundary conditions and composed of two identical cells of particles. Each cell of the chain comprises exactly one heavy particle succeeded by a group of N light particles of identical masses (i.e., 1: N dimer chain). We show that resonant inter-particle interaction occurring on the optical branch (in the limit of a strong mass mismatch $\varepsilon = m/M \ll 1$) leads to a very interesting dynamical transitions undergone by the entire chain. Here, we would like to bring the main features of the regimes corresponding to the optical and acoustic branches. Thus, the regimes of the acoustic branch are manifested by an in-phase motion of each pair of neighboring elements. Moreover, the most significant amount of system energy is carried by the heavy particles. In contrast to the case of acoustic branch, regimes of the optical branch are manifested by the anti-phase motion of each pair of neighboring elements, as well as the low-amplitude oscillations of the heavy particles. Thus, in the limit of a strong mass mismatch $\varepsilon \ll 1$, it is easy to show that for any particular regime corresponding to the acoustic branch energy carried by the light particles is negligibly small $(O(\varepsilon))$ in comparison with the heavy ones. However, in the same limit of a high mass mismatch, regimes of an optical branch are characterized by a negligibly small fraction of energy carried by the heavy particles in comparison with the light ones.

We demonstrate the existence of peculiar regimes belonging to the optical branch and leading to a complete energy transport between the groups of light particles when the heavy ones remain nearly stationary. Applying the method of limiting phase trajectories, we find a threshold value of the parameter of nonlinearity above which energy imparted on one of the groups remains permanently localized within the same group and this without leaking to the other parts of the chain. Moreover, as it will be shown below, the aforementioned regime of strong energy exchanges between the groups of light particles exists for an arbitrary number of elements included in each one of the groups.

This result highly differs from those obtained for a homogeneous chain (Manevitch and Smirnov 2010) where the system is required to include sufficient number of particles to ensure the formation of beating phenomenon between the two halves of the chain. Thus, dynamics of the dimer chain reveals a new mechanism of formation of highly non-stationary regime leading to a massive transport of energy between the groups of light elements. As it will become clear from the analysis brought below, this mechanism of formation of strong beats is fully governed by the mass ratio between the light and heavy particles. Thus, decreasing the mass ratio, one effectively creates the closely spaced pairs of modes situated on the optical branch. These pairs are responsible for resonant interactions between the groups of light particles leading to complete energy exchanges.

It is worthwhile emphasizing that formation of beats between the groups of light particles is not the only possible mechanism of a massive energy transport in the FPU dimer chain. Indeed, one can easily show that operating on the acoustic branch under assumption of a strong mass mismatch ($\varepsilon \to 0$) the FPU dimer chain effectively reduces to a homogeneous FPU chain (i.e., where the elements of the reduced, chain correspond to the heavy particles of the dimer chain). Thus, according to Manevitch and Smirnov (2010), inclusion of a sufficient number of heavy particles in the chain will allow for a complete energy transport solely between the two groups of heavy particles (i.e., two halves of the chain) when the energy remaining in the light ones is negligibly small. Here, we note once again that in scope of the present study we are focusing solely on the regimes corresponding to the optical branch and revealing the all new mechanism of energy transport between the light elements.

The results of the current study are by no means limited to serve the academic purposes but have the far-going applications in the real engineering fields such as bridge dynamics, vibration isolation, and shock absorption. For instance, as it was discussed in Manevitch and Oshmyan (1999), a one-dimensional system which consists of the discrete, light masses linked by a beam with elastic supports is an adequate dynamic model in a wide field of technical applications such as aeronautics, ship building, civil, nuclear, and rocket engineering. As it is evident from the study pursued in the paper, the introduction of several heavy masses in the structure of mechanical system (delimiting the light ones) may induce the very peculiar and at the same time important dynamical regimes leading to energy localization on a single group of light masses, thus preventing the penetration of unwanted vibrations or shock waves to the second part of the structure. This is a highly important feature for the purposes of energy isolation and shock absorption.

In the first section, we present the dynamical system with the necessary physical and asymptotic adoptions. We demonstrate the mechanism of formation of intense energy exchanges between the light particles in a simple linear, (1:1) dimer chain. Sections 4.3.2, 4.3.3 are fully devoted to the analysis of the FPU (1:1) dimer chain. The main objective of Sect. 4.3.3 is to analytically estimate a threshold value for the parameter of nonlinearity leading to the dynamical transitions from the regime of a near complete energy exchanges between the light particles to a strong energy localization on one of them. In Sect. 4.3.4, we bring the generalization of the results obtained for (1:1) dimer chain to a (1:N) dimer chain. In the same section, we show that aforementioned regimes of strong energy exchanges and localization are manifested in terms of interacting groups of light particles rather than single ones, giving rise to N—possible modes of interaction situated on the optical branch of the chain.

4.3.1 The Model

In the present section, we consider a nonlinear cyclic, FPU dimer chain comprising 2 heavy particles and $N > 1$ light particles, such that each group of N light particles is delimited by the two heavy ones from both ends (Fig. 4.14). The system under consideration is described by the following set of equations

$$\ddot{x}_1 = x_{2N+2} - 2x_1 + x_2 + \varepsilon\alpha\tilde{F}_1$$
$$\varepsilon\ddot{x}_2 = x_1 - 2x_2 + x_3 + \varepsilon\alpha\tilde{F}_2$$

$$\cdots$$

$$\varepsilon\ddot{x}_{N+1} = x_N - 2x_{N+1} + x_{N+2} + \varepsilon\alpha\tilde{F}_{N+1}$$
$$\ddot{x}_{N+2} = x_{N+1} - 2x_{N+2} + x_{N+3} + \varepsilon\alpha\tilde{F}_{N+2} \tag{4.33}$$
$$\varepsilon\ddot{x}_{N+3} = x_{N+2} - 2x_{N+3} + x_{N+4} + \varepsilon\alpha\tilde{F}_{N+3}$$

$$\cdots$$

$$\varepsilon\ddot{x}_{2N+2} = x_{2N+1} - 2x_{2N+2} + x_1 + \varepsilon\alpha\tilde{F}_{2N+2}$$

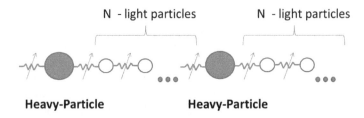

Heavy-Particle   Heavy-Particle

Fig. 4.14 Scheme of the model under consideration

where $x_k = x_{k+2N+2}$, $\tilde{F}_j = \{x_{j-1} - x_j\}^3 - \{x_j - x_{j+1}\}^3$ is a nonlinear part of the inter-particle, interaction force, x_i is the displacement of the ith particle of the chain, $\varepsilon = m/M$ is the normalized mass ratio between the light and heavy particles of the chain and is defined as a small system parameter $(0 < \varepsilon \ll 1)$, and $\alpha \sim O(1)$ is a parameter of stiffness nonlinearity. In the course of the present study, we are interested in demonstration of the peculiar intermodal, resonant interactions leading to the intensive energy exchanges between the light particles as well as strong energy localization on one of them.

4.3.2   Intensive Energy Exchanges: Linear Case $(N = 1, \alpha = 0)$

Let us start with the illustration of the idea of intensive energy exchanges between the light particles from the simplest, linear case $(N = 1, \alpha = 0)$. The linear subsystem of (1.138) reads

$$
\begin{aligned}
\ddot{x}_1 &= x_4 - 2x_1 + x_2 \\
\varepsilon \ddot{x}_2 &= x_1 - 2x_2 + x_3 \\
\ddot{x}_3 &= x_2 - 2x_3 + x_4 \\
\varepsilon \ddot{x}_4 &= x_3 - 2x_4 + x_1
\end{aligned}
\tag{4.34}
$$

It is important to note that in the framework of the current study we solely focus on the resonant interactions between the particles having their eigenfrequencies situated on the optical branch which for the regular $(N = 1)$, linear $(\alpha = 0)$ dimer chain is defined by a well-known relation (Brillouin 1946)

$$
\omega_n^2 = \frac{(1+\varepsilon)}{\varepsilon} + \left\{ \left(\frac{1+\varepsilon}{\varepsilon}\right)^2 - \frac{4}{\varepsilon}\sin\left(\frac{\pi n}{2}\right) \right\}^{1/2}, \quad n = 0, 1
\tag{4.35}
$$

where n defines a particular mode of the optical branch and is fully dependent on the periodicity of the chain. Dimer chain considered in the present section

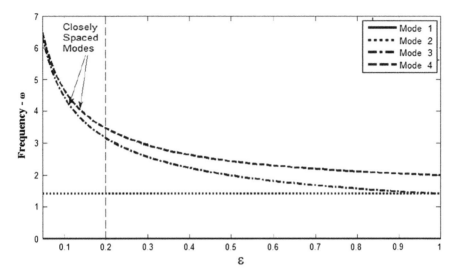

Fig. 4.15 Evolution of the four eigenfrequencies of (4.35) versus the ε parameter

comprises exactly two cells, each one containing exactly one heavy and one light particles, and therefore, n assumes only two values, namely $n = 0, 1$.

System (4.35) has the following set of eigenvectors and eigenvalues

$$
\mathbf{M} = \begin{pmatrix} 1 & -1 & 0 & -\varepsilon \\ 1 & 0 & -1 & 1 \\ 1 & 1 & 0 & -\varepsilon \\ 1 & 0 & 1 & 1 \end{pmatrix}, \quad \mathbf{\Lambda} = \left(0, 2, 2\varepsilon^{-1}, 2\varepsilon^{-1}(1+\varepsilon)\right) \tag{4.36}
$$

where $\mathbf{M}$ is the modal matrix, and $\mathbf{\Lambda}$ is the set of the corresponding eigenvalues.

In Fig. 4.15, we plot the evolution of all the four eigenfrequencies with respect to the variation of ε parameter in the range of $\varepsilon \in (0, 1]$.

As one may infer from the diagram presented in Fig. 4.15, the two highest modes of system (4.32a, b) become closely spaced in the limit of $\varepsilon \to 0$. In the present section, we show that resonant interaction of these two modes leads to the complete energy exchanges between the light particles. Thus, initial energy primarily supplied to one of the light particles of (4.32a, b) will be completely exchanged (in the recurrent fashion) with the second light particle of the chain. Amount of energy stored on the heavy particles is negligibly small in comparison with that of light ones.

To understand this peculiar resonant exchange between the light particles, it is convenient to introduce modal coordinates,

$$
\mathbf{x} = \mathbf{Mq} \tag{4.37}
$$

where, $\mathbf{x} = \begin{bmatrix} x_1 & x_2 & x_3 & x_4 \end{bmatrix}$, $\mathbf{q} = \begin{bmatrix} q_1 & q_2 & q_3 & q_4 \end{bmatrix}$

Thus, after some trivial algebraic manipulations, one obtains the following set of linear oscillators each one of which corresponds to a certain mode

$$
\begin{aligned}
&\ddot{q}_1 = 0 \\
&\ddot{q}_2 + 2q_2 = 0 \\
&\ddot{q}_3 + \left(\frac{2}{\varepsilon}\right)q_3 = 0 \\
&\ddot{q}_4 + \left(\frac{2(1+\varepsilon)}{\varepsilon}\right)q_4 = 0
\end{aligned}
\tag{4.38}
$$

Rescaling of system (6) with respect to time $\tau = \sqrt{\frac{2}{\varepsilon}}t$ yields

$$
\begin{aligned}
&q_1'' = 0 \\
&q_2'' + \varepsilon q_2 = 0 \\
&q_3'' + q_3 = 0 \\
&q_4'' + (1+\varepsilon)q_4 = 0
\end{aligned}
\tag{4.39}
$$

It is quite evident from the rescaled system (4.40) that considering ε as a small system parameter ($\varepsilon \ll 1$) the two highest modes (q_3, q_4) become closely spaced creating the appropriate conditions for internal resonant interaction in the chain leading to the recurrent energy exchanges between the different parts of the chain. It is important to emphasize that this peculiar mechanism of resonant interaction between certain distinct elements of the chain cannot be observed in the modal coordinates, as the modal systems (4.38) and (4.39) are decoupled. Thus, in order to properly illustrate the resonant interaction between the two highest modes, we introduce the new set of coordinates, referred to as coordinates of "effective particles"

$$
\begin{aligned}
&\eta = q_3 + q_4 \\
&v = q_3 - q_4
\end{aligned}
\tag{4.40}
$$

Rewriting the last two equations of (4.39) in terms of the new coordinates, the so-called coordinates of "effective particles" yield

$$
\begin{aligned}
&\ddot{\eta} + \eta = \varepsilon\left(\frac{v - \eta}{2}\right) \\
&\ddot{v} + v = \varepsilon\left(\frac{\eta - v}{2}\right)
\end{aligned}
\tag{4.41}
$$

The motivation behind this transformation to the new set of coordinates of "effective particles" is simply to gain back the information regarding the particular behavior of particles (e.g., energy exchanges between certain groups of particles in the chain) which is completely obscured in the modal representation (4.38), (4.39). In fact, (4.41) describes the system of the two weakly coupled linear oscillators (coherent domains) exhibiting strong beating phenomenon, leading to the recurrent energy exchanges between them. Thus, initially exciting one of the effective particles of (4.41), the energy will be completely transferred from one particle to another and will wander in a recurrent fashion.

To give a clear physical interpretation of the approach, we look again into the definition of these coordinates (4.40) as well as on their correspondence to the mode shapes of the linear system (4.39). In fact, the first effective particle (η) is formed by summation of both the interacting modes. Thus, a simple summation of the two corresponding eigenvectors yields

$$V_\eta^T = (V_3 + V_4)^T = [-\varepsilon \quad 0 \quad -\varepsilon \quad 2]^T \tag{4.42}$$

The second effective particle (v) is formed by subtraction of the modes leading to the following vector

$$V_v^T = (V_3 - V_4)^T = [\varepsilon \quad -2 \quad \varepsilon \quad 0]^T \tag{4.43}$$

From the observation of the two vectors corresponding to the effective particles (V_η^T, V_ξ^T,), it is clear that the physical meaning of the first effective particle (η) is the energy localization on the second light particle of the chain when that of the second effective particle (v) is the energy localization on the first light particle of the chain. Therefore, complete energy exchanges between the effective particles governed by system (4.41) simply mean recurrent transfers of energy from one light particle of the chain to another. In Fig. 4.16, we illustrate the time series of the response of all the four elements of the chain, namely the two light and two heavy particles.

As it is clear from the results of Fig. 4.16, the significant amount of energy is mainly stored on the light particles of the chain and recurrently wanders from one light particle to another. We also note that initial conditions are chosen in a way to initially excite one of the "effective particles," namely $\eta(0) = I_0, \eta'(0) = v(0) = v'(0) = 0$ which finally results in a well-known, pure beating phenomenon exhibited by the weakly coupled effective particles (4.41) with the closely spaced natural frequencies. In Fig. 4.17, we illustrate the kinetic energy distribution among all the particles of the chain for the two distinct time snapshots. The first snapshot (Fig. 4.17a) corresponds to the very initial moment where all the initial energy is localized on the first light particle. The second snapshot (Fig. 4.17b) corresponds to the moment when the entire system energy gets almost completely transferred to the second light particle.

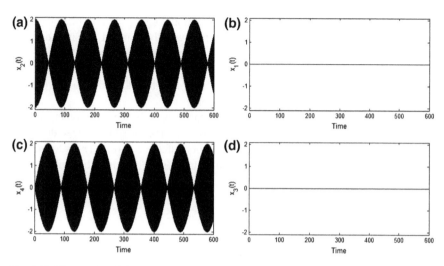

Fig. 4.16 Time series of the response recorded on each one of the particles: **a** first light particle, **b** first heavy particle, **c** second light particle, and **d** second heavy particle. System parameters: $\varepsilon = 0.01$. Initial conditions, $x_1(0) = -\varepsilon, \dot{x}_1(0) = 0$, $x_2(0) = 2, \dot{x}_2(0) = 0$, $x_3(0) = -\varepsilon, \dot{x}_3(0) = 0$, $x_4(0) = 0, \dot{x}_4(0) = 0$

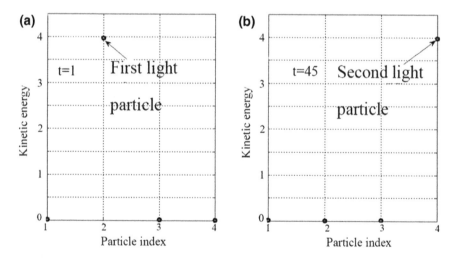

Fig. 4.17 Kinetic energy recorded for each one of the particles in the chain **a** $t = 0$ and **b** $t = 44.54$. System parameters: $\varepsilon = 0.01$. Initial conditions: $x_1(0) = -\varepsilon, \dot{x}_1(0) = 0$, $x_2(0) = 2, \dot{x}_2(0) = 0$, $x_3(0) = -\varepsilon, \dot{x}_3(0) = 0$, $x_4(0) = 0, \dot{x}_4(0) = 0$

4.3.3 Complete Energy Exchanges and Localization: Nonlinear Case $(N = 1, \alpha > 0)$

In the present section, we study the effect of the parameter of nonlinearity (α) on energy transfer and find the threshold value beyond which the regime of intensive energy exchanges between the light particles ceases to exist resulting in permanent energy localization on one of the light particles.

Again, we start the analysis from transferring system (4.32a, b) to modal coordinates of the underlying linear subsystem

$$
\begin{aligned}
\ddot{q}_1 &= \varepsilon\alpha\{V^{-1}\}_{1j}F_j \\
\ddot{q}_2 + 2q_2 &= \varepsilon\alpha\{V^{-1}\}_{2j}F_j \\
\ddot{q}_3 + \left(\frac{2}{\varepsilon}\right)q_3 &= \varepsilon\alpha\{V^{-1}\}_{3j}F_j \\
\ddot{q}_4 + \left(\frac{2(1+\varepsilon)}{\varepsilon}\right)q_4 &= \varepsilon\alpha\{V^{-1}\}_{4j}F_j
\end{aligned}
\tag{4.44}
$$

where V^{-1} is the inverse of the modal matrix V, and F_j is given by

$$
F_j = \begin{cases} \{x_{j-1} - x_j\}^3 - \{x_j - x_{j+1}\}^3, & j\text{-odd} \\ \frac{1}{\varepsilon}\left(\{x_{j-1} - x_j\}^3 - \{x_j - x_{j+1}\}^3\right), & j\text{-even} \end{cases}
\tag{4.45}
$$

We note that in expressions (4.44), (4.45), x_j are considered as a linear combination of the modal coordinates, $x_j = V_{jk}q_k$. Again, we are interested in studying the resonant interaction of the two highest modes of (4.44), and therefore, further analytical treatment concerns only the last two equations of (4.44). Rescaling system (1.149) with respect to time $\tau = \sqrt{\frac{2}{\varepsilon}}t$ yields

$$
\begin{aligned}
q_3'' + q_3 &= \frac{\varepsilon^2\alpha}{2}\{V^{-1}\}_{3j}F_j(\mathbf{q}) \\
q_4'' + (1+\varepsilon)q_4 &= \frac{\varepsilon^2\alpha}{2}\{V^{-1}\}_{4j}F_j(\mathbf{q})
\end{aligned}
\tag{4.46}
$$

As it is clear from the previous discussion, we are considering only the resonant interaction of the two highest modes with the closely spaced frequencies (q_3, q_4). In fact, no additional intermodal, resonant interaction is possible in the system under consideration and this is because the frequencies of the rest two modes (q_1, q_2) are fairly distant from the resonant frequency under consideration. Therefore, assuming that only the pair (q_3, q_4) is initially excited and the level of excitation is of order

$O(1)$, then the magnitude of the rest two modes is bounded by $O(\varepsilon)$ ($q_1, q_2 \sim O$ (ε)). Taking this into account, we expand the RHS of (14) retaining the terms up to $O(\varepsilon)$

$$
\begin{aligned}
\frac{\varepsilon^2 \alpha}{2} \{V^{-1}\}_{3j} F_j(\mathbf{q}) &= \frac{-\varepsilon \alpha}{4} \left(2(q_3 - q_4)^3 + 2(q_3 + q_4)^3\right) + O(\varepsilon^2) \\
\frac{\varepsilon^2 \alpha}{2} \{V^{-1}\}_{4j} F_j(\mathbf{q}) &= \frac{\varepsilon \alpha}{4} \left(2(q_3 - q_4)^3 - 2(q_3 + q_4)^3\right) + O(\varepsilon^2)
\end{aligned}
\tag{4.47}
$$

Substituting (4.47) into (4.46), we obtain

$$
\begin{aligned}
q_3'' + q_3 &= \frac{-\varepsilon \alpha}{4} \left(2(q_3 - q_4)^3 + 2(q_3 + q_4)^3\right) + O(\varepsilon^2) \\
q_4'' + (1 + \varepsilon)q_4 &= \frac{\varepsilon \alpha}{4} \left(2(q_3 - q_4)^3 - 2(q_3 + q_4)^3\right) + O(\varepsilon^2)
\end{aligned}
\tag{4.48}
$$

Rewriting (4.48) in terms of coherent domains (4.40) yields

$$
\begin{aligned}
\eta'' + \eta &= \frac{\varepsilon}{2} \{v - \eta\} - \varepsilon \alpha \eta^3 + O(\varepsilon^2) \\
v'' + v &= -\frac{\varepsilon}{2} \{v - \eta\} - \varepsilon \alpha v^3 + O(\varepsilon^2)
\end{aligned}
\tag{4.49}
$$

Additional rescaling of time brings system (4.49) to the more convenient form

$$
\begin{aligned}
\eta'' + \eta &= \frac{\varepsilon}{2} v - \frac{\varepsilon \alpha}{1 + \varepsilon/2} \eta^3 + O(\varepsilon^2) \\
v'' + v &= \frac{\varepsilon}{2} \eta - \frac{\varepsilon \alpha}{1 + \varepsilon/2} v^3 + O(\varepsilon^2)
\end{aligned}
\tag{4.50}
$$

As it was shown earlier in Manevitch et al. (2010), there is a threshold value of the parameter of nonlinearity $\alpha = \alpha_{\mathrm{cr}}$ beyond which strong energy exchange between the oscillators of (4.50) ceases to exist, leading to the energy localization on one of them. In terms of the system under consideration, the transition from the regime of strong energy exchanges (beating phenomenon) to localization simply means that the main portion of energy initially supplied to one of the light particles can barely be transferred to another light particle and remains permanently entrapped on the first one. To make this point more clear, we briefly repeat the analysis of Manevitch et al. (2010). To this extent, we perform the following change of variables

$$
\psi_1 = \eta' + i\eta, \quad \psi_2 = v' + iv
\tag{4.51}
$$

Substitution of (19) into (18) yields

$$\psi_1' - i\psi_1 = \frac{\varepsilon}{2}\left\{\frac{\psi_2 - \psi_2^*}{2i}\right\} - \varepsilon\alpha\left\{\frac{\psi_1 - \psi_1^*}{2i}\right\}^3$$

$$\psi_2' - i\psi_2 = \frac{\varepsilon}{2}\left\{\frac{\psi_1 - \psi_1^*}{2i}\right\} - \varepsilon\alpha\left\{\frac{\psi_2 - \psi_2^*}{2i}\right\}^3$$

(4.52)

Following the method of multiple scales, we assume the following expansion

$$\psi_i = \psi_{i0} + \varepsilon\psi_{i1} + O(\varepsilon^2), \frac{d}{dt} = \frac{\partial}{\partial\tau_0} + \varepsilon\frac{\partial}{\partial\tau_1}\cdots, \quad i = 1, 2 \qquad (4.53)$$

After some lengthy, straightforward calculations (Manevitch et al. 2010), we arrive at the modal system of equations describing the slow evolution

$$\varphi_{10}' + \frac{i}{4}\varphi_{20} - \frac{3i\alpha}{8}\varphi_{10}|\varphi_{10}|^2 = 0$$

$$\varphi_{20}' + \frac{i}{4}\varphi_{10} - \frac{3i\alpha}{8}\varphi_{20}|\varphi_{20}|^2 = 0$$

(4.54)

where $\psi_{10} = \varphi_{10}\exp(it), \psi_{20} = \varphi_{20}\exp(it)$

System (4.54) is fully integrable and possesses two integrals of motion:

$$H = \frac{1}{4}\left(\varphi_2\varphi_1^* + \varphi_2^*\varphi_1\right) - \frac{3}{16}\alpha\left(|\varphi_1|^4 + |\varphi_2|^4\right)$$

$$I = |\varphi_1|^2 + |\varphi_2|^2$$

(4.55)

Accounting for (4.55), it is convenient to transform from the complex coordinates $(\varphi_{10}, \varphi_{20})$ to the new angular coordinates (θ, Δ),

$$\varphi_{10} = \sqrt{I}\cos\theta\exp(i\delta_1), \varphi_{20} = \sqrt{I}\sin\theta\exp(i\delta_2), \quad \Delta = \delta_1 - \delta_2 \qquad (4.56)$$

such that the first integral of (4.55) takes the following form

$$H^* = \sin(2\theta)\{\cos\Delta + k\sin(2\theta)\}, \quad k = \frac{3I\alpha}{8} \qquad (4.57)$$

In Fig. 4.18, we present several phase portraits governed by (4.50) for the different values of k. As it was shown in Manevitch et al. (2010), the existence of the complete energy exchanges between the interacting particles (4.49) can be fully explained by means of a unique phase trajectory (on the θ, Δ plane) the so-called limiting phase trajectory (LPT) (Manevitch et al. 2010). The LPTs present on Fig. 4.18a–c and clearly correspond to the complete energy exchanges between the

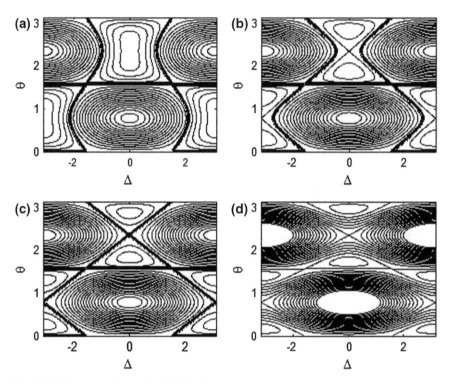

Fig. 4.18 Phase portraits on the (θ, Δ) plane **a** $k = 0.55$, **b** $k = 0.9$, **c** $k = 1$, and **d** $k = 1.5$. Limiting phase trajectory is denoted with the *bold, solid line*

effective particles. In fact, the correspondence of the LPTS to the complete energy exchanges between the interacting particles can be more evident when looking to the latest coordinate transformation brought in (4.56). Thus, the amplitude of the envelopes related to the response of each one of the particles is: $|\varphi_{10}| = \sqrt{I} \cos \theta, |\varphi_{20}| = \sqrt{I} \sin \theta$. Therefore, to assure the full energy exchange between the particles, one may seek for the special phase trajectories (which do not include any fixed point) on the θ, Δ plane, such that θ variable performs the excursions back and forth between 0 and $\pi/2$. Interestingly enough, there is a unique trajectory on the θ, Δ phase plane which exhibits such behavior. This trajectory is exactly LPT mentioned above.

Results of Fig. 4.18 show that LPTs exist only below a certain threshold value of $k = k_{cr}$ above which the LPT ceases to exist, and therefore, there is no possible way for a complete transfer from one effective particle to another. Recent works on the subject (Manevitch et al. 2010) have shown that this threshold value can be easily evaluated from the integral of motion (4.57). To show that we note that limiting phase trajectories (LPTs) correspond to the case of $H^* = 0$. This brings to the following non-trivial relation between θ and Δ

$$\cos \Delta + k \, \sin(2\theta) = 0 \qquad (4.58)$$

Obviously enough, (4.58) has a closed loop solution for the values of $k < k_{cr} = 1$. Thus, for the k values above the threshold $(k > k_{cr} = 1)$, no LPT is possible which brings in turn to the annihilation of the regime of strong energy exchanges between the effective particles and results in the energy localization on one of them.

Using this criterion, we compute the critical value of nonlinearity α, above which the regime of strong energy exchange is annihilated.

$$\alpha_{cr} = \frac{8}{3I} \qquad (4.59)$$

As a numerical verification of this prediction, we excite one of the light particles of (4.44) choosing the parameter of nonlinearity slightly below the threshold $(\alpha < \alpha_{cr})$ (Fig. 4.19) and slightly above the threshold $(\alpha > \alpha_{cr})$ (Fig. 4.20).

Results of Figs. 4.19 and 4.20 suggest a very good correspondence between the analytical predictions of the threshold value ($\alpha_{cr} = 0.6667$—for the initial conditions chosen) with that of numerical simulations. In fact, in our simulations, we found out that the transition from the regime of strong energy exchanges to localization happens for $\alpha = \alpha_{cr} = 0.65$ which is extremely close to the theoretical prediction.

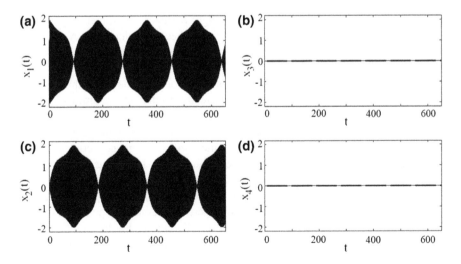

Fig. 4.19 Time series of the response recorded on each one of the particles: **a** first light particle, **b** first heavy particle, **c** second light particle, and **d** second heavy particle. Initial conditions: $x_1(0) = -\varepsilon, \dot{x}_1(0) = 0, x_2(0) = 2, \dot{x}_2(0) = 0, x_3(0) = -\varepsilon, \dot{x}_3(0) = 0, x_4(0) = 0, \dot{x}_4(0) = 0$. System parameters: $\varepsilon = 0.01, \alpha = 0.65 < \alpha_{cr} = 0.6667, I = 4$

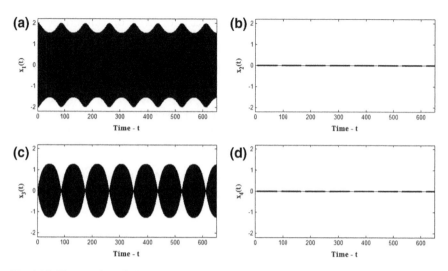

Fig. 4.20 Time series of the response recorded on each one of the particles: **a** first light particle, **b** first heavy particle, **c** second light particle, and **d** second heavy particle. Initial conditions, $x_1(0) = -\varepsilon, \dot{x}_1(0) = 0, \ x_2(0) = 2, \dot{x}_2(0) = 0, \ x_3(0) = -\varepsilon, \dot{x}_3(0) = 0, \ x_4(0) = 0, \dot{x}_4(0) = 0$. System parameters: $\varepsilon = 0.01, \ \alpha = 0.67 > \alpha_{cr} = 0.6667, \ I = 4$

4.3.4   Extension to the Higher Number of Light Particles $(N > 1, \alpha > 0)$

In the present section, we illustrate an extension of the simple, cyclic dimer chain (4 particles periodicity) with alternating heavy light particles (1:1 dimer) to the more complicated, periodic dimer chains comprising 2 heavy particles and $N > 1$ light particles, such that each group of N light particles is delimited by the two heavy ones from both ends (4.32a, b).

The primary goal of the present section is to demonstrate the applicability of the method of limiting phase trajectories to correctly predict the regimes of strong energy exchanges and localization when dealing with the group of light particles rather than single light particle included in each cell of the chain. In the present section, we concentrate solely on the regimes with a frequency content corresponding to the optical branch of the dimer chain (4.32a, b). Linear normal modes ($\alpha = 0$) of (1) related to the optical branch can be easily differentiated and classified. Indeed, let us consider a pair of linear modes $(\mathbf{V}_{0i}, \mathbf{V}_{\varepsilon i})$, where $\mathbf{V}_{0i}$ corresponds to the vibrational mode with the two heavy particles remaining permanently stationary and $\mathbf{V}_{\varepsilon i}$ is related to the $O(\varepsilon)$ small oscillations of the heavy particles in comparison with the light ones. Because of the symmetry of the system under consideration (4.32a, b), the motion of heavy particles for any vibrational mode satisfies the following relation

$$x_1(t) = X_1^{(i)} \exp(j\omega_i t), x_{N+2}(t) = X_{N+2}^{(i)} \exp(j\omega_i t), X_1^{(i)} = (-1)^{p(i)} X_{N+2}^{(i)} \quad (4.60)$$

where $x_1(t), x_{N+2}(t)$ describe the displacement of the first and the second heavy particles, and $X_1^{(i)}, X_{N+2}^{(i)}$ define their amplitudes, accordingly. It is also important to note that there are only two possible motions of the heavy particles, either in-phase or out-of-phase, and it strictly depends on the mode, under consideration. Therefore, we introduce additional parameter $p(i)$ which returns the values of 0 (in-phase motion) or 1 (out-of-phase motion) depending on the mode under consideration (index i is used as a label for the pair of modes on the optical branch, possessing closely spaced frequencies). As it will become clear from the further analysis for each linear, eigenmode $\mathbf{V}_{0i}$ of the optical branch (with the corresponding eigenvalue λ_{0i}), there exists a mode $\mathbf{V}_{\varepsilon i}$ such that the following asymptotical relation holds

$$\begin{aligned} \lambda_{\varepsilon i} &= \lambda_{0i} + \sigma\varepsilon + O(\varepsilon^2) \\ \mathbf{V}_{\varepsilon i} &= \mathbf{V}_{0i} + \mathbf{v}\varepsilon + O(\varepsilon^2) \end{aligned} \quad (4.61)$$

where σ denotes the parameter of the frequency detuning between the ith pair of modes with the closely spaced frequencies.

In fact, further analysis concerns solely the dynamics on the optical branch. Therefore, one may find it convenient to transform (1) to the anti-continuum limit,

$$\begin{aligned} x_1'' &= \varepsilon(x_{2N+2} - 2x_1 + x_2) + \varepsilon^2 \alpha \tilde{F}_1 \\ x_2'' &= x_1 - 2x_2 + x_3 + \varepsilon\alpha \tilde{F}_2 \\ &\qquad \ldots \\ x_{N+1}'' &= x_N - 2x_{N+1} + x_{N+2} + \varepsilon\alpha \tilde{F}_{N+1} \\ x_{N+2}'' &= \varepsilon(x_{N+1} - 2x_{N+2} + x_{N+3}) + \varepsilon^2 \alpha \tilde{F}_{N+2} \\ x_{N+3}'' &= x_{N+2} - 2x_{N+3} + x_{N+4} + \varepsilon\alpha \tilde{F}_{N+3} \\ &\qquad \ldots \\ x_{2N+2}'' &= x_{2N+1} - 2x_{2N+2} + x_1 + \varepsilon\alpha \tilde{F}_{2N+2} \end{aligned} \quad (4.62)$$

where the differentiation is performed with respect to a new timescale $\tau = \varepsilon^{-1/2} t$. Let us start with the analysis of different modes of a linear subsystem of (30) ($\alpha = 0$). Thus, the modal solutions of a linear subsystem of (4.62) ($\tau = \varepsilon^{-1/2} t$) are sought in the following form

$$x_k = X_k \exp(j\omega\tau) \quad (4.63)$$

Substituting (4.63) into (4.62) and accounting for (4.60), we obtain the following linear system corresponding to a single cell of (4.62) (i.e., single heavy particle, succeeded by N—light ones),

$$
\begin{pmatrix}
2\varepsilon - \omega_i^2 & -\varepsilon & 0 \dots 0 & -\varepsilon \\
-1 & 2 - \omega_i^2 & -1 & \\
0 & \ddots & \ddots & \ddots \\
\vdots & & & \\
(-1)^{p(i)} & & -1 & 2 - \omega_i^2
\end{pmatrix}
\begin{pmatrix}
X_1 \\
\vdots \\
\vdots \\
\vdots \\
\vdots \\
\vdots \\
X_{N+1}
\end{pmatrix}
=
\begin{pmatrix}
0 \\
\vdots \\
\vdots \\
\vdots \\
\vdots \\
\vdots \\
0
\end{pmatrix}
\tag{4.64}
$$

It is easy to see that the problem of finding the set of eigenfrequencies corresponding to the family of eigenmodes with the stationary heavy particles ($\lambda_{i0} = \omega_{i0}^2$) is effectively reduced to the homogeneous chain of light particles with the fixed ends. Thus, these eigenfrequencies can be easily computed from the following characteristic polynomial

$$
P_0(\omega) = Det
\begin{pmatrix}
2 - \omega^2 & -1 & & & \\
-1 & 2 - \omega^2 & -1 & & \\
& \ddots & \ddots & \ddots & \\
& & & & -1 \\
& & & -1 & 2 - \omega^2
\end{pmatrix}
= 0
\tag{4.65}
$$

We denote the ith solution of (33) as $\lambda_{i0} = \omega_{i0}^2$. In fact, all the solutions of (1.170) can be represented in the well-known explicit form Gregory and Karney (1969), Elliott (1953),

$$
\lambda_{i0} = \omega_{i0}^2 = 2\left(1 - \cos\left(\frac{i\pi}{N+1}\right)\right)
\tag{4.66a}
$$

with their corresponding eigenvectors

$$
V_k^{(i)} = \sin\left(\frac{ik\pi}{N+1}\right), \quad k = 1, \dots, N
\tag{4.66b}
$$

Note that the eigenmode of the entire linear subsystem ($\alpha = 0$) of (30) corresponding to λ_{i0} reads

$$(\mathbf{V}_{0i})_j = \begin{cases} 0, & j = 1, N+2 \\ V_{j-1}^{(i)}, & 1 < j < N+2 \\ (-1)^{r(i)} V_{j-N-2}^{(i)}, & N+2 < j \leq 2N+1 \end{cases} \tag{4.67}$$

where $r(i)$ assumes the values of 0 or 1 depending on the mode under consideration. We are at a point to describe the second type of the eigenmodes and eigenvalues defined in (4.60). Clearly enough, the eigenvalue $\lambda_{i\varepsilon} = \omega_{i\varepsilon}^2$ is the solution of the following characteristic polynomial derived from (4.62) and (4.64)

$$P_\varepsilon(\omega) = \mathrm{Det} \begin{pmatrix} 2\varepsilon - \omega^2 & -\varepsilon & 0...0 & -\varepsilon \\ -1 & 2-\omega^2 & -1 & \\ 0 & \ddots & \ddots & \ddots \\ \vdots & & & \\ (-1)^{p(i)} & & -1 & 2-\omega^2 \end{pmatrix} = \left(2\varepsilon - \omega^2\right) P_0(\omega) + \varepsilon\left(\Delta_{12} + (-1)^{N+1}\Delta_{1(N+1)}\right)$$

$$\tag{4.68}$$

Here, the expression for the determinant is developed with respect to the first row; Δ_{12} and $\Delta_{1(N+1)}$ are the minors of the matrix.

In fact, we are interested in the eigenvalues corresponding to the optical branch and therefore $\lambda_\varepsilon \sim O(1)$ which is a solution of the characteristic polynomial (4.65). Let as rewrite the characteristic equation in the more compact form

$$P_0(\omega) + \varepsilon P_1(\omega) = 0 \tag{4.69}$$

Using implicit function theorem, it can be shown that

$$\omega_{i\varepsilon}^2 = \lambda_{i\varepsilon} = \lambda_{i0} + \varepsilon\sigma + O(\varepsilon^2) \tag{4.70}$$

Our next goal is to find the eigenvectors $\mathbf{V}_{\varepsilon i}$ corresponding to (4.70). Deriving an explicit expression for these eigenvectors is a formidable task. However, recalling that on the optical branch the motion of the heavy particles is of the order $O(\varepsilon)$ (this can be directly seen from (4.64) assuming the amplitude of the light particles to be of order $O(1)$), we depict the motion of the heavy particles as the following [see (4.62)]

$$X_1 = \varepsilon A, X_{N+2} = \varepsilon(-1)^{p(i)}A \tag{4.71}$$

Accounting for (4.71) in (4.64), we derive a linear system which synchronizes the motion of the light particles with the motion of heavy ones

$$
\begin{pmatrix}
2-\omega_{i\varepsilon}^2 & -1 & & \\
-1 & 2-\omega_{i\varepsilon}^2 & -1 & \\
& \ddots & & \ddots \\
& & -1 & 2-\omega_{i\varepsilon}^2
\end{pmatrix}
\begin{pmatrix}
X_2 \\
\vdots \\
\vdots \\
X_{N+1}
\end{pmatrix}
=
\begin{pmatrix}
\varepsilon A \\
\vdots \\
\vdots \\
\varepsilon(-1)^{p(i)} A
\end{pmatrix}
\tag{4.72}
$$

Substituting (4.70) into (4.72) yields

$$
\left(
\begin{pmatrix}
2-\omega_{i0}^2 & -1 & & \\
-1 & 2-\omega_{i0}^2 & -1 & \\
& \ddots & & \ddots \\
& & -1 & 2-\omega_{i0}^2
\end{pmatrix}
-\varepsilon\sigma\mathbf{I}
\right)
\begin{pmatrix}
X_2 \\
\vdots \\
\vdots \\
X_{(N+1)}
\end{pmatrix}
=
\begin{pmatrix}
\varepsilon A \\
0 \\
\vdots \\
\varepsilon(-1)^{p(i)} A
\end{pmatrix}
\tag{4.73}
$$

It is clear from (4.83) that for any mode related to the optical branch the amplitude of the light particles is of order $O(1)$. Therefore, applying the perturbation theory, we arrive at the solution of (41) corresponding to the optical branch which can be written in the following form

$$
X_k = \sin\left(\frac{i(k-1)\pi}{N+1}\right) + O(\varepsilon), \quad k = 2,\dots,N+1
\tag{4.74}
$$

The motion of the light particles included in the second cell of the dimer chain is easily derived

$$
X_k = (-1)^{p(i)} \sin\left(\frac{i(k-1)\pi}{N+1}\right) + O(\varepsilon), \quad k = N+3,\dots,2N+1
\tag{4.75}
$$

Note that the eigenmodes of the entire linear subsystem of (4.62) corresponding to $\lambda_{i\varepsilon}$ read

$$
(\mathbf{V}_{\varepsilon i})_j =
\begin{cases}
O(\varepsilon), & j = 1, N+2 \\
V_{j-1}^{(i)} + O(\varepsilon), & 1 < j < N+2 \\
(-1)^{p(i)} V_{j-N-2}^{(i)} + O(\varepsilon), & N+2 < j \leq 2N+1
\end{cases}
\tag{4.76}
$$

Further study primarily concerns the resonant interaction of effective particles $(\eta_i \xi_i)$ related to the different pairs of eigenmodes $(\mathbf{V}_{0i}, \mathbf{V}_{\varepsilon i})$ with the corresponding eigenfrequencies $(\lambda_{i0}, \lambda_{i\varepsilon})$. Before proceeding with the direct analysis of the regimes exhibited by (4.62), we make additional very important observation concerning the orthogonality of each pair of modes $(\mathbf{V}_{0i}, \mathbf{V}_{\varepsilon i})$. Thus, the following property holds

$$\mathbf{V}_{0i}^{\mathrm{T}}\mathbf{V}_{\varepsilon i} = 0 \qquad (4.77)$$

This property can be easily derived by observing that the motion of particles for both the modes is in-phase on one of the cells of the dimer chain and out-of-phase on the second one (see Appendix for details). Thus, the result of a scalar multiplication of the considered eigenmodes is zero.

Let us rewrite system (1.167) in the matrix form

$$\ddot{\mathbf{x}}(t) + \mathbf{K}(\varepsilon)\mathbf{x}(t) = \varepsilon\mathbf{F}_{\mathrm{nlin}}(\mathbf{x}(t)) \qquad (4.78)$$

where $\mathbf{F}_{\mathrm{nlin}} = \begin{bmatrix} \varepsilon^2\alpha\tilde{F}_1 & \varepsilon\alpha\tilde{F}_2 & \cdots & \varepsilon\alpha\tilde{F}_{N+1} & \cdots & \varepsilon^2\alpha\tilde{F}_{N+2} & \varepsilon\alpha\tilde{F}_{N+3} & \cdots & \varepsilon\alpha \\ \tilde{F}_{2N+2} \end{bmatrix}^{\mathrm{T}}$

Arguing as before, we transform (46) into modal coordinates of the linear subsystem ($\alpha = 0$) by assigning

$$\mathbf{x}(t) = \mathbf{M}\mathbf{q}(t) \qquad (4.79)$$

where $\mathbf{M}$ is the modal matrix and $\mathbf{q} = [q_1(t), \ldots, q_{2N+2}(t)]^{\mathrm{T}}$ is a vector of modal coordinates. Inserting (4.79) into (4.78), we obtain

$$\mathbf{M}\ddot{\mathbf{q}}(t) + \mathbf{K}\mathbf{M}\mathbf{q}(t) = \varepsilon\mathbf{F}_{\mathrm{nlin}}(\mathbf{M}\mathbf{q}(t)) \qquad (4.80)$$

Before proceeding with the further analysis, let us bring additional convention for labeling the modes. In fact, in the present section, we concentrate each time on a single pair of modes $\mathbf{V}_{0i}, \mathbf{V}_{\varepsilon i}$. However, $(0i)$, (εi) are not the real indices, but rather labels which are a somewhat problematic when used in the matrix or vector analysis. Therefore, each time we need to provide a real index for these pair of modes, we use the following notation, $\mathbf{V}_{0i} = \mathbf{V}_l, \mathbf{V}_{\varepsilon i} = \mathbf{V}_{l+1}$.

Multiplying (48) by $\mathbf{V}_{0i}, \mathbf{V}_{\varepsilon i}$ separately, we have

$$\ddot{q}_{0i} + \omega_{0i}^2 q_{0i} = \varepsilon\mathbf{V}_{0i}^{\mathrm{T}}\mathbf{F}_{\mathrm{nlin}}(\mathbf{M}\mathbf{q}(t)) - \sum_{\substack{j=1, \\ j\neq l, \\ j\neq l+1,}}^{2N+2} (\gamma_{0j}\ddot{q}_j + \beta_{0j}q_j)$$

$$\ddot{q}_{\varepsilon i} + \omega_{\varepsilon i}^2 q_{\varepsilon i} = \varepsilon\mathbf{V}_{\varepsilon i}^{\mathrm{T}}\mathbf{F}_{\mathrm{nlin}}(\mathbf{M}\mathbf{q}(t)) - \sum_{\substack{j=1, \\ j\neq l, \\ j\neq l+1,}}^{2N+2} (\gamma_{\varepsilon j}\ddot{q}_j + \beta_{\varepsilon j}q_j) \qquad (4.81)$$

$$\gamma_{0j} = \left(\mathbf{V}_{0i}^T\right)_k \mathbf{M}_{kj}, \quad \beta_{0j} = \left(\mathbf{V}_{0i}^T\right)_k (\mathbf{K}\mathbf{M})_{kj},$$

$$\gamma_{\varepsilon j} = \left(\mathbf{V}_{\varepsilon i}^T\right)_k \mathbf{M}_{kj}, \quad \beta_{\varepsilon j} = \left(\mathbf{V}_{\varepsilon i}^T\right)_k (\mathbf{K}\mathbf{M})_{kj}$$

where the indices $l, l+1$ correspond to the modes $\mathbf{V}_{0i}, \mathbf{V}_{\varepsilon i}$, respectively, and $q_{0i}, q_{\varepsilon i}$ are their modal coordinates, $\omega_{\varepsilon i}^2 = \omega_{0i}^2 + \varepsilon\sigma$. Einstein summation convention is used in the expressions for γ_{ij}, β_{ij} (49). In scope of the present paper, we are interested in the analysis of resonant interaction of the effective particles in the vicinity of (1:1) resonance. Therefore, applying the initial excitation only for the two interacting modes, namely $\mathbf{V}_{0i}$ and $\mathbf{V}_{\varepsilon i}$, and assuming that the number of light particles is such that the spectrum provided by (4.66a) is well separated $\left(\frac{\pi}{\varepsilon} \gg N\right)$ (i.e., negating the possibility of additional internal resonances in the system), the following asymptotical estimation can be made

$$q_j, \ddot{q}_j \sim O(\varepsilon) \quad \forall j \neq l, l+1 \tag{4.82}$$

Moreover, focusing on a certain resonant interaction occurring in the vicinity of ω_{0i}^2, we can completely neglect the contribution of the epsilon order small, non-resonant terms given in (4.82). This can be also verified by performing a simple multiple scales expansion of (4.81) in the vicinity of main (1:1) resonance, which will result in the modulation equations describing the dynamics of the system without containing the non-resonant terms of (4.82). In scope of the present paper, we do not bring the multiple scales expansion, but rather get rid of the non-resonant terms in (4.81) for the sake of brevity.

Thus, system (4.81) reduces to the following form, containing only the resonant term in the right-hand side of (4.81)

$$\begin{aligned}
\ddot{q}_{0i} + \omega_{0i}^2 q_{0i} &= \varepsilon \mathbf{V}_{0k}^{\mathrm{T}} \mathbf{F}_{\mathrm{nlin}}(\mathbf{V}_{0i} q_{0i} + \mathbf{V}_{\varepsilon i} q_{\varepsilon i}) \\
\ddot{q}_{\varepsilon i} + \omega_{\varepsilon i}^2 q_{\varepsilon i} &= \varepsilon \mathbf{V}_{\varepsilon k}^{\mathrm{T}} \mathbf{F}_{\mathrm{nlin}}(\mathbf{V}_{0i} q_{0i} + \mathbf{V}_{\varepsilon i} q_{\varepsilon i})
\end{aligned} \tag{4.83}$$

where

$$\mathbf{F}_{\mathrm{nlin}} = \begin{bmatrix} \varepsilon^2 \alpha \tilde{F}_1 & \varepsilon \alpha \tilde{F}_2 & \cdots & \varepsilon \alpha \tilde{F}_{N+1} & \cdots & \varepsilon^2 \alpha \tilde{F}_{N+2} & \varepsilon \alpha \tilde{F}_{N+3} & \cdots & \varepsilon \alpha \tilde{F}_{2N+2} \end{bmatrix}^{\mathrm{T}}$$

$$\tilde{F}_j = \left(\Gamma^0_{(j-1)j} q_k + \Gamma^\varepsilon_{(j-1)j} q_{k+1}\right)^3 - \left(\Gamma^0_{(j)j+1} q_k + \Gamma^\varepsilon_{j(j+1)} q_{k+1}\right)^3$$

$$\Gamma^0_{(j-1)j} = \left((\mathbf{V}_{0i})_{j-1} - (\mathbf{V}_{0i})_j\right), \quad \Gamma^\varepsilon_{(j-1)j} = \left((\mathbf{V}_{\varepsilon i})_{j-1} - (\mathbf{V}_{\varepsilon i})_j\right)$$

$$\Gamma^0_{(j)j+1} = \left((\mathbf{V}_{0i})_j - (\mathbf{V}_{0i})_{j+1}\right), \quad \Gamma^\varepsilon_{j(j+1)} = \left((\mathbf{V}_{\varepsilon i})_j - (\mathbf{V}_{\varepsilon i})_{j+1}\right)$$

$$(\mathbf{V}_{0i})_j = (\mathbf{V}_{0i})_{j+2N+2}, \quad (\mathbf{V}_{\varepsilon i})_j = (\mathbf{V}_{\varepsilon i})_{j+2N+2}$$

$$\tag{4.84}$$

Accounting for (4.67), (4.76), it is easy to see that

$$\begin{aligned}
\Gamma^\varepsilon_{(j-1)j} &= \Gamma^0_{(j-1)j} + O(\varepsilon), \quad j = 1, \ldots, N+1 \\
\Gamma^\varepsilon_{(j-1)j} &= -\Gamma^0_{(j-1)j} + O(\varepsilon), \quad j = N+2, \ldots, 2N+2
\end{aligned} \tag{4.85}$$

System (4.85) can be further simplified by explicitly performing all the scalar multiplications of the right-hand side of (4.85)

$$\ddot{q}_{0i} + \omega_{0i}^2 q_{0i} = \varepsilon\alpha \left(\sum_{j=2}^{N+1} V_{j-1}^{(i)} \tilde{F}_j + \sum_{j=N+3}^{2N+2} (-1)^{r(i)} V_{j-N-2}^{(i)} \tilde{F}_j \right) + O(\varepsilon^2)$$

$$\ddot{q}_{\varepsilon i} + \omega_{\varepsilon i}^2 q_{\varepsilon i} = \varepsilon\alpha \left(\sum_{j=2}^{N+1} V_{j-1}^{(i)} \tilde{F}_j + \sum_{j=N+3}^{2N+2} (-1)^{p(i)} V_{j-N-2}^{(i)} \tilde{F}_j \right) + O(\varepsilon^2)$$

(4.86)

where $r(i)$ is the binary function assuming the values of 0 and 1 depending on the interacting pair (i—index) of modes of the optical branch. Exact definitions of $r(i)$ and $p(i)$ are given in Appendix (4.100, 4.102). As it is clear from the arguments brought in the appendix, $r(i)$ and $p(i)$ are correlated functions, satisfying $r(i) = 1 - p(i)$.

Accounting for (4.84), (4.85) yields

$$\tilde{F}_j = \begin{cases} \left(\left(\Gamma_{(j-1)j}^0 \right)^3 - \left(\Gamma_{(j)j+1}^0 \right)^3 \right) \eta_i^3 + O(\varepsilon), & j = 2, \dots, N+1 \\ \left(\left(\Gamma_{(j-1)j}^0 \right)^3 - \left(\Gamma_{(j)j+1}^0 \right)^3 \right) \xi_i^3 + O(\varepsilon), & j = N+3, \dots, 2N+2 \end{cases}$$

(4.87)

where $\eta_i = q_{0i} + q_{\varepsilon i}$, $\xi_i = q_{0i} - q_{\varepsilon i}$ are the coordinates of effective particles corresponding to the pair of modes $(\mathbf{V}_{0i}, \mathbf{V}_{\varepsilon i})$. Transforming (4.86) into coordinates of effective particles leads to the following form

$$\ddot{\eta}_i + \omega_{0i}^2 \eta_i + \frac{\varepsilon\sigma(\eta_i - \xi_i)}{2} = 2\varepsilon\tilde{\alpha}_i \eta_i^3 + O(\varepsilon^2)$$

$$\ddot{\xi}_i + \omega_{0i}^2 \xi_i - \frac{\varepsilon\sigma(\eta_i - \xi_i)}{2} = 2\varepsilon\tilde{\alpha}_i \xi_i^3 + O(\varepsilon^2)$$

$$\tilde{\alpha}_i = \alpha \sum_{j=2}^{N+1} V_{j-1}^{(i)} \left(\left(\Gamma_{(j-1)j}^0 \right)^3 - \left(\Gamma_{(j)j+1}^0 \right)^3 \right)$$

(4.88)

It is quite easy to see that (56) admits exactly the same form of (1.154) analyzed in the previous section. We also note that in the derivation of (1.153), we used the fact that $(r(i) \neq p(i))$ (Appendix).

Arguing exactly as in the previous section, we find a threshold value for the parameter of nonlinearity $\tilde{\alpha}_i$ above which energy gets localized on a single group of light particles exhibiting oscillations of the type governed by the ith pair of modes $(\mathbf{V}_{0i}, \mathbf{V}_{\varepsilon i})$.

$$\tilde{\alpha}_{i\text{CR}} = \frac{4\sigma}{3I} \Rightarrow \alpha_{i\text{CR}} = \frac{4\sigma}{3I} \left(\sum_{j=2}^{N+1} V_{j-1}^{(i)} \left(\left(\Gamma_{(j-1)j}^0 \right)^3 - \left(\Gamma_{(j)j+1}^0 \right)^3 \right) \right)^{-1}$$

(4.89)

To illustrate numerical evidences of energy localization and transfer between the groups of light particles, we choose $N = 3$ and show the transitions between the regimes for each pair of modes situated on the optical branch.

Let us start with the consideration of the pair of modes corresponding to the lowest frequency of the optical branch. In fact, this mode corresponds to the in-phase motion of the light particles in each cell of the dimer chain.

The corresponding eigenvectors and eigenvalues read

$$\mathbf{V}_{01} = \begin{bmatrix} 0 & -1 & -\sqrt{2} & -1 & 0 & 1 & \sqrt{2} & 1 \end{bmatrix}/\sqrt{8}, \quad \lambda_{01} = \sqrt{2}\left(\sqrt{2}-1\right)$$

$$\mathbf{V}_{\varepsilon 1} = \begin{bmatrix} 0 & 1 & \sqrt{2} & 1 & 0 & 1 & \sqrt{2} & 1 \end{bmatrix}/\sqrt{8} + O(\varepsilon), \quad \lambda_{\varepsilon 1} = \sqrt{2}\left(\sqrt{2}-1\right) + \varepsilon\sigma + O(\varepsilon^2)$$

$$\sigma = \left(\frac{\sqrt{2}+1}{\sqrt{2}}\right)$$

$$(4.90)$$

Using the expression (4.89), we compute the threshold value of the parameter of nonlinearity above which the regime of strong, recurrent energy exchanges between the groups of light particles terminates and energy gets localized on one of the groups.

$$\alpha_{1_{CR}} = \frac{4\sigma}{3I}(0.0322)^{-1} \qquad (4.91)$$

Exciting one of the corresponding effective particles (i.e., $\mathbf{x}(0) = \mathbf{V}_{01} + \mathbf{V}_{\varepsilon 1}$, $q_{01}(0) = q_{\varepsilon 1}(0) = 1$, $\eta_1(0) = 2, \xi_1(0) = 0, I = 4$), we plot (Fig. 4.21) the total energies $E_1(t), E_2(t)$ for each one of the groups of light particles for the two aforementioned regimes, namely regime of strong energy exchange between the

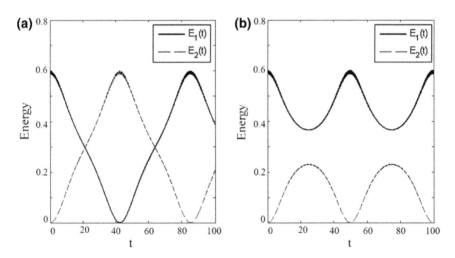

Fig. 4.21 Total energy stored on each group of light particles: **a** $\alpha_1 = 17$, **b** $\alpha_1 = 19$ Threshold values: theoretical—$\alpha_{1_{CR}} = 17.7$, numerical: $\alpha_{1_{CR}} \cong 18.4$, $I = 4$

groups $(\alpha < \alpha_{1_{CR}})$ and a regime of energy localization on one of the groups $(\alpha > \alpha_{1_{CR}})$. The expressions for $E_1(t), E_2(t)$ are given by

$$E_1(t) = \varepsilon \sum_{j=2}^{4} \frac{(\dot{x}_j)^2}{2} + \sum_{j=2}^{5} \left(\frac{(x_{j-1} - x_j)^2}{2} + \varepsilon \alpha \frac{(x_{j-1} - x_j)^4}{4} \right)$$

$$E_2(t) = \varepsilon \sum_{j=6}^{8} \frac{(\dot{x}_j)^2}{2} + \sum_{j=6}^{9} \left(\frac{(x_{j-1} - x_j)^2}{2} + \varepsilon \alpha \frac{(x_{j-1} - x_j)^4}{4} \right)$$

(4.92)

where $x_9 = x_1$. We note that the kinetic energies of the heavy particles are of $O(\varepsilon^2)$ small and therefore are omitted from the consideration in (4.93).

We continue with the illustration of the regimes of resonant energy interaction between the effective particles corresponding to the intermediate frequency of the optical branch. The corresponding eigenvectors and eigenvalues read

$$\mathbf{V}_{02} = [0 \quad -1/2 \quad 0 \quad 1/2 \quad 0 \quad -1/2 \quad 0 \quad 1/2], \quad \lambda_{02} = 2$$
$$\mathbf{V}_{\varepsilon 2} = [0 \quad 1/2 \quad 0 \quad -1/2 \quad 0 \quad -1/2 \quad 0 \quad 1/2], \quad \lambda_{\varepsilon 2} = 2 + \varepsilon\sigma + O(\varepsilon^2)$$
$$\sigma = 1$$

(4.93)

Using the expression (4.89), we compute the threshold value of the parameter of nonlinearity above which the regime of strong energy exchange between the groups of light particles ceases and energy gets localized on one of the groups.

$$\alpha_{2_{CR}} = \frac{4\sigma}{3I}(0.25)^{-1}$$

(4.94)

Exciting one of the corresponding effective particles (i.e., $\mathbf{x}(0) = \mathbf{V}_{02} + \mathbf{V}_{\varepsilon 2}$, $q_{02}(0) = q_{\varepsilon 2}(0) = 1$, $\eta_2(0) = 2, \xi_2(0) = 0, I = 4$), we plot (Fig. 4.22) the total energies $E_1(t), E_2(t)$ for each one of the groups of light particles given by (4.92).

As a final example, we consider the interaction of effective particles corresponding to the pair of modes with the highest frequency of the optical branch. In fact, these modes correspond to the out-of-phase motion of the light particles included in each cell of the dimer chain. The corresponding eigenvectors and eigenmodes read

$$\mathbf{V}_{03} = [0 \quad -1 \quad \sqrt{2} \quad -1 \quad 0 \quad 1 \quad -\sqrt{2} \quad 1]/\sqrt{8}, \quad \lambda_{03} = 2 + \sqrt{2}$$
$$\mathbf{V}_{\varepsilon 3} = [0 \quad 1 \quad -\sqrt{2} \quad 1 \quad 0 \quad 1 \quad -\sqrt{2} \quad 1]/\sqrt{8} + O(\varepsilon), \quad \lambda_{\varepsilon 3} = 2 + \sqrt{2} + \varepsilon\sigma + O(\varepsilon^2)$$
$$\sigma = \left(\frac{\sqrt{2} - 1}{\sqrt{2}} \right)$$

(4.95)

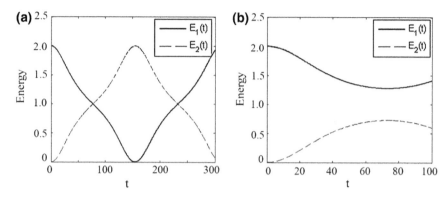

Fig. 4.22 Total energy stored on each group of light particles: **a** $\alpha = 1.3$ and **b** $\alpha = 1.4$. Threshold values: theoretical—$\alpha_{2_{CR}} = 1.33$, $\alpha_{2_{CR}} \cong 1.34$, $I = 4$

Again, using the expression (4.89), we compute the threshold value of the parameter of nonlinearity above which the regime of strong energy exchange between the groups of light particles is canceled and energy gets localized on one of the groups.

$$\alpha_{3_{CR}} = \frac{4\sigma}{3I}(1.0928)^{-1} \tag{4.96}$$

Exciting one of the corresponding effective particles (i.e., $\mathbf{x}(0) = \mathbf{V}_{03} + \mathbf{V}_{\varepsilon 3}$, $q_{03}(0) = q_{\varepsilon 3}(0) = 1$, $\eta_3(0) = 2$, $\xi_3(0) = 0, I = 4$), we plot (Fig. 4.23) the total energies $E_1(t), E_2(t)$ for each one of the groups of light particles defined in (4.92).

As it is evident from the results of Figs. 4.21, 4.22 and 4.23, the theoretical prediction of the threshold value ($\alpha_{i_{CR}}$) is in a very good agreement with that obtained from numerical simulation. Thus, near complete energy exchanges between the groups of light particles corresponding to the different modes of the optical branch can be observed in Figs. (4.21a, 4.22 and 4.23a). Breakdown in the regime of complete energy exchanges and localization on a single group of light particles is evident from Figs. 4.21b, 4.22b and 4.23b.

Thus, in the present study, we have shown that highly non-stationary regime of a strong energy transfer can also be realized in the short, periodic (1:N) FPU dimer chains and is manifested by recurrent energy exchanges between the groups of light particles (i.e., heavy particles remain almost immobile). The mechanism of formation of the near complete energy exchanges between the light particles of the chain has been explained first for the simple, linear dimer chain of period 4. Therefore, concentrating on the two highest modes of the linear system, we first show that in the limit of a strong mass mismatch ($\varepsilon \to 0$), these modes become closely spaced resulting in the possibility of strong beats exhibited by the two light particles. As it is also clear from the analysis and discussion brought in Sect. 4.3, this highly

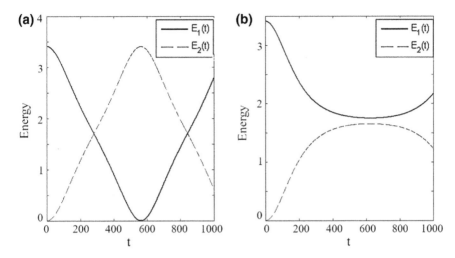

Fig. 4.23 Total energy stored on each group of light particles: **a** $\alpha = 0.08$ and **b** $\alpha = 0.09$. Threshold values: theoretical—$\alpha_{3_{CR}} = 0.0893$, numerical: $\alpha_{3_{CR}} \cong 0.09$, $N = 4$

non-stationary regime cannot be described in terms of modal coordinates of the linear subsystem as the diagonalized system of modal oscillators is fully decoupled and therefore no intermodal interaction can be inferred from it. However, using the concept of "effective" particles previously introduced for the homogeneous FPU chain (Manevitch and Smirnov 2010), we define new set of coordinates resulting in the symmetric linear system of weakly coupled oscillators ("effective" particles) describing the recurrent energy wandering between the light particles of the original linear system. The concept of "effective" particles is farther extended to the non-linear case. Again, focusing on the resonant interaction on the optical branch, we effectively describe the dynamics of a nonlinear dimer chain of period 4 by a symmetric system of the two weakly coupled, Duffing oscillators. Fixing the level of initial excitation, we apply the method of limiting phase trajectories and then find a threshold value of the parameter of nonlinearity below which strong energy exchange between the groups of light particles of the chain holds and above which energy gets permanently localized on one of the light particles. In the last section of the paper, we consider the extended version of the nonlinear dimer chain where each single light particle is replaced with the group of light particles. We show that the analysis carried out in Sect. 4 can be efficiently applied to the higher dimensional model and again estimates the transition between the regime of a complete energy transfer between the group of light particles to the permanent energy localization on one of the groups. We note that for the case of (1:N) dimer chain, the mechanism of a near complete energy exchange (between the groups of light particles) as well as energy localization (on one of the groups) can be realized for each vibrational mode of the optical branch and for any number of light particles included in the cell. Basically, the fact that near complete energy exchange holds for any number of light particles included in the cell differs from the results obtained for a homogeneous chain (Manevitch and Smirnov 2010) where the length of the chain should be sufficiently large (i.e., at least 24 particles in the chain) so that the modes considered become closely spaced. As it will become clear from the results of the current study, a closeness of the modes of the dimer chain is fully governed by a mass ratio parameter. Therefore, in the limit of low mass ratios, we show the formation of the pairs of modes with closely spaced frequencies which are responsible for the peculiar regimes of intense energy transfer between the groups of light particles. Consequently, dynamics of dimer chain reveals a new mechanism of formation of inter-chain resonances without invoking the well-known phenomenon of densification of eigenfrequencies driven by the increasing number of particles (Manevitch and Smirnov 2010). Analytical predictions derived in the paper are in an extremely close correspondence with the results of numerical simulations.

Appendix

Let us show that the pair of modes $\mathbf{V}_{0i}, \mathbf{V}_{\varepsilon i}$ is orthogonal

$$\mathbf{V}_{0i}^{\mathrm{T}}\mathbf{V}_{\varepsilon i} = 0 \tag{4.97}$$

From the consideration of symmetry, the modes have the following form

$$\mathbf{V}_{0i} = \begin{bmatrix} 0 & X_2^0 & \cdots & X_{N+1}^0 & 0 & (-1)^{r(i)}X_2^0 & \cdots & (-1)^{r(i)}X_{N+1}^0 \end{bmatrix}$$
$$\mathbf{V}_{\varepsilon i} = \begin{bmatrix} X_1^\varepsilon & X_2^\varepsilon & \cdots & X_{N+1}^\varepsilon & (-1)^{p(i)}X_1^\varepsilon & (-1)^{p(i)}X_2^\varepsilon & \cdots & (-1)^{p(i)}X_{N+1}^\varepsilon \end{bmatrix}$$
$$\tag{4.98}$$

Using perturbation theory, we have shown that

$$X_j^\varepsilon = X_j^0 + O(\varepsilon) \tag{4.99}$$

where $X_j^0 \sim O(1)$.

Therefore,

$$\mathrm{sgn}(X_j^\varepsilon) = \mathrm{sgn}(X_j^0), j = 2, \quad \ldots N+1 \tag{4.100}$$

Note that (A4) can be violated when dealing with the mode possessing the nodal points on some of the light particles (i.e., $X_j^0 = 0, j \in [2,\ldots,N]$). However, in this case, the product $X_j^0 X_j^\varepsilon = 0$ and therefore the disparity in signs do not affect the orthogonality of the modes.

For each mode $\mathbf{V}_{0i}$, we note the two possibilities for the choice of $r(i)$. To this end, let us consider the signs of X_2^0 and X_{N+1}^0. In case X_2^0 and X_{N+1}^0 have identical signs, the immobile, heavy masses can be balanced by the oscillating light neighbors only if $r(i) = 1$.

Thus,

$$r(i) = \begin{cases} 1, & \text{sgn}(X_2^0) = \text{sgn}(X_{N+1}^0) \\ 0, & \text{sgn}(X_2^0) = -\text{sgn}(X_{N+1}^0) \end{cases} \tag{4.101}$$

As for the second mode $\mathbf{V}_{\varepsilon i}$, it is obvious that operating on the optical branch of the chain, each oscillating heavy particle should be out-of-phase with its light neighbors. In other words, both the light particles neighboring to any of heavy masses should move in phase. Thus,

$$p(i) = \begin{cases} 1, & \text{sgn}(X_2^\varepsilon) = \text{sgn}(X_{N+1}^\varepsilon) \\ 0, & \text{sgn}(X_2^\varepsilon) = -\text{sgn}(X_{N+1}^\varepsilon) \end{cases} \tag{4.102}$$

From (4.100)–(4.102), it is evident that

$$r(i) \neq p(i) \tag{4.103}$$

Thus, the motion of particles for both the modes $\mathbf{V}_{0i}, \mathbf{V}_{\varepsilon i}$ is in-phase on one cell of the dimer chain and out-of-phase on the second one. Accounting for (4.98) and (4.103), it is easy to see that

$$\mathbf{V}_{0i}^{\mathrm{T}}\mathbf{V}_{\varepsilon i} = 0 \tag{4.104}$$

References

Ablowitz, M.J., Ladik, J.E.: Nonlinear differential-difference equations and Fourier analysis. J. Math. Phys. **17**, 1011 (1976)

Akhmediev, N.N., Ankiewicz, A.: Solitosns: Nonlinear Pulses and Beams. Chapman and Hall, London (1992)

Berman, G.P., Izrailev, F.M.: The Fermi–Pasta–Ulam problem: fifty years of progress. Chaos **15**, 15104 (2005)

Berman, G.P., Kolovsky, A.R.: The limit of stochasticity for a one-dimensional chain of interacting oscillators. Sov. Phys. JETP **60**, 1116 (1984)

Bogolyubov, N.N., Mitropol'skii Yu.A.: Asymptotic Methods in the Theory of Nonlinear Oscillations, Gordon & Breach, Delhi (1961)

Brillouin, L.: Wave Propagation in Periodic Structures. Dover Publications (1946)

Budinsky, N., Bountis, T.: Stability of nonlinear modes and chaotic properties of 1D Fermi–Pasta–Ulam lattices. Physica D **8**, 445 (1983)

Burlakov, V.M., Darmanyan, S.A., Pyrkov, V.N.: Modulation instability and recurrence phenomena in anharmonic lattices. Phys. Rev. B **54**, 3257 (1996)

Chechin, G.M., Novikova, N.V., Abramenko, A.A.: Bushes of vibrational modes for Fermi–Pasta–Ulam chains. Physica D **166**, 208 (2002)

Claude, Ch., Kivshar, YuS, Kluth, O., Spatschek, K.H.: Moving localized modes in nonlinear lattices. Phys. Rev. B **47**, 14228 (1993)

Dash, P.C., Patnaik, K.: Nonlinear wave in a diatomic Toda lattice. Phys. Rev. A, **23** (1981)

Dauxois, T., Peyrard, M.: Physics of Solitons. Cambridge University Press, Cambridge (2006)

Dauxois, T., Khomeriki, R., Piazza, F., Ruffo, S.: The anti-FPU problem. Chaos **15**, 15110 (2005)

Dodd, R.K., Eilbeck, J.C., Gibbon, J.D., Morris, H.S.: Solitons and Nonlinear Wave Equations. Academic Press Inc., London (1982)

Elliott, J.F.: The characteristic roots of certain real symmetric matrices. Mater thesis, University of Tennessee (1953)

Feng, B.-F.: An integrable three particle system related to intrinsic.localized modes. J. Phys. Soc. Jpn. **75**, 014401 (2006)

Fermi, E., Pasta, J., Ulam, S.: studies of the nonlinear problems. In: Segre, E. (ed.) Collected Papers of Enrico Fermi, p. 978. University of Chicago Press, Chicago (1965)

Flach, S., Gorbach, A.V.: Discrete breathers—Advances in theory and applications. Phys. Rep. **467**, 1 (2008)

Flach, S., Willis, C.R.: Discrete breathers. Phys. Rep. **295**, 181 (1998)

Flach, S., Ivanchenko, M.V., Kanakov, O.I.: q-Breathers and the Fermi-Pasta-Ulam problem. Phys. Rev. Lett. **95**, 64102 (2005)

Gallavotti, G. (ed.): The Fermi-Pasta-Ulam Problem: A Status Report, vol. 728. Springer, Berlin (2008) (Springer Series Lect. Notes Phys.)

Gregory, R.T., Karney, D.: A collection of matrices for testing computational algorithm. Wiley-Interscience (1969)

Hajnal, D., Schilling, R.: Delocalization-localization transition due to anharmonicity. Phys. Rev. Lett. **101**, 124101 (2008)

Henrici, A., Kappeler, T.: Results on normal forms for fpu chains. Commun. Math. Phys. **278**, 145 (2008a)

Henrici, A., Kappeler, T.: Commun. Math. Phys. **278**, 145 (2008b)

James, G., Kastner, M.: Bifurcations of discrete breathers in a diatomic Fermi–Pasta–Ulam chain. Nonlinearity **20**, 631 657 (2007)

James, G., Noble, P.: Breathers on di-atomic Fermi-Pasta-Ulam lattices. Physica D **196**, 124–171 (2004)

Jensen, S.M.: The nonlinear coherent coupler. IEEE J Quantum Electron QE **18**, 1580–1583 (1982)

Khusnutdinova, K.R.: Non-linear waves in a double row particle system. Vestnik MGU Math Mech **2**, 71–76 (1992)

Khusnutdinova, K.R., Pelinovsky, D.E.: On the exchange of energy in coupled Klein-Gordon equations. Wave Motion **38**, 1–10 (2003)

Kivshar, Y.S., Peyrard, M.: Modulational instabilities in discrete lattices. Phys. Rev. A **46**, 3198 (1992)

Kosevitch, A.M., Kovalev, A.C.: Introduction to Nonlinear Physical Mechanics. Naukova Dumka, Kiev (1989). (in Russian)

Kosevich, YuA, Manevich, L.I., Savin, A.V.: Wandering breathers and self-trapping in weakly coupled nonlinear chains: classical counterpart of macroscopic tunneling quantum dynamics. Phys. Rev. E **77**, 046603 (2008)

Kovaleva, A., Manevitch, L.I., Manevitch E.L.: Intense energy transfer and superharmonic resonance in a system of two coupled oscillators. Phys. Rev. E **81**(5) (2010)

Kruskal, M.D., Zabusky, N.J.: Stroboscopic-perturbation procedure for treating a class of nonlinear wave equations. J. Math. Phys. **5**, 231 (1964)

Landau, L.D., Lifshitz, E.M.: Course of Theoretical Physics, vol. 1. Pergamon, Oxford, UK (1960)

Lichtenberg, A.J., Livi, R., Oettini, M., Ruffo, S.: Dynamics of oscillator chains. In: Gallavotti, G. (ed.) The Fermi-Pasta-Ulam Problem: A status report, vol. 728. Springer, Berlin (2008)

Manevitch, L.I.: The description of localized normal modes in a chain of nonlinear coupled oscillators using complex variables. Nonlinear Dyn. **25**, 95 (2001)

Manevich, L.I.: In: Proceedings of the 8th Confence on Dynamical Systems: Theory and Applications, Lodz, 12–15 Dec 2005 (Poland, 2005), vol. 1, pp. 119–136 (2005)

Manevitch, L.I.: In: Awrejcewicz, J., Olejnik, P. (eds.) 8th Conference on Dynamical Systems-Theory and Applications, DSTA-2005, p. 289 (2005)

Manevitch, L.I.: New approach to beating phenomenon in coupled nonlinear oscillatory chains. Arch Appl Mech. **301** (2007)

Manevitch, L.I.: Vibro-impact models for smooth non-linear systems. In: Ibrahim, R.A., Babitsky, V.I., Okuma, M. (eds.) Lecture Notes in Applied and Computational Mechanics, vol. 44, pp. 191–202. Springer, Berlin (2009)

Manevitch, L.I., Gendelman, O.V.: Tractable Models of Solid Mechanics. Springer, Berlin (2011)

Manevich, A.I., Manevich, L.I.: The Mechanics of Nonlinear Systems with Internal Resonances. Imperial College Press, London (2005)

Manevitch, L.I., Musienko, A.I.: Limiting phase trajectories and energy exchange between an anharmonic oscillator and external force. Nonlinear Dyn. **58**, 633–642 (2009)

Manevitch, L.I., Oshmyan, V.G.: An asymptotic study of the linear vibrations of a stretched beam with concentrated masses and discrete elastic supports. J. Sound Vib. **223**(5), 679–691 (1999)

Manevitch, L.I., Smirnov, V.V.: In: Indeitsev, D.A. (ed.) Advanced Problem in Mechanics, APM-2007, p. 289. IPME RAS, St.Petersburg (Repino), Russia (2007)

Manevitch, L.I., Smirnov, V.V.: In: Awrejcewicz, J., Olejnik, P., Mrozowski, J. (eds.) 9th Conference on Dynamical Systems Theory and Applications, DSTA-2007, p. 301 (2007)

Manevitch, L.I., Smirnov, V.V., 9th Conference on dynamical systems theory and applications. In: Awrejcewicz, J., Olejnik, P., Mrozowski, J. (eds.) 301, DSTA-2007, L'od'z, Poland 17–20 Dec 2007 (2007)

Manevich, L.I., Smirnov, V.V.: Discrete breathers and intrinsic localized modes in small FPU systems. In: Proceedings of APM 2007: 293, St-Peterburg (2007)

Manevitch, L.I., Smirnov, V.V.: Solitons in Macromolecular Systems. Nova Publ, New-York (2008)

Manevitch, L.I., Smirnov, V.V.: Limiting phase trajectories and the origin of energy localization in nonlinear oscillatory chains. Phys. Rev. E **82** (2010)

Manevich, L.I., Mikhlin, Yu.V., Pilipchuk, V.V.: The Method of Normal Oscillations for Essentially Nonlinear Systems. Nauka, Moscow (1989) (in Russian)

Manevitch, L.I., Mikhlin, YuV, Pilipchuk, V.N.: The Normal Vibrations Method For Essentially Nonlinear Systems. Nauka Publ, Moscow (1989). (in Russian)

Manevitch, L.I., Kovaleva, A.S., Manevitch, E.L.: Limiting phase trajectories and resonance energy transfere in a system of two coupled oscillators. Math Problems Eng, vol. 2010. doi:10. 1155/2010/760479 (2010)

Manevitch, L.I., Kovaleva, A.S., Shepelev, D.S.: Non-smooth approximations of the limiting phase trajectories for the duffing oscillator near 1:1 resonance. Accepted for publication in Physica D (2011)

Mokrosst, F., Buttner, H.: Thermal conductivity in the diatomic toda lattice. J. Phys. C: Solid State Phys. **16**, 4539–4546 (1983)

Nayfeh A.H., Mook, D.T.: Nonlinear Oscillations. Wiley, New York (1979)

Ovchinnikov, A.A., Erikhman, N.S., Pronin, K.A.: Vibrational-Rotational Excitations in Nonlinear Molecular Systems. Springer, Berlin (2001)

Poggi, P., Ruffo, S.: Exact solutions in the FPU oscillator chain. Physica D **103**, 251 (1997)

Rabinovich, M.I., Trubetskov, D.I.: Introduction to the Theory of Oscillations and Waves. Nauka, Moscow (1984). [in Russian]

Rink, B.: Symmetry and resonance in periodic FPU chains. Commun. Math. Phys. **218**, 665 (2001)

Rink, B., Verhulst, F.: Near-integrability of periodic FPU-chains. Phys. A **285**, 467 (2000)

Rosenberg, R.M.: On nonlinear vibrations of systems with many degrees of freedom. Adv. Apl. Mech. **9**, 156 (1966)

Sandusky, K.W., Page, J.B.: Interrelation between the stability of extended normal modes and the existence of intrinsic localized modes in nonlinear lattices with realistic potentials. Phys. Rev. B **50**, 866 (1994)

Scott A.: Nonlinear Science: Emergence and Dynamics of Coherent Structures. Oxford University Press, Oxford (2003)

Smirnov V.V., Manevitch L.I.: Limiting phase trajectories and dynamic transitions in nonlinear periodic systems,Acoust. Phys. **57**, 271–276 (2011)

Starosvetsky, Y., Manevitch L.I.: Nonstationary regimes in a Duffing oscillator subject to biharmonic forcing near a primary resonance. Phys. Rev. E **83** (2011)

Starosvetsky, Y., Manevitch, L.I.: On intense energy exchange and localization in periodic FPU dimer chains,Physica D **264**, 66–79 (2013)

Toda, M.: Theory of Nonlinear Lattices, vol. 20. Springer, Berlin (1989)

Uzunov, I.M., Muschall, R., Gölles, M., Kivshar, Yuri S., Malomed, B.A., Lederer, F.: Pulse switching in nonlinear fiber directional couplers. Phys. Rev. E **51**, 2527–2537 (1995)

Vakakis, A.F.: Advanced Nonlinear Strategies for Vibration Mitigation and System Identification, 1st Ed. Springer, Berlin (2010)

Vakakis, A.F., Manevitch, L.I., Mikhlin, YuV, Pilipchuk, V.N., Zevin, A.A.: Normal Modes and Localization in Nonlinear Systems. Wiley, New York (1996)

Chapter 5
Localized Nonlinear Excitations and Inter-chain Energy Exchange

The description of the nonlinear media in the framework of the quasi-one-dimensional models assumes the existence of the essential anisotropy of the media properties, for example, the considerable difference between coupling constants along and transversely the chains in the polymeric crystals. Under these assumptions, the dislocations motion in the Frenkel–Kontorova model and the domain wall creation in the φ^4 one are considered as a plane wave along the transversal direction. However, there are some physical problems where the transfer of the excitation energy in this direction is the subject of the interest. First of all, one should keep in mind the polymeric crystals, where the transition of the localized excitation between the neighbor chains corresponds to the motion of the point-like defect transversely the crystal (Manevitch and Smirnov 2008). The nonlinear optical waveguides are the typical quasi-one-dimensional systems with the transversal transfer of the localized pulses (Akhmediev and Ankiewicz 1997). Finally, the wandering of the excitation energy between strands of the DNA may be interesting from the viewpoint of the biology as well as the nanoelectronics. This phenomenon is important also for description of the processes of the charge transport along the DNA.

The inter-chain energy exchange by the breathers was first studied both analytically and numerically in Kosevich et al. (2008). A principal possibility of this phenomenon has been shown. In the present chapter, we consider the problem of inter-chain energy exchange using the LPT concept. The weakly coupled nonlinear chains are regarded as the continuum systems with various degrees of nonlinearity (linear chains, weakly nonlinear chains, and chains with nonlinearity compared with coupling).

© Springer Nature Singapore Pte Ltd. 2018
L.I. Manevitch et al., *Nonstationary Resonant Dynamics of Oscillatory Chains and Nanostructures*, Foundations of Engineering Mechanics,
DOI 10.1007/978-981-10-4666-7_5

5.1   Linear Chains with Weak Coupling

Let us consider a system of weakly coupled Fermi–Pasta–Ulam (FPU) chains with potential energy containing the terms of fourth order alongside with parabolic ones. The respective Hamilton function is:

$$H = \sum_n \left\{ \sum_{j=1,2} \left[\left(\frac{dq_{n,j}}{dt} \right)^2 + \frac{c^2}{2} \left(q_{n+1,j} - q_{n,j} \right)^2 + \frac{\beta}{4} \left(q_{n+1,j} - q_{n,j} \right)^4 \right] + \varepsilon \frac{\gamma}{2} \left(q_{n,1} - q_{n,2} \right)^2 \right\}$$

(5.1)

where $q_{n,j}$ is dimensionless displacement of particle "n"-th in "j"-th chain, c, β, and γ are dimensionless parameters of interaction, and ε is a small coefficient of inter-chain coupling. It is easy to show that a modulation of particle displacements at the right edge of the spectrum of linearized system $(u_{n,j} = (-1)^n q_{n,j})$ leads to the following continuum equations for the envelope functions φ^4:

$$\frac{\partial^2 u_j}{\partial \tau^2} + \frac{\partial^2 u_j}{\partial x^2} + u_j + 16\beta u_j^3 - \varepsilon \gamma u_{3-j} = 0$$
$$\tau = \omega t, \quad \omega^2 = 4 + \varepsilon \gamma$$

(5.2)

It is convenient to use the complex variables:

$$\Psi_j = \frac{1}{\sqrt{2}} \left(\frac{\partial u_j}{\partial \tau} + i u_j \right), \quad \bar{\Psi}_j = \frac{1}{\sqrt{2}} \left(\frac{\partial u_j}{\partial \tau} - i u_j \right)$$

(5.3)

The equations of motion (5.2) are converted to form:

$$i \frac{\partial}{\partial \tau} \Psi_j + \Psi_j + \frac{1}{2} \frac{\partial^2}{\partial x^2} (\Psi_j - \bar{\Psi}_j) - 4\beta (\Psi_j - \bar{\Psi}_j)^3 - \varepsilon \frac{\gamma}{2} (\Psi_{3-j} - \bar{\Psi}_{3-j}) = 0 \quad (5.4)$$

Let us assume that the parameter of nonlinearity β is equal to zero. Now, we consider two linear chains with linear coupling. Using the multi-scale expansion in ε:

$$\Psi_j = \varepsilon(\psi_j + \varepsilon \psi_{j,1} + \varepsilon^2 \psi_{j,2} + \ldots)$$
$$\tau_0 = \tau, \quad \tau_1 = \varepsilon \tau, \quad \tau_2 = \varepsilon^2 \tau$$
$$\xi = \varepsilon x$$

(5.5)

We obtain the following equations of different orders by small parameter ε: ε^1:

$$i \partial_{\tau_0} \psi_j + \psi_j = 0$$
$$\psi_j = \chi_j e^{i\tau_0}$$

(5.6)

ε^2:

$$i\partial_{\tau_0}\psi_{j,1} + i\partial_{\tau_1}\psi_j + \psi_j - \tfrac{\gamma}{2}(\psi_{3-j} - \bar{\psi}_{3-j}) = 0$$
$$\psi_{j,1} = \chi_{j,1}e^{i\tau_0}$$
$$i\partial_{\tau_0}\chi_{j,1} + i\partial_{\tau_1}\chi_j - \tfrac{\gamma}{2}(\chi_{3-j} - \bar{\chi}_{3-j}e^{-2i\tau_0}) = 0$$

$$(5.7)$$

The last equations lead to following important relationships

$$i\partial_{\tau_1}\chi_j - \tfrac{\gamma}{2}\chi_{3-j} = 0$$
$$\chi_{j,1} = \tfrac{\gamma}{4}\bar{\chi}_{3-j}e^{-2i\tau_0}$$

$$(5.8)$$

Now, we can get the solution of Eqs. (5.8) in the form:

$$\chi_1 = \frac{1}{\sqrt{2}}\left[X_1 \cos\left(\frac{\gamma}{2}\tau_1\right) - iX_2 \sin\left(\frac{\gamma}{2}\tau_1\right)\right]$$
$$\chi_2 = \frac{1}{\sqrt{2}}\left[X_2 \cos\left(\frac{\gamma}{2}\tau_1\right) - iX_1 \sin\left(\frac{\gamma}{2}\tau_1\right)\right]$$

$$(5.9)$$

ε^3:

$$i\partial_{\tau_0}\psi_{j,2} + i\partial_{\tau_1}\psi_{j,1} + i\partial_{\tau_2}\psi_j + \tfrac{1}{2}\partial_\xi^2(\psi_j - \bar{\psi}_j) - \tfrac{\gamma}{2}(\psi_{3-j,1} - \bar{\psi}_{3-j,1}) = 0$$
$$i\partial_{\tau_0}\chi_{j,2} + i\partial_{\tau_1}\chi_{j,1} + i\partial_{\tau_2}\chi_j + \tfrac{1}{2}\partial_\xi^2(\chi_j - \bar{\chi}_j e^{-2i\omega\tau_0}) - \tfrac{\gamma}{2}(\chi_{3-j,1} - \bar{\chi}_{3-j,1}e^{-2i\omega\tau_0}) = 0$$

$$(5.10)$$

It is easy to view that after integrating with respect to fast time τ_0 with using last relation (5.8), the equations for different chains turn out to be uncoupled. The main point in analysis of Eqs. (5.10) are that the unknown functions depend on "intermediate" time τ_1. The adequate procedure to remove this dependence is the averaging over time τ_1. After simple algebraic manipulations, we obtain:

$$i\partial_{\tau_2}X_1 + \frac{1}{2}\partial_\xi^2 X_1 - \frac{\gamma^2}{8}X_1 = 0$$
$$i\partial_{\tau_2}X_2 + \frac{1}{2}\partial_\xi^2 X_2 - \frac{\gamma^2}{8}X_2 = 0,$$

$$(5.11)$$

where X_1 and X_2 are the functions of slow time τ_2. Equations (5.11) have solution in the form of plane wave.

$$X_j = A_j \exp(i(k\xi - \omega\tau_2))$$

$$(5.12)$$

with dispersion relation

$$\omega = \frac{1}{2}\left(k^2 + \left(\frac{\gamma}{2}\right)^2\right).$$

It is important that the structure of Eqs. (5.11) allow the wave localizing on one chain only. This case corresponds to full energy exchange between the chains, if the solution (5.12) is considered as initial condition for first of Eqs. (5.8) in the "intermediate" time τ_1. It is obvious that the plane waves (5.12) migrate from one chain to other in accordance with Eqs. (5.9). Figure 5.1 shows an example of full exchange between the chains for the initial conditions $A_1 = 0.10$ and $A_2 = 0.0$.

5.2 Nonlinear Chains

If the parameter β in Eqs. (5.4) is not zero, we can study how the nonlinearity influences on the process of energy exchange. Using the series like (5.5) by small parameter ε, we obtain the weak nonlinearity asymptotics, because the order of coupling terms is equal ε^2, while the nonlinear terms give contribution $\sim \varepsilon^3$. Thus, Eqs. (5.6)–(5.9) are the same as for linear and nonlinear systems, but Eqs. (5.11) resulting from averaging is changed:

$$i\partial_{\tau_2}X_1 + \frac{1}{2}\partial_\xi^2 X_1 - \frac{\gamma^2}{8}X_1 + \frac{3\beta}{8}\left(3|X_1|^2X_1 + 2|X_2|^2X_1 - X_2^2\bar{X}_1\right) = 0$$
$$i\partial_{\tau_2}X_2 + \frac{1}{2}\partial_\xi^2 X_2 - \frac{\gamma^2}{8}X_2 + \frac{3\beta}{8}\left(3|X_2|^2X_2 + 2|X_1|^2X_2 - X_1^2\bar{X}_2\right) = 0$$

$$(5.13)$$

Equations (5.13) describe the pair of nonlinear oscillatory chains with nonlinear coupling contrary to the initial system with the linear coupling. It is interesting that the structure of nonlinear terms is similar to that of the case of small FPU system (Manevich and Smirnov 2007). These equations admit both anharmonic plane wave solution and solution in the form of localized vibrations (breathers). The plane wave solution has the form:

$$X_j(\xi, \tau_2) = A_j \exp(-i(\omega\tau_2 - k\xi)) \tag{5.14}$$

with dispersion relations

$$\omega = \frac{1}{2}\left(k^2 + \left(\frac{\gamma}{2}\right)^2\right) - 6\beta\left(3A_j^2 + A_{3-j}^2\right) \tag{5.15}$$

Like the case of linear chains, Eqs. (5.13) admit a wave solution, localized on one chain only. This solution leads to full energy exchange between chains. Figure 5.2 shows an example of small-amplitude anharmonic plane wave in the weakly nonlinear system.

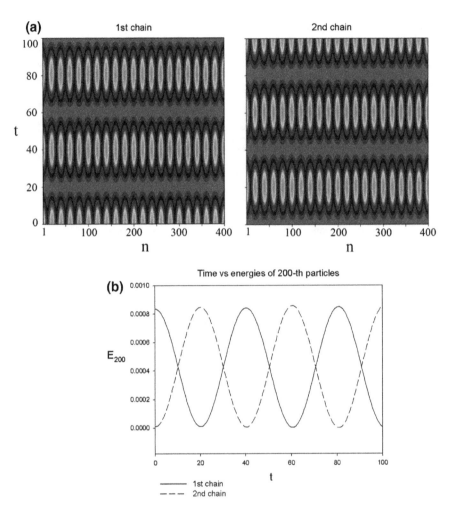

Fig. 5.1 **a** "Map" of total energy of linear chains—*bright bands* correspond to high energy value, *dark bands*—correspond to low energy, *t*—time, *n*—number of particle in the chain; **b** energy of 200-th particles in the different chains versus time. The plane wave was initiated in left chain at the time $t = 0$ only

The analysis of the phase plane in the terms of polar coordinates (Kosevich and Kovalyov 1989; Manevich 2007) does not show bifurcation both in-phase and anti-phase stationary points. The solution in the form of plane wave has the phase shift which is equal to $\pi/2$. Thus, this trajectory corresponds to the limiting phase trajectory (LPT) in the case of two nonlinear oscillators.

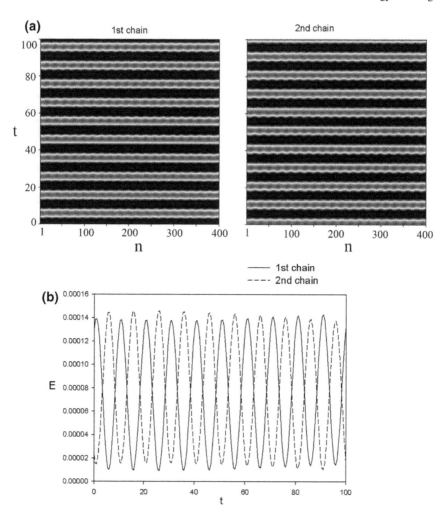

Fig. 5.2 Plane wave with full energy exchange in the weakly nonlinear system. **a** Energy "map"—*bright* and *dark bands* correspond to high and low energy values, respectively. The plane wave was initiated in right chain at $t = 0$ only. **b** The energy profiles of 200-th particles in both chains versus time

Let us consider a localized solution of Eqs. (5.13):

$$X_j(\xi, \tau_2) = A_j(\xi - v\tau_2)\exp[-i(\omega\tau_2 - q\xi)], \quad j = 1, 2. \tag{5.16}$$

Here, A_j are real functions. Substitution of this form into the equations of motion gives the relation between wave number q and velocity v:

$$v = -q.$$

The equations for amplitudes A_j can be written as follows:

$$A_j'' + \left(\omega - \frac{q^2}{2} - \frac{\gamma^2}{8}\right)A_j + 6\beta\left(3A_j^3 + A_{3-j}^2 A_j\right) = 0, \tag{5.17}$$

where primes denote differentiation with respect to argument. Let us suppose that $A_1 = kA_2$. The conditions of compatibility of Eqs. (5.17) lead to following values of k: $k = +1$; $k = -1$; and $k = 0$. The last value has a principal importance because of existence of full energy exchange. Thus, the solution of Eqs. (5.13) describing the localized oscillations have the following form:

$$X_1(\xi, \tau_2) = \sqrt{\frac{\left(q^2 + (\gamma/2)^2\right) - 2\omega}{6(3 + \kappa^2)\beta}} \exp[-i(\omega\tau_2 - q\xi)]$$

$$\times \, sch\left[\sqrt{\frac{\left(q^2 + (\gamma/2)^2\right) - 2\omega}{2}}(\xi - q\tau_2 - \xi_0)\right] \tag{5.18}$$

$$X_2(\xi, \tau_2) = \kappa X_1(\xi, \tau_2), \quad \kappa = 0, \pm 1.$$

It is worth mentioning that the shape of small-amplitude solution is formed in the timescale that is more slower than the characteristic time of energy transfer between different chains. This statement is valid for both linear and nonlinear systems. An example of energy transfer in the case of moving breather is shown in Fig. 5.3.

5.2.1 Chains with Nonlinearity, Compatible with Coupling

Let us return to Eqs. (5.4) for the case of the amplitudes providing compatibility of nonlinear and coupling terms by parameter ε. If their values reach a magnitude $\sqrt{\varepsilon}$, the expansion (5.5) turns out to be invalid. Therefore, we will use multi-scale expansion:

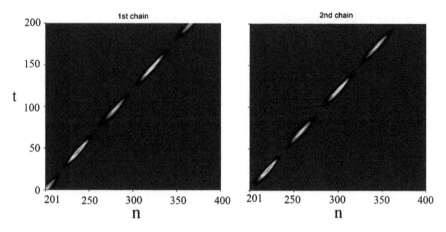

Fig. 5.3 Energy "map" for moving small-amplitude breather in the case of weak nonlinearity

$$\Psi_j = \sqrt{\varepsilon}(\psi_j + \varepsilon\psi_{j,1} + \varepsilon^2\psi_{j,2} + \dots)$$
$$\tau_0 = \tau, \quad \tau_1 = \varepsilon\tau, \quad \tau_2 = \varepsilon^2\tau \qquad , \qquad (5.19)$$
$$\xi = \sqrt{\varepsilon}x$$

That leads to following equations for different orders of small parameter ε:
$\varepsilon^{1/2}$:

$$i\partial_{\tau_0}\psi_j + \psi_j = 0$$
$$\psi_j = \chi_j e^{i\tau_0} \qquad\qquad (5.20)$$

$\varepsilon^{3/2}$:

$$i\partial_{\tau_0}\psi_{j,1} + i\partial_{\tau_1}\psi_j + \psi_j + \tfrac{1}{2}\partial_\xi^2(\psi_j - \bar\psi_j) - \tfrac{\gamma}{2}(\psi_{3-j} - \bar\psi_{3-j}) - 4\beta(\psi_j - \bar\psi_j)^3 = 0$$
$$\psi_{j,1} = \chi_{j,1}e^{i\tau_0}$$
$$i\partial_{\tau_0}\chi_{j,1} + i\partial_{\tau_1}\chi_j + \tfrac{1}{2}\partial_\xi^2(\chi_j - \bar\chi_j e^{-2i\tau_0}) - \tfrac{\gamma}{2}(\chi_{3-j} - \bar\chi_{3-j}e^{-2i\tau_0}) - 4\beta(\chi_j e^{i\tau_0} - \bar\chi_j e^{i\tau_0})^3 e^{-i\tau_0} = 0$$

$$(5.21)$$

Integrating last of Eqs. (5.21) with respect to "fast" time τ_0, we get two coupled equations:

$$i\partial_{\tau_1}\chi_j + \frac{1}{2}\partial_\xi^2\chi_j - \frac{\gamma}{2}\chi_{3-j} + 12\beta|\chi_j|^2\chi_j = 0 \qquad (5.22)$$

First, we can see that there are two symmetric solutions of Eqs. (5.22). In the class of localized soliton-like solutions, they have the form:

in-phase solution

$$\chi_1(\xi,\tau_1) = \frac{1}{4}\sqrt{\frac{2\omega+q^2+\gamma}{3\beta}}sch\left(\frac{1}{4}\sqrt{\frac{2\omega+q^2+\gamma}{6\beta}}(\xi+q\tau_1)\right)$$
$$\times \exp(i(\omega\tau_1 - q\xi)),$$
$$\chi_1(\xi,\tau_1) = \chi_2(\xi,\tau_1)$$

(5.23)

anti-phase solution

$$\chi_1(\xi,\tau_1) = \frac{1}{4}\sqrt{\frac{2\omega+q^2-\gamma}{3\beta}}sch\left(\frac{1}{4}\sqrt{\frac{2\omega+q^2-\gamma}{6\beta}}(\xi+q\tau_1)\right)$$
$$\times \exp(i(\omega\tau_1 - q\xi)),$$
$$\chi_1(\xi,\tau_1) = -\chi_2(\xi,\tau_1)$$

(5.24)

The Hamilton function, corresponding to Eqs. (5.22), is

$$h = -\frac{\gamma}{2}(\chi_1\bar{\chi}_2 + \bar{\chi}_1\chi_2) + \frac{1}{2}\left(|\partial_\xi\chi_1|^2 + |\partial_\xi\chi_2|^2\right)$$
$$+ 6\beta\left(|\chi_1|^4 + |\chi_1|^4\right)$$

(5.25)

Similar to symmetric solutions (5.23)–(5.24), we can suppose that localized soliton-like solutions of Eqs. (5.22) can be represented in the form:

$$\chi_j = A(\xi)X_j(\tau_1),$$

(5.26)

where a space-dependent amplitude A has the same profile for both chains. Thus, integrating Eqs. (5.25) with respect to space variable ξ, we get the "energy" of the system as a function of time variable:

$$H = -\frac{\gamma}{2}N(X_1\bar{X}_2 + \bar{X}_1X_2) + \frac{1}{2}\mu N\left(|X_1|^2 + |X_2|^2\right) + 6\beta\nu N^2\left(|X_1|^4 + |X_2|^4\right),$$

(5.27)

where new parameters are defined by soliton profile:

$$N = \int A^2 d\xi, \quad \mu = \int (\partial_\xi A)^2 d\xi \Big/ \int A^2 d\xi,$$
$$\nu = \int A^4 d\xi \Big/ \left(\int A^2 d\xi\right)^2$$

(5.28)

In such a case, we get an analog of two nonlinear oscillators, described by functions X_j. Both beating with full energy exchange and confinement of initial excitation in the one of chain can be observed when the value of "occupation number" N grows. It was shown that the process of energy exchange is defined by trajectories, closed to LPT and pertinent to attractive area of one of two stationary points of the system.

There are two stationary points of Eqs. (5.24) at a small "occupation number" N and four ones exist if N is large enough. It is easy to see from analysis of phase plane in the terms of "polar variables":

$$X_1 = \cos \theta \, e^{i\delta_1}, \quad X_2 = \sin \theta \, e^{i\delta_2}.$$

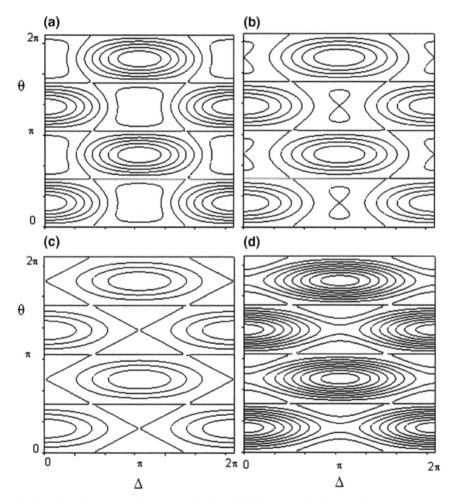

Fig. 5.4 Transformation of "phase plane" of Eq. (5.24) in terms of polar variables. θ characterizes the amplitude ratio and $\Delta = \delta_1 - \delta_2$—the phase shift. The occupation number N increases from (**a**) to (**d**) fragments: **a** $\kappa < 0.5$, **b** $0.5 < \kappa < 1$, **c** $\kappa = 1.0$, and **d** $\kappa > 1.0$ (see text)

The parameter controlling a structure of phase plane is $\kappa = 6\beta\nu N/\gamma$. Four typical cases are shown in Fig. 5.4. Like the case of two weakly coupled oscillators, if the parameter κ is smaller than 0.5, only two stationary points exist: in-phase ($\delta = 0$, $\theta = \pi/4$) and out-of-phase ($\delta = \pi$, $\theta = \pi/4$) ones (see Fig. 5.4). Closed trajectories near the LPT describe full energy exchange. At $\kappa = 0.5$, out-of-phase mode becomes unstable one that leads to separatrix creation (Fig. 5.4b). So, if we start from the state near the new asymmetric modes, we can not transfer energy effectively from one chain to another one. However, a possibility of full exchange along the LPT is well preserved. The total prohibition of the energy exchange appears when κ becomes more than unit. Then, the separatrix coincides with LPT and trajectories closed around out-of-phase mode are broken (Fig. 5.4c). So, full confinement of excitation on the one of the chains occurs. The main conclusion is that the full energy exchange is possible till the parameter κ do not exceed unit (Fig. 5.4d). After that only partial exchange can occur near the asymmetric modes. The computer simulation data, an example of which is shown in Fig. 5.5, demonstrate a confinement of initial excitations in the one chain at $\kappa \sim 2$.

Analytical and numerical studies of the wandering excitation both in linear and nonlinear chains coupled by weak linear interaction show an existence of two asymptotic limits of energy transfer between different chains. First of them is characterized by quick energy transfer in comparison with processes of excitation formation. In such a case, the waves in the different chains exhibit the phase shift which is equal $\pi/2$. It means that respective trajectory is closed to the LPT. On the contrary the excitations with large amplitudes can show both full energy exchange near LPT and partial exchange near stationary points up to full confinement of excitation in one of the chains.

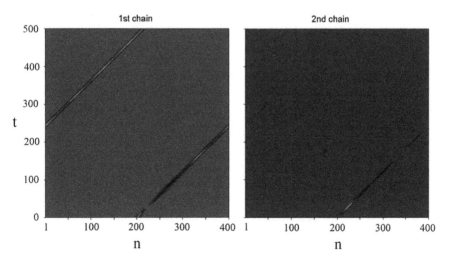

Fig. 5.5 Confinement of the breather in the first chain. Breather was initiated in the first chain at $t = 0$. After exchange with the second chain, the breather returns to the "parent" chain

References

Akhmediev, N.N., Ankiewicz, A..: Solitons, Nonlinear Pulses and beams, Chapman and Hall, London (1997)

Kosevich, A.M., Kovalyov, A.S.: Introduction to Nonlinear Physical Mechanics. Naukova Dumka, Kiev (1989) (in Russian)

Kosevich, Y.A., Manevich, L.I., Savin, A.V.: Wandering breathers and self-trapping in weakly coupled nonlinear chains: classical counterpart of macroscopic tunneling quantum dynamics. Phys. Rev. E **77**, 046603 (2008)

Manevich, L.I.: New approach to beating phenomenon in coupled nonlinear oscillatory chains. Arch. Appl. Mech. **77**, 301–312 (2007)

Manevich, L.I., Smirnov, V.V.: Discrete breathers and intrinsic localized modes in small FPU systems. In: Proceedings APM 2007, p. 293. St. Petersburg (2007)

Manevitch, L.I., Smirnov, V.V.: Solitons in Macromolecular Systems. Nova Publ, New York (2008)

Part II
Extensions to Non-conservative Systems

In this part of the book, we show that the LPT concept which was initially developed for conservative nonlinear models can be efficiently extended. This extension allows taking into account the external periodic forcing and damping as well as feedback which leads to existence of self-sustained oscillations. It is significant that, contrary to conservative models, in this case even the asymptotic equations in the main approximation are not integrable. Therefore, the LPTs as fundamental non-stationary solutions provide an unique possibility to understand and describe analytically a wide class of non-stationary resonance processes (similarly to description of the stationary processes on non-conservative systems in the frameworks of the NNMs concept). Moreover, the LPT concept turns out to be useful for understanding and analytical description of such significant processes as autoresonance in the nonlinear systems with variable parameters.

Chapter 6
Duffing Oscillators

In this chapter, we illustrate the role of LPTs in the analysis of nonlinear non-stationary oscillations in non-stationary systems by a simple example of a periodically forced single-degree-of-freedom (SDOF) Duffing oscillator. The main difference from the conservative system described in the preliminary section is that we deal here with resonance energy flow from the source of energy instead of internal (intermodal) resonance.

In the first section of this chapter, we define the stationary states and the LPTs for this model and then employ the LPT concept to describe salient features of the non-stationary dynamics of the forced Duffing oscillator near 1:1 resonance. It is shown that LPTs can be considered as borderlines between different types of trajectories, associated with maximum targeted energy transfer from the source of energy to the oscillator and related, depending on the parameters, to quasi-linear, moderately nonlinear, and strongly nonlinear regimes of oscillations, while the steady (stationary) oscillations of the non-autonomous system play the role of NNMs. We also extend the notion of the LPT to oscillations with maximum energy in non-dissipated and dissipated oscillators subjected to biharmonic external excitations.

The main features of the dynamical behavior of the model under consideration are highlighted with the help of explicit asymptotic solutions. Similar to previous chapters, we make an extensive use of a special asymptotic technique based on the complexification of the dynamic and using the non-smooth transformations. Considering the case of biharmonic excitation leads to revealing the qualitatively new strongly nonlinear phenomenon. We consider also the case of super-harmonic resonance.

© Springer Nature Singapore Pte Ltd. 2018
L.I. Manevitch et al., *Nonstationary Resonant Dynamics of Oscillatory Chains and Nanostructures*, Foundations of Engineering Mechanics,
DOI 10.1007/978-981-10-4666-7_6

6.1   Duffing Oscillator with Harmonic Forcing Near 1:1 Resonance

6.1.1   Main Equations and Definitions

We investigate a dimensionless weakly nonlinear oscillator subject to a periodic excitation in the neighborhood of 1:1 resonance. The equation of motion is given by

$$\frac{d^2u}{d\tau_0^2} + 2\varepsilon\gamma\frac{du}{d\tau_0} + u + 8\alpha\varepsilon u^3 = 2\varepsilon F \sin(1+\varepsilon s)\tau_0, \tag{6.1}$$

where γ, α, F, s are positive parameters, and $\varepsilon > 0$ is a small parameter of the system. The oscillator is assumed to be initially at rest; this assumption is equivalent to initial conditions $u = 0$, $v = du/d\tau_0 = 0$ at $\tau_0 = 0^+$. Below, we write $\tau_0 = 0$ instead of $\tau_0 = 0^+$, except as otherwise noted. The trajectory satisfying these initial conditions is a *limiting phase trajectory* (LPT) corresponding to maximum possible (under given parameters of the system) energy transfer from the source of energy to the oscillator. This extension of the LPT concept onto forced oscillations was made in (Manevitch and Musienko 2009) and then in (Manevitch et al. 2011a, b, c).

To construct an explicit asymptotic solution, we make use of the multiple scales analysis (Nayfeh and Mook 2004). This approach is especially convenient in neighborhoods of resonances in the framework of the LPT concept.

Similar to previous chapters, the first step is to introduce the complex-valued variables

$$Y = (v + iu)e^{-i\tau_0}, \quad Y^* = (v - iu)e^{i\tau_0}, \quad i = \sqrt{-1}, \tag{6.2}$$

where asterisk denotes complex conjugate. Substituting the representation (6.2) into (6.1) yields the following alternative (still exact) equation of motion:

$$\frac{dY}{d\tau_0} = 3i\varepsilon\alpha|Y|^2]Y - \gamma Y - i\varepsilon F e^{i\tau_1} + i\varepsilon G_0(\tau_0, \tau_1, Y, Y^*), \quad Y(0) = 0, \tag{6.3}$$

where $\tau_1 = \varepsilon\sigma\tau_0$ is the leading-order slow timescale. The coefficient G_0 is given by

$$G_0(\tau_0, \tau_1, Y, Y^*) = (Fe^{-i\tau_1} + i\gamma Y^* - 3\alpha Y^*|Y|^2)e^{-2i\tau_0} - \alpha Y^3 e^{2i\tau_0} + (Y^*)^3 e^{-4i\tau_0}. \tag{6.4}$$

Equation (6.3) is exact, as it is derived from the original equation of motion without omitting any terms in the process. To obtain an approximate solution, we apply the multiple scales method (Nayfeh and Mook 2004) to Eq. (6.3). Since Eq. (6.3) depends on two timescales, the fast timescale τ_0 and the leading-order slow timescale $\tau_1 = \varepsilon\sigma\tau_0$, the solution is sought in the form of the expansion:

$$Y(\tau_0, \tau_1, \varepsilon) = \phi^{(0)}(\tau_1) + \varepsilon\phi^{(1)}(\tau_0, \tau_1) + O(\varepsilon^2), \tag{6.5}$$

with the slow leading-order term $\varphi^{(0)}(\tau_1)$. After substituting (6.5) into (6.3) and eliminating the sum of non-oscillating terms from the resulting equation, we obtain the following equation for the slow variable $\varphi^{(0)}(\tau_1)$:

$$s\frac{d\phi^{(0)}}{d\tau_1} + \gamma\phi^{(0)} - 3i\alpha|\phi^{(0)}|^2\phi^{(0)} = -iFe^{i\tau_1}, \quad \phi^{(0)}(0) = 0 \tag{6.6}$$

This equation represents the approximation of the *slow flow dynamics* of the system; i.e., it governs approximately the slow evolution of the complex amplitude with time. Rescaling of the variables and the parameters

$$\phi^{(0)}(\tau_1) = \Lambda\phi(\tau_1)e^{i\tau_1}, \quad \Lambda = \left(\frac{s}{3\alpha}\right)^{1/2}, \quad f = \frac{F}{s}\Lambda = F\sqrt{\frac{3\alpha}{s^3}}, \quad \gamma_1 = \frac{\gamma}{s} \tag{6.7}$$

reduces Eq. (6.6) to the form

$$\frac{d\phi}{d\tau_1} + \gamma_1\phi + i(1 - |\phi|^2)\phi = if, \quad \phi(0) = 0. \tag{6.8}$$

After introducing a polar decomposition of ϕ in terms of a real amplitude and a real phase, $\phi = \alpha\varepsilon^{i\Delta}$, Eq. (6.8) is rewritten as follows:

$$\frac{da}{d\tau_1} + \gamma_1 a = -f\sin\Delta$$
$$a\frac{d\Delta}{d\tau_1} = -a + a^3 - f\cos\Delta, \tag{6.9}$$

with initial condition $a(0) = 0$. The second initial condition, corresponding to $\Delta(0)$, will be derived below.

It now follows from (6.2), (6.5) and (6.7) that

$$u(\tau_0, \varepsilon) = \Lambda a(\tau_1)\sin(\tau_0 + \Delta(\tau_1) + \tau_1) + O(\varepsilon)$$
$$v(\tau_0, \varepsilon) = \Lambda a(\tau_1)\cos(\tau_0 + \Delta(\tau_1) + \tau_1) + O(\varepsilon) \tag{6.10}$$

It is well known that the difference between a precise solution of Eq. (6.1) and its approximation (6.10) is of $O(\varepsilon)$ in large time intervals $\tau_0 \sim O(1/\varepsilon)$ (Nayfeh and Mook 2004). However, a refined analysis (Mirkina (Kovaleva) 1977) shows that the interval of convergence depends on the properties of higher approximations and may tend to infinity. Moreover, relatively large values of ε in particular problems do not necessarily imply that the derived analytical approximations will be poor at larger times; numerical examples are given in (Vakakis et al. 1996).

6.1.2   *Stationary States, LPTs, and Critical Parameters*

We start the analysis from the non-dissipative system

$$\frac{da}{d\tau_1} = -f \sin \Delta$$
$$a \frac{d\Delta}{d\tau_1} = -a + a^3 - f \cos \Delta. \tag{6.11}$$

with initial condition $a(0) = 0$. It is easy to prove that system (6.11) is integrable, yielding the following integral of motion

$$H = a \left(\frac{1}{4} a^3 - \frac{1}{2} a - f \cos \Delta \right) \tag{6.12}$$

Since $a(0) = 0$, then $H = 0$ on the LPT. This implies that there exist two branches of the LPT, and the branch $a \equiv 0$ for any Δ corresponds to an instant change of the phase shift (Fig. 6.1); the non-trivial branch solves the cubic equation

$$\frac{1}{4} a^3 - \frac{1}{2} a - f \cos \Delta = 0. \tag{6.13}$$

Equality (6.13) determines the second initial condition $a(0^+) = 0$, $\cos\Delta(0^+) = 0$. Suppose that $da/d\tau_1 > 0$ at $\tau_1 = 0^+$; under this assumption, $\Delta(0^+) = -\pi/2$. Hence, initial conditions $a = 0$, $\Delta = -\pi/2$ at $\tau_1 = 0$ correspond to the LPT of system (6.11).

Our purpose is to find critical values of the parameter f dictating different types of the dynamical behavior. Since any stable orbit encircles a corresponding stationary point, the first step is to find the steady states of Eq. (6.11) (they can be considered as the analogues of NNMs in conservative systems) from the equations $da/d\tau_1 = 0$, $d\Delta/d\tau_1 = 0$, or,

$$-a + a^3 - f \, \mathrm{sgn}(\cos \Delta) = 0, \quad \cos \Delta = \pm 1. \tag{6.14}$$

Due to periodicity, only two stationary points $\Delta = 0$ and $\Delta = -\pi$ may be considered. The corresponding stationary states are denoted by C_+ and C_-, respectively. We analyze the roots of the algebraic equation (6.14) through the properties of its discriminant (see Korn and Korn 2000 for more details)

$$D_2 = 4 - 27 f^2.$$

If $D_2 < 0$, then Eq. (6.14) has three different real roots; if $D_2 > 0$, then there exists a single real and two complex conjugate roots; if $D_2 = 0$, two real roots merge (Korn and Korn 2000). The latter condition determines the critical value

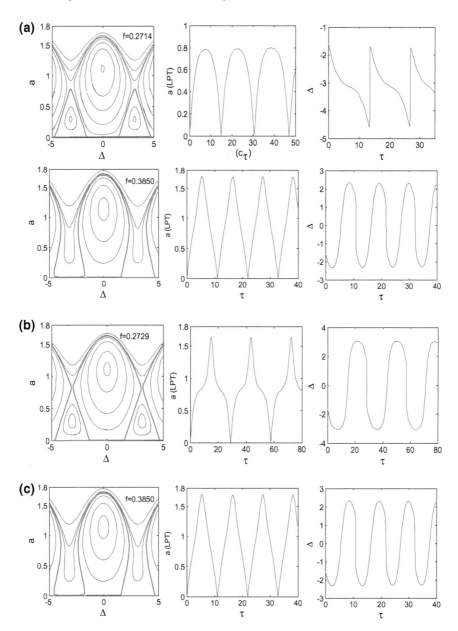

Fig. 6.1 Phase portraits, slow amplitudes, and phases of system (6.11): **a** weakly nonlinear oscillations, $f = 0.2714$; **b** moderately nonlinear oscillations, $f = 0.2729$; **c** strongly nonlinear oscillations, $f = 0.385$. The LPTs are depicted by *red color*

$$f_2 = 2/\sqrt{27} \approx 0.3849. \tag{6.15}$$

A straightforward investigation of Eq. (6.14) proves that the system has a single stable center C_+: $(0, a_+)$ if $f > f_2$ (Fig. 6.1c). Note that the parameter $f = F\sqrt{3\alpha/s^3}$ reflects the effect of all parameters on the system dynamics, and the condition $f > f_2$ implies not only strong nonlinearity or large excitation amplitude but also intense excitation of the oscillator with small frequency detuning.

If $f < f_2$, then there exist two stable centers C_-: $(-\pi, a_-)$, C_+: $(0, a_+)$ and an intermediate unstable hyperbolic point O: $(-\pi, a_0)$ (Fig. 6.1a, b).

In both cases, the LPT begins at $a = 0$, $\Delta = -\pi/2$, but its direction depends on the value of f. Note that the stable center C_- and the small LPT near C_- exist provided Eq. (6.13) is solvable at $\Delta = -\pi$. In order to find a critical value $f_1 < f_2$ ensuring the transition from small to large oscillations, we analyze the discriminant of Eq. (6.13) at $\Delta = -\pi$ given by

$$D_1 = 16(2 - 27f^2).$$

If $D_1 = 0$, the hyperbolic point in the axis $\Delta = -\pi$ coincides with the maximum point of the small LPT defined by the conditions $da/d\tau_1 = 0$, $\sin \Delta = 0$. Thus, the critical value f_1 is given by the condition $D_1 = 0$, or

$$f_1 = \sqrt{2/27} \approx 0.2722. \tag{6.16}$$

The threshold f_1 corresponds to a boundary between small and large oscillations: At $f = f_1$, the LPT of small oscillations coalesces with the separatrix going through the homoclinic point on the axis $\Delta = -\pi$. This implies that the transition from small to large oscillations occurs due to loss of stability of the LPT of small oscillations. At $f = f_2$, the stable center on the axis $\Delta = -\pi$ vanishes due to the coalescence with the homoclinic point, and only a single stable center remains on the axis $\Delta = 0$.

Conditions $f < f_1$, $f_1 < f < f_2$, and $f > f_2$ specify *quasi-linear*, *moderately nonlinear*, and *strongly nonlinear* dynamical behavior, respectively. These definitions are consistent with the plots presented in Fig. 6.1.

Figure 6.1 clearly demonstrates the "limiting" property of the LPTs in the time-invariant system. It is seen that the LPT represents an outer boundary for a set of closed trajectories encircling the stable center in the phase plane (Δ, a). In particular, this property proves that motion along the LPT possesses the maximum amplitude and therefore the maximum energy among all other closed trajectories characterizing oscillatory motion.

It is important to note that the value f_2 is independent on the initial conditions (see, e.g., Kovaleva and Manevitch 2013; Neishtadt 1975) as a boundary between the regimes with one or three stationary points. However, the boundary f_1 between small and large oscillations strictly depends on the choice of the initial point. This boundary was revealed in (Manevitch 2007; Manevitch and Musienko 2009; Manevitch et al. 2011a, b, c).

Now, we exclude the phase $\Delta(\tau_1)$ from (6.11) and examine the amplitude $a(\tau_1)$ on the LPT. It follows from (6.11) that

$$\frac{d^2a}{d\tau_1^2} = -f \cos \Delta \frac{d\Delta}{d\tau_1}$$

on the LPT. Taking into account that, by virtue of (6.13), $f \cos\Delta = a(a^2 - 2)/4$, we obtain the following second-order equation for the LPT:

$$\frac{d^2a}{d\tau_1^2} + \frac{dU}{da} = 0 \tag{6.17}$$

with initial conditions $a(0) = 0$, $v(0) = da/d\tau_1 = f$. The potential $U(a)$ is defined by the condition of the energy conservation, namely $E = \frac{1}{2}v^2 + U(a) = \frac{1}{2}v^2(0)$. It follows from (6.11) and (6.13) that $v = -f \sin\Delta = \pm[f^2 - 1/16 \, a^2(a^2 - 2)^2]^{1/2}$, and therefore,

$$U(a) = \frac{a^2(a^2 - 2)^2}{32}$$

$$u(a) = \frac{dU}{da} = \frac{a}{4}\left(\frac{a^2}{2} - 1\right)\left(\frac{3a^2}{2} - 1\right) \tag{6.18}$$

The function $U(a)$ and the phase portraits of oscillator (6.17) in the plane (a, v) are shown in Fig. 6.2. It is seen that, dependent on the initial energy, motion varies from small to large oscillations with observable deceleration near the saddle points and then to motion with an almost constant velocity up to the reflection from the wall of the potential well (Fig. 6.2). In the latter case, $a(\tau_1)$ tends to a sawtooth

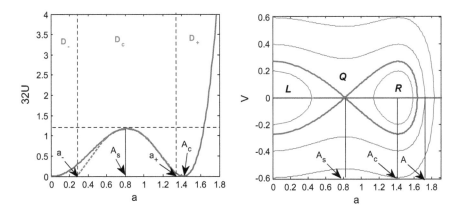

Fig. 6.2 Normalized potential and phase portraits of system (6.17) corresponding to different initial energy; A_s the saddle point; A_c the stable center; A the amplitude of oscillations; *red dashed lines* in the left plot represent the quadratic approximation of the potential in the domain D_C

Fig. 6.3 Functions $\tau(\phi)$ and $e(\phi)$

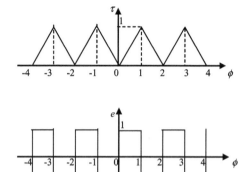

function (Fig. 6.3), and the phase portrait of system (6.17) is consistent with motion of a free particle between two rigid walls (Sect. 1.1).

The coordinates of the saddle point A_s and the stable center A_c are defined by the equality $u(a) = 0$, that is, $A_s = \sqrt{2/3} \approx 0.816, A_c = \sqrt{2} \approx 1.414$. The amplitude of oscillations A may be calculated by Eq. (6.13) at $\Delta = 0$ and $\Delta = \pi$, that is, $A|$ $A^2 - 2| = 4f$; the half-period $T(A)$ is calculated by formula (Sagdeev et al. 1988)

$$T(A) = \int_0^A \frac{da}{\sqrt{f^2 - 2U(a)}} \tag{6.19}$$

6.1.3 Non-smooth Approximations of Strongly Nonlinear Oscillatory Modes

The maximum attention will be paid to highly energetic regimes with large amplitudes of oscillations occurring at $f > f_2$. In this case, motion along the LPT is characterized by a sawtooth envelope $a(\tau_1)$ (Fig. 6.1c). This underlines the fact that important essentially nonlinear phenomena (such as this one) may be missed when resorting to perturbation techniques based on linear (harmonic) generating functions, whose range of validity is restricted to stationary (or almost stationary) and non-stationary, but non-resonance processes. The similarity of the LPT to a sawtooth function admits an approximation of strongly nonlinear highly energetic smooth oscillations by a non-smooth response of a free particle moving with constant velocity between two rigid walls $a = 0$ and $a = A$, where A is a maximum amplitude of oscillations (see also Fig. 1.3 in Sect. 1.1).

Referring to the method of non-smooth transformations (Pilipchuk 2010), we introduce the pair of non-smooth basic functions $\tau(\phi)$ and $e(\phi) = d\tau/d\phi$, $\phi = \Omega\tau_1$ by formulas

$$\tau(\phi) = \frac{2}{\pi} |\arcsin\left(\sin\frac{\pi\phi}{2}\right)|, \quad e(\phi) = \frac{d\tau}{d\phi} = \text{sgn}[\sin(\pi\phi)]$$

$$\frac{d}{d\tau_1} = \Omega[e(\phi)\frac{\partial}{\partial\tau} + \frac{\partial}{\partial\phi}]$$

(6.20)

where $\Omega = 1/T$; the half-period T will be defined below. Plots of non-smooth functions (6.20) are given in Fig. 6.3.

It was shown in earlier works (Manevitch et al. 2011a, b, c) that solutions of (6.11) can be expressed through non-smooth functions as

$$a(\tau_1) = X(\tau), \quad \Delta(\tau_1) = e(\phi)Y(\tau),$$

$$\frac{d}{d\tau_1} = \Omega\left(e\frac{\partial}{\partial\tau} + \frac{\partial}{\partial\phi}\right)$$

(6.21)

While the functions $\Delta(\tau_1)$, $da/d\tau_1$, and $d\Delta/d\tau_1$ are formally continuous at $\phi = 2n - 1$, $\tau = 1$ and discontinuous at $\phi = 2n$, $\tau = 0$, Eq. (6.21) yields discontinuity at both $\phi = 2n - 1$, $\tau = 1$ and $\phi = 2n$, $\tau = 0$. Singularity at $\phi = 2n - 1$ vanishes if $Y(\tau) = 0$ at $\tau = 1$. Formally, one can impose this smoothening condition in order to eliminate singular terms from the resulting equations, but, as shown below, this condition holds by virtue of the dynamical equations.

To derive the equations for X, Y, we insert (6.21) into (6.11) and then separate the terms with and without the coefficient e. This yields the set of equations

$$\frac{dX}{d\tau} = -f\sin Y$$

$$\frac{dY}{d\tau} + X - X^3 = -f\cos Y$$

(6.22)

with initial conditions $X = 0$, $Y = -\pi/2$ at $\tau = 0^+$. It is obvious that

$$\frac{dx}{d\tau} = 0 \text{ at } Y = 0.$$

It now follows that equality $Y(1) = 0$ has a simple physical meaning: It means that X_1 is maximal at $Y = 0$, $\tau = 1$.

In order to assess the amplitude and the period of oscillations, we reduce (6.22) to the second-order form similar to (6.17). Transformations of the same sort as above reduce (6.22) to the second-order equation

$$\Omega\frac{d^2X}{d\tau^2} + u(X) = 0$$

(6.23)

with initial conditions $X = 0$, $dX/d\tau = f/\Omega$ at $\tau = 0^+$; the function $u(X)$ is given by (6.18). Using the vibro-impact approximations, we calculate successive iterations by formulas

$$X = x_0 + x_1 + \ldots, \quad Y_2 = y_0 + y_1 + \ldots, \tag{6.24}$$

where it is assumed that $|x_1(\tau)| \ll |x_0(\tau)|$, $|y_1(\tau| \ll |y_0(\tau)|$ in the interval $0 \le \tau \le 1$. The leading-order approximation x_0 is chosen as the solution of the equation of free particles moving between the walls, namely $d^2 x_0/d\tau^2 = 0$, with initial conditions $x_0 = 0$, $dx_0/d\tau = f/\Omega$ at $\tau = 0^+$. This yields the following non-smooth approximations:

$$x_0(\phi) = a_0(\phi) = A_0 \tau(\phi), \quad v_0(\phi) = A_0 \mathrm{sgn}(\sin 2\phi),$$
$$y_0(\phi) = \Delta_0(\phi) = -\frac{\pi}{2} e(\phi), \quad \phi = \Omega_0 \tau_1, \quad \Omega_0 = {}^1\!/_{T_0} \tag{6.25}$$

By construction, the inverse transformation $a_0 = A_0 \tau(\Omega_0 \tau_1)$ produces the saw-tooth periodic solution associated with the vibro-impact process. The generating half-period $T_0 = 1/\Omega_0$ is defined as $T_0 = A_0/f$, where A_0 is calculated after the substitution of the latter equality into expression (6.19).

The first-order term x_1 is governed by the equations

$$\frac{d^2 x_1}{d\tau^2} = -\Omega_0^{-2} u(a_0), \quad x_1(\tau) = -\Omega_0^{-2} \int_0^\tau (\tau - \xi) u(A_0 \xi) d\xi$$

Integration by parts together with formulas (6.24) gives

$$a_1(\tau) = a_0(\tau) + x_1(\tau) = A_0 \tau - \frac{A_0 \tau^3}{8\Omega_1^2} \left[\frac{(A_0 \tau)^4}{28} - \frac{(A_0 \tau)^2}{5} + \frac{1}{3} \right], \ldots$$
$$\Delta_1(\tau) = \Delta_0(\tau) + y_1(\tau) = e(\varphi) \left[-\frac{\pi}{2} + \frac{\tau}{\Omega_1} \left(-\frac{1}{2} + \frac{(A_0 \tau)^2}{4} \right) \right] \tag{6.26}$$

Fig. 6.4 Exact solution $a(\tau_1)$ (*solid*) and approximations $a_0(\tau_1)$ (*dash*) and $a_1(\tau_1)$ (*dash-dot*) for Eq. (6.17) with $f = 0.385$

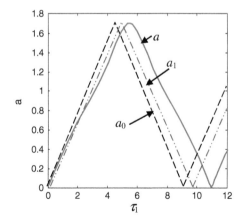

[a detailed derivation of (6.26) can be found in Manevitch et al. (2011a, b, c)]. Note that the solution (6.26) is constructed as a function of τ, but the inverse transformation $\tau \rightarrow \tau_1$ by formulas (6.20) automatically yields the solution periodic in τ_1 (Fig. 6.4). A more detailed consideration of this approach as well as an analysis of numerical solutions can be found in Manevitch et al. (2011a, b, c).

6.1.4 Analysis with Taking into Account the Energy Dissipation

We apply now the LPT concept to examine high-energy oscillations of a weakly damped oscillator. The system with parameters $f = 0.385$, $\gamma_1 = 0.05$ is considered as an example. In Fig. 6.5, one can observe two stages of motion: In the initial interval $0 \leq \tau_1 \leq \tau^*$, the trajectory is close to the LPT of the conservative system ($\gamma_1 = 0$), whereas in the second interval $\tau_1 > \tau^*$, motion is similar to quasi-linear oscillations and then converges to stationary oscillations of constant amplitude; an instant τ^* corresponds to the first maximum of the amplitude $a(\tau_1)$. Thus, the first part of the trajectory may be approximated by the previously obtained segment of the LPT for the non-dissipative system; the matching point is $a(\tau^*) = A$ at $\tau^* = T_0 = A/f$, with amplitude A calculated form (6.19).

In the second interval of motion, the trajectory of the dissipative system tends to the steady-state O: (a_0, Δ_0) as $\tau_1 \rightarrow \infty$. The point O: (a_0, Δ_0) is defined by the equality

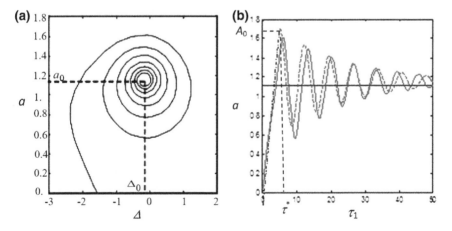

Fig. 6.5 Numerical results for the weakly dissipative nonlinear oscillator with parameters $f = 0.385$, $\gamma 1 = 0.05$: **a** phase portrait; **b** plots of a($\tau 1$): the exact (numerical) solution (*solid line*); segment (6.25) (*dotted-dashed line*); segment (6.27) (*dashed line*)

$$a^2 \left[(1-a)^2 + \gamma_1^2 \right] = f^2,$$

or, for sufficiently small γ_1,

$$\gamma_1 a_0 = f \sin \Delta_0, \quad a_0 - a_0^3 = -f \cos \Delta_0$$
$$\Delta_0 \approx -\gamma_1 a_0/f + O(\gamma_1^3), \quad a_0(1 - a_0^2) = -f + O(\gamma_1^2)$$

Let $\xi = a - a_0$, $\delta = \Delta - \Delta_0$ be deviations from the steady state satisfying the matching conditions $a_0 + \xi = A$, $d\xi/d\tau_1 = 0$ at $\tau_1 = T_0$.

Assuming negligible contributions of nonlinearity in small oscillations near O, the system linearized near O can be considered. The linearized equations are given by

$$\frac{d\xi}{d\tau_1} + \gamma_1 \xi + f\delta = 0, \quad \frac{d\delta}{d\tau_1} - k - \xi + \gamma_1 \delta = 0$$

where $k = 2a_0 + f/a_0^2$. This yields

$$\xi(\tau_1) = c_0 e^{-\gamma_1(\tau_1 - T_0)} \cos \kappa (\tau_1 - T_0),$$
$$\delta(\tau_1) = r c_0 e^{-\gamma_1(\tau_1 - T_0)} \sin \kappa (\tau_1 - T_0), \quad \tau_1 > T_0 \tag{6.27}$$

where $c_0 = A - a_0$, $\kappa = (fk)^{1/2}$, $r = \kappa/f$.

Figure 6.5 demonstrates a good agreement between the numerical solution of Eq. (6.9) (solid line) and its approximations. Despite a certain discrepancy in the initial interval, the numerical and analytical solutions closely approach as τ_1 increases. This simplest matched approximation is sufficient to describe complicated near-resonance dynamics. Note that the simplicity of the obtained solutions, which results from the attention to the physical properties of the system and effective treatment of the LPT theory, is contrasted with the daunting complexity of the traditional analysis of transient nonlinear processes.

6.2 Duffing Oscillator Subjected to Biharmonic Forcing Near the Primary Resonance

Analytical investigation of non-stationary processes in the Duffing oscillator subjected to biharmonic forcing under conditions of a primary resonance is carried out in this section. First, we employ the LPT methodology to investigate a non-dissipative system with a biharmonic excitation. We demonstrate that the presence of an additional harmonic with a slowly changing frequency entails recurrent transitions from one type of LPT to another one. Next, we investigate the occurrence of relaxation oscillations in a lightly damped system. It is also demonstrated that the mechanism of relaxations may be approximated and explained through the existence of the LPTs characterized by a strong energy exchanges between a single oscillator and an external source of energy. It is shown that the results of analytical approximations and numerical simulations are in a quite satisfactory agreement.

6.2.1 Equations of Fast and Slow Motion

We investigate dimensionless weakly nonlinear oscillator subjected to a biharmonic excitation in the neighborhood of 1:1 resonance. The equation of motion is given by

$$\frac{d^2u}{d\tau_0^2} + 2\varepsilon\gamma\frac{du}{d\tau} + u + 8\alpha\varepsilon u^3 = 2\varepsilon[\tilde{F}_1\sin(1+\varepsilon s_1)\tau_0 + \tilde{F}_2\sin(1+\varepsilon s_2)\tau_0] \quad (6.28)$$

where $s_2 = (1 + \varepsilon\sigma)s_1$. External forcing in this system consists of two distinct harmonic components with close frequencies. We will refer to this type of excitation as a biharmonic (quasi-periodic) one.

We recall that the maximum energy transfer from the source of energy to the oscillator happens if the oscillator is initially at rest, that is, $u = 0$, $v = du/d\tau_0 = 0$ at $\tau_0 = 0$. These initial conditions define the above-introduced LPT of the oscillator with biharmonic forcing.

To derive an analytical solution of Eq. (6.28), we invoke the complex-valued transformation $(u, v) \rightarrow (Y, Y^*)$ and then apply the multiple scales decomposition $Y = \varphi^{(0)} + \varepsilon\varphi^{(1)} + \cdots$. Taking into account that the excitation directly depends on the fast timescale τ_0, the slow timescale $\tau_1 = \varepsilon s\tau_0$, and the super-slow timescale $\tau_2 = \varepsilon\tau_1$, the leading-order slow term $\varphi^{(0)}$ and the expansion of the full time-derivative are presented as

$$\varphi^{(0)} = \varphi^{(0)}(\tau_1, \tau_2), \qquad \frac{d\varphi^{(0)}}{d\tau_1} = \frac{\partial\varphi^{(0)}}{\partial\tau_1} + \varepsilon\frac{\partial\varphi^{(0)}}{\partial\tau_2} \quad (6.29)$$

Therefore, the leading-order equation for φ_0 includes the slow and super-slow timescales:

$$\frac{d^2u}{d\tau_0^2} + 2\varepsilon\gamma\frac{du}{d\tau} + u + 8\alpha\varepsilon u^3 = 2\varepsilon[\tilde{F}_1\sin(1+\varepsilon s_1)\tau_0 + \tilde{F}_2\sin(1+\varepsilon s_2)\tau_0]$$

$$s\frac{\partial\varphi^{(0)}}{\partial\tau_1} + \gamma\varphi^{(0)} - 5i\alpha|\varphi^{(0)}|^2\varphi^{(0)} = -\tilde{F}_1 e^{i\tau_1} - i\tilde{F}_2 e^{i(\tau_1+\sigma\tau_2)}, \quad \varphi^{(0)}(0) = 0$$

$$(6.30)$$

Introducing rescaling of the variables and the parameters

$$\delta(\tau_2) = \sigma\tau_2, \quad \varphi^{(0)}(\tau_1) = \Lambda\varphi(\tau_1)e^{i\tau_1}, \quad \Lambda = \sqrt{s/3\alpha},$$
$$F_j = F_j/s\Lambda = F_j\sqrt{3\alpha/s^3}, \quad \gamma_1 = \gamma/s, \quad j = 1,2$$

we transform (6.30) into a more convenient form similar to (6.6)

$$\frac{\partial\phi}{\partial\tau_1} + \gamma_1\phi + i\phi(1 - |\phi|^2) = -i(F_1 + F_2\exp(i\delta(\tau_2))), \quad \phi(0) = 0. \qquad (6.31)$$

6.2.2 LPTs of Slow Motion in a Non-dissipative System

In this section, we consider a non-dissipative model with the "frozen" phase δ. The equation of motion is given by

$$\frac{\partial\phi}{\partial\tau_1} + i\phi(1 - |\phi|^2) = -i(F_1 + F_2\exp(i\delta)), \quad \varphi(0) = 0. \qquad (6.32)$$

After introducing a polar decomposition $\varphi = ae^{i\Delta}$, we transform (6.32) into the equations for the real-valued amplitude $a \geq 0$ and phase Δ

$$\frac{\partial a}{\partial\tau_1} = -F_1\sin\Delta + F_2\sin(\delta - \Delta) = 0, \quad a(0) = 0$$
$$a\frac{\partial\Delta}{\partial\tau_1} = -a + a^3 - F_1\cos\Delta - F_2\cos(\delta - \Delta) \qquad (6.33)$$

As in Sect. 6.1, we investigate the system dynamics through the analysis of LPTs. It is easy to prove that system (6.33) possesses the integral of motion (with respect to the timescale τ_1)

$$K = a\left(\frac{a^3}{4} - \frac{a}{2} - F_1\cos\Delta - F_2\cos(\delta - \Delta)\right), \qquad (6.34)$$

and $K = 0$ on the LPT. As in the time-invariant case, the LPT has two branches: the first branch is trivial ($a = 0$), while the second branch satisfies the cubic equation similar to (6.13)

$$g(a, \Delta) = \frac{1}{4}a^3 - \frac{1}{2}a - F_1\cos\Delta - F_2\cos(\delta - \Delta) = 0 \qquad (6.35)$$

Note that the above equation determines the phase $\Delta(a)$ on the LPT for any δ. Since the point $a(0) = 0$ belongs to the LPT, we obtain from (6.35) the following expression for the initial phase Δ_0 corresponding to a non-trivial branch of the LPT:

$$
\begin{aligned}
F_1 \cos \Delta_0 + F_2 \cos(\delta - \Delta_0) &= 0, \\
\Delta_0 &= -\arctan \frac{F_1 + F_2 \cos \delta}{g_2 \sin \delta} + n\pi.
\end{aligned}
\tag{6.36}
$$

In the next step, we study bifurcations of the slow motion. We begin with the analysis of the stationary (in τ_1) points of system (6.33). By letting $da/d\tau_1 = 0$, $d\Delta/d\tau_1 = 0$, one obtains the following algebraic equations for the stationary points (a_s, Δ_s):

$$
\begin{aligned}
F_1 \sin \Delta - F_2 \sin(\delta - \Delta) &= 0 \\
a - a^3 + f &= 0
\end{aligned}
\tag{6.37}
$$

where $f = \pm(F_1 \cos\Delta_s + F_2 \sin(\delta - \Delta_s))$. It follows from the first equation in (6.37) that the coordinates Δ_s of the stationary points in the phase plane are given by

$$
\Delta_s = \arctan \frac{F_2 \sin\delta}{F_1 + F_2 \cos\delta} + n\pi.
\tag{6.38}
$$

It is important to note that the second equation in (6.37) formally coincides with Eq. (6.14) and thus determines quasi-stationary points depending on the "frozen" phase δ. This implies that one can consider the critical values (6.15) and (6.16) of the parameter f as the boundaries separating different types of motion.

As in Sect. 6.1, we obtain two critical relationships determining locations of the centers and the shape of the phase orbits of system. We recall that the critical value $f_1 = \sqrt{2/27}$ characterizes the boundary between quasi-linear oscillations with relatively small amplitude ($|f| < f_1$) and moderately nonlinear oscillations with larger amplitude at $f_1 < |f| < f_2$. The threshold $f_2 = 2/\sqrt{27}$ corresponds to the transition from moderately nonlinear ($f_1 < |f| < f_2$) to strongly nonlinear ($|f| > f_2$) regimes with large amplitudes and energy.

It is worth mentioning that, in virtue of (6.37) and (6.38), the reduced forcing amplitude f directly depends on the parameters F_1, F_2, δ, and a proper choice of these parameters may provide zero forcing (in τ_1) such that $\partial a/\partial \tau_1 = 0$ in (6.33). In this case, the bifurcation analysis presented in this section becomes unacceptable, as it formally corresponds to an unforced response of the nonlinear oscillator.

6.2.3   Super-Slow Dynamics

Next, we consider the super-slow evolution of the LPT due to monotonous variations of the parameter $\delta(\tau_2)$. It was mentioned that in the system under investigation the parameter f is δ-dependent. Furthermore, since the phase $\delta(\tau_2)$ monotonously varies with respect to the super-slow timescale τ_2, all bifurcation parameters are

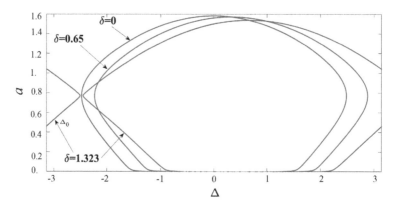

Fig. 6.6 Evolution of the LPT of the second type (strongly nonlinear oscillations) up to the transition to the LPT of the first type (moderately nonlinear oscillations)

also time-varying. This also means that global changes in the system dynamics may arise as the time increases. We show that at $\delta = 0$ there exists the LPT of the second type (strongly nonlinear high-energy oscillations with a single stable center in the phase plane) which, after a certain time interval, will bifurcate to the LPT of the first type (moderately nonlinear oscillations) due to the parametric switching from $|f(\delta(\tau_2))| > f_2$ to $|f(\delta(\tau_2))| < f_2$.

As shown in Fig. 6.6, the LPT starting at $\delta = 0$ undergoes global bifurcations. It is evident that, as the parameter δ increases, the LPT slowly changes, and the left and right corner points of the LPT meet at the saddle point at $\delta_{cr} = 1.99$. The coincidence of the corner points brings about the global bifurcation that finally results in the disappearance of strongly nonlinear regime and the transition to the LPT of moderately nonlinear regime (cf. Fig. 6.1).

The sequence of LPTs arising right after the bifurcation is illustrated in Fig. 6.7 for several values of δ. It is clearly seen that an increase in phase detuning $\delta(\tau_2)$ is equivalent to deviations of the excitation frequency from resonance, which, in turn, entails a decreasing amplitude and passage from one type of oscillations to another one.

One may observe in Fig. 6.7 that the aforementioned bifurcation leads to the transition from a strongly nonlinear regime to a moderately nonlinear one. The LPT gradually evolves with the variation of δ in time until it reaches a certain critical point beyond which another transition from the moderately to strongly nonlinear regime is observed. These transitions occur recurrently in large time intervals.

We now estimate analytically the critical values δ_{cr}, at which the transitions may occur. First, we use Eq. (6.35) to find the values of the phase Δ and the amplitude a_{cr}, at which $\partial\Delta/\partial a = 0$. The critical amplitude a_{cr} is defined as follows:

Fig. 6.7 Formation of the
LPTs of the first type and
their super-slow evolution till
the time point of additional
transition to the LPT of the
second type. The *numbers*
near curves show the values
of δ

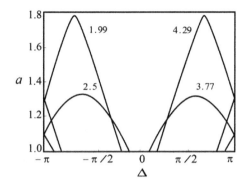

$$\frac{\partial g(a, \Delta(a))}{\partial a}\Big|_{\frac{\partial \Delta}{\partial a}=0} = \frac{3}{2}a^2 - 1 = 0, \quad a_{cr} = \sqrt{2/3}. \tag{6.39}$$

Substituting $a = a_{cr}$ into (6.35) yields:

$$2F_1\cos\Delta + 2F_2\cos(\delta - \Delta) = \frac{1}{2}a_{cr}^3 - a_{cr}. \tag{6.40}$$

Now, we find the critical value δ_{cr} at which two stationary points collide. To this
end, we derive the condition for Eq. (6.40) to have a single solution. This can be
achieved by equating the amplitude of the left-hand side "oscillating part" (with
respect to Δ) to that of the right-hand side. This implies that the critical value
$\delta = \delta_{cr}$ satisfies the following equations:

$$\delta_{cr} = \pm\arccos\frac{\left(\frac{a_{cr}^3}{2} - a_{cr}\right)^2 - 4\left(F_1^2 + F_2^2\right)}{8F_1F_2}, \quad a_{cr} = \sqrt{\frac{2}{3}} \tag{6.41}$$

6.2.4 Relaxation Oscillations in a Lightly Damped System

In this section, we demonstrate the possibility of relaxation oscillations in the
lightly damped system (6.31) and a connection between the trajectories of relaxation oscillations and the LPTs.

We recall that relaxation oscillations are characterized by the occurrence of
recurrent segments of fast and slow motion. In order to highlight the super-slow
motion, we rewrite Eq. (6.31) as a singular equation

$$\varepsilon\frac{\partial\phi}{\partial\tau_2} + \gamma_1\phi + i\phi(1 - |\phi|^2) = -i(F_1 + F_2\exp(i\delta(\tau_2))), \quad \phi(0) = 0. \tag{6.42}$$

The limit as $\varepsilon \to 0$ gives the following algebraic equation

$$i\Phi(1 - |\Phi|^2) + \gamma_1 \Phi = -i(F_1 + F_1 e^{i\sigma\tau_2}). \tag{6.43}$$

with trajectories depending on the super-slow time τ_2. It follows from (6.42), (6.43) that $\Phi(\tau_2) = \lim \varphi(\tau_1, \tau_2)$ as $\varepsilon \to \infty$. Thus, $\Phi(\tau_2)$ can be interpreted as a quasi-stationary value of the complex amplitude φ_0 with "frozen" parameter τ_2.

The transformation $|\Phi| = \sqrt{Z}$ reduces Eq. (6.43) to the real-valued form:

$$\gamma_1^2 Z + (1 - Z)^2 Z = Q(\tau_2), \tag{6.44}$$

where $Q(\tau_2) = F_1^2 + F_2^2 + 2F_1 F_2 \cos \sigma\tau_2$ represents the quadratic amplitude of excitation. The function $|\Phi| = \sqrt{Z}$ obviously evaluates the amplitude of oscillations $a = |\varphi_0|$ as $\tau_1 \to \infty$.

The plot of Φ versus Q is presented in Fig. 6.8. It is seen that the plot may be folded at certain parameters. This folded structure contains two stable branches and one unstable branch (Fig. 6.8).

It was shown in earlier works(Andronov et al. 1966; Gendelman et al. 2006, 2008; Guckenheimer et al. 2005, 2006; Starosvetsky and Gendelman, 2008a, b: 2, c: 3, 2009a, b: 2; Theocharis et al. 2010; Szmolyan and Wechselberger 2004) that relaxation oscillations are characterized by consequent jumps from one stable branch to another one, accompanied by a super-slow evolution on each of the stable branches. To find the conditions of jumps, expression (6.44) is differentiated with respect to Z. Equating the derivative to zero, we obtain the equation

$$3Z^2 - 4Z + \gamma_1^2 + 1 = 0 \tag{6.45}$$

Fig. 6.8 LPT of the second type slightly before the bifurcation occurs

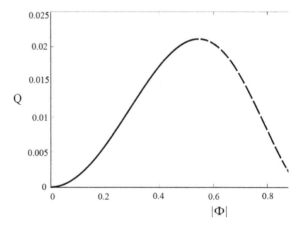

which determines extreme points. It is easy to deduce that a real-valued solution $Z > 0$ exists if $\gamma_1^2 < 1/3$. Therefore, above critical damping $\gamma_{cr} = \sqrt{1/3}$, relaxation is impossible.

It follows from Eq. (6.44) that motion on the stable branches is fully governed by the forcing amplitudes F_1, F_2. It is easy to deduce that if $F_1 \neq 0$, $F_2 = 0$, then there exist only stationary points on the stable branches of the manifold. If $F_1 \neq 0$, $F_2 \neq 0$, then there are three possibilities: The first two types of motion correspond to continuous oscillations on each of the stable branches; the third type corresponds to the aforementioned relaxation oscillations.

The conditions of the occurrence of relaxation oscillations can be found from (6.44) and (6.45). Solutions of Eq. (6.45) define the values of Z corresponding to the fold points:

$$Z_{1,2} = \frac{2}{3}\left(1 \pm \sqrt{1 - \frac{3}{4}(\gamma_1^2 + 1)}\right). \tag{6.46}$$

Substituting (6.46) into (6.44), we find the corresponding excitation amplitude Q

$$Q_j = \gamma_1^2 Z_j + (1 - Z_j)^2 Z_j, \quad j = 1, 2, \tag{6.47}$$

Now, one may formulate the following necessary conditions ensuring the regime of relaxation:

$$\begin{aligned} F_1^2 + F_1^2 + 2F_1F_2 > Q_1 \\ F_1^2 + F_1^2 - 2F_1F_2 < Q_2 \end{aligned} \tag{6.48}$$

The effect of relaxation is illustrated in Figs. 6.9 and 6.10. In Fig. 6.9, one can compare numerical results and analytical approximations for the amplitude of oscillations on the stable branches of the manifold provided conditions (6.48) are not fulfilled ($F_1 = F_2 = 0.065$).

As shown in Fig. 6.9, there exists an initial phase of relaxation oscillations with a super-slow change of the amplitude corresponding to the upper stable branch, and then, there is a single jump to the lower stable branch, wherein the system continues evolving. No additional relaxation oscillations are possible for this case, since conditions (6.48) are not fulfilled. If the forcing parameters satisfy (6.48), then there exist relaxation oscillations on the upper and lower stable branches (Fig. 6.10).

The super-slow flow analysis does not demonstrate a global dynamical picture of the lightly damped system. To make the analysis complete, we need to study the slow flow dynamics, which corresponds to the phase of relaxations from one stable branch of the super-slow manifold to another one followed by quasi-linear damped oscillations. We study the dynamics of a weakly damped oscillator, in which $0 < \gamma < \gamma_{cr}$. As in the previous sections, it is rather natural to assume that during the relaxation period T^* (the time required for the trajectory emanating from the fold of

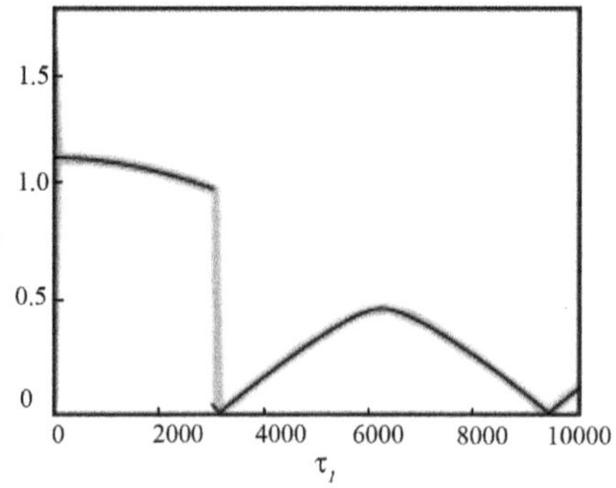

Fig. 6.9 Comparison of numerical and analytical results in the absence of the stable relaxation regime. *Gray line* Analytical approximation and *black line* Numerical simulation

Fig. 6.10 Comparison of numerical and analytical results for the case of relaxation oscillations. *Gray line* Analytical approximation and *black line* Numerical simulation

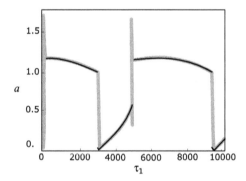

a super-slow surface to reach its first peak) the weakly damped trajectory runs sufficiently close to the non-dissipative one, and motion during this period is very close to motion along the corresponding LPT of the non-dissipative system.

In order to illustrate the correspondence of the LPT of strongly nonlinear oscillations to the initial phase of relaxation ($0 < t < T^*$), we plot the LPT corresponding to the value of δ at the point of relaxation on the phase plane of the damped system (Fig. 6.11).

It is evident from Fig. 6.11 that the damped trajectory corresponding to a jump from the lower branch of the super-slow manifold wraps up the LPT of the corresponding non-dissipative system, thus reaching the maximum value in the close vicinity of the LPT. This also means that the LPT of the non-dissipative system predicts fairly well the maximum amplitude of oscillations. It is also clear from Fig. 6.11 that after reaching the peak of the response, the amplitude of oscillations diminishes and the response becomes quasi-linear. As in Sect. 6.1, this response may be described with the help of the model linearized around the upper stable branch.

We now calculate the critical values of δ in the lower folds, corresponding to the minimal relaxation amplitude. It follows from (6.46) that the amplitude of the lower fold equals to:

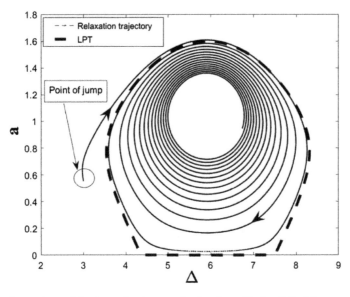

Fig. 6.11 Comparison of the damped trajectory of relaxation with the LPT of the corresponding Hamiltonian system

$$Z_1 = \frac{2}{3}\left(1 - \sqrt{1 - \frac{3}{4}(\gamma_1^2 + 1)}\right). \tag{6.49}$$

Thus, according to (6.44), one obtains

$$Q_1 = F_1{}^2 + F_2{}^2 + 2F_1F_2\cos\delta_{cr},$$
$$\delta_{cr} = \pm\arccos\left(\frac{Q_1 - F_1{}^2 - F_2{}^2}{2F_1F_2}\right). \tag{6.50}$$

The amplitudes of oscillations and the sequences of the LPTs in the corresponding undamped system at the points of jump are demonstrated in Fig. 6.12. As one may observe, the LPTs are reconstructed accordingly and provide fairly good estimations for jumps.

As mentioned above, at the initial stage of motion, wherein the effect of damping is negligibly small, the trajectory of a lightly damped system is close to the LPT of a corresponding non-dissipated system (Fig. 6.11), but at the second stage, after reaching the first peak of the relaxation phase with the coordinates $a(T^*) = a_m$, $\Delta(T^*) = \Delta_m$, motion of a lightly damped system may be described fairly well by linearizing (6.45) near the upper stable branch of the super-slow flow manifold. Thus, assuming small deviations near the upper stable branch, one may suggest the following approximation:

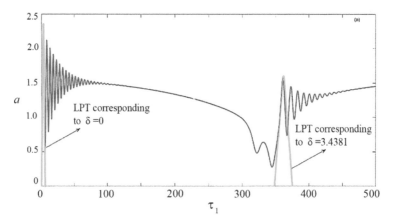

Fig. 6.12 Sequence of jumps of the LPT for relaxation oscillations; the actual response is marked with the *green solid line*

$$\varphi(\tau_1, \tau_2) = \Phi(\tau_2) + \phi(\tau_1, \tau_2) \tag{6.51}$$

Substituting (6.51) into (6.31), we obtain the following linearized equation:

$$\frac{\partial \varphi}{\partial \tau_1} + i(1 - 2|\Phi|^2)\varphi + \gamma\varphi - i\Phi^2\varphi = 0. \tag{6.52}$$

Equation (6.52) describes the last stage of damped oscillations near the stable attractor. Initial conditions are taken at the point $a(T^*) = a_m$, $\Delta(T^*) = \Delta_m$.

Comparison of analytical approximations with the results of numerical simulation is given in Fig. 6.13. It is seen that the interval of the slow timescale oscillations can be analytically approximated to a quite satisfactory extent.

We now return to the consideration of the lightly damped system. From Fig. 6.8, it is seen that the upper fold point is slightly distant from the horizontal axis. Arguing as above, one can state that the value of damping parameter dictates the distance of the upper fold from the horizontal axis. At the same time, for sufficiently light damping the critical parameter δ_{cr} (corresponding to a point of jump) may be related to the non-dissipative case of the LPT of the first type.

Thus, if the jump from the upper stable branch to the lower one occurs far away from the LPT of the first kind (which is most likely to happen when forcing is small enough and thus hardly affects the damped response of large amplitude), its trajectory may be roughly approximated by the equation of free oscillations

$$\frac{\partial \varphi}{\partial \tau} + i\varphi - i|\varphi|^2\varphi + \gamma\varphi \cong 0 \tag{6.53}$$

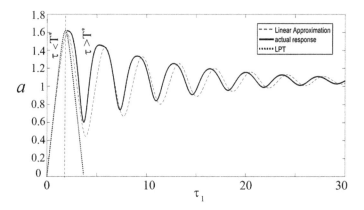

Fig. 6.13 Comparison of analytical approximations and numerical simulations

This immediately yields the exponential decay of the response amplitude:

$$|\varphi| = \sqrt{Z_2}\,\exp(-\gamma\tau) \tag{6.54}$$

We underline that Eq. (6.54) is valid only if the response amplitude far exceeds the amplitude of forcing. Therefore, if the trajectory of relaxation starts far away from the LPT of the first type, then the response is of simple exponential decay type. A comparison of the analytical approximation (6.54) with the actual (numerical) response is presented in Fig. 6.14. As shown in Fig. 6.14, relaxation oscillations may be approximated by the response of the damped unforced weakly nonlinear oscillator.

6.3   Super-Harmonic Resonance

In this section, we analyze energy exchange in a system subject to super-harmonic resonance. We show that the energy imparted in the system is partitioned among the principal and super-harmonic modes, but energy exchange can arise due to super-harmonic oscillations. Using the LPT concept, we construct approximate analytical solutions describing intense irreversible energy transfer in a harmonically excited Duffing oscillator. Numerical simulations confirm the accuracy of the analytic approximations.

6.3.1   Equations of Motion

We consider the equation similar to (6.1)

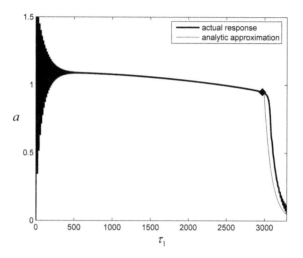

Fig. 6.14 Comparison of an analytical approximation of relaxation with the result of numerical simulation. Actual system response is denoted with the *bold solid line*; analytical approximation is denoted with *thin line*

$$\frac{d^2y}{dt^2} + 2\varepsilon\gamma\frac{dy}{dt} + y + 8\varepsilon\alpha y^3 = 2F\,\sin\left(\frac{1}{3} + \varepsilon s\right)t \tag{6.55}$$

with initial conditions $y = 0$; $dy/dt = 0$ at $t = 0$. The solution of (6.55) can be represented as the sum $y = y_0 + u$, in which y_0 is defined as a partial solution of the linear equation

$$\frac{d^2y_0}{dt^2} + y_0 = 2F\,\sin\omega_\varepsilon\tau_0 \atop y_0 = \Lambda\,\sin\omega_\varepsilon\tau_0, \quad \Lambda = 9F/4, \quad \omega_\varepsilon = 1/3 + \varepsilon s \tag{6.56}$$

From (6.55) and (6.56) we obtain the equation for the super-harmonic component

$$\frac{d^2u}{dt^2} + u + 2\varepsilon\gamma\frac{du}{dt} + 8\varepsilon\alpha(u + y_0)^3 = 0, \tag{6.57}$$

with initial conditions $u = 0$; $v = du/dt = -\Lambda/3$ at $t = 0$. In Eq. (6.57), principal harmonics of frequency $1/3$ are of $O(\varepsilon)$; this implies that the generating solution of (6.57) includes only super-harmonic components of frequency 1.

The non-stationary dynamics of system (6.57) is analyzed with the help of the multiple scales procedure demonstrated in Sect. 6.1. The change of variables (6.2) transforms Eq. (6.57) into the complex-valued equation

$$\frac{d\psi}{dt} - i\psi + \varepsilon\gamma(\psi + \psi^*) + i\varepsilon\alpha[(\psi - \psi^*) + \Lambda(e^{i\omega_\varepsilon\tau_0} - e^{-i\omega_\varepsilon\tau_0})]^3 = 0,$$
$$\psi(0) = -\Lambda/3 \tag{6.58}$$

As in Sect. 6.1, the approximate solution of (6.58) is sought by the multiple scales method with further selection of resonance terms. Introducing the representation

$$\psi(t, \varepsilon) = \psi_0(\tau_0, \tau_1) + \varepsilon\psi_1(\tau_0, \tau_1) + \cdots$$
$$\frac{d}{d\tau_0} = \frac{\partial}{\partial\tau_0} + 3s\varepsilon\frac{\partial}{\partial\tau_1},$$

with fast and slow timescales $\tau_0 = t$ and $\tau_1 = 3\varepsilon st$, respectively, and then reproducing necessary transformations of Sect. 6.1, one obtain the leading-order solution (6.5) with the slow envelope $\varphi(\tau_1)$ satisfying the equation

$$\frac{\partial\varphi}{\partial\tau_1} + \gamma_1\varphi - i\alpha[|\varphi|^2\varphi + 2\Lambda^2\varphi_0 - \frac{1}{3}\Lambda^3 e^{i\tau_1}] = 0, \quad \varphi(0) = -\frac{\Lambda}{3} \tag{6.59}$$

where $\gamma_1 = \gamma/3\ s$, $\alpha_1 = \alpha/s$. The transformation $\varphi(\tau_1) = \Lambda\varphi_0(\tau_1)e^{i\tau_1}$ reduces (6.37) to the two-parameter form

$$\frac{\partial\varphi_0}{\partial\tau_1} + \gamma_1\varphi_0 + i[(1 - 2\lambda^2) - \lambda^2|\varphi_0|^2]\varphi_0 = \frac{1}{3}\lambda^2, \quad \varphi_0(0) = -\frac{1}{3} \tag{6.60}$$

where $\lambda^2 = \alpha_1\Lambda^2 = 81\alpha F^2/16s$. Finally, the representation $\varphi_0 = ae^{i\Delta}$ transforms Eq. (6.60) into the system

$$\frac{da}{d\tau_1} = -\gamma a - \beta\sin\Delta$$
$$a\frac{d\Delta}{d\tau_1} = a(-\zeta + 3\beta\alpha^2) - \beta\cos\Delta \tag{6.61}$$

where $\beta = \lambda^2/3$, $\zeta = 1 - 2\lambda^2$; initial conditions are given by $a(0) = 1/3$, $\Delta(0) = -\pi$. System (6.61) may be analyzed in the same way as (6.10). Note that the asymptotic representations of the solutions for Eqs. (6.55) and (6.57) are given by

$$u(\tau_0, \varepsilon) = \Lambda a(\tau_1)\sin(\tau_0 + \Delta(\tau_1) + \tau_1) + O(\varepsilon)$$
$$y(\tau_0, \varepsilon) = \Lambda[a(\tau_1)\sin(\tau_0 + \Delta(\tau_1) + \tau_1) + \sin 1/3(\tau_0 + \tau_1)] + O(\varepsilon) \tag{6.62}$$

6.3.2 Super-Harmonic Resonance in the Non-dissipative System

Now, we consider the non-dissipative oscillator with $\gamma = 0$, namely

$$
\frac{da}{d\tau_1} = -\beta \sin \Delta
$$
$$
a\frac{d\Delta}{d\tau_1} = a(-\zeta + 3\beta\alpha^2) - \beta \cos \Delta
$$
(6.63)

The dynamical analysis is performed using the procedures developed in Sect. 6.2. First, we find stationary points of (6.63) from the equations

$$
a(-\zeta + 3\beta\alpha^2) = \beta \cos \Delta_i, \quad \Delta_1 = 0, \quad \Delta_2 = -\pi
$$

Respective stationary points are denoted as C_+ and C_-. The phase portraits corresponding to different values of the parameter $\beta(\lambda)$ are similar to the ones shown in Fig. 6.11. Using previous arguments, we obtain the following expressions for the discriminants and critical parameters, which determine the boundaries between quasi-linear, moderately nonlinear, and strongly nonlinear systems:

$$
D_1 = \beta^2 - \frac{2\zeta^3}{81\beta} = 0, \quad \lambda^2 - \sqrt[3]{\frac{2}{3}}(1 - 2\lambda^2) = 0, \quad \lambda_1 = 0.564
$$
$$
D_1 = \frac{\beta^2}{4} - \frac{\zeta^3}{81\beta} = 0, \quad \lambda^2 - \sqrt[3]{\frac{4}{3}}(1 - 2\lambda^2) = 0, \quad \lambda_1 = 0.586
$$
(6.64)

As in the case of 1:1 resonance, the LPT of small oscillations ($\lambda < \lambda_1$) encircles the stable center in the axis $\Delta = -\pi$; if $\lambda_1 < \lambda < \lambda_2$, the LPT of large oscillations encircles the stable center in the axis $\Delta = 0$; if $\lambda > \lambda_2$, then there exists a single stable center in the axis $\Delta = 0$. Inequalities

$$
\lambda < \lambda_1, \quad \lambda_1 < \lambda < \lambda_2, \quad \lambda > \lambda_2
$$
(6.65)

determine, respectively, *quasi-linear, moderately nonlinear,* and *strongly nonlinear* oscillations.

We now examine *strongly nonlinear* oscillations ($\lambda > \lambda_2$). We recall that system (6.63) conserves the integral of motion

$$
h = -a(\zeta a + 3\beta a^3/2 - 2\beta \cos \Delta)
$$
(6.66)

identifying the phase trajectories in the plane (a, Δ).

Considering initial conditions $a = 1/3$, $\Delta = -\pi$, we obtain

$$h = \frac{1}{3}\left(-\frac{\zeta}{3} + \frac{37\beta}{18}\right). \tag{6.67}$$

Taking into account expressions (6.66) and (6.67), we exclude the variable Δ and thus reduce (6.63) to a single second-order equation. From (6.66) and (6.67), we obtain

$$\cos\Delta(a) = \frac{1}{2\beta}\left(\frac{3\beta a^3}{2} - \zeta a - \frac{h}{a}\right), \quad \frac{d\Delta}{d\tau_1} = \omega(a),$$

$$\omega(a) = \frac{1}{2}\left(\frac{9\beta a^2}{2} - \zeta\right) + \frac{h}{2a^2} \tag{6.68}$$

and therefore, system (6.63) can be rewritten as the second-order equation

$$\frac{d^2 a}{d\tau_1^2} + u(a) = 0, \quad \tau_1 = 0: \quad a = \frac{1}{3}; \quad \frac{da}{dt} = v = 0 \tag{6.69}$$

where

$$u(a) = \omega(a)\beta\cos\Delta = \frac{a}{4}\left[\left(\frac{3}{2}\beta a^2 - \zeta\right)\left(\frac{9}{2}\beta a^2 - \zeta\right) + 3\beta h - \frac{h^2}{a^4}\right].$$

Initial conditions are given by $a = 1/3$, $v = 0$ at $\tau_1 = 0$. Equation (6.69) corresponds to the conservative oscillator with the potential

$$U(a) = \int_0^a u(x)dx = \frac{a^2}{8}\left(\frac{3\beta a^2}{2} - \zeta^2\right) + \frac{3}{8}\beta h a^2 - \frac{h}{8a^2} \tag{6.70}$$

Note that oscillator (6.70) preserves the energy $E = v^2/2 + U(a) = E_0$, where $E_0 = U(1/3)$. Thus, the amplitude of oscillations A is defined by the condition $U(A) = E_0$, $v = 0$; the half-period of oscillations $T(A)$ is given by (cf. 6.18)

$$T(A) = \int_0^A \frac{da}{\sqrt{E_0 - 2U(a)}}. \tag{6.71}$$

Exact solutions of Eqs. (6.63) and (6.68) cannot be found in closed form. We compare an exact (numerical) solution with explicit analytical approximations obtained by the following iterative procedure:

$$\frac{da_{i+1}}{d\tau_1} = -\beta \sin \Delta_i; \quad \frac{d\Delta_i}{d\tau_1} = \omega(a_i), \quad i = 0, 1$$

$$a_i(0) = 1/3, \quad \Delta_i = -\pi \tag{6.72}$$

$$a_0 = a(0) = 1/3, \quad \omega_0 = \omega(a_0)$$

This implies that

$$\frac{da_1}{d\tau_1} = -\beta \sin \Delta_0, \quad \frac{d\Delta_0}{d\tau_1} = \omega_0,$$

$$a_1(0) = 1/3, \quad \Delta_0(0) = -\pi \tag{6.73}$$

It follows from (6.72) and (6.73) that the leading-order approximation takes the form

$$a_1(\tau_1) = 1/3 + (\beta/\omega_0)[1 - \cos(\omega_0\tau_1)], \quad \Delta_0(\tau_1) = -\pi + \omega_0\tau_1 \tag{6.74}$$

We take the parameters of numerical simulation:

$$F = 1, \quad \alpha = 0.27, \quad s = 1/3, \quad \varepsilon = 0.007, \quad \gamma = 0 \quad \text{or} \quad \gamma = 0.2 \tag{6.75}$$

corresponding to $\Lambda = 9/4$, $\lambda = 0.9 > \lambda_2$, $\beta = 1.35$, $\zeta = 0.3$. It is easy to find that $\omega_0 \approx 3.49$, $T_0 = \pi/\omega_0 \approx 0.9$, and $A = a_0(T_0) \approx 1.1$. As shown in Fig. 6.15, the discrepancy between analytical solution and numerical simulation does not exceed 10%.

Figure 6.16 compares numerical (exact) results for the non-dissipated Duffing Eq. (6.55) with analytical approximations (6.62) and (6.74).

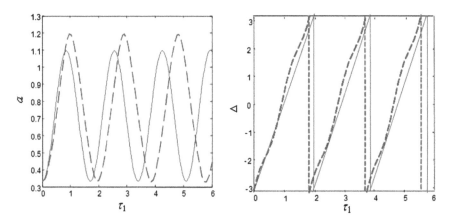

Fig. 6.15 Temporal behavior of $a(\tau_1)$ (*left*); principal value of $\Delta(\tau_i)$ in the interval $(-\pi, \pi]$ (*right*): *dash* Numerical solutions of (6.41) and *solid* Analytic approximations (6.52)

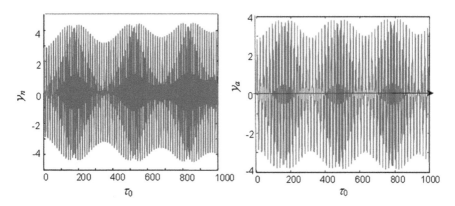

Fig. 6.16 Numerical solution of Eq. (6.55) at $\gamma = 0$ (*left*) and its analytical approximation (*right*) in the fast timescale

Figure 6.16 demonstrates a fairly good agreement of the exact (numerical) solution and its explicit analytical approximation, despite slight irregularity of high-frequency components in the right plot.

Transient dynamics of the dissipative oscillator

Figure 6.17 depicts the typical dynamics of a strongly nonlinear oscillator with weak dissipation ($\gamma = 0.2$, $\varepsilon = 0.007$). Note that the plots in Figs. 6.16 and 6.17 agree with the underlying assumption: Motion of the damped system is similar to motion of the undamped system up to an instant of the first maximum of the slow envelope; then, the damped motion approaches stationary oscillations independent of initial conditions. This assumption motivates the approximation procedure.

We denote by $a\gamma(\tau_1)$ the solution of Eq. (6.61) with initial conditions (6.16); by $a\gamma(\tau_1*)$, its first maximum at $\tau_1 = \tau_1*$ (Fig. 6.18); and by $a(\tau_1)$, the solution of (6.63) in the absence of damping ($\gamma = 0$). As mentioned above, $a\gamma(\tau_1)$ is partitioned into two segments: On the interval $[0, \tau_1*]$, $a\gamma(\tau_1)$ is considered as being close to the solution $a(\tau_1)$ of the undamped system (6.63), while on the interval $\tau_1 \geq \tau_1*$, the solution $a\gamma(\tau_1)$ is similar to smooth decaying oscillations of the dissipative system.

If $\gamma \neq 0$, the steady-state O: $(a\gamma^0, \Delta\gamma^0)$ is determined by the equality

$$a^2 \left[\left(\zeta - 3\beta a^2 \right)^2 + \gamma^2 \right] = \beta^2. \tag{6.76}$$

or, for sufficiently small γ,

$$\gamma a_\gamma^0 = -\beta \sin \Delta_\gamma^0, \quad \zeta a_\gamma^0 - 3\beta (a_\gamma^0)^3 = -\beta \cos \Delta_\gamma^0$$
$$\Delta_\gamma^0 \approx -\gamma a_\gamma^0 / \beta + O(\gamma^3), \quad a_\gamma^0 [\zeta - 3\beta (a_\gamma^0)^2] = -\beta + O(\gamma^2) \tag{6.77}$$

Fig. 6.17 Numerical solution
of (6.55); $\gamma = 0.2$

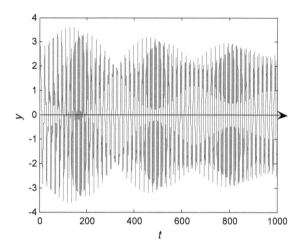

Fig. 6.18 Numerical and
analytical solutions: *solid
line* Numerical solution of
(6.61); *dotted line* Segment
(6.74); and *starred
line* segment (6.80); $\gamma = 0.2$

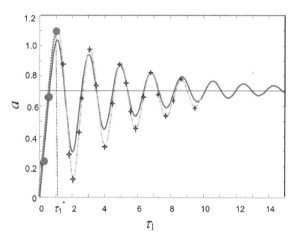

For the system with parameters (6.75), we have $a\gamma^0 \approx 0.73$ and $\Delta\gamma^0 \approx -0.54$. In
addition, we note that the contribution of nonlinear force in oscillations near O is
relatively small. Under this assumption, in the interval $\tau_1 \geq \tau_1{}^*$ one can consider
the system linearized near the steady-state $a\gamma^0$, $\Delta\gamma^0$, namely

$$\frac{d\xi}{d\tau_1} + \beta\eta = -\gamma\xi, \quad \frac{d\eta}{d\tau_1} - \frac{k_1}{a_\gamma^0}\xi = -\gamma\eta, \tag{6.78}$$

where $\xi = a\gamma - a\gamma^0$, $\eta = \Delta\gamma - \Delta\gamma^0$, and $k_1 = 9\beta(a\gamma^0)^2 - \zeta$. The matching conditions
at $\tau_1 = \tau_1{}^*$ are given by

$$a_\gamma^0 + \xi = a(\tau_1^*) = A_0, \qquad \frac{d\xi}{d\tau_1} = 0 \qquad (6.79)$$

If $k_1 > 0$, then we obtain from (6.78)

$$
\begin{aligned}
\xi(\tau_1) &= c_0 e^{-\gamma(\tau_1 - \tau_1^*)} \cos \kappa(\tau_1 - \tau_1^*), \\
\eta(\tau_1) &= r c_0 e^{-\gamma(\tau_1 - \tau_1^*)} \sin \kappa(\tau_1 - \tau_1^*), \quad \tau_1 - \tau_1^* > 0
\end{aligned}
\qquad (6.80)
$$

where $c_0 = A_0 - a\gamma^0$, $\kappa^2 = \Phi k_1 / a\gamma^0 > 0$, and $r = \kappa/F$. Figure 6.18 demonstrates a good agreement between the numerical solution of Eq. (6.61) (solid line) and approximations found from matching segment (6.74) (dot line) with solution (6.80) of the linearized systems (dot-dash line) at the point τ_1^*. Despite a certain discrepancy in the initial interval of motion, the numerical and analytical solutions approach closely to the steady-state $a\gamma^*$ as τ_1 increases. This implies that a simplified model (6.74) and (6.80) suffices to describe the complicated resonance dynamics.

References

Andronov, A.A., Vitt, A.A., Khaikin, S.E.: Theory of Oscillators. Dover, New York (1966)

Gendelman, O.V., Gourdon, E., Lamarque, C.H.: Quasiperiodic energy pumping in coupled oscillators under periodic forcing. J. Sound Vib. **294**, 651–662 (2006)

Gendeman, O.V., Starosvetsky, Y., Feldman, M.: Attractors of harmonically forced linear oscillator with attached nonlinear energy sink I: description of response regimes. Nonlinear Dyn. **51**, 31–46 (2008)

Guckenheimer, J., Hoffman, K., Weckesser, W.: Bifurcations of relaxation oscillations near folded saddles. Int. J. Bifurcat. Chaos **15**, 3411 (2005)

Guckenheimer, J., Wechselberger, M., Young, L.S.: Chaotic attractors of relaxation oscillators. Nonlinearity **19**, 701–720 (2006)

Korn, G.A., Korn, T.M.: Mathematical handbook for scientists and engineers, 2nd edn. Dover Publications, New York (2000)

Kovaleva, A., Manevitch, L.I.: Resonance energy transport and exchange in oscillator arrays. Phys. Rev. E **88**, 022904 (2013)

Manevitch, L.I.: New approach to beating phenomenon in coupling nonlinear oscillatory chains. Arch. Appl. Mech. **77**, 301–312 (2007)

Manevitch, L.I., Musienko, A.I.: Limiting phase trajectories and energy exchange between anharmonic oscillator and external force, Nonlinear Dyn. **58**, 633–642 (2009)

Manevitch, L.I., Kovaleva, A., Manevitch, E.L., Shepelev, D.S.: Limiting phase trajectories and non-stationary resonance oscillations of the duffing oscillator, part 1. A non-dissipative oscillator. Commun. Nonlinear Sci. Numer. Simul. **16**, 1089–1097 (2011a)

Manevitch, L.I., Kovaleva, A., Manevitch, E.L., Shepelev, D.S.: Limiting phase trajectories and non-stationary resonance oscillations of the Duffing oscillator. part 2. A dissipative oscillator. Commun. Nonlinear Sci. Numer. Simulat. **16**, 1098–1105 (2011b)

Manevitch, L.I., Kovaleva, A., Shepelev, D.S.: Non-smooth approximations of the limiting phase trajectories for the Duffing oscillator near 1:1 resonance. Physica D **240**, 1–12 (2011c)

Mirkina (Kovaleva), A.S.: On a modification of the averaging method and estimates of higher approximations. J. Appl. Math. Mech. **41**, 901–909 (1977)

Nayfeh, A.H., Mook, D.T.: Nonlinear Oscillations. Wiley-VCH, Weinheim (2004)

Neishtadt, A.I.: Passage through a separatrix in a resonance problem with slowly varying parameter. J. Appl. Math. Mech. **39**, 594–605 (1975)

Pilipchuk, V.N.: Nonlinear dynamics: between linear and impact limits. Springer, Berlin (2010)

Raghavan, S., Smerzi, A., Fantoni, S., Shenoy, S.R.: Coherent oscillations between two weakly coupled Bose-Einstein condensates: Josephson effects, π-oscillations, and macroscopic quantum self-trapping. Phys. Rev. A **59**, 620–633 (1999)

Sagdeev, R.Z., Usikov, D.A., Zaslavsky, G.M.: Nonlinear physics: from the pendulum to turbulence and chaos. Harwood Academic Publishers, New York (1988)

Starosvetsky, Y., Gendelman, O.V.: Strongly modulated response in forced 2DOF oscillatory system with essential mass and potential asymmetry. Physica D **237**, 1719–1733 (2008a)

Starosvetsky, Y., Gendelman, O.V.: Attractors of harmonically forced linear oscillator with attached nonlinear energy sink II: optimization of a nonlinear vibration absorber. Nonlinear Dynam **51**, 47–57 (2008b)

Starosvetsky, Y., Gendelman, O.V.: Dynamics of a strongly nonlinear vibration absorber coupled to a harmonically excited two-degree-of-freedom system. J. Sound Vibr **312**, 234–256 (2008c)

Starosvetsky, Y., Gendelman, O.V.: Interaction of nonlinear energy sink with a two degrees of freedom linear system: Internal resonance. J. Sound Vibr **329**, 1836–1852 (2009a)

Starosvetsky, Y., Gendelman, O.V.: Vibration absorption in systems with a nonlinear energy sink: nonlinear damping. J. Sound Vibr **324**, 916–939 (2009b)

Szmolyan, P., Wechselberger, M.: Relaxation oscillations in R^3J. Diff. Eqns **200**, 69–104 (2004)

Theocharis, G., Boechler, N., Kevrekidis, P.G., Job, S., Porter, M.A., Daraio, C.: Intrinsic energy localization through discrete gap breathers in one-dimensional diatomic granular crystals. Phys. Rev. E **820 56604** (1–11) (2010)

Vakakis, A.F., Manevitch, L.I., Mikhlin, Yu.V., Pilipchuk, V.N., Zevin, A.A.: Normal Modes and Localization in Nonlinear Systems. Wiley, New York (1996)

Vakakis, A.F., Gendelman, O., Bergman, L.A., McFarland, D.M., Kerschen, G., Lee, Y.S.: Nonlinear targeted energy transfer in mechanical and structural systems. Springer, Berlin (2008)

Chapter 7
Non-conventional Synchronization of Weakly Coupled Active Oscillators

In this chapter, we describe a new type of self-sustained oscillations associated with the phenomenon of synchronization. Conventional studies of synchronization in the model of two weakly coupled Van der Pol oscillators considered their synchronization in the regimes close to nonlinear normal modes (NNMs). In Kovaleva et al. (2013) and Manevitch et al. (2013). It was shown for the first time that in the important case of the hard excitation, an alternative synchronization mechanism can emerge even if the conventional synchronization becomes impossible. We identify this mechanism as an appearance of the dynamic attractor with the complete periodic energy exchange between the oscillators and then show that it can be interpreted as a dissipative analogue of highly intensive beats in a conservative system. This type of motion is therefore opposite to the NNM-type synchronization in which, by definition, no energy exchange can occur. The analytical study is based on the LPT concept but, in contrast to the conservative systems, in the present case the LPT can be regarded as an attractor. Finally, it is shown that within the LPT approach, the localized mode represents an attractor in the range of model parameters wherein the LPT as well as the in-phase and anti-phase NNMs become unstable.

The chain of coupled dissipative oscillators described by the Van der Pol (VdP) or Van der Pol-Duffing (VdP-D) equations represents one of the fundamental models of nonlinear dynamics (Pikovsky et al. 2001; Verhulst 2005) with numerous applications in different fields of physics and biophysics (Chakraborty and Rand 1988; Landa 1989; Pikovsky et al. 2001; Newel and Nazarenko 2001; Rompala et al. 2007). In the continuum limit, it can be reduced to the complex Ginsburg–Landau equation (Akhmediev and Ankiewicz 1997; Malomed 1994; Mihalache et al. 2008), which admits periodic or localized solutions in several significant cases (Landa 1996; Malomed 1994; Mihalache et al. 2008; Pikovsky et al. 2001). The simplest discrete model that combines two nonlinear dissipative oscillators was considered in series of books and papers. The main attention was paid, as a rule, to synchronization of the oscillators in the regimes close to NNMs (Pikovsky et al. 2001; Rompala and Rand 2007; Newel and Nazarenko 2001; Vakakis et al. 1996;

© Springer Nature Singapore Pte Ltd. 2018 187
L.I. Manevitch et al., *Nonstationary Resonant Dynamics of Oscillatory Chains and Nanostructures*, Foundations of Engineering Mechanics,
DOI 10.1007/978-981-10-4666-7_7

Landa 1996; Chakraborty and Rand 1988). We have noted above that, in the conservative case, another type of motion characterized by the complete energy exchange between oscillators can play the fundamental role. This dynamical regime, opposite to NNMs by its physical meaning, was defined as the limiting phase trajectory (LPT). This chapter analytically demonstrates that the LPT corresponding to the complete energy exchange between weakly coupled oscillators appears to be an attractor of the alternative to this one of the NNM types. Revealing a new class of attractors may serve as the starting point for deeper understanding and further applications of the synchronization phenomenon.

7.1 Main Equations

We begin with a system of two linearly coupled Lienard oscillators, separately described by the equation

$$\frac{d^2u}{dt^2} + f(u)\frac{du}{dt} + g(u) = 0 \qquad (7.1)$$

It is known that such an oscillator possesses stable periodic attractors (limit cycles), whenever functions f and g satisfy the conditions of the well-known Poincare–Bendixson theorem [see, e.g., Landa (1996)]. Usually, f is an even whereas g is an odd analytical function represented in the form of the truncated power series

$$f(u) = a_0 + a_2u^2 + a_4u^4 + O(u^6); \;\; g(u) = b_1u + b_3u^3 + O(u^5) \qquad (7.2)$$

The signs of the coefficients are responsible for the qualitative dynamical behaviors of the model. If $a_0 < 0$, $a_2 > 0$, $a_4 = 0$, and $b_1 > 0$, the oscillator takes the VdP-D form. In this case, the equilibrium point ($u = 0$, $du/dt = 0$) represents an unstable focus. As a result, the energy flows into the system near the equilibrium for any small magnitude of initial perturbations. However, it would be physically reasonable to assume the existence of a threshold of the initial excitation above which the global dynamical process can be triggered. Within the above expansion for $f(u)$, the only way to introducing the threshold is to choose the coefficient as follows: $a_0 > 0$, $a_2 < 0$, $a_4 > 0$. This choice provides local stability of the equilibrium point so that the energy in-flow into the system is triggered only above certain nonzero level of excitation, when $f(u)$ becomes negative (Landa 1996).

7.1.1 Coupled Active Oscillators

Our principal goal is to demonstrate that the presence of the energy threshold allows revealing a new type of synchronization in the system of two weakly coupled active oscillators. We denote the numerical and scaling factors as follows: $a_0 = 2\varepsilon\gamma$, $a_2 = -8\varepsilon g$, $a_4 = 16\varepsilon d$, $b_1 = 1$, $b_2 = 8\varepsilon\alpha$, and the strength of coupling $2\varepsilon\beta$ where $0 < \varepsilon \ll 1$ is a small parameter of the system. As a result, the model under consideration takes the form

$$\frac{d^2 u_1}{d\tau_o^2} + u_1 + 2\beta\varepsilon(u_1 - u_2) + 8\varepsilon\alpha u_1^3 + 2\varepsilon(\gamma - 4g u_1^2 + 8d u_1^4)\frac{du_1}{d\tau_0} = 0$$
$$\frac{d^2 u_2}{d\tau_o^2} + u_2 + 2\beta\varepsilon(u_2 - u_1) + 8\varepsilon\alpha u_2^3 + 2\varepsilon(\gamma - 4g u_2^2 + 8d u_2^4)\frac{du_2}{d\tau_0} = 0 \tag{7.3}$$

As in Sect. 2.1, we introduce the change of variables $Y_r = (v_r + iu_r)e^{-i\tau_0}$, $r = 1$, 2, such that $Y_r(\tau_0, \tau_1, \varepsilon) = \varphi_r^{(0)}(\tau_1) + \varepsilon\varphi_r^{(0)}(\tau_0, \tau_1) + O(\varepsilon^2)$, $\tau_1 = \varepsilon t$ and then conclude that the leading order slow terms $\varphi_r^{(0)}(\tau_1)$ can be presented in the form (1.4): $\varphi_1^{(0)}(\tau_1) = a(\tau_1)\exp(i\beta\tau_1)$, $\varphi_2^{(0)}(\tau_1) = b(\tau_1)\exp(i\beta\tau_1)$ where the slow complex amplitudes a, b are given by

$$\frac{da}{d\tau_1} - 3ia|a|^2 a + \left(\gamma - g|a|^2 + d|a|^4\right)a + i\beta b = 0,$$
$$\frac{db}{d\tau_1} - 3ia|b|^2 b + \left(\gamma - g|b|^2 + d|b|^4\right)b + i\beta a = 0. \tag{7.4}$$

Finally, rewriting a, b in the polar form

$$a = R_1 \exp(i\delta_1), \quad b = R_2 \exp(i\delta_2) \tag{7.5}$$

and considering the phase difference $\Delta = \delta_1 - \delta_2$ as a new variable, we derive the following system of real-valued equations:

$$\frac{dR_1}{d\tau_1} + \gamma R_1 - g R_1^3 + d R_1^5 + \beta R_2 \sin \Delta = 0,$$
$$\frac{dR_2}{d\tau_1} + \gamma R_2 - g R_2^3 + d R_2^5 - \beta R_1 \sin \Delta = 0,$$
$$R_1 R_2 \frac{d\Delta}{d\tau_1} + 3\alpha R_1 R_2 (R_2^2 - R_1^2) + \beta(R_2^2 - R_1^2)\cos \Delta = 0. \tag{7.6}$$

7.2 NNMs and LPTs Symmetries

System (7.6) is non-integrable, but it possesses the discrete symmetry, that is, it preserves its form under the following coordinate replacement:

$$
\begin{array}{lll}
\text{(a)} & R_1 \rightarrow R_2;\ R_2 \rightarrow R_2; & \Delta \rightarrow \Delta \\
\text{(b)} & R_1 \rightarrow -R_2;\ \ R_2 \rightarrow -R_1; & \Delta \rightarrow \Delta
\end{array}
\tag{7.7}
$$

The symmetries (a) and (b) provide the existence of the in-phase ($R_1 = R_2, \Delta = 0$) and the out-of-phase ($R_1 = R_2$; $\Delta = \pi$) NNMs, respectively. In a general case, Eqs. (7.6) do not explicitly reveal any other simple symmetry, discrete or continuous, except of the temporal translation. However, if any non-trivial continuous symmetry exists under certain conditions, it can be found in the framework of the Lie group theory (Ovsyannikov 1982), by manipulating the infinitesimal differential operator of the dynamical system (7.6), $X = X_0 + X_1$ where

$$
X_0 = \xi(R_1, R_2, \Delta)\frac{\partial}{\partial \tau_1} + \eta(R_1, R_2, \Delta)\frac{\partial}{\partial R_1} + \zeta(R_1, R_2, \Delta)\frac{\partial}{\partial R_2} + \varsigma(R_1, R_2, \Delta)\frac{\partial}{\partial \Delta},
$$

and X_1 is the first continuation of the operator X_0, whose components are given by time derivatives in system (5.6). Following the technique of Ovsyannikov (1982) and considering the partial differential equations for the components of operator X, we reveal the existence of the rotation group in the plane (R_1, R_2) with the invariant $I = N = R_1^2 + R_2^2$ under certain additional conditions. For the rotational symmetry to take place, the parameters of system (5.6) must satisfy the relation $g^2 = 9\gamma d/2$, while the initial conditions provide a certain excitation level given by the constant $N = 2\,g/3d$. In this case, introducing the coordinate transformation $R_1 = \sqrt{N}\cos\theta$ and $R_2 = \sqrt{N}\sin\theta$, we obtain the system

$$
\begin{aligned}
\frac{d\theta}{d\tau_2} &= \tfrac{1}{2}\left(\sin\Delta - \lambda\sin 4\theta\right), \\
\sin 2\theta\,\frac{d\Delta}{d\tau_2} &= \cos 2\theta\cos\Delta + 2k\sin 4\theta,
\end{aligned}
\tag{7.8}
$$

where $\tau_2 = \beta\tau_1$, and the parameters $k = 3\alpha N/2\beta$ and $\lambda = N^2 d/8\beta$ characterize nonlinearity and dissipation relatively to coupling of the generators, respectively.

7.3 Analysis of the Phase Plane and Analytical Solutions

Figure 7.1 depicts the system behavior for different combinations of the nonlinearity and dissipation parameters. First, we consider the case $\lambda = 0$, when system (7.8) is conservative and the NNMs are stable (Fig. 7.1a). Two branches of the LPT are associated with complete energy exchange between the generators. In order to avoid the conservative type bifurcation of the NNMs occurring at $k = \frac{1}{2}$, we choose the parameter of nonlinearity between 0 and 1/2. When the parameter λ is relatively small, the system has two unstable focuses corresponding to unstable NNMs of the original system (7.1) (Fig. 7.1b–d). The focuses transform into unstable nodes when $\lambda^2 > 1 - 2k$ (Fig. 7.1e, f). If the dissipative parameter λ does not exceed the value $0.5\left(1 + \sqrt{1 - 4k^2}\right)$, the only attractor of the system is the LPT with intensive energy exchange between the oscillators.

In Fig. 7.1, one can see that the phase shift between the oscillators on the LPT remains near $\pm\pi/2$ (Fig. 7.1b), while the sign changes almost instantly as the system approaches the LTP attractor. Therefore, the oscillators become

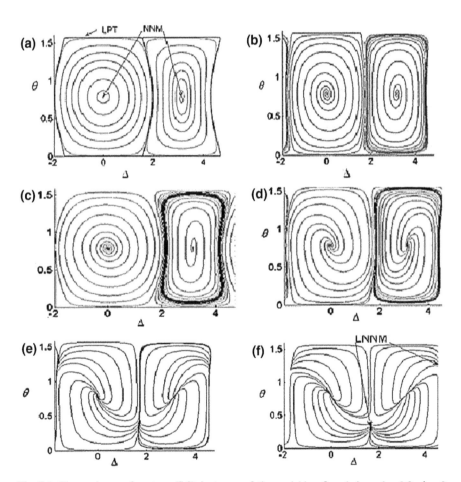

Fig. 7.1 Phase planes of system (7.8) in terms of the variables θ and Δ: **a** $k = 0.2$, $\lambda = 0$ (conservative system), **b** $k = 0.2$, $\lambda = 0.1$; **c** $k = 0.44$, $\lambda = 0.1$ **d** $k = 0.2$, $\lambda = 0.5$; **e** $k = 0.1$, $\lambda = 0.9$; **f** $k = 0.1$, $\lambda = 0.99$. The results are in a good agreement with simulations of the original system (7.1)

synchronized in a non-conventional way that can be qualified as the "LPT-type synchronization". Furthermore, if $\lambda \geq 0.5\left(1 + \sqrt{1 - 4k^2}\right)$, the LPT becomes unstable, and the attractor is a stationary point corresponding to the localized NNM with energy predominantly trapped on one of the two oscillators (Fig. 7.1f).

It is important to note that the evolution of the LPT leading to the transition from energy exchange to energy localization turns out to be independent on the evolution of the stationary points and occurs "later" than the transformation from an unstable focus to an unstable node (Fig. 7.1e).

In the range of intensive energy exchange, one can obtain an analytical solution of system (7.8) by using the saw-tooth transformations (6.20):

$$\theta = A\tau + \frac{\lambda}{4}[\cos(4A\tau) - 1]e + \ldots, \quad \Delta = \pi - \left[\frac{\pi}{2} - 2k \sin(2a\tau)\right]e + \ldots \quad (7.9)$$

where the basis functions τ and e are defined by formulas (6.20) and depicted in Fig. 6.3. Note that numerical solutions shown in Fig. 7.2a, b appear to be in good agreement with analytical solutions (7.9) demonstrated in Fig. 7.2c, d.

The behavior of the system before and after the transition (in the parametric space) from non-conventional synchronization on the LPT to synchronization on the localized NNM is illustrated in Fig. 7.3a, b, in terms of the original variables. The amplitudes $R_j = \sqrt{u_j^2 + v_j^2}, j = 1, 2$, presented in Fig. 7.3, are obtained by numerical integration of the original system (7.3). Non-conventional synchronization on the LPT far from the localization threshold is shown in Fig. 7.3c. Also, we

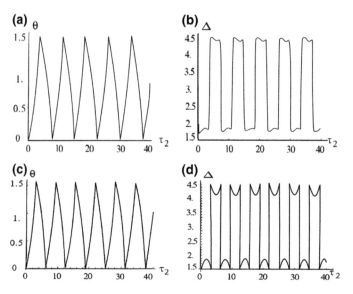

Fig. 7.2 Numerical (exact) solutions θ (**a**) and Δ (**b**) and their non-smooth approximations (**c**) and (**d**)

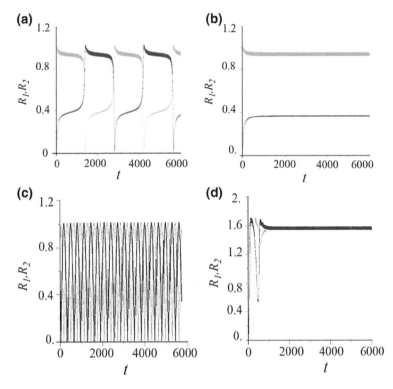

Fig. 7.3 Transitions from energy exchange to energy localization in system (7.1) of two coupled generators. The amplitudes $R_j = \sqrt{u_j^2 + v_j^2}$, $j = 1, 2$ are obtained by numerical integration of the original system (7.3) (R_1—*orange line*, R_2—*blue line*) **a** energy exchange just before the transition to localization for the parameters $k = 0.1$, $\lambda = 0.98$; **b** the behavior just over the localization threshold, in which most part of energy is localized on one of the oscillators for the parameters $k = 0.1$, $\lambda = 0.99$; **c** non-conventional synchronization corresponding to the parameter set of Fig. 7.1b, far from the localization threshold; and **d** conventional synchronization on out-of-phase NNM when the parameters of generators do not satisfy the accepted symmetry conditions

present for comparison the plot demonstrating the well-studied conventional synchronization, realized on the out-of-phase NNM (Fig. 7.3d). In the latter case, the set of parameters is taken far enough from the accepted symmetry conditions.

It is worth noticing that the presented scenario seems to be quite general in the area of model parameters wherein the conventional synchronization is impossible (it is confirmed by the analytical estimates and numerical simulations presented in Manevitch et al. (2013) and Kovaleva et al. (2013). However, the considered case allows the detailed analytical description and reveals a new important type of synchronization. Besides, one can see how the complete energy exchange between different parts of the systems described by the LPT goes over to the energy localization. After prediction of the domain of the dissipative parameters wherein the LPT-type synchronization exists, such type of synchronization can be found experimentally in the physical, chemical, and biological systems modeled by a pair of coupled generators.

References

Akhmediev, N.N., Ankiewicz, A.: Solitons: Nonlinear Pulses and Beams. Chapman & Hall, London (1997)

Chakraborty, T., Rand, R.H.: The transition from phase locking to drift in a system of two weakly coupled Van der Pol oscillators. Int. J. Nonlinear Mech. **23**, 369–376 (1988)

Kovaleva, M.A., Manevitch, L.I., Pilipchuk, V.N.: New type of synchronization for auto-generator with hard excitation. J. Exp. Theor. Phys. **116**, 369–377 (2013)

Landa, P.S., Duboshinskiĭ, Ya.B.: Self-oscillatory systems with high-frequency energy sources. Sov. Phys. Usp. **32**, 723 (1989)

Landa, P.S.: Nonlinear Oscillations and Waves in Dynamical Systems. Kluwer Academic Publishers, Dordrecht (1996)

Malomed, B.A.: Waves and solitary pulses in a weakly inhomogeneous Ginzburg-Landau equations. Phys. Rev. E **50**, 4249–4252 (1994)

Manevitch, L.I., Kovaleva, M.A., Pilipchuk, V.N.: Non-conventional synchronization of weakly coupled active oscillators, Eur. Lett. **101**, 50002 (1–5) (2013)

Mihalache, D., Mazilu, D., Lederer, F., Kivshar, Y.S.: Spatiotemporal surface Ginzburg-Landau solitons. Phys. Rev. A **77**, 043828 (1–6) (2008)

Newell, A.C., Nazarenko, S., Biven, L.: Wave turbulence and intermittency. Physica D **152**(153), 520–550 (2001)

Ovsyannikov, L.V.: Group Analysis of Differential Equations. Academic Press, New York (1982)

Pikovsky, A., Rosenblum, M., Kurths, J.: Synchronization: A Universal Concept in Nonlinear Sciences. Cambridge University Press (2001)

Rompala, K., Rand, R., Howland, H.: Dynamics of three coupled van der Pol oscillators, with application to circadian rhythms. Commun. Nonlinear Sci. Numer. Simul. **12**, 794–803 (2007)

Vakakis, A.F., Manevitch, L.I., Mikhlin, Y.V., Pilipchuk, V.N., Zevin, A.A.: Normal Modes and Localization in Nonlinear Systems. Wiley, New York (1996)

Verhulst, F.: Invariant manifolds in dissipative dynamical systems. Acta Appl. Math. **87**, 229–244 (2005)

Chapter 8
Limiting Phase Trajectories
and the Emergence of Autoresonance
in Anharmonic Oscillators

As mentioned above, resonance energy transfer represents one of the most effective ways of the response enhancement for a broad range of physical and engineering systems. In this chapter, the notion of resonance energy transfer is extended to the oscillators subjected to harmonic forcing with a slowly varying frequency. We investigate capture into resonance of a Klein–Gordon chain of identical linearly coupled Duffing oscillators excited by a harmonic force with a slowly varying frequency applied at an edge of the chain.

We begin with the consideration of a single Duffing oscillator. It is important to underline that properties of the LPTs in the oscillator subjected to a periodic excitation with a slowly varying frequency do not directly correspond to the properties of the LPTs in a forced oscillator discussed in Sect. 6.1. In the case under consideration, the LPT can also be identified as a trajectory corresponding to a maximum possible irreversible energy transfer from the source of energy to the oscillator. However, the amplitude of the LPT does not represent a slow periodic process with a constant mean value; it either converges to a certain stationary level at large times or can be considered as small oscillations near a *slowly increasing mean value*. The ability of a nonlinear oscillator to stay captured into resonance due to variance of its structural or/and excitation parameters is known as *autoresonance* (AR).

An idea of AR or "resonance under the action of a force produced by the system's itself" was first suggested in Andronov et al. (1966). The occurrence of AR may be informally explained by example of the Duffing oscillator. Let the oscillator be initially at rest and subjected to an external force of small amplitude and constant frequency equal to the frequency of linear oscillations. It is known that an increase in the natural frequency due to continuous growth of the amplitude of nonlinear oscillations results in breakup of resonance in the oscillator with constant parameters. However, an oscillator may remain persistently captured into resonance with its drive if the driving frequency, being initially close or equal to the natural frequency of the oscillator, varies slowly in time to be consistent with the slowly changing frequency of the oscillator. It is important to underline that the emergence

© Springer Nature Singapore Pte Ltd. 2018   195
L.I. Manevitch et al., *Nonstationary Resonant Dynamics of Oscillatory Chains
and Nanostructures*, Foundations of Engineering Mechanics,
DOI 10.1007/978-981-10-4666-7_8

of AR leads to persistently growing mean amplitude of oscillations, and thus, this process may be employed to attain the required energy level.

After first studies for the purposes of the particles' acceleration (McMillan 1945; Veksler 1944), AR has become a very active field of research. Theoretical approaches, experimental evidence, and applications of AR in different fields of natural science, from plasmas to planetary dynamics, are reported in numerous papers (Friedland); additional theoretical and computational results can be found in (Ben-David et al. 2006; Chacon 2005; Kalyakin 2008); recent advances in this field are discussed, e.g., in (Friedland; Murch et al. Shalibo et al. 2012).

It was noticed (e.g., in Bohm 1947) that the physical mechanism behind autoresonance can be interpreted as adiabatic nonlinear phase-locking between the system and the driving signal. However, to the best of authors' knowledge, the mechanism causing the transition from bounded oscillations to AR has not been reported in the literature.

In Sect. 8.1, we discuss the emergence and stability of AR in a single Duffing oscillator subjected to a harmonic forcing with time-varying frequency. The asymptotic analysis of resonant oscillations is developed with the emphasis on the calculation of quasi-steady states. This analysis leads to the definition of the parametric domain, in which stable AR with persistently growing mean amplitude can occur.

We demonstrate that AR occurs due to the loss of stability of the LPT corresponding to quasi-linear oscillations of small amplitudes. It is shown that a critical parameter, which determines a boundary between the oscillations with small and large amplitudes in the time-invariant system, may be treated as a lower threshold of the emergence of autoresonance in the oscillator with slowly varying parameters. Furthermore, it is demonstrated that the threshold parameter numerically obtained, e.g., in Friedland (2008), Marcus et al. (2004), is inacceptable in the problem examined. Conditions of the transition from small to large oscillations in the Duffing oscillator is used to derive the critical sweeping rate. The obtained analytical estimates are proved to be very close to the results of numerical simulations. Note that the Duffing system is chosen for illustrative purposes. The qualitative features of the results hold true for a large class of nonlinear oscillators with slowly time-varying frequencies.

In Sect. 8.2, the results obtained for a single oscillator are extended to the nonlinear Klein–Gordon chain consisting of $n \geq 2$ identical linearly coupled Duffing oscillators. Parametric criteria, which guarantee the emergence and stability of AR in the entire chain, are established. It is shown that an increase in the number of oscillators in the chain does not change the conditions of capture into resonance in comparison with a two-particle cell. Furthermore, the emergence of AR in the chain entails the asymptotic equipartition of energy among all oscillators at large times.

8.1   Autoresonance in a SDOF Nonlinear Oscillator

The occurrence of autoresonance may be informally explained by an example of capture into resonance of the Duffing oscillator. Let us assume that the oscillator is initially at rest and subjected to an external force of small amplitude ε and constant frequency ω_0 equal to the frequency of linear oscillations. It is known that an increase in the natural frequency due to continuous growth of the amplitude of nonlinear oscillations results in breakup of resonance in the oscillator with constant parameters. However, if slowly decreasing stiffness counterbalances an effect of increasing amplitudes and sustains the value of the natural frequency near its initial value ω_0, the oscillator remains captured into resonance with the external force. After first studies for the purposes of particle acceleration (McMillan 1945; Veksler 1944), AR has become a very active field of research. Theoretical approaches, experimental evidence, and applications of AR in different fields of natural science, from plasmas to planetary dynamics, are reported in numerous papers (Friedland); additional theoretical and computational results can be found in (Ben-David et al. 2006; Bohm 1947; Chacon 2005; Kalyakin 2008); recent advances in this field are discussed, e.g., in (Andronov et al. 1966; Friedland; Murch et al. 2011; Neishtadt 1975; Shalibo et al. 2012).

In this section, we study the dynamics of the Duffing oscillator with slowly varying linear stiffness subjected to a periodic excitation with constant frequency. The equation of motion is given by

$$\frac{d^2 u}{d\tau_0^2} + (1 - \varepsilon \zeta(\tau))u + 8\varepsilon a u^3 = 2\varepsilon F \, \cos \tau_0, \tag{8.1}$$

where $\varepsilon > 0$ is a small parameter of the system, $\zeta(\tau) = s + b\tau$, $\tau = \varepsilon\tau_0$ is the slow timescale. As in the previous sections, initial conditions $u = 0$, $v = du/d\tau_0 = 0$ at $\tau_0 = 0$ determine the LPT of system (8.1) corresponding to the maximum possible energy transfer from the source of energy to the oscillator. We consider the case of $s > 0$, $\alpha > 0$.

Asymptotic solutions of Eq. (8.1) for small ε can be obtained in the same way as in Sect. 6.1. First, we introduce the change of variables $Y = (v + iu)e^{-i\tau_0}$, where the amplitude Y is constructed in the form of the multiple scales expansion with a slow main term: $Y(\tau_0, \tau, \varepsilon) = \phi^{(0)}(\tau) + \varepsilon\phi^{(1)}(t, \tau) + O(\varepsilon^2)$. Using rescaling

$$\tau_1 = s\tau; \; \varphi = \Lambda^{-1}\varphi^{(0)}; \; \Lambda = (s/3\alpha)^{1/2}; f = F/s\Lambda = F\sqrt{3\alpha/s^3}; \; \beta = b/s^2 \tag{8.2}$$

and repeating the reasoning of Sect. 6.1, a two-parameter equation similar to (6.8) is derived:

$$\frac{d\phi}{d\tau_1} + i\left(1 + \beta\tau_1 - |\phi|^2\right)\phi \quad \text{if} \quad \phi(0) = 0. \tag{8.3}$$

The polar representation $\varphi = ae^{i\Delta}$ transforms Eq. (8.3) into the system

$$\frac{da}{d\tau_1} = -f \sin \Delta,$$
$$\frac{d\Delta}{d\tau_1} = -(1 + \beta\tau_1) + a^2 - a^{-1}f \cos \Delta. \tag{8.4}$$

with initial conditions $a(0) = 0$, $\Delta(0) = -\pi/2$. It now follows from (8.2) to (8.4) that

$$u(\tau_0, \varepsilon) = \Lambda a(\tau_1) \sin(\tau_0 + \Delta(\tau_1) + \tau_1) + O(\varepsilon),$$
$$v(\tau_0, \varepsilon) = \Lambda a(\tau_1) \cos(\tau_0 + \Delta(\tau_1) + \tau_1) + O(\varepsilon) \tag{8.5}$$

8.1.1 Critical Parameters

For better understanding of the occurrence of unbounded modes, we first consider the underlying *time-invariant* system

$$\frac{da}{d\tau_1} = -f \sin \Delta,$$
$$\frac{d\Delta}{d\tau_1} = -1 + a^2 - a^{-1}f \cos \Delta. \tag{8.6}$$

with initial conditions $a(0) = 0$, $\Delta(0) = -\pi/2$ corresponding to the LPT.
  It was proved in Sect. 6.1 that there exist two critical relationships

$$f_1 = \sqrt{2/27} \approx 0.2721; \quad f_2 = 2\big/\sqrt{27} \approx 0.3849 \tag{8.7}$$

which define the boundaries between different types of the dynamical behavior. Conditions $f < f_1$, $f_1 < f < f_2$, and $f > f_2$ characterize *quasi-linear, moderately nonlinear*, and *strongly nonlinear* dynamics, respectively.
  Now, we extend the above-mentioned results to system (8.4). Figure 8.1 present the results of numerical simulations for system (8.4) with detuning $\beta\tau_1$ and the following values of the system parameters: $\beta = \pm 0.07$, $f = 0.34$ ($f_1 < f < f_2$).
  It is seen in Fig. 8.1 that the solution of Eq. (8.4) is very close to the LPT of the time-independent system during the first half-cycle of oscillations. If $\beta > 0$, detuning $\beta\tau_1$ increases with an increase in τ_1, thereby shifting the system to the domain of small oscillations; if $\beta < 0$, detuning decreases with an increase in τ_1; in the latter case, the system passes through resonance with large amplitude of oscillations. This implies that the LPT of the time-invariant system ($\beta = 0$) determines the maximum achievable energy level.

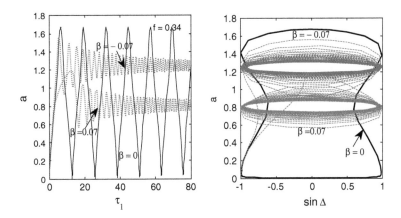

Fig. 8.1 Plots of $a(\tau_1)$ and phase portraits of system (8.4) with parameters $\beta = \pm0.07$, $f = 0.34$. Plots of the LPTs for the corresponding time-invariant systems ($\beta = 0$, *black solid lines*) are shown for comparison

The projection of the trajectory $a(\tau_1)$ onto the phase plane represents the spiral orbit with an attracting focus ($a = a_0$, $\sin \Delta = 0$),where $a_0 = \lim a(\tau_1)$ as $\tau_1 \to \infty$.

Figure 8.2 depicts the emergence of AR from stable bounded oscillations under the change of rate $\beta > 0$. As seen in Fig. 8.2, under very slow sweep, the transition from bounded oscillations to AR takes place if the parameter f is close to the critical value f_1. In particular, this means that the peak of the LPT in the time-invariant system ($\beta = 0$) determines a minimum energy level achievable in the process of capture into resonance.

It is shown in Fig. 8.2 that at $f = 0.274$ the transition occurs at $\beta \approx 0.001$; the difference between f and $f_1 = 0.2721$ is less than 1%; at $f \approx 0.28$, the transition takes place at $\beta \approx 0.006$; the difference between f and f_1 is less than 2.7%. On the other hand, for $f = 0.34$, the critical rate $\beta \approx 0.061$; the difference between f and f_1 is about 20%. This implies that the inequality $f > f_1$ can be interpreted as *the necessary condition* of the emergence of AR.

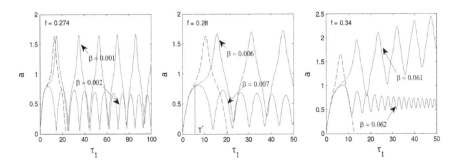

Fig. 8.2 Transitions to AR in system (8.4) with different parameters f and β; the cycle of oscillations in the time-independent system (*black line*) is demonstrated for comparison

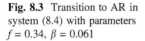

Fig. 8.3 Transition to AR in system (8.4) with parameters $f = 0.34$, $\beta = 0.061$

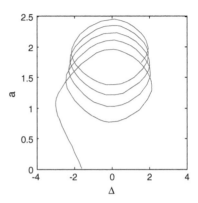

Figure 8.3 elucidates the nature of AR oscillations. It is seen that in the first half-cycle of oscillations the amplitude $a(\tau_1)$ is close to the LPT of a moderately nonlinear mode of motion (see Fig. 8.1b). Then, the shape of the trajectory changes, and it turns into quasi-linear oscillations near an upward quasi-steady center $(\bar{a}(\tau_1), 0)$ with a slowly increasing value of $\bar{a}(\tau_1)$. The quasi-stationary amplitude $\bar{a}(\tau_1)$. is calculated below.

The obtained numerical results motivate the derivation of an analytical threshold between bounded and unbounded oscillations. In order to evaluate the critical rate β corresponding to the transition from bounded to unbounded oscillations, we employ the fact that for sufficiently small τ_1, the solution $a(\tau_1)$ of system (8.4) is very close to the LPT of the time-invariant system (8.6). We recall that the LPT of the moderately nonlinear ($f_1 < f < f_2$) time-invariant systems has a distinctive inflection at an instant $\tau_1 = T^*$ (Fig. 8.2). We introduce the parameter $\tilde{f}(\tau_1) = f/(1 + \beta\tau_1)^{3/2}$ such that $\tilde{f}(0) = f > f_1$. Numerical results in Figs. 8.1 and 8.2 indicate that an adiabatically varying system in which $\tilde{f}(0) > f_1$ gets captured into the domain of small oscillations provided $\tilde{f}(T^*) < f_1$. Under this assumption, the critical rate is given by

$$\beta^* = (T^*)^{-1} \left[(f/f_1)^{2/3} - 1 \right]. \tag{8.8}$$

If $\beta < \beta^*$, the system admits AR. In order to check the correctness of equality (8.8), we calculate the critical rate β^* in the system with linear-in-time detuning ($n = 1$). First, the instant T^* will be found from the obtained numerical results. In the next step, the analytical estimate of T^* and the respective value β^* will be derived.

We recall that the point of inflection is determined by the conditions $da/d\tau_1 \neq 0$, $d^2a/d\tau_1^2 = 0$. It follows from (8.6) that the latter condition corresponds to $d\Delta/d\tau_1 = 0$, i.e., the envelope $a(\tau_1)$ achieves the point of inflection when the phase Δ achieves its minimum (Fig. 8.4).

As seen in Fig. 8.4, $T^* \approx 6.5$ for $f = 0.274$; this yields $\beta^* \approx 0.00075$, while computational results give $0.001 < \beta < 0.002$. Then, $T^* \approx 5$ and $\beta^* \approx 0.004$ for $f = 0.28$, while the numerical simulation gives $0.006 < \beta < 0.007$. Note that for

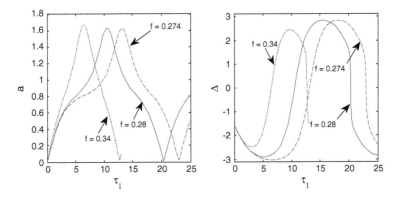

Fig. 8.4 Envelopes $a(\tau_1)$ and phases $\Delta(\tau_1)$ for different values of the parameter f

$f = 0.28$, the threshold parameter $\mu_{th} = |\beta_{th}|^{3/4} f = 0.41$ yields $\beta_{th} = (f/\mu_{th})^{4/3} = 0.467$, which is vastly larger than the real threshold rate. In a similar way, we find that for $f = 0.34$, the critical rate $\beta^* = 0.053$, while the numerical simulation gives $0.061 < \beta < 0.062$. Note that at $f = 0.34$, the inflection of the curve $a(\tau_1)$ is practically indistinguishable, but the phase has the distinct minimum at $T^* \approx 3$ (Fig. 8.4).

It is important to note that formula (8.8) defines the critical rate for systems with both linear-in-time and nonlinear-in-time detuning laws. For example, in the case of quadratic detuning $\beta \tau_1^2$ and $f = 0.28$, we find $\beta^* = 0.0008$; at the same time, the numerical simulation gives $0.001 < \beta < 0.002$ (Fig. 8.4).

An analytical derivation of both an instant T^* and a point of inflection a^*, Δ^* is built upon the results obtained in Sect. 6.1. We recall that the amplitude $a(\tau_1)$ corresponding to the LPT of conservative system (8.6) satisfies the second-order Eq. (6.17), namely

$$\frac{d^2 a}{d\tau_1^2} + \frac{dU}{da} = 0 \tag{8.9}$$

with potential (6.18). The inflection point is defined by the condition $dU/da = 0$ at $a = a^*$ or

$$a^* = \sqrt{2/3} = 0.8165 \tag{8.10}$$

The value of the phase Δ^* at inflection is defined by the equality $d\Delta/d\tau_1 = 0$. Using (8.6) and (8.10), it is easy to derive that $\cos \Delta^* = a^*[(a^*)^2 - 1]/f$, where $a^*[(a^*)^2 - 1] = -f_1$, and thus,

$$\cos \Delta^* = -f_1/f. \tag{8.11}$$

Since the maximum of $U(a)$ is also defined by the condition $dU/da = 0$, the potential barrier passes through the point of inflection $a = a^*$. It follows then that the time up to inflection T^* equals the time τ_1^* needed to reach the potential barrier. Using the same arguments as in Sect. 6.1, one can find that

$$T^* = \tau_1 \approx 3 \ \ln\frac{(f^2 - f_1^2)^{1/2}}{f - f_1} = \frac{3}{2} \ \ln\frac{f + f_1}{f - f_1}. \tag{8.12}$$

For example, in the case of $f = 0.274$ and $f = 0.28$, we obtain the approximations $\tau^* \approx 7.48$ and $\tau_1^* \approx 5.9$, which exceed the corresponding numerical values T^* (Fig. 8.4) for 15%. It follows then that the substitution of τ_1^* for T^* into (8.8) gives the rate $\beta_1^* < \beta^*$. Therefore, detuning rate $\beta < \beta_1^* < \beta^*$ allows the occurrence of autoresonance.

8.1.2 Numerical Evidence of Capture into Resonance

The obtained numerical results allow for the representation of the complex amplitude $\phi(\tau_1)$ as $\phi(\tau_1) = \bar{\phi}(\tau_1) + \tilde{\phi}(\tau_1)$, where the terms $\bar{\phi}(\tau_1)$ and $\tilde{\phi}(\tau_1)$ denote a quasi-stationary value of $\varphi(\tau_1)$ and rather small fast fluctuations near $\bar{\varphi}$ respectively. The state $\bar{\varphi}$ can be approximately calculated as a stationary point of Eq. (8.3) with "frozen" detuning ζ_0. We thus obtain

$$(\zeta_0 = |\bar{\phi}|^2)\bar{\phi} = -f, \quad \bar{\phi} \approx \pm\sqrt{\zeta_0} + f/2\zeta_0$$
$$\bar{\phi} \approx \pm\sqrt{\zeta_0}, \quad \bar{a} = |\bar{\phi}| \approx \sqrt{\zeta_0} \quad \text{if} \quad |f/2\zeta_0| \ll 1 \tag{8.13}$$

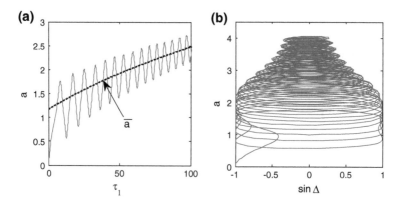

Fig. 8.5 Capture into resonance: **a** convergence of $a(\tau_1)$ to a monotonically increasing backbone $\bar{a}(\tau_1)$; **b** phase-locking at $\tau \to \infty$

The quasi-stationary amplitude $\bar{a} \approx \sqrt{\zeta_0}$ corresponds to the backbone curve (Andronov 1966; Hayfeh and Mook 2004) and expresses a relationship between the amplitude and the frequency of free oscillations (Fig. 7.5). The rapidly oscillating component $\tilde{\phi}_0(\tau)$ can be approximately computed from the linearized equation

$$\frac{\mathrm{d}\tilde{\phi}}{\mathrm{d}\tau} + 2i\zeta_0(\tau)\mathrm{Re}\left(\tilde{\phi}\right) = \mp \frac{\beta}{2\sqrt{\zeta_0(\tau)}} \tag{8.14}$$

Capture into resonance in the system with parameters $\beta = 0.05$, $f = 0.34$ is depicted in Fig. 8.5. Numerical results demonstrate the convergence of the amplitude $a(\tau_1)$ to a monotonically increasing backbone curve and phase-locking at $\tau_1 \to \infty$.

8.2 Autoresonance Versus Localization in Weakly Coupled Oscillators

In this section, we extend the notion of the resonance energy transfer along the LPT on a 2DOF system consisting of a passive linear oscillator weakly coupled with a nonlinear (Duffing) actuator excited by an external force.

We recall that the analysis of passage through resonance was first concentrated on the basic nonlinear oscillator, but then, the developed methods and approaches were extended to two- or three-dimensional systems. Examples in this category are excitations of continuously phase-locked plasma waves (Barth and Friedland 2013), particle transport in a weak external field with slowly changing frequency (Dodin and Flisch 2012; Zelenyi et al. 2013), control of nanoparticles (Kivshar 1993), etc. Some particular results (Barth and Friedland 2013; Friedland) demonstrated that external forcing with a slowly varying frequency applied to a pair of coupled nonlinear oscillators generates AR in both oscillators. We show that this conclusion is not universal; in particular, it does not hold for a pair of weakly coupled linear and nonlinear (Duffing) oscillators considered in this section.

In this section, we examine passage through resonance in two classes of systems. The first class includes the systems, in which a periodic force with constant frequency acts on the Duffing oscillator with slowly time-decreasing linear stiffness; in the systems of the second class, the time-invariant nonlinear oscillator is excited by a force with slowly increasing frequency. In both cases, stiffness of the linear oscillator and linear coupling remains constant, and the entire system is initially captured into resonance.

We demonstrate that AR in the nonlinear actuator may entail oscillations with growing amplitude in the coupled oscillator only in the system of the first class with constant excitation frequency, whereas in the system of the second class, the most part of energy remains localized on the excited oscillator, and a portion of energy transferred to the linear oscillator is insufficient to provide growing oscillations.

This means that the systems that seem to be almost identical exhibit different dynamical behavior caused by their different resonance properties.

It is important to note that the change of frequency of the forcing field is usually considered as an effective tool for producing the desired resonance dynamics, and the failure of this approach in a multi-degrees-of-freedom system has not been discussed thus far in the literature.

8.2.1 Energy Transfer in a System with Constant Excitation Frequency

The equations of motion of two coupled oscillators are given by

$$m_1 \frac{d^2 u_1}{dt^2} + c_1 u_1 + c_{10}(u_1 - u_0) = 0,$$
$$m_0 \frac{d^2 u_0}{dt^2} + C(t)u + ku^3 + c_{10}(u_0 - u_1) = A \cos \omega t, \tag{8.15}$$

where u_0 and u_1 denote absolute displacements of the nonlinear and linear oscillators, respectively; m_0 and m_1 are their masses; c_1 denotes stiffness of the linear oscillator; c_{10} is the linear coupling coefficient; k is the coefficients of cubic non-linearity; $C(t) = c_0 - (k_1 + k_2 t)$, $k_{1,2} > 0$; A and ω are the amplitude and the frequency of the periodic force. The system is initially at rest, that is, $u_k = 0$, $v_k = du_k/dt = 0$ at $t = 0$, $k = 0, 1$. We recall that these initial conditions determine the LPT of system (8.15) associated with maximum possible energy transfer from the source of energy to the oscillator.

As in the previous sections, we define the small parameter $2\varepsilon = c_{10}/c_1 \ll 1$. Considering weak nonlinearity and taking into account resonance properties of the system, we denote

$$\begin{aligned} c_1/m_1 = c_0/m_0 = \omega^2; & \quad A = \varepsilon m \omega^2 F \\ k_1/c_0 = 2\varepsilon s; & \quad k_2/c_0 = 2\varepsilon^2 b\omega; \quad k/c_0 = 8\varepsilon\alpha; \\ c_{10}/c_1 = 2\varepsilon\lambda_1; & \quad c_{10}/c_0 = 2\varepsilon\lambda_0 \end{aligned} \tag{8.16}$$

and then reduce the equations of motion to the form:

$$\frac{d^2 u_1}{d\tau_0^2} + u_1 + 2\varepsilon\lambda_1(u_1 - u_0) = 0,$$
$$\frac{d^2 u_0}{d\tau_0^2} + (1 - 2\varepsilon\zeta(\tau))u_0 + 2\varepsilon\lambda_0(u_0 - u_1) + 8\varepsilon a u_0^3 = 2\varepsilon F \sin \tau_0, \tag{8.17}$$

where $\tau_0 = \omega t$ is the fast timescale, and $\tau = \varepsilon\tau_0$ is the leading-order slow timescale; $(\tau) = \sigma + \beta\tau$. The asymptotic analysis of system (8.17) is analogous to the analysis performed in Sects. 8.1 and 2.6. As in the previous sections, we introduce the following transformations:

$$v_k + iu_k = Y_k e^{i\tau_0}; \quad Y_k(\tau_0, \tau, \varepsilon) = \varphi_k^{(0)}(\tau) + \varepsilon\varphi_k^{(0)}(\tau_0, \tau) + \varepsilon^2 + \cdots$$
$$\tau_1 = s\tau; \quad \varphi_k = \Lambda^{-1}\varphi_k^{(0)}; \quad \Lambda = (s/3\alpha)^{1/2}, \quad k = 0, 1 \tag{8.18}$$

and then perform the separation of the fast and slow timescales. As a result, we obtain the following system for the leading-order slow complex amplitudes $\varphi_0(\tau_1)$ and $\varphi_1(\tau_1)$:

$$\frac{d\phi_1}{d\tau_1} - i\mu_1(\phi_1 - \phi_0) = 0, \phi_1(0) = 0,$$
$$\frac{d\phi_0}{d\tau_1} - i\mu_0(\phi_0 - \phi_1) + i(\zeta_0(\tau_1) - |\phi_0|^2)\phi_0 \quad \text{if} \quad \varphi_0(0) = 0. \tag{8.19}$$

with coefficients $f = F/s\Lambda$, $\beta = b/s^2$, $\mu_k = \lambda_k/s$, and detuning $\zeta_0(\tau_1) = 1 + \beta\tau_1$. For more details concerning the asymptotic analysis of similar systems, one can refer to Kovaleva and Manevitch (2012), Kovaleva et al. (2010).

The study of tunneling in weakly coupled oscillator (Kovaleva et al. 2010) proved that an asymptotic solution of the nonlinear equation can be obtained separately in the following cases: either coupling is weak enough to provide the condition $\mu_1\mu_2 \ll \beta$, or the mass of the attached oscillator is much less than the mass of the actuator, $m_1 \ll m_0$. For brevity, only the first case is discussed below. Under this assumption, the term proportional to μ_0 can be removed from (8.19) in the main approximation. The resulting truncated system

$$\frac{d\phi_1}{d\tau_1} - i\mu_1\phi_1 = -i\mu_1\phi_0, \phi_1(0) = 0,$$
$$\frac{d\phi_0}{d\tau_1} + i(\zeta_0(\tau_1) - |\phi_0|^2)\phi_0 \quad \text{if} \quad \phi_0(0) = 0. \tag{8.20}$$

includes an independent nonlinear equation for the complex amplitude φ_0 and a linear equation for φ_1, in which φ_0 plays the role an external excitation. The effect of weak coupling on motion of the nonlinear oscillator may be considered in subsequent iterations (Kovaleva and Manevitch 2010).

The response $\varphi_1(\tau)$ can be defined from (8.19) or (8.20) by the expression:

$$\phi_1(\tau_1) = -i\mu_1 \int_0^{\tau_1} e^{i\mu_1(\tau - s)}\phi_0(s)ds. \tag{8.21}$$

As in Sect. 8.1, the solution of the nonlinear equation is represented as $\phi(\tau_1) = \bar{\phi}(\tau_1) + \tilde{\phi}(\tau_1)$, where $\bar{\phi}(\tau_1)$ and $\tilde{\phi}(\tau_1)$ denote a quasi-stationary value of $\varphi(\tau_1)$ and fast fluctuations near $\bar{\phi}(\tau_1)$, respectively. Since the contribution from fast fluctuations $\tilde{\phi}_0(\tau)$ to integral (8.21) is relatively small compared to the contribution from the slowly varying component $\bar{\phi}_0$, we employ approximation (8.13) to obtain:

$$\phi_1(\tau_1) \approx -i\mu_1 e^{-i\mu_1\tau_1} J(\tau_1), \quad J(\tau_1) = \int_0^{\tau_1} e^{i\mu_1 s}\overline{\phi}_0(s)ds, \qquad (8.22)$$

where $\bar{\phi}_0(\tau_1) = \sqrt{1 + \beta\tau_1}$. Integration by parts gives

$$J(\tau_1) = -i\mu_1^{-1}\left[e^{i\mu_1\tau_1}\bar{\phi}_0(\tau_1) - \bar{\phi}_0(0)\right] - \Phi(\tau_1),$$
$$\Phi(\tau_1) = \frac{\beta}{2}\int_0^{\tau_1}\frac{e^{i\mu_1 s}}{\sqrt{1+\beta s}}ds. \qquad (8.23)$$

It follows from (8.23) that $\Phi(\tau_1) = \sqrt{\beta}F(\tau_1)$, where $F(\tau_1)$ is a Fresnel-type integral bounded for any $\tau_1 > 0$. Hence, $\phi_1(\tau_1) = \bar{\phi}_1(\tau_1) + \tilde{\phi}_1(\tau_1) + O(\sqrt{\beta})$, where

$$\bar{\phi}_1(\tau_1) = \bar{\phi}_0(\tau_1), \tilde{\phi}_1(\tau_1) \approx -\bar{\phi}(0)e^{i\mu_1\tau_1}. \qquad (8.24)$$

Although equality $\bar{\phi}_1(\tau_1) = \bar{\phi}_0(\tau_1)$ can be directly obtained from (8.19), transformations (8.21)–(8.24) formally demonstrate the equality of the averaged amplitudes for both oscillators, as well as the occurrence of growing oscillations in the linear oscillator.

Let a_k and a_k^{tr} ($k = 0, 1$) denote real-valued amplitudes oscillations in the full system (8.19) and the truncated system (6.20), respectively. Plots of $a_k(\tau_1)$ and $a_k^{tr}(\tau_1)$ in the systems with parameters

$$\beta = 0.05, \quad \mu_0 = 0.02, \quad \mu_1 = 0.15, \quad f = 0.34 \qquad (8.25)$$

are presented in Fig. 2.36. It is important to note that a single oscillator with parameters $\mu_0 = 0$, $\beta^* \approx 0.06$, $f = 0.34$ admits AR (see Sect. 8.1).

It is seen that the amplitudes a_k (solid lines) and a_k^{tr} (dotted lines) calculated for the full system (8.19) and the truncated system (8.20), respectively, are close to each other. Dashed lines depict the identical backbone curves. Furthermore, Fig. 8.6a shows that, as in the case of a SDOF oscillator, in the first half-period of oscillations, the amplitudes a_0 and a_0^{tr} are close to the LPT of the time-invariant system, but then, motion turns into small fast oscillations near the backbone curve $\bar{a}_0 \approx \sqrt{\zeta_0}$.

Figure 8.6b demonstrates that irregular oscillations of the linear oscillator at the early stage are then transformed into regular oscillations near the backbone curve. In the system with coupling $\mu_1 = 0.15$, we obtain the period of oscillations $T_1 = 2\pi/\mu_1 \approx 25.12$, and the amplitude of fluctuations $|\tilde{\phi}_1(\tau_1)| = |\bar{\phi}(0)| = 1$; both these values are close to the parameters in Fig. 8.6b.

Although both oscillators possess gradually increasing amplitudes, the nature of the emerged processes is different. Figure 8.6a depicts AR in the nonlinear

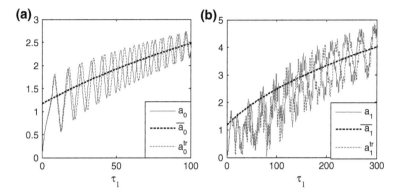

Fig. 8.6 Amplitudes and phases of oscillations in the full and truncated systems with parameters (8.25): **a** amplitudes of the nonlinear oscillators; **b** amplitudes of the linear oscillators; *dashed lines* represent quasi-stationary amplitudes

oscillator with the trajectory corresponding to the LPT of large oscillations. At the same time, Fig. 8.6b demonstrates forced oscillations, in which the response of the nonlinear oscillator acts as an external force.

8.2.2   Energy Localization and Transport in a System with a Slowly Varying Forcing Frequency

In this section, we briefly analyze energy transport in coupled oscillators with constant parameters and a slowly changing forcing frequency. The equations of motion are reduced the form similar to (8.17)

$$\frac{d^2 u_1}{d\tau_0^2} + u_1 + 2\varepsilon\lambda_1(u_1 - u_0) = 0,$$
$$\frac{d^2 u_0}{d\tau_0^2} + u + 2\varepsilon\lambda(u_0 - u_1) + 8\varepsilon a u_0^3 = 2\varepsilon F \sin(\tau_0 + \theta(\tau)), \qquad (8.26)$$
$$\frac{d\theta}{d\tau} = s + b\tau.$$

Transformations (8.18) together with the change of variables $\tau_1 = s\tau$, $\phi_j = \varphi_j e^{-i\theta}$ yield the following dimensionless equations for the slow complex amplitudes $\phi_j(\tau_1)$:

$$\frac{d\varphi_1}{d\tau_1} + i\zeta_0(\tau_1)\varphi_1 - i\mu_1(\varphi_1 - \varphi_0) = 0, \quad \varphi_1(0) = 0,$$
$$\frac{d\varphi_0}{d\tau_1} - i\mu_0(\varphi_0 - \varphi_1) + i(\zeta_0(\tau) - |\varphi_0|^2)\varphi \quad \text{if} \quad \varphi_0(0) = 0. \qquad (8.27)$$

Where $\zeta_0(\tau_1) = 1 + \beta\tau_1$. Note that Eq. (8.27) are similar to (8.19), but the time-dependent coefficient $\zeta_0(\tau_1)$ is now involved in both equations. Details of transformations are provided in Kovaleva (2015), Kovaleva and Manevitch (2016).

As in the previous section, the sought amplitude of the nonlinear oscillator is presented as $\varphi_0 = \bar\varphi_0 + \widetilde{\varphi_0}$, where $\bar\varphi_0(\tau_1)$ and $\widetilde{\varphi_0}(\tau_1)$ denote the quasi-stationary amplitude and small fast fluctuations near $\bar\varphi_0(\tau_1)$, respectively. It is easy to show that the state $\bar\varphi_0(\tau_1)$ and the backbone curve $\bar{a}_0 = |\bar\phi_0|$ are defined by relations (8.13).

After calculating the nonlinear response $\phi_0(\tau_1)$, the solution $\phi_1(\tau_1)$ can be directly found from the first Eq. (8.27). Ignoring the effect of small fast fluctuations, we obtain

$$\varphi_1(\tau_1) \approx -i\frac{\mu_1}{2\beta}e^{-iS(\tau_1)/2\beta}K(\tau_1), \quad S(\tau_1) = (1+\beta\tau_1)^2,$$

$$K(\tau_1) = K_0(\tau_1) - K_0(1), \quad K_0(\tau_1) = \int_0^{S(\tau_1)} e^{iz/2\beta}z^{-1/4}dz. \tag{8.28}$$

Although the expression for $K_0(\tau_1)$ cannot be found in closed form, the limiting value $K_0(\infty)$ can be explicitly evaluated and equals $K_0(\infty) = (2\beta)^{4/3}\Gamma(\frac{3}{4})e^{3i\pi/8}$, where Γ is the gamma function (Gradshtein and Ryzhik 2000). Hence, $a_1(\tau_1) \to a_{1\infty} = \mu_1(2\beta)^{1/3}\Gamma(\frac{3}{4})$ as $\tau_1 \to \infty$. This result indicates that AR in the nonlinear actuator is unable to generate oscillations with permanently growing energy in the attached oscillator, but the transferred energy suffices to sustain linear oscillations with bounded amplitude.

Figure 8.7 depicts the amplitudes of oscillations $a_0(\tau_1) = |\phi_0(\tau_1)|$ and $a_1(\tau_1) = |\phi_1(\tau_1)|$ calculated from Eq. (8.27). Figure 8.7a shows that the amplitude of nonlinear oscillations is very close to its analogue presented in Fig. 8.6a but the amplitude of linear oscillations in Fig. 8.7b drastically differs from the amplitude of oscillations with growing energy in Fig. 8.6b. The shape of the amplitude $a_1(\tau_1)$ is

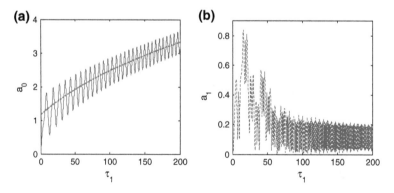

Fig. 8.7 Amplitudes of oscillations: **a** nonlinear oscillations; *dashed line* denotes the backbone curve $\bar{a}_0$; **b** linear oscillations

similar to the resonance curve with a noticeable peak in the initial stage of motion, but then, motion turns into small oscillations with near the limiting amplitude close to the theoretically predicted value $a_{1\infty} \approx 0.1$.

8.2.3 Energy Transfer in a System with Slow Changes of the Natural and Excitation Frequencies

A key conclusion from the obtained results is that in the system of with slowly changing excitation frequency, the energy transferred from the nonlinear oscillator is insufficient to produce oscillations with growing energy in the attached linear oscillator. The different dynamical behavior can be considered as a consequence of different resonance properties of the systems. In the system with a constant frequency of external forcing but slowly varying parameters of the nonlinear actuator, both oscillators are captured in resonance. If the forcing frequency slowly increases and the parameters of the system remain constant, resonance in the nonlinear oscillator is still sustained by increasing amplitude, while the frequency of the linear oscillator falls into the domain beyond the resonance. This implies that the slow change of the linear stiffness of the actuator can be considered as a parameter controlling the occurrence of high-energy oscillations with growing amplitude in the linear oscillator.

As an illustrating example, we consider a system with slowly changing linear stiffness of the actuator. The system dynamics is described by the equations

$$\frac{\mathrm{d}^2 u_1}{\mathrm{d}\tau_0^2} + u_1 + 2\varepsilon\lambda_1(u_1 - u_0) = 0,$$

$$\frac{\mathrm{d}^2 u_0}{\mathrm{d}\tau_0^2} + (1 - 2\varepsilon\xi(\tau))u_0 + 2\varepsilon\lambda_0(u_0 - u_1) + 8\varepsilon a u_0^3 = 2\varepsilon F \sin(\tau_0 + \theta(\tau)), \quad (8.29)$$

$$\frac{\mathrm{d}\theta}{\mathrm{d}\tau} = \zeta(\tau),$$

where $\zeta(\tau) = s + b_1\tau$, $\xi(\tau) = b_3\tau^3$, $\tau = \varepsilon\tau_0$; all other coefficients are defined by relations (8.16). We recall that if $\xi(\tau) = 0$, then AR may appear only the nonlinear oscillator. We will show that slow changes in both natural and excitation frequencies of the actuator may sustain growing oscillations in the coupled linear oscillator.

Transformations (8.18), together with the change of variables $\tau_1 = s\tau$, $\phi_j = \varphi_j e^{-i\theta}$, result in the following equations for the slow complex amplitudes $\phi_j(\tau_1)$:

$$\frac{\mathrm{d}\varphi_1}{\mathrm{d}\tau_1} + i\zeta_1(\tau_1)\varphi_1 - i\mu_1(\varphi_1 - \varphi_0) = 0, \quad \varphi_1(0) = 0,$$

$$\frac{\mathrm{d}\varphi_0}{\mathrm{d}\tau_1} + i[\zeta_0(\tau_1) - |\varphi_0|^2]\varphi_0 - i\mu_0(\varphi_0 - \varphi_1) \quad \text{if} \quad \varphi_0(0) = 0, \quad (8.30)$$

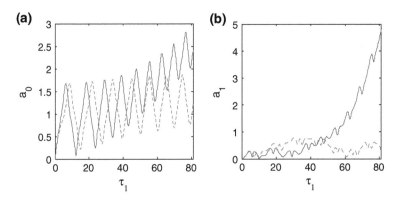

Fig. 8.8 Amplitudes of oscillations of the actuator (**a**) and the linear oscillator (**b**); *solid lines* corresponds to system (8.30); *dashed lines* correspond to the time-independent actuator ($\beta_3 = 0$)

where

$$\tau_1 = s\tau, \quad \zeta_0(\tau) = \zeta_1(\tau) + \xi_1(\tau), \quad \zeta_1(\tau_1) = 1 + \beta_1\tau_1,$$
$$\xi_1(\tau_1) = \beta_3\tau_1^3, \quad \beta_1 = b_1/s^2, \beta_3 = b_3/s^4;$$

other parameters are defined in (8.19) and (8.27). Note that details of converting the full system (8.29) into the equations for slow complex amplitudes (8.30) can be found in (Kovaleva 2015).

It is easy to obtain from (8.30) that the quasi-steady states $\overline{\phi}_j$ can be evaluated as

$$\begin{aligned}
|\overline{\phi}_0(\tau_1)| &\approx [\zeta_1(\tau_1) + \xi_1(\tau_1)]^{1/2} \sim O(\tau^{3/2}), \\
|\overline{\phi}_1(\tau_1)| &\approx \mu_1\left[\zeta_1(\tau_1) + \xi_1(\tau_1)\right]^{1/2}/(\zeta_1(\tau_1) - \mu_1) \sim O(\tau^{1/2}).
\end{aligned} \tag{8.31}$$

Expressions (8.31) imply simultaneous (but not equal) growth of backbone curves and thus suggest growing energy of oscillations of both oscillators provided linear stiffness of the actuator varies with rate $\xi_1(\tau)$ exceeding the rate $\zeta_1(\tau_1)$ of the change of the forcing frequency.

We illustrate these conclusions by the results of numerical simulations for the system with parameter $\beta_1 = 10^{-3}$, $\beta_3 = 10^{-5}$, $f = 0.34$, $\mu_0 = 0.01$, $\mu_1 = 0.15$ (Fig. 8.8).

It is seen that an additional slow change of the actuator stiffness entails an increase in the nonlinear response, thus enhancing energy transfer and making it sufficient to sustain growing oscillations of the linear oscillator. A thorough theoretical analysis of this model is omitted but the obtained results motivate further analytical investigation of feasible energy transfer in the multi-dimensional arrays (see, e.g., Kovaleva 2015).

8.3 Autoresonance in Nonlinear Chains

It follows from Sect. 8.1 that AR may potentially serve as an effective tool to excite and control the high-energy regime in a single oscillator. However, the behavior of coupled oscillators may drastically differ from the dynamics of a single oscillator. In particular, capture into resonance may not exist, or AR in the excited oscillator may be insufficient to enhance the response of the attachment. In this section, these effects are investigated for a nonlinear Klein–Gordon chain consisting of identical linearly coupled Duffing oscillators, one of which is subjected to periodic forcing with a slowly varying frequency. It is shown that capture into resonance of the entire chain may occur if both forcing amplitude and coupling stiffness exceed certain threshold values, but detuning rate is small enough.

8.3.1 The Model

The dynamics of the chain is described by the following equations:

$$
\begin{aligned}
&\frac{d^2 u_1}{dt^2} + \omega^2 u_1 + \gamma u_r^3 + \kappa(u_1 - u_2) = A \cos \theta(t),\\
&\frac{d^2 u_r}{dt^2} + \omega^2 u_r + \gamma u_r^3 + \kappa(2u_r - u_{r+1} - u_{r-1}) = 0, \quad r = 2, \ldots, n-1,\\
&\frac{d^2 u_n}{dt^2} + \omega^2 u_n + \gamma u_r^3 + \kappa(u_n - u_{n-1}) = 0,\\
&\frac{d\theta}{dt} = \omega + \zeta(t), \zeta(t) = k_1 + k_2 t.
\end{aligned}
\tag{8.32}
$$

Here and below, the variable u_r denotes the position of the rth oscillator; $\omega^2 = c/m$, m and c being the mass and the linear stiffness of each oscillator; $\gamma > 0$ is the cubic stiffness coefficient; the coefficient κ represents the stiffness of linear coupling between the oscillators. The first oscillator is subjected to periodic forcing with amplitude A and time-dependent frequency $\Omega(t) = \omega + \zeta(t)$, where $\zeta(t) = k_1 + k_2 t$; the parameters $k_1 > 0$ and $k_2 > 0$ denote the initial constant detuning and the detuning rate, respectively. The array under consideration may be considered an example of a microelectromechanical system (MEMS) with a broad spectrum of applications.

Now we reduce Eq. (8.32) to the form more convenient for further analysis. First, assuming small initial frequency detuning, we introduce the small parameter $\varepsilon = k_1/\omega$, $0 < \varepsilon \ll 1$. Then, taking into consideration resonance properties of the oscillators, we define the following rescaled parameters:

$$
\begin{aligned}
&\tau_0 = \omega t, \quad \tau_1 = \varepsilon \tau_0,\\
&\gamma = 8\varepsilon \alpha \omega^2, \quad A = 2\varepsilon F \omega^2, \quad \kappa = 2\varepsilon k \omega^2, \quad k_2 = 2\varepsilon^2 \beta \omega^2.
\end{aligned}
\tag{8.33}
$$

In these notations, the equations of motion are reduced to the form

$$\frac{d^2 u_1}{d\tau_0^2} + u_1 + 8\varepsilon\alpha u_1^3 + 2\varepsilon k(u_1 - u_2) = 2\varepsilon F \sin\theta(\tau_0, \varepsilon),$$

$$\frac{d^2 u_r}{dt^2} + u_r + 8\varepsilon\alpha u_r^3 + 2\varepsilon k(2u_r - u_{r+1} - u_{r-1}) = 0, \quad r = 2, \ldots, n-1,$$

$$\frac{d^2 u_n}{d\tau_0^2} + u_n + 8\varepsilon\alpha u_n^3 + 2\varepsilon k(u_n - u_{n-1}) = 0,$$

$$\frac{d\theta}{d\tau_0} = 1 + \varepsilon\zeta_0(\tau_1), \zeta_0(\tau_1) = 1 + \beta\tau_1,$$

(8.34)

If the system is initially at rest, then $\theta = 0$, $u_r = 0$, $v_r = du_r/d\tau_0 = 0$ at $\tau_0 = 0$ ($r = 1, \ldots, n$). We recall that zero initial conditions identify the *limiting phase trajectory* (LPT) of the oscillator.

As in the previous section, asymptotic solutions of (8.34) for small ε are derived with the help of the multiple timescale formalism. First, we introduce the dimensionless complex conjugate vector envelopes Ψ and Ψ^* with components Ψ_r, Ψ_r^* and related dimensionless parameters f, μ by formulas similar to (8.35)

$$\Psi_r = \Lambda^{-1}(v_r + iu_r)e^{-i\theta}, \Psi_r^* = \Lambda^{-1}(v_r - iu_r)e^{i\theta}, \Lambda = (1/3\alpha)^{1/2},$$
$$f = F/\Lambda, \mu = k/\Lambda,$$

(8.35)

where $r = 1, \ldots, n$. It follows from (8.35) that the real-valued dimensionless amplitudes and the phases of oscillations are defined as $\tilde{a}_r = |\Psi_r|$ and $\tilde{\Delta}_r = \arg\Psi_r$, respectively.

Substituting (8.35) into (8.34), we derive the following equations for the envelopes Ψ_r:

$$\frac{d\Psi_1}{d\tau_0} = -i\varepsilon[(\zeta_0(\tau_1) - |\Psi_1|^2)\Psi_1 - \mu(\Psi_1 - \Psi_2) + f + G_1],$$

$$\frac{d\Psi_r}{d\tau_0} = -i\varepsilon[(\zeta_0(\tau_1) - |\Psi_r|^2)\psi_r - \mu(2\Psi_r - \Psi_{r-1} - \Psi_{r+1}) + G_r], \quad 2 \leq r \leq n-1,$$

$$\frac{d\Psi_n}{d\tau_0} = -i\varepsilon[(\zeta_0(\tau_1) - |\Psi_n|^2)\Psi_n - \mu(\Psi_n - \Psi_{n-1}) + G_n],$$

(8.36)

subject to zero initial conditions. The parameters $G_1, \ldots G_n$ involve fast harmonics with coefficients depending on Ψ and Ψ^*, but explicit expressions of these parameters uninvolved in further analysis.

Asymptotic approximations to the solutions of Eq. (8.36) are constructed with the help of the multiple timescale approach. First, the asymptotic decomposition similar to (8.37) is introduced:

$$\Psi_r(\tau_0, \tau_1, \varepsilon) = \psi_r(\tau_1) + \varepsilon \psi_r^{(1)}(\tau_0, \tau_1) + O(\varepsilon^2), \quad r = 1, \ldots, n. \qquad (8.37)$$

Standard transformations yield the following averaged equations for the slowly varying envelopes $\psi_r(\tau)$:

$$\frac{d\psi_1}{d\tau_1} = -i[(\zeta_0(\tau_1) - |\psi_1|^2)\psi_1 - \mu(\psi_1 - \psi_2) + f]$$

$$\frac{d\psi_r}{d\tau_1} = -i[(\zeta_0(\tau_1) - |\psi_r|^2)\psi_r - \mu(2\psi_r - \psi_{r-1} - \psi_{r+1})], \quad 2 \leq r \leq n-1,$$

$$\frac{d\psi_n}{d\tau_1} = -i[(\zeta_0(\tau_1) - |\psi_n|^2)\psi_n - \mu(\psi_n - \psi_{n-1})]$$

$$(8.38)$$

with zero initial conditions. Note that the averaged Eq. (8.38) involve three independent coefficients instead of six parameters in (8.32).

The change of variables $\psi_r = a_r e^{i\Delta_r}$ yields the following equations for the real-valued dimensionless amplitudes a_r and the phases Δ_r:

$$\frac{da_1}{d\tau_1} = -\mu a_2 \sin(\Delta_1 - \Delta_2) - f \sin \Delta_1,$$

$$a_1 \frac{d\Delta_1}{d\tau_1} = \mu[a_1 - a_2 \cos(\Delta_1 - \Delta_2)] - \zeta_0(\tau_1)a_1 + a_1^3 - f \cos \Delta_1,$$

$$\frac{da_r}{d\tau_1} = \mu[a_{r-1} \sin(\Delta_{r-1} - \Delta_r) + a_{r+1} \sin(\Delta_{r+1} - \Delta_r)],$$

$$a_r \frac{d\Delta_r}{d\tau_1} = \mu[2a_r - a_{r-1} \cos(\Delta_{r-1} - \Delta_r) - a_{r+1} \cos(\Delta_{r+1} - \Delta_r)] - \zeta_0(\tau_1)a_r + a_r^3,$$

$$r \in [2, n-1],$$

$$\frac{da_n}{d\tau_1} = \mu a_{n-1} \sin(\Delta_{n-1} - \Delta_n),$$

$$a_n \frac{d\Delta_n}{d\tau_1} = \mu[a_n - a_{n-1} \cos(\Delta_{n-1} - \Delta_n)] - \zeta_0(\tau_1)a_n + a_n^3$$

$$(8.39)$$

with initial amplitudes $a_r(0) = 0$ and indefinite initial phases $\Delta_r(0)$, $r = 1,\ldots,n$. To overcome this uncertainty, one needs to solve the nonsingular complex-valued Eq. (8.38), and then, calculate the real-valued amplitudes and the phases by formulas $a_r = |\psi_r|$, $\Delta_r = \arg(\psi_r)$. Numerical results presented below have been obtained from regular Eq. (8.38).

The accuracy of asymptotic approximations for systems with slowly varying parameters has been discussed, e.g., in Sanders et al. (2007). Recall that the errors of approximation $|\tilde{a}_r(\tau, \varepsilon) - a_r(\tau)| \to 0$ as $\varepsilon \to 0$ in the time interval of interest, which is, at least, of order $O(1/\beta)$.

8.3.2 Quasi-steady States

Quasi-state values of the amplitudes and the phase of oscillations can be calculated from the following equations:

$$P_r = \frac{da_r}{d\tau_1} = 0, Q_r = \frac{d\Delta_r}{d\tau_1} = 0, \quad r = 1, \ldots, n. \tag{8.40}$$

The equation $P_n = 0$ implies $\sin\left(\overline{\Delta}_n - \overline{\Delta}_{n-1}\right) = 0$. Inserting this equality into the equation $P_{n-1} = 0$, we then have $\sin\left(\overline{\Delta}_{n-1} - \overline{\Delta}_{n-2}\right) = 0$. Repeating this procedure for each equation $P_r = 0$, we find that $\sin\left(\overline{\Delta}_r - \overline{\Delta}_{r-1}\right) = 0$, $r > 1$. Finally, the equation $P_1 = 0$ yields $\sin\overline{\Delta}_1 = 0$. This means that either $\overline{\Delta}_r = 0 \pmod{2\pi}$ or $\overline{\Delta}_r = \pi \pmod{2\pi}$, $r = 1, \ldots, n$. One can conclude that, in analogy to a single oscillator, the phases $\overline{\Delta}_r = 0$ $(r = 1, \ldots, n)$ correspond to the stable AR, while the phases $\overline{\Delta}_r = \pi$ $(r = 1, \ldots, n)$ are unstable. Substituting $\overline{\Delta}_r = 0$ into the conditions $Q_r = 0$, one obtains the following equations for the quasi-stationary amplitudes:

$$\begin{aligned}
&\mu(\bar{a}_1 - \bar{a}_2) - \zeta_0(\tau_1)\bar{a}_1 + \bar{a}_1^3 - f = 0 \\
&\mu(2\bar{a}_r - \bar{a}_{r-1} - \bar{a}_{r+1}) - \zeta_0(\tau_1)\bar{a}_r + \bar{a}_r^3 = 0, \quad r \in [2, n-1], \\
&\mu(\bar{a}_n - \bar{a}_{n-1}) - \zeta_0(\tau_1)\bar{a}_n + \bar{a}_n^3 = 0.
\end{aligned} \tag{8.41}$$

Maximal quasi-steady solutions corresponding to AR in the entire chain are given by

$$\begin{aligned}
&\bar{a}_1(\tau_1) \approx \sqrt{\zeta_0(\tau_1)} + [f/2\zeta_0(\tau_1)] + O(\mu f/\zeta_0(\tau_1)), \\
&\bar{a}_r(\tau_1) \approx \sqrt{\zeta_0(\tau_1)} + O(\mu f/\zeta_0^r(\tau_1)), \quad r = 2, \ldots, n.
\end{aligned} \tag{8.42}$$

The slow functions $\bar{a}_1(\tau_1), \ldots, \bar{a}_n(\tau_1)$ can be interpreted as the backbone curves.

It follows from (8.42) that the higher-order corrections may be ignored; furthermore,

$$\bar{a}_r(\tau_1) \rightarrow \bar{a}(\tau_1) = \sqrt{\zeta_0(\tau_1)}, \quad \tau_1 \rightarrow \infty, \quad r = 1, \ldots, n. \tag{8.43}$$

Formulas (8.42) and (8.43) clearly indicate that the energy initially placed in the first oscillator tends to *equipartition* among all oscillators with the limiting amplitudes of oscillations $\bar{a}(\tau)$. This conclusion is illustrated below by the results of numerical simulations for the chains with different number of particles.

8.3.3 Parametric Thresholds

Note that solutions (8.42) formally exist for arbitrary values of structural and excitation parameters. We establish the parametric thresholds, which allow the emergence of AR in the entire chain.

In order to simplify an analytical framework and elucidate the interpretation of the results, we assume that the parameters $f \sim o(1)$, $\mu \sim o(1)$ but $\beta \ll 1$. These assumptions agree with the earlier obtained theoretical and numerical results (Kovaleva and Manevitch 2013a, b, c) as well as with the numerical results presented below.

As remarked above, resonance in the forced oscillator represents a necessary condition for the emergence of resonance in the passive attachment. This means that the first parametric boundary can be found assuming small oscillations of the attachment. Assuming $a_r = \varepsilon^{1/2} \tilde{a}_r$, $r \geq 2$, the equations of the first oscillator are given by

$$\frac{da_1}{d\tau_1} = -f \sin \Delta_1,$$
$$a_1 \frac{d\Delta_1}{d\tau_1} = -(\zeta_0(\tau_1) - \mu)a_1 + a_1^3 - f \cos \Delta_1 \quad (8.44)$$

with initial conditions $a_1(0) = 0, \Delta_1(0) = -\pi/2$. Equation (8.44) describe the slow dynamics of a single Duffing oscillator with an additional linear spring of stiffness μ. Thus, the results derived in Sect. 8.1 can be directly applied to (8.44). In particular, this implies that in the first half-cycle of oscillations, the solution of Eq. (8.44) is very close to the LPT of the underlying *time-invariant* system

$$\frac{da_1}{d\tau_1} = -f \sin \Delta_1,$$
$$a_1 \frac{d\Delta_1}{d\tau_1} = -(1 - \mu)a_1 + a_1^3 - f \cos \Delta_1 \quad (8.45)$$

with initial conditions $a_1(0) = 0$, $\Delta_1(0) = -\pi/2$ corresponding to the LPTs. Reproducing the transformation of Sect. 2.2, one can conclude that the parametric thresholds that determine the boundaries between small and large oscillations of the oscillator (8.45) are given by

$$(a)\, f_{1\mu} = f_1 \sqrt{(1 - \mu)^3}, \quad (b)\, f_{2\mu} = f_2 \sqrt{(1 - \mu)^3}, \quad (8.46)$$

where $f_1 = \sqrt{2/27}$, $f_2 = 2/\sqrt{27}$. As in the "classical" oscillator (Sect. 2.2), the transition from small (non-resonant) to large (resonant) oscillations for the oscillator being initially at rest occurs due to the loss of stability of the LPT of small oscillations at $f = f_{1\mu}$. At $f = f_{2\mu}$, the stable center at $\Delta_1 = -\pi$ vanishes, and only a single stable center corresponding to nonlinear resonance remains on the axis $\Delta_1 = 0$.

If the actuator is captured into resonance, resonant oscillations in the attachment can occur if the coupling response is large enough to transfer the required amount of energy. This implies that the coupling parameter μ cannot be negligibly small. In order to evaluate the lower bound of the coupling strength μ, we consider the time-invariant analog of Eq. (8.40) with the parameter $\zeta_0 = 1$. It is easy to verify that the stationary phases are given by the equalities $\sin \Delta_1 = 0$, $\sin(\Delta_r - \Delta_{r-1}) = 0$, $r \in [2, n]$, and the solutions $\overline{\Delta}_r = 0$ are stable. If $\overline{\Delta}_r = 0$, then the corresponding stationary amplitudes satisfy the equations

$$
\begin{aligned}
(1 - a_1^2)a_1 - \mu(a_1 - a_2) + f &= 0, \\
(1 - a_r^2)a_r - \mu(2a_r - a_{r-1} - a_{r+1}) &= 0, \quad r \in [2, n-1], \\
(1 - a_n^2)a_n - \mu(a_n - a_{n-1}) &= 0.
\end{aligned}
\tag{8.47}
$$

It is easy to prove that maximal solutions of (8.47) are expressed as

$$
\bar{a}_1 = 1 + f/2 + O(\mu f); \quad \bar{a}_r = 1 + O(\mu^{r-1} f), \quad r \geq 2
\tag{8.48}
$$

Solutions (8.48) formally exist even if the coefficient μ is insufficient to produce the coupling response needed to sustain resonance in the attached oscillators. Our purpose is to define the coefficient of coupling strength μ, which yields the coupling response sufficient to sustain resonance in an arbitrary rth oscillator in the chain under the condition of resonance in the preceding oscillators. We begin with the analysis of resonance in the last oscillator. Ignoring higher-order corrections, we rewrite the nth equation in (8.47) in the form

$$
a_n^3 - (1 - \mu)a_n = \varphi_n, \quad \varphi_n = \mu \bar{a}_{n-1},
\tag{8.49}
$$

where $\bar{a}_{n-1} = 1$. The roots of Eq. (8.49) are analyzed through the properties of the discriminant $D_n = 27\varphi_n^2 - 4(1 - \mu)^3$ (Korn and Korn 2000). If $D_n < 0$, then Eq. (8.49) has 3 different real roots; if $D_n = 0$, two real roots merge; if $D_n > 0$, there exists a single real and two complex conjugate roots. The condition $D_n > 0$ is easily transformed into the inequality $\mu > f_{2\mu}$, or $\mu > \mu_{cr} = 0.25$.

Next, we analyze the behavior of the rth oscillator assuming resonance in the preceding oscillator and small oscillations of the subsequent oscillator. Under these assumptions, the equation for the amplitude a_r is rewritten as

$$
a_r^3 - (1 - 2\mu)a_r = \varphi_r, \varphi_r = \mu \bar{a}_{r-1}, \quad r \in [2, n-1]
\tag{8.50}
$$

The roots of Eq. (8.50) are analyzed through the properties of the discriminant $D_r = 27\varphi_r^2 - 4(1 - 2\mu)^3$. It is easy to prove that $D_r > 0$ at $\mu > 0.189$ for all attached oscillators from $r = 2$ to $r = n - 1$. It now follows that an admissible parametric domain for a multi-particle chain is determined by the conditions

$$f > f_{1\mu} = f_1 \sqrt{(1 - \mu)^3}, \quad \mu > \mu_{cr} = 0.25. \tag{8.51}$$

Conditions (8.51) are illustrated in Fig. 8.9. It follows from (8.51) that all oscillators with the parameters from the domain D_1 below $f_{1\mu}$ execute small quasi-linear oscillations; if the parameters f, μ lie within the dotted domain D_0, then the entire chain is captured into resonance; if the parameters belong to the shaded domain D, then the actuator is captured into resonance but the dynamics of the attachment should be investigated separately.

It is important to note that expressions (8.51) have been derived from the conditions of resonance for the time-invariant oscillator. However, numerical examples demonstrate that slow variations of the forcing frequency weakly affect the conditions of the emergence of AR in a multi-particle chain.

The emergence of AR also depends on the critical detuning rate β^*, at which the transition from bounded to unbounded oscillations occurs. As seen in Fig. 8.2b, the LPT of the time-invariant oscillator has a noticeable inflection at $\tau 1 = T^*$, and the transition from small to large oscillations takes place at a time instant close to T^*. Considering $\zeta_0(\tau)$ as a "frozen" parameter, we deduce that AR in the oscillator (8.44) may occur if

$$f > f_{1\mu}^* = f_1 (1 + \beta T^* - \mu)^{3/2} > f_{1\mu},$$
$$\beta < \beta^* = (T^*)^{-1} [(f/f_1)^{2/3} - (1 - \mu)]. \tag{8.52}$$

An approximate analytic expressions of the instant T^* as well as the points of inflection $a(T^*)$, $\Delta(T^*)$ can be derived in the same way as in Sect. 8.1. To improve the accuracy of calculations, in practice, it is convenient to employ the values of T^* found from the numerical simulation.

Fig. 8.9 Parametric thresholds (8.51)

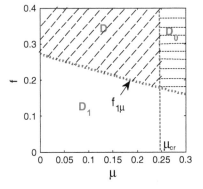

8.3.4   Numerical Results

In this section, we demonstrate the numerical results that confirm the effect of parameters on the formation and persistence of AR in the Klein–Gordon chain. We begin with the basic two-particle model. Equation (8.39) for two coupled oscillators are given by

$$
\begin{aligned}
\frac{da_1}{d\tau_1} &= -\mu a_2 \sin(\Delta_1 - \Delta_2) - f \sin \Delta_1, \\
a_1 \frac{d\Delta_1}{d\tau_1} &= -\mu[a_2 \cos(\Delta_1 - \Delta_2) - a_1] - \zeta_0(\tau_1)a_1 + a_1^3 - f \cos \Delta_1, \\
\frac{da_2}{d\tau_1} &= \mu a_1 \sin(\Delta_1 - \Delta_2), \\
a_2 \frac{d\Delta_2}{d\tau_1} &= -\mu[a_1 \cos(\Delta_1 - \Delta_2) - a_2] - \zeta_0(\tau_1)a_2 + a_2^3.
\end{aligned}
\tag{8.53}
$$

Figure 8.10 depicts the amplitudes of oscillations in the chain with parameters $\mu = 0.08$, $\beta = 0.01$ under the change of forcing amplitude from $f = 0.235$ to $f = 0.3$. Figure 8.10a demonstrates that the occurrence of small oscillations or AR with growing mean amplitude depends on the forcing amplitude. It is easy to obtain from (8.51) that $f_{1\mu} = 0.24$ at $\mu = 0.08$. Figure 8.10b shows that in the first cycle of oscillations, the amplitude $a_1(\tau)$ is close to the LPT of the time-independent oscillator with the same parameters (the LPTs of the time-invariant oscillator are depicted by dashed lines). The instant $T^* \approx 3.8$ corresponds to the point of inflection of the "large" LPT. It follows from (8.42) that the corresponding detuning rate $\beta^* \approx 0.037$, and thus, the admissible values of detuning rate are determined by the condition $\beta < 0.037$.

Figure 8.11 confirms that the coupling response may be insufficient to sustain growing oscillations in the attachment even in the presence of AR in the actuator. This effect is illustrated for two-particle arrays excited by forcing with parameters

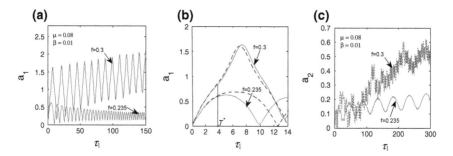

Fig. 8.10 Amplitudes of oscillations of the actuator (**a, b**) and the attachment (**c**) in the cell subjected to an external forcing with amplitude $f = 0.235 < f_{1\mu}$ or $f = 0.3 > f_{1\mu}$, respectively; LPTs of the time-invariant ($\beta = 0$) actuator are depicted by the *dashed lines* in plot (**b**)

$f = 0.3$, $\beta = 0.01$. It is clear that at $\mu = 0.25$, both oscillators exhibit AR, but at $\mu = 0.08$, the excited oscillator is captured into resonance while the passive attachment executes non-resonant oscillations with small amplitude. Localization of energy in the actuator accompanied by small oscillations of the attachment is illustrated in Fig. 8.11b.

The quasi-steady amplitudes depicted in Fig. 8.11 are close to curves (8.42). It is evident that the backbone curves for both oscillators in Fig. 8.11a are nearly identical at large times.

We recall that the frequency of a nonlinear oscillator changes as its amplitude changes, and the oscillator stays captured in resonance with its drive if the driving frequency varies slowly in time to be consistent with the frequency of the oscillator. The growth of detuning rate entails crossing the resonance domain without capture (Arnold et al. 2006). Figure 8.12 depicts this effect in the coupled oscillators with

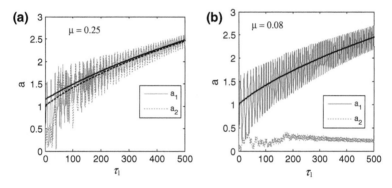

Fig. 8.11 Response amplitudes $a_1(\tau)$ and $a_2(\tau)$: **a** AR in both oscillators at $\mu = 0.25$; **b** AR in the actuator and small-amplitude oscillations in the attachment at $\mu = 0.08$. The *solid* and *dashed bold lines* in plots (**a**) and (**b**) depict backbones (8.42) for the 1st and 2nd oscillator, respectively

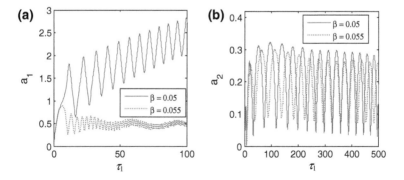

Fig. 8.12 Influence of rate β on the emergence of AR: **a** transitions from AR to small oscillations in the actuator with the growth of detuning rate; **b** small oscillations in the attached oscillator

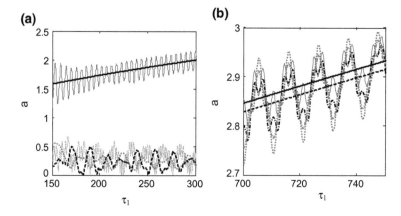

Fig. 8.13 Amplitudes of oscillations in the four-particle chain: **a** AR in the actuator and **b** small oscillations of the attached oscillators at $\mu = 0.2$; **c** AR with monotonically growing backbones for all oscillators at $\mu = 0.23$. The *solid* and *dashed bold lines* in plots (**a**) and (**c**) depict backbone curves (8.42) for the actuator and all attached oscillators, respectively

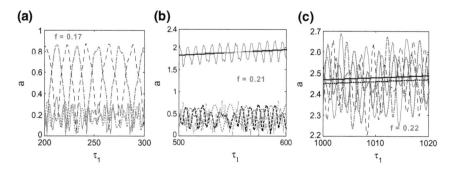

Fig. 8.14 Response amplitudes of the four-particle chain: **a** small oscillations of the entire chain at $f = 0.17 < f_{1\mu}$; **b** energy localization in the excited oscillator and small oscillations of the attachment at $f = 0.21 > f_{1\mu}$; **c** AR in the entire chain at $f = 0.22 > f_{1\mu}$; *solid lines* correspond to a_1; *dashed-dotted lines* to a_2; *dashed lines* to a_3; and *dotted lines* to a_4. The *solid* and *dashed bold lines* in Fig. 8.14b, c depict backbone curves (8.42) for the actuator and all attached oscillators, respectively

parameters $f = 0.3$, $\mu = 0.08$. From Fig. 8.12a, it is seen that even the excited oscillator escapes from resonance with an increase in the rate β.

Now, we illustrate the formation of stable AR in the multi-particle chains with $n > 2$. First, we investigate the effect of linear coupling on the dynamics of the four-particle chain. Figure 8.13 demonstrates that a decrease in the coupling stiffness μ results in escape of the attached oscillators from resonance in a four-particle chain subjected to an external force with amplitude $f = 0.4$ and detuning rate $\beta = 0.01$. Figure 8.13a, b shows that in the weakly coupled chain with

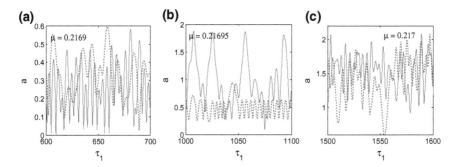

Fig. 8.15 Amplitudes $a_1(\tau)$ (*solid lines*) and $a_8(\tau)$ (*dotted lines*) of the eight-particle chain: **a** small oscillations of the entire chain at $\mu = 0.2169$; **b** energy localization in the excited oscillator against small oscillations of the attachment at $\mu = 0.21695$; and **c** AR in the entire chain at $\mu = 0.217$

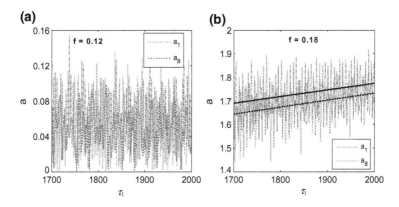

Fig. 8.16 Dependence of the response amplitudes $a_1(\tau)$ and $a_8(\tau)$ from the forcing amplitude f: **a** small oscillations of the entire chain at $f = 0.12$; **b** AR in the entire chain at $f = 0.18$. The *solid* and *dashed bold* lines depict backbones (8.32) for the actuator and the attached oscillators, respectively

$\mu = 0.2 < \mu_{cr}$, the energy is localized in the excited oscillator, but the attachment exhibits small non-resonant oscillations. Further increase in the coefficient μ enhances the coupling response and entails AR in the entire chain (Fig. 8.13c). Note that the both pairs of the parameters ($f = 0.4$, $\mu = 0.2$ and $f = 0.4$, $\mu = 0.23$) belong to the domain D (Fig. 8.9). This implies that the behavior of the chain cannot be predicted beforehand and requires a numerical study.

Figure 8.14 depicts the response amplitudes of the four-particle chain with parameters $\mu = 0.25$, $\beta = 0.005$ but with different amplitudes of external forcing.

It follows from (8.36) that $f_{1\mu} = 0.176$ at $\mu = 0.25$. Figure 8.14a depicts small oscillations of the chain at $f = 0.17 < f_{1\mu}$. An increase in the forcing amplitude entails energy localization in the excited oscillator against small oscillations of the attachment at $f = 0.21$ (Fig. 8.14b) and capture into resonance of the entire chain at $f = 0.22$ (Fig. 8.14c). The initial segments of chaotic motion and transitions to regular oscillations are withdrawn from consideration.

Figures 8.15 and 8.16 demonstrate the results of numerical simulations for the eight-particle chain subjected to forcing with parameters $f = 0.25$, $\beta = 0.001$. For brevity, only the dynamics of the first (excited) and last (eight) oscillators are illustrated. Figure 8.15 indicates that an increase in the coupling strength μ leads to the transformations of small oscillations (Fig. 8.15a) into large oscillations of the actuator (Fig. 8.15b) and then into AR in the entire chain (Fig. 8.15c).

Figure 8.16 illustrates the transitions from small oscillations to AR with an increase in the forcing amplitude f in the chain with parameters $\mu = 0.25$, $\beta = 0.001$.

References

Andronov, A.A., Vitt, A.A., Khaikin, S.E.: Theory of Oscillators. Dover, New York (1966)

Arnold, V.I., Kozlov, V.V., Neishtadt, A.I.: Mathematical Aspects of Classical and Celestial Mechanics, 3rd edn. Springer, Berlin (2006)

Barth, I., Friedland, L.: Two-photon ladder climbing and transition to autoresonance in a chirped oscillator. Phys. Rev. A **87**(1–4), 053420 (2013)

Ben-David, O., Assaf, M., Fineberg, J., Meerson, B.: Experimental study of parametric autoresonance in Faraday waves. Phys. Rev. Lett. **96**(1–4), 154503 (2006)

Bohm, D., Foldy, L.L.: Theory of the synchro-cyclotron. Phys. Rev. **72**, 649–661 (1947)

Chacón, R.: Energy-based theory of autoresonance phenomena: application to duffing-like systems. Europhys. Lett. **70**, 56–62 (2005)

Dodin, I.Y., Fisch, N.J.: Adiabatic nonlinear waves with trapped particles. III. Wave dynamics. Phys. Plasmas **19**, 012104 (2012)

Friedland L.: http://www.phys.huji.ac.il/ ~ lazar/

Friedland, L.: Efficient capture of nonlinear oscillations into resonance. J. Phys. A: Math. Theor. **41**(1–8), 415101 (2008)

Gradshteyn, I.S., Ryzhik, I.M.: Tables of Integrals, Series, and Products, 6th edn. Academic Press, San Diego (2000)

Kalyakin, L.A.: Asymptotic analysis of autoresonance models. Russ. Math. Surv. **63**, 791–857 (2008)

Kivshar, Y.S.: Intrinsic localized modes as solitons with a compact support. Phys. Rev. E **48**, R43–R45 (1993)

Korn, G.A., Korn, T.M.: Mathematical Handbook for Scientists and Engineers, 2nd edn. Dover Publications, New York (2000)

Kovaleva A.: Resonance energy transport in an oscillator chain. http://arXiv:1501.00552. (2015)

Kovaleva, A., Manevitch, L.I.: Classical analog of quasilinear Landau-Zener tunneling. Phys. Rev. E. **85**(1–8), 016202 (2012)

Kovaleva, A., Manevitch, L.I.: Resonance energy transport and exchange in oscillator arrays. Phys. Rev. E **88**(1–10), 022904 (2013a)

Kovaleva, A., Manevitch, L.I.: Emergence and stability of autoresonance in nonlinear oscillators. Cybern. Phys. **2**, 25–30 (2013b)

Kovaleva, A., Manevitch, L.I.: Limiting phase trajectories and emergence of autoresonance in nonlinear oscillators. Phys. Rev. E **88**(1–6), 024901 (2013c)

Kovaleva, A., Manevitch, L.I.: Autoresonance energy transfer versus localization in weakly coupled oscillators. Phys D Nonlinear Phenom **320**, 1–8 (2016)

Kovaleva, A., Manevitch, L.I., Manevitch, E.L.: Intense energy transfer and superharmonic resonance in a system of two coupled oscillators. Phys. Rev. E **81**(1–12), 056215 (2010)

Marcus, G., Friedland, L., Zigler, A.: From quantum ladder climbing to classical autoresonance. Phys. Rev. A **69**(1–5), 013407 (2004)

McMillan, E.M.: The synchrotron—a proposed high energy particle accelerator. Phys. Rev. **68**, 143–144 (1945)

Murch, K.W., Vijay, R., Barth, I., Naaman, O., Aumentado, J., Friedland, L., Siddiqi, I.: Quantum fluctuations in the chirped pendulum. Nat. Phys. **7**, 105–108 (2011)

Nayfeh, A.H., Mook, D.T.: Nonlinear Oscillations. Wiley-VCH, Weinheim (2004)

Neishtadt, A.I.: Passage through a separatrix in a resonance problem with slowly varying parameter. J. Appl. Math. Mech. **39**, 594–605 (1975)

Sanders, J.A., Verhulst, F., Murdock, J.: Averaging Methods in Nonlinear Dynamical Systems, 2nd edn. Springer, Berlin (2007)

Shalibo, Y., Rofe, Y., Barth, I., Friedland, L., Bialczack, R., Martinis, J.M., Katz, N.: Quantum and classical chirps in an anharmonic oscillator. Phys. Rev. Lett. **108**(1–5), 037701 (2012)

Veksler, V.I.: Some new methods of acceleration of relativistic particles. ComptesRendus (Dokaldy)de l'Academie Sciences de l'URSS **43**, 329–331 (1944)

Zelenyi, L.M., Neishtadt, A.I., Artemyev, A.V., Vainchtein, D.L., Malova, H.V.: Quasiadiabatic dynamics of charged particles in a space plasma. Physics-Uspekhi **56**, 347–394 (2013)

Part III
Applications

The final part of this book contains several important and extremely interesting applications. Besides the classical problem of the forced pendulum, which is solved for the arbitrary amplitudes for the first time, we consider a number of discrete systems, which concern the nonlinear sinks, the metamaterials with intrinsic degrees of freedom, and the locally supported chains. Using the effective semi-inverse method allows exceeding the frontiers of the small-amplitude approximation even in the problems where the small parameter is not presented in the initial formulation of the puzzle. By such a manner, we demonstrate the fundamental properties of the LPT concept not only for the discrete systems, but for the objects with distributed parameters such as the carbon nanotubes.

Chapter 9
Targeted Energy Transfer

This chapter presents the analytical and numerical study of energy transport in a system of n linear impulsively loaded oscillators (a primary linear system), in which the nth oscillator is coupled with an essentially nonlinear attachment—the nonlinear energy sink (NES). It was noted [see, e.g., (Vakakis et al. 2008 and references therein; Manevitch et al. 2007; Vaurigaud et al. 2011; Kovaleva and Manevitch 2013)] that the most effective energy exchange arises when the response of the primary system without an attachment is almost monochromatic; that is, all oscillators exhibit harmonic oscillations with an identical dominant frequency, close to that of a linear normal mode.

We suggest an order-reduction procedure, which allows the separated dynamical analysis for the edge oscillator coupled with the NES and the remaining linear part of the system. Using simplifications based on the reduced two degree-of-freedom (2 DOF) model, we depict the admissible domain of parameters and derive a closed-form approximate solution adequately describing the strongly non-stationary processes in the entire system in terms of LPTs.

9.1 The Model

We study an array of n linear oscillators (a primary linear system), in which the nth oscillator is weakly coupled to an essentially nonlinear attachment (NES). For simplicity, we assume unidirectional motion of the primary system such that each oscillator moves in line or in parallel with the NES. In this case, the system dynamics is governed by the equations

© Springer Nature Singapore Pte Ltd. 2018
L.I. Manevitch et al., *Nonstationary Resonant Dynamics of Oscillatory Chains and Nanostructures*, Foundations of Engineering Mechanics,
DOI 10.1007/978-981-10-4666-7_9

$$\frac{d^2 X_r}{dt^2} + \sum_{k=1}^{n} c_{rk} X_k = 0, \quad r = 1, 2, \ldots, n-1,$$

$$\frac{d^2 X_n}{dt^2} + \sum_{k=1}^{n} c_{nk} X_k - M_n^{-1}(KX^3 + 2\chi X) = 0, \quad (9.1)$$

$$\frac{d^2 X}{dt^2} + \frac{d^2 X_n}{dt^2} + m^{-1}(KX^3 + 2\chi X) = 0,$$

where $X_r (r = 1, 2, \ldots, n)$ and $X = X_{n+1} - X_n$ refer to the absolute displacements of the rth oscillator and the relative displacement of the NES, respectively; $c_{rk} = C_{rk}/M_r$, where M_r is the mass of the rth oscillator, C_{rk} is stiffness of linear coupling between the rth and kth oscillators; m is the attached mass, 2χ and K denote the coefficients of linear and cubic stiffness of the NES with domination of the nonlinear contribution (in particular, the linear stiffness of NES can be equal to zero).

The selected initial conditions correspond to impulses imposed onto the linear oscillators with the system being initially under conditions: $X_r = 0$, $dX_r/dt = P_r$, $r = 1, 2, \ldots, n$; $X = 0$, $dX/dt = -P_n$ at $t = +0$. In what follows, the sign "+" is omitted.

9.2   Analytical Study

Let $\Omega_1, \ldots, \Omega_n$ be the natural frequencies of the decoupled primary linear system. It is supposed $\Omega_r \neq \Omega_k$ if $r \neq k$. The resonance interaction between the linear system and the NES may exist only if motion of the linear oscillator is close to harmonic oscillations with a natural frequency, and the NES is well-tuned at the same frequency (it is possible due to domination of the nonlinear contribution into elastic force arising in the NES). We assume that this condition holds for the frequency Ω_1.

We develop an order-reduction procedure that makes the essential mathematics of the problem as clear as possible. Assuming a lightweight attachment, we define the small parameter of the system as $\varepsilon = (m/M_n)^{1/2} \ll 1$. Then, we define the dimensionless time $\tau_0 = \Omega_1 t$ and transform the parameters and the initial conditions as follows:

$$K/m\Omega_1^2 = \kappa^{-2}, \quad \chi/m\Omega_1^2 = \varepsilon\sigma, \quad c_{rk}/\Omega_1^2 = \sigma_{rk}, \quad P_r/\kappa\Omega_1 = \varepsilon V_r \quad (9.2)$$

It can be shown that the relative displacement X is of $O(1)$, while the displacements X_r are of $O(\varepsilon)$. For convenience, the dimensionless variables are redefined as follows:

$$u = K^{-1}X, \quad u_r = \kappa^{-1}X_r = \varepsilon\zeta_r, \tag{9.3}$$

in which the variable ζ_r is of $O(1)$. Substituting (9.2) and (9.3) into Eq. (9.1), we obtain the dimensionless equations

$$\frac{d^2\zeta_r}{d\tau_0^2} + \sum_{k=1}^{n} \sigma_{rk}\zeta_k = 0, \quad r = 1, 2, \ldots, n-1,$$

$$\frac{d^2\zeta_n}{d\tau_0^2} + \sum_{k=1}^{n} \sigma_{nk}\zeta_k - \varepsilon(u^3 + 2\varepsilon\sigma u) = 0, \tag{9.4}$$

$$\frac{d^2u}{d\tau_0^2} + \varepsilon\frac{d^2\zeta_n}{d\tau_0^2} + u + \varepsilon\mu(u^3 - u) + 2\varepsilon\sigma u = 0, \quad \mu = 1/\varepsilon.$$

With initial conditions $\zeta_r = 0$, $d\zeta_r/dr_0 = V_r$; $u = 0$, $du/d\tau_0 = -\varepsilon V_n$ at $\tau_0 = 0$. The dimensionless natural frequencies of the primary linear subsystem are defined as $\omega_k = \Omega_k/\Omega_1$; the resonance frequency $\omega_1 = 1$. The resonance condition implies that the parenthetical expression with factor $\varepsilon\mu$ in the last equation is relatively small compared to all other terms of order 1, while the term $\mu(u^3 - u)$ is of $O(1)$ [see detailed discussion in Kovaleva and Manevitch (2013)].

The selected mode of oscillations is extracted with the help of the Laplace transform for the first group of equation in (9.4). The equations for the Laplace transform $Z_r(s)$ are given by

$$s^2 Z_r(s) + \sum_{k=1}^{n} \sigma_{rk}Z_k(s) = V_r, \quad r = 1, 2, \ldots, n-1,$$

$$s^2 Z_n(s) + \sum_{k=1}^{n} \sigma_{nk}Z_k(s) = \varepsilon R(s) + V_n, \tag{9.5}$$

where $Z_r(s)$ is the Laplace transforms of $\zeta_r(\tau_0)$, $R(s)$ is the Laplace transform of the nonlinear coupling $r = u^3 + 2\varepsilon\sigma u$ considered as a function of time τ_0, and s designates the Laplace transform variable. It follows from (9.5) that

$$Z_r(s) = D^{-1}(s)\left[\sum_{j=1}^{n} l_{rj}(s)V_j + \varepsilon l_{rn}(s)R(s)\right], \quad r = 1, \ldots, n. \tag{9.6}$$

It was proved [see, e.g., (Kolovsky 1999; Meirovitch 2001)] that the characteristic polynomial of the non-dissipative oscillators takes the form $D(s) = \prod_{k=1}^{n}(s^2 + \omega_k^2)$, while $l_{rj}(s)$ are the polynomials of order less than n in s^2. It now follows that Eq. (9.5) can be rewritten as

$$(s^2 + w_1^2)Z_r(s) = \sum_{j=1}^{n} \Lambda_{rj}(s)V_j + \varepsilon\Lambda_m(s)(s)R(s), \quad r = 1,\ldots,n,$$

$$\Lambda_{rj}(s) = l_{rj}(s)/D_1(s), \quad D_1(s) = \prod_{k=2}^{n}(s^2 + \omega_k^2).$$

Inverse Laplace transforming yields the equations with the highlighted resonance substructure

$$\frac{d^2\zeta_r}{d\tau_0^2} + \sum_{k=1}^{n-1}\sigma_{rk}\zeta_k = -\sigma_m\zeta_n, \quad r = 1,2,\ldots,n-1, \tag{9.7}$$

$$\frac{d^2\zeta_n}{d\tau_0^2} + \zeta_n + \varepsilon\mu f_n(\tau_0) + \varepsilon^2 q_n(\tau_0) - \varepsilon(u^3 + 2\varepsilon\sigma u) = 0,$$

$$\frac{d^2 u}{d\tau_0^2} + \varepsilon\frac{d^2\zeta_n}{d\tau_0^2} + u + \varepsilon\mu(u^3 - u) + 2\varepsilon\sigma u = 0, \tag{9.8}$$

where the functions f_n and q_n are defined by the equalities

$$f_n(\tau_0) = -\sum_{k=1}^{n-1}\sigma_{nk}w_k(\tau_0), \quad w_k(\tau_0) = \sum_{j=2}^{n}a_{kj}V_j\sin\omega_k\tau_0,$$

$$q_n(\tau_0) = \sum_{k=2}^{n}b_k\psi_k(\tau_0), \quad \psi_k(\tau_0) = \int_0^{\tau_0}\sin\omega_k(\tau_0 - z)r(u(z))dz,$$

with easily calculated coefficients a_{kj} and b_k; the sum $w_k(\tau_0)$ represents the partial response of the kth oscillator in the primary linear system (excluding the harmonic of frequency 1). The coefficient $\varepsilon\mu \equiv 1$ implies that $|f_n(\tau_0)| \ll 1$ but $\mu|f_n(\tau_0)| \sim O(1)$. Note that if the primary linear system exhibits nearly harmonic oscillations of frequency $\omega_1 = 1$, then the contribution of the higher harmonics in the system dynamics is insignificant and the function $f_n(\tau_0)$ is necessarily small.

9.3   Selection of Resonance Terms and Principal Asymptotic Approximation

As in the previous sections, the selection of the resonance terms is performed with the help of the complexification-averaging procedure. To this end, the following complex-valued variables are introduced:

$$\varphi_n = (\mathrm{d}\zeta_n/\mathrm{d}\tau_0 + i\zeta_n)e^{-i\tau_0}, \quad \varphi = (\mathrm{d}u/\mathrm{d}\tau_0 + iu)e^{-i\tau_0}, \tag{9.9}$$

where the functions φ and φ_n are sought in the form of the timescale series

$$\begin{aligned}
\varphi_n(\tau_0, \tau_1) &= \varphi_n^{(0)}(\tau_1) + \varepsilon\varphi_n^{(1)}(\tau_0, \tau_1) + \cdots, \\
\varphi(\tau_0, \tau_1) &= \varphi^{(0)}(\tau_1) + \varepsilon\varphi^{(1)}(\tau_0, \tau_1) + \cdots, \quad \tau_1 = \varepsilon\tau_0
\end{aligned} \tag{9.10}$$

Substituting (4.9) and (4.10) into (4.8) and separating the resonant terms, we obtain the following equations to the main approximations $\varphi_n^{(0)}(\tau_1)$, $\varphi^{(0)}(\tau_1)$:

$$\begin{aligned}
&\frac{\mathrm{d}\varphi_n^{(0)}}{\mathrm{d}\tau_1} + \frac{i}{2}\varphi^{(0)} = 0, \quad \varphi_n^{(0)}(0) = V_n, \\
&\frac{\mathrm{d}\varphi^{(0)}}{\mathrm{d}\tau_1} + i\mu\left(\frac{1}{2} - \frac{3}{8}|\varphi^{(0)}|^2\right)\varphi^{(0)} - i\sigma\varphi^{(0)} + \frac{i}{2}\varphi_n^{(0)} = 0, \quad \varphi^{(0)}(0) = 0.
\end{aligned} \tag{9.11}$$

It has been demonstrated (Kovaleva and Manevitch 2013) that Eq. (9.11) is identical to the equations of the two-state atomic tunneling (Raghavan et al. 1999), thereby confirming a direct mathematical analogy between quantum and classical transitions.

For brevity, we denote $V_n = v$. The transformations $\varphi_n^{(0)} = v\cos\theta e^{i\delta_1}$ and $\varphi^{(0)} = v\sin\theta e^{i\delta_2}$ lead to the equations for the real-valued variables θ and $\Delta = \delta_1 - \delta_2$

$$\begin{aligned}
&\frac{\mathrm{d}\theta}{\mathrm{d}\tau_1} = \frac{1}{2}\sin\Delta, \\
&\sin 2\theta\frac{\mathrm{d}\Delta}{\mathrm{d}\tau_1} = (\cos\Delta + 2k\sin 2\theta)\cos 2\theta - g\sin 2\theta,
\end{aligned} \tag{9.12}$$

with initial conditions $\theta(0) = 0$, $\Delta(0) = \pi/2$. The parameters k and g are defined as

$$k = \frac{3\mu}{32}v^2, \quad g = \sigma - \frac{\mu}{2}\left(1 - \frac{3}{8}v^2\right). \tag{9.13}$$

As shown in recent studies (Manevitch et al. 2007), the trajectory with initial conditions $\theta = 0$ and $\Delta = \pi/2$ represents an outer boundary for a set of closed trajectories encircling the stable center in the phase plane (θ, Δ). It is referred to as the limiting phase trajectory (LPT) of system (9.12). We recall that motion along the LPT ensures the maximum possible energy exchange between two coupled oscillators.

Note that system (9.12) conserves the integral of motion $K = (\cos\Delta + k\sin 2\theta)\sin 2\theta + g\cos 2\theta = \text{const.}$ Since $\theta(0) = 0$, $\Delta(0) = \pi/2$ on the LPT, we obtain

$$H = (\cos \Delta + k \sin 2\theta) \sin 2\theta + g \cos 2\theta = g. \tag{9.14}$$

Once $\theta(\tau_1)$ and $\Delta(\tau_1)$ are found, the main approximations $\zeta_n^{(0)}$ and $u^{(0)}$ to the solution of system (9.8) can be calculated by (9.9) and (9.10). As a result, we obtain

$$\zeta_n^{(0)}(\tau_0, \tau_1) = -v \cos \theta(\tau_1) \sin(\tau_0 - \delta_1(\tau_1)),$$
$$u^{(0)}(\tau_0, \tau_1) = -v \sin \theta(\tau_1) \sin(\tau_0 - \delta_2(\tau_1)). \tag{9.15}$$

It follows from (9.15) that Eq. (9.12) suffices to approximately describe resonance energy transfer provided the phases δ_1 and δ_2 are defined. We recall that $\zeta_n^{(0)} = v \sin \delta_1(0) = 0$, $u^{(0)} = 0$ at $\tau_0 = 0$. Hence, one can set $\delta_1 = \Delta + \pi/2$, $\delta_2 = \pi/2$, or $\delta_1(0) = \pi$. It now follows that the leading-order approximations to the solutions ζ_n and u of system (9.8) are given by:

$$\zeta_n^{(0)}(\tau_0, \tau_1) = v \cos \theta(\tau_1) \cos(\tau_0 - \Delta(\tau_1)), \quad u^{(0)}(\tau_0, \tau_1) = v \sin \theta(\tau_1) \cos \tau_0. \tag{9.16}$$

The leading-order approximations $\zeta_r^{(0)}(r \le n-1)$ can be easily found from Eq. (9.7). An example is given below. Formally, expressions (9.16) are independent of $\zeta_r(r \le n-1)$. However, further approximations take into account the higher frequency components f_n, q_n that reflect the effect of additional degrees of freedom on the motion of the nth oscillator and the NES.

It was shown in the previous section that the parametric boundary between the energy localization on the excited oscillator and strong energy exchange in system (9.8) is expressed as $|k+g| = 1$ or, by the above definition,

$$\left| \frac{\mu}{2} \left(\frac{9}{16} v^2 - 1 \right) + \sigma \right| = 1. \tag{9.17}$$

The condition $|k+g| < 1$ corresponds to localization of energy on the initially excited oscillators, while $|k+g| > 1$ corresponds to energy transfer. We note that these conditions reflect the essential difference between linear and nonlinear resonances; namely, the NES resonates with a mode of the primary system only above a certain energy threshold. The obtained conditions are illustrated in Fig. 9.1. The parameters of the numerical simulation are given by

$$\varepsilon = 0.333, \quad \sigma = 0.158, \quad \mu = 3.16 \tag{9.18}$$

It now follows from (9.17) and (9.18) that the critical (dimensionless) value of the impulse $v^* = 1.59$. Plots of $\theta(\tau_1)$ and $\Delta(\tau_1)$ for $v < v^*$ and $v > v^*$ are shown in Fig. 9.1.

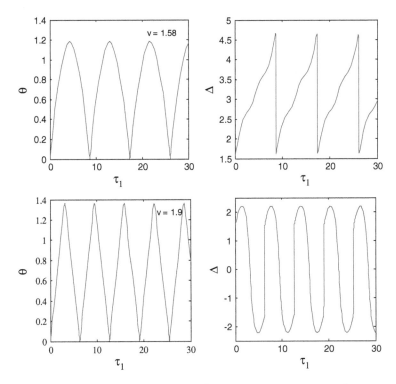

Fig. 9.1 Plots of $\theta(\tau_1)$ and $\Delta(\tau_1)$ for $v = 1.58 < v^*$ (*upper panel*) and $v = 1.9 > v^*$ (*bottom panel*)

As shown in Fig. 9.1, in the regime of intense energy exchange, $v > v^* \theta(\tau_1)$ is close to the sawtooth function (9.19). This implies the leading-order approximation in the form

$$
\theta^{(0)}(\tau_1) = \theta_M \tau(\phi), \quad \Delta^{(0)}(\tau_1) = \frac{\pi}{2}\frac{d\tau}{d\phi}, \quad \phi = 2\frac{\tau_1}{T_0},
$$

$$
\tau(\phi) = \frac{2}{\pi}|\arcsin[\sin(\pi\phi/2)]|, \quad e(\phi) = \tau_\varphi(\phi) = \mathrm{sign}[\sin(\pi\phi/2)],
$$

(9.19)

where θ_M and T_0 denote the magnitude and the period of functions (9.19), respectively. Plots of $\tau(\phi)$ and $d\tau/d\phi$ are shown in Fig. 6.3 (Part II).

Note that in the interval $0 \leq \tau_1 \leq T_0/2$ the sawtooth function $\theta_0(\tau_1) = 2\theta_M \tau_1/T_0$. Then, it follows from (9.12) that $d\theta_0/d\tau_1 = 1/2$ at $\tau_1 = 0$, and therefore, $2\theta_M/T_0 = 1/2$, $T_0 = 4\theta_M$. It is important to note that the amplitude θ_M can be analytically found using the integral of motion (9.14). Since the maximum θ_M is achieved at $\Delta = 0$, the maximum θ_M satisfies the equation

$$(1 + k \sin 2\theta) \sin 2\theta_M - 2g \sin^2 \theta_M = 0 \tag{9.20}$$

Note that equality (9.20) defines an exact maximum θ_M of $\theta(\tau_1)$, while the corresponding period of the slow envelope $T_0 = 4\theta_M$ gives an approximation of the exact period T based on the sawtooth approximation. For example, if $v = 1.9 > v^*$, then $\theta_M = 1.39$, $T_0 = 4\theta_M = 5.56$, while the exact period of beat oscillations $T \approx 6.3$ (Fig. 9.1); the difference between the approximate and exact values is about 12%. We recall that the envelope of the NES oscillations is defined as $a = v|\sin\theta|$. The obtained estimates define the maximum of the slow envelope as $a_M = v|\sin\theta_M|$. Finally, the approximate solution can be represented as

$$
\begin{aligned}
u_n^{(0)}(\tau_0, \tau_1) &= \varepsilon v \cos\theta^{(0)}(\tau_1) \cos(\tau_0 - \Delta^{(0)}(\tau_1)), \\
u^{(0)}(\tau_0, \tau_1) &= v \sin\theta^{(0)}(\tau_1) \sin\tau_0.
\end{aligned}
\tag{9.21}
$$

The approximate solution may be improved through successive iterations. The rth iterations $\theta^{(r)}(\tau_1), \Delta^{(r)}(\tau_1)$ are defined by the following equations:

$$
\begin{aligned}
\frac{d\theta^{(r)}}{d\tau_1} &= \frac{1}{2}\sin\Delta^{(r)}, \\
\frac{d\Delta^{(r)}}{d\tau_1} &= \cos\Delta^{(r)}\cot 2\theta^{(r-1)} + 2k\cos 2\theta^{(r-1)} - g, \quad r = 1, 2, \ldots \\
\theta^{(r)}(0) &= 0, \quad \Delta^{(r)}(0) = \pi/2.
\end{aligned}
\tag{9.22}
$$

Fig. 9.2 Plots of the exact solution $\theta(\tau_1)$ (*red solid line*) and iterations $\theta_0(\tau_1)$ (*black dashed line*), $\theta_1(\tau_1)$ (*blue dotted line*)

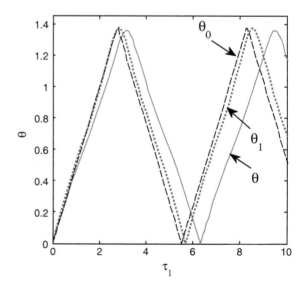

From Fig. 9.2, it is seen that the first iteration $\theta^{(1)}(\tau_1)$ (dotted line) only slightly improves the initial sawtooth approximation $\theta^{(0)}(\tau_1)$ (dashed line). This implies that a simple function $\theta^{(0)}(\tau_1)$ suffices to adequately describe the real process.

9.4   3 DOF Oscillators with the NES

As an example, we consider a 4 DOF systems consisting of three identical linear oscillators excited by equal impulses and nonlinearly coupled with an attachment (Fig. 9.3). Note that this model is chosen for illustrative purposes, but the qualitative features of the results hold true for more complicated systems with a single NES.

Absolute displacements of the oscillators are denoted by $X_r(r = 1,...,4)$; the equal masses of the oscillators are denoted by M; the attached mass is denoted by m, the coefficient of linear stiffness is denoted by C; and the nonlinear and linear stiffness coefficients of the NES are denoted by K and 2χ, respectively. In this notation, the equations of motion are given by

$$\frac{d^2 X_1}{dt^2} + \lambda^2(2X_1 - X_2) = 0,$$
$$\frac{d^2 X_2}{dt^2} + \lambda^2(2X_2 - X_1 - X_3) = 0,$$

(9.23)

$$\frac{d^2 X_3}{dt^2} + \lambda^2(X_3 - X_2) = M^{-1}(KX^3 + 2\chi X),$$
$$\frac{d^2 X}{dt^2} + \frac{d^2 X_3}{dt^2} = -m^{-1}(KX^3 + 2\chi X),$$

where $X = X_4 - X_3$, and $\lambda^2 = C/M$. Initial conditions at $t = 0$ correspond to equal impulses of strength P applied to linear oscillators, i.e., $X_r = 0$, $dX_r/dt = P$, $r = 1$, 2, 3; $X = 0$, $dX/dt = -P$.

The characteristic equation of the primary linear system takes the form

$$D(s) = (s^2 + 2\lambda^2)(s^2 + \lambda^2) - \lambda^2(2s^2 + 3\lambda^2) = 0$$

(9.24)

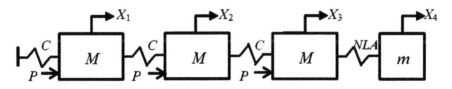

Fig. 9.3 Identical oscillators coupled with the NES

The roots $s_1^2 = -0.2\lambda^2$, $s_2^2 = -1.5\lambda^2$, and $s_3^2 = -3.2\lambda^2$ correspond to the frequencies $\Omega_1 = 0.445\lambda$, $\Omega_2 = 1.22\lambda$, and $\Omega_3 = 1.79\lambda$. As in the previous section, we assume that $m/M = \varepsilon^2 \ll 1$. The introduction of the timescale $\tau_0 = \Omega_1 t$ and rescaling (9.2) and (9.3) reduces system (9.23) to the dimensionless form with the nonlinear coefficient to be unity, that is

$$\frac{d^2 u_1}{d\tau_0^2} + 5(2u_1 - u_2) = 0,$$

$$\frac{d^2 u_2}{d\tau_0^2} + 5(2u_2 - u_1 - u_3) = 0,$$

$$\frac{d^2 u_3}{d\tau_0^2} + 5(u_3 - u_2) - \varepsilon^2(u^3 + 2\varepsilon\sigma u) = 0,$$

$$\frac{d^2 u}{d\tau_0^2} + \frac{d^2 u_3}{d\tau_0^2} + u^3 + 2\varepsilon\sigma u = 0.$$

(9.25)

With initial conditions $u_r = u = 0$, $du_r/d\tau_0 = \varepsilon v$, and $du/d\tau_0 = -\varepsilon v$ at $\tau_0 = 0$. As indicated previously, $\varepsilon v = \kappa P/\Omega_1$, and $\kappa = \sqrt{K/m\Omega_1^2}$. The dimensionless natural frequencies of the primary linear system are calculated as $\omega_1 = 1$, $\omega_2 = 2.71$, and $\omega_3 = 3.98$.

It was mentioned earlier that the resonance interaction with intense energy transfer may exist only when the linear oscillations of the primary system are nearly monochromatic. Formally, this condition can be verified by solving for the linear system, but the numerical simulations give straightforward results. Figure 9.4 demonstrates numerical results for system (9.25) and its linear counterpart with parameters (9.18) and $v = 1.61 > v^*$.

In Fig. 9.4, it is clearly seen that all oscillators in the primary linear systems are nearly monochromatic with frequency close to 1, even though the effect of higher harmonics is distinguished in the displacement u_1 of the first oscillator. At the same time, in the nonlinear system, the nearly harmonic oscillations turn into nonlinear beating but with the clearly pronounced higher frequency components in the displacement u_1 of the first oscillator.

From Fig. 9.5, it is seen that leading-order approximations $u^{(0)}$ and $u_3^{(0)}$ are close to the exact (numerical) solutions u and u_3 of the full system (9.25); the discrepancy between the periods and maximum values of the envelope of beating and its analytical approximation $a = v|\sin\theta|$ is about 12 and 10%, respectively. We recall that the function $\theta(\tau_0)$ has been found analytically in the end of Part II. Therefore, the developed asymptotic procedure provides an effective tool for obtaining a closed-form approximate solution adequately describing the transient dynamics of the entire system.

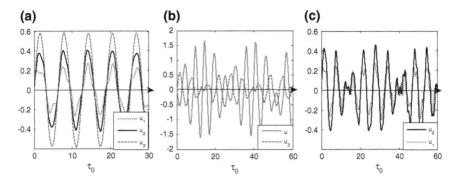

Fig. 9.4 Displacements $u_r(\tau_0)$ and $u(\tau_0)$ of the linear (**a**) and nonlinear (**b, c**) oscillators, with parameters $\varepsilon^2 = 0.1$, $\sigma = 0.158$, $\nu = 1.61$, $\nu^* = 1.59$

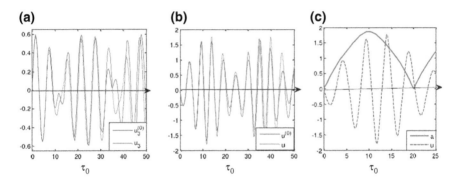

Fig. 9.5 Exact and approximate solutions of (9.25)

Now, we calculate the approximate solutions of system (9.25). Recalling that $u_r = \varepsilon \zeta_r$, we obtain from (9.21) that

$$
\begin{aligned}
u^{(0)}(\tau_0, \tau_1) &= a(\tau_1)\cos\tau_0 \\
u_3^{(0)}(\tau_0, \tau_1) &= a_3(\tau_1)\cos(\tau_0 - \Delta^{(0)}(\tau_1)),
\end{aligned}
\tag{9.26}
$$

where $a(\tau_1) = \nu \sin\theta^{(0)}(\tau_1)$, and $a_3(\tau_i) = \varepsilon\nu\cos\theta^{(0)}(\tau_1)$. It follows from (4.26) that the maximum amplitudes $a_{3M} = \varepsilon\nu\sin\theta_M = 0.52$, $a_M = \nu\sin\theta_M = 1.65$ are close to the exact (numerical) results presented in Fig. 9.5.

The leading-order approximations to beat oscillations of the first and second oscillators can be found from the first pair of equations in (9.25), in which u_3 is changed to $u_3^{(0)}$. Considering the slow timescale τ_1 as a "frozen" parameter, we obtain the leading-order approximations to beat oscillations as $u_r^{(0)}(\tau_0, \tau_1) = a_r(\tau_i)\cos(\tau_0 - \Delta^{(0)}(\tau_1))$ with slow envelopes $a_r(\tau_i) = a_{rM}\cos\theta^{(0)}(\tau_1)$, $r = 1, 2$. The maximum amplitudes $a_{1M} = 0.162$ and $a_{2M} = 0.34$ are close to the maximums of the processes plotted in Fig. 9.4.

9.5   Transient Dynamics of the Dissipative System

To provide the efficiency of NES, we have to pass it in the resonance conditions a significant part of the energy from the linear subsystem and then to prevent a reverse energy flow. This can be achieved due to dissipation leading to decrease of the amplitude and consequently of the frequency also. Therefore, in this section, we analyze the effect of NES in the weakly dissipative system. For brevity, we consider the 2 DOF systems consisting of the linear oscillator coupled with the NES. The equations of motion are given by

$$M\frac{d^2X_1}{dt^2} + CX_1 - \left(KX^3 + 2\chi X + H\frac{dX}{dt}\right) = 0,$$

$$m\left(\frac{d^2X}{dt^2} + \frac{d^2X_1}{dt^2}\right) + \left(KX^3 + 2\chi X + H\frac{dX}{dt}\right) = 0. \tag{9.27}$$

We recall that oscillations of the dissipative system vanish at rest O: ($X_1 = X_2 = 0$, $dX_1/dt = dX_2/dt = 0$) as $t \to \infty$. This implies that the effect of dissipation, whatever small it might be, must be considered in the approximate solution; otherwise, the convergence to O is lost.

As in previous sections, we consider the small parameter $\varepsilon = (m/M)^{1/2}$ and then introduce the dimensionless time $\tau_0 = \Omega t$, $\Omega = (C/M)^{1/2}$, the dissipation parameter $H/m\omega_1 = 2\varepsilon\eta$, and the rescaled variables $u_{1,2}$ and u (see (9.2), (9.3)). The resulting dimensionless equations of the weakly dissipative system can be written as follows:

$$\frac{d^2u_1}{d\tau_0^2} + u_1 - \varepsilon^2\left[c(u_2 - u_1)^3 + 2\varepsilon\sigma(u_2 - u_1) + \varepsilon\eta\frac{d}{d\tau_0}(u_2 - u_1)\right] = 0,$$

$$\frac{d^2u_2}{d\tau_0^2} + \left[c(u_2 - u_1)^3 + 2\varepsilon\sigma(u_2 - u_1) + \varepsilon\eta\frac{d}{d\tau_0}(u_2 - u_1)\right] = 0. \tag{9.28}$$

Where $u_2 = u + u_1$ represents the absolute NES displacement. Initial conditions are defined as $u_1 = u_2 = 0$; $du_1/d\tau_0 = \varepsilon v$, $du_2/d\tau_0 = 0$ at $\tau_0 = 0$. Figure 9.6 presents the results of numerical simulation with parameters (9.18) and, additionally, $v = 1.67$, and $\eta = 0.333$.

From Fig. 9.6, it is seen that motion of the dissipative system is separated into two parts: In the first interval $0 \le \tau_0 \le \tau_0^*$, where τ_0^* corresponds to the first maximum of the slow envelope of beat oscillations, motion is close to strongly nonlinear undamped oscillations described well by LPT, but then the system approaches the rest state O_1 with an exponentially decreasing amplitude of oscillations. Different frequencies of damping oscillations and dissipation rates for the processes u_1, u_2, and $u = u_2 - u_1$ are observed. Note that the difference in dissipation rates stems from the difference in masses of coupled bodies.

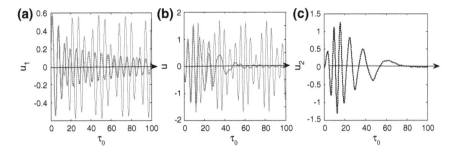

Fig. 9.6 Displacements $u_1(\tau_0)$ (plot (**a**)) and $u(\tau_0)$ (plot (**b**)) of the dissipative (*dashed lines*) and non-dissipative (*solid lines*) systems, respectively; plot (**c**) depicts the absolute NES displacement $u_2(\tau_0)$ in the dissipative system

In the first interval $0 \le \tau_0 \le \tau_0^*$, the transient dynamics can be analyzed with the help of the previously developed procedure. Since the contribution of the nonlinear force near the rest state is negligible, at large times system (9.28) can be approximated by a system linearized near the rest state O_1. Linearization of Eq. (9.28) near O yields the system

$$\frac{d^2\tilde{u}_1}{d\tau_0^2} + \tilde{u}_1 - 2\varepsilon^3\left[\sigma(\tilde{u}_2 - \tilde{u}_1) + \eta\frac{d}{d\tau_0}(\tilde{u}_2 - \tilde{u}_1)\right] = 0,$$

$$\frac{d^2\tilde{u}_2}{d\tau_0^2} - 2\varepsilon^3\left[\sigma(\tilde{u}_2 - \tilde{u}_1) + \eta\frac{d}{d\tau_0}(\tilde{u}_2 - \tilde{u}_1)\right] = 0, \quad \tau_0 \gg \tau_0^*,$$

(9.29)

The characteristic values of the linearized system (9.29) are given by: $\lambda_{1,2} \approx \pm i - \varepsilon^3\eta$; $\lambda_{3,4} \approx \pm i\sqrt{2\varepsilon\sigma} - \varepsilon\eta$. It now follows that the logarithmic decrements for the first and second oscillators are different and equal to $\varepsilon^3\eta$ and $\varepsilon\eta$, respectively; the corresponding frequencies of oscillations are equal to 1 and $\sqrt{2\varepsilon\sigma}$. Besides, the following matching constraints at $\tau_0^* = \tau_1^*/\varepsilon$ are imposed:

$$\tilde{u}_1 = u_1, \quad \tilde{u} = u, \quad \frac{d\tilde{u}_1}{d\tau_0} = \frac{du_1}{d\tau_0}, \frac{d\tilde{u}}{d\tau_0} = \frac{du}{d\tau_0},$$

(9.30)

where $u_1(\tau_0^*)$ and $u_1(\tau_0^*)$ are calculated from Eq. (9.28) in the first interval $[0, \tau_0^*)$. By definition, $u_1(\tau_0^*) \approx 0$, while $|u|$ is close to the maximum u^*, that is, $d\tilde{u}/d\tau_0 = du/d\tau_0 \approx 0$ at τ_0^*. This implies that solutions of Eq. (9.28) for $\tau_0 > \tau_0^*$ can be approximated as

$$u_1(\tau_0) \approx e^{-\varepsilon^3\eta(\tau_0 - \tau_0^*)}v_1^* \sin(\tau_0 - \tau_0^*) + \varepsilon^2 \dots,$$

$$u(\tau_0) \approx u^* e^{-\varepsilon\eta(\tau_0 - \tau_0^*)} \cos\left[\sqrt{2\varepsilon\sigma}(\tau_0 - \tau_0^*)\right] - u_1(\tau_0) \dots,$$

(9.31)

Fig. 9.7 Precise (numerical) solution $u = u_{num}$ of Eq. (9.28) (*solid blue line*) and its analytical approximation (9.31) in the first (*red dashed line*) and second (*black dotted line*) intervals

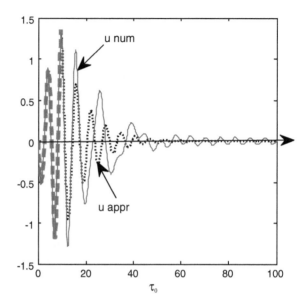

where $v_1^* = du_1/d\tau_0$ at $\tau_0 = \tau_0^*$. Different effects of dissipation on the linear and nonlinear oscillators result from the difference in the masses of oscillators. Figure 9.7 allows the comparison of the exact (numerical) solution $u(\tau_0)$ of system (9.28) with its analytical approximation (9.31). Note that due to different rates of decay, the slow-frequency components in $u(\tau_0)$ become negligibly small and thus $u(\tau_0) \approx -u_1(\tau_0)$ for $\tau_0 \gg \tau_0^*$. This conclusion is confirmed by the numerical results presented in Figs. 9.6 and 9.7.

9.6 Reduction to a Model of the Single Oscillator

This section presents further simplification of the model through the reduction of the multi-dimensional equations to an equation of a single oscillator. For illustrative purposes, the non-dissipative $(H = 0)$ two-mass system analogous to (9.27) is considered. Arguing as above, we introduce the small parameter $\varepsilon = (m/M)^{1/2}$ and define the dimensionless time $\tau_0 = \Omega t$, $\Omega = (C/M)^{1/2}$ and the rescaled variables (9.2), (9.3). The resulting dimensionless equations take the form similar to (9.8)

$$\frac{d^2\zeta_1}{d\tau_0^2} + \zeta_1 - \varepsilon(cu^3 + 2\varepsilon\sigma u) + \cdots = 0,$$

$$\frac{d^2u}{d\tau_0^2} + \varepsilon\frac{d^2\zeta_1}{d\tau_0^2} + u + \varepsilon\mu(cu^3 - u) + 2\varepsilon\sigma u = 0,$$

(9.32)

(higher-order non-resonant excitations in the first equation are omitted). Initial conditions are given by $\zeta_1 = 0$, $d\zeta_1/d\tau_0 = v$; $u = 0$, $du/d\tau_0 = \varepsilon v$ at $\tau_0 = 0$. At the next step, we introduce the complex envelopes analogous to (9.9) and then consider the multiple scales decomposition (9.10). The resulting equations for the main approximations $\varphi_1^{(0)}$ and $\varphi^{(0)}$ coincide with (9.11), namely:

$$
\begin{aligned}
&\frac{d\varphi_1^{(0)}}{d\tau_1} + \frac{i}{2}\varphi^{(0)} = 0, \quad \varphi_1^{(0)}(0) = v, \\
&\frac{d\varphi^{(0)}}{d\tau_1} + i\mu\left(\frac{1}{2} - \frac{3}{8}\left|\varphi^{(0)}\right|^2\right)\varphi^{(0)} - i\sigma\varphi^{(0)} + \frac{i}{2}\varphi_1^{(0)} = 0, \quad \varphi^{(0)}(0) = 0.
\end{aligned}
\tag{9.33}
$$

The elimination of $\varphi_1^{(0)}$ from the first equation reduces (9.33) to the integro-differential equation for the complex envelope $\varphi_0(\tau_1)$

$$
\varphi_1^{(0)}(\tau_1) = v - \frac{i}{2}I(\tau_1); \quad I(\tau_1) \int_0^{\tau_1} \varphi^{(0)}(s)ds
$$

$$
\frac{d\varphi^{(0)}}{d\tau_1} + i\mu\left(\frac{1}{2} - \frac{3}{8}\left|\varphi^{(0)}\right|^2\right)\varphi^{(0)} - i\sigma\varphi^{(0)} = -\frac{i}{2}\left(v - \frac{i}{2}I(\tau_1)\right), \quad \varphi^{(0)}(0) = 0.
\tag{9.34}
$$

The right-hand side of (9.34) expresses the cumulative effect of the exciting impulse and the coupling response on the dynamical behavior of the NES. As in previous studies, we assume that the contribution of the coupling response in the system dynamics is less than the effect of the initial impulse, and thus, Eq. (9.34) may be approximately solved through successive iterations. The initial iteration $\phi^{(0)}$ is defined from the truncated equation, in which the term $I(\tau_1)$ is ignored, namely

$$
\frac{d\phi^{(0)}}{d\tau_1} + i\left[\mu\left(\frac{1}{2} - \frac{3}{8}\left|\phi^{(0)}\right|^2\right) - \sigma\right]\phi^{(0)} = -\frac{i}{2}v, \quad \phi^{(0)}(0) = 0.
\tag{9.35}
$$

It follows from (9.34), (9.35) that the initial iteration adequately approximates the process $\varphi^{(0)}(\tau_1)$ if the integral $I^{(0)}(\tau_1)\int_0^{\tau_1}\phi^{(0)}(s)ds$ satisfies the condition $\left|I^{(0)}(\tau_r)\right| \ll 2v$. Formally, this inequality may be verified when the solution of Eq. (9.35) is found.

It is easy to deduce that Eq. (9.35) is identical to the equation of the slow complex envelope for a single oscillator subjected to harmonic forcing with amplitude $v/2$. The change of variables $\phi^{(0)} = a^{(0)}e^{i\Delta^{(0)}}$ leads to the following equations for the amplitude $a^{(0)}$ and the phase $\Delta^{(0)}$:

$$\frac{da^{(0)}}{d\tau_1} = -\frac{1}{2} v \sin \Delta^{(0)}$$

$$a^{(0)} \frac{d\Delta^{(0)}}{d\tau_1} = \left[\mu \left(-\frac{1}{2} + \frac{3}{8} (a^{(0)})^2 \right) + \sigma \right] a^{(0)} - \frac{1}{2} v \cos \Delta^{(0)}. \qquad (9.36)$$

With initial conditions $a^{(0)}(0) = 0, \Delta^{(0)}(0) = -\pi/2$. The first iteration $\phi^{(1)}(\tau_1)$ satisfies the equation

$$\frac{d\phi^{(1)}}{d\tau_1} - i\sigma\phi^{(1)} + i\mu \left(-\frac{3c}{8} \left| \phi^{(1)} \right|^2 \phi^{(1)} + \frac{1}{2}\phi^{(1)} \right) = -\frac{i}{2} \left[v_0 - \frac{i}{2} I^{(0)}(\tau_1) \right], \quad (9.37)$$

with $\phi^{(1)}(0) = 0$. Once $\phi^{(0)}(\tau_1)$ is found, the coupling response $I^{(0)}(\tau_1)$ is expressed as an explicit function of time. Computations can be simplified if the real and imaginary components Re $\phi^{(1)}$ and Im $\phi^{(1)}$ are found separately from Eq. (9.37) and then the slow envelope $a^{(1)} = \left| \phi^{(1)} \right|$ is calculated. Successive iterations can be constructed in a similar way.

For computational purposes, the numerical values of the system parameters selected in Sect. 9.5 are used. It is shown in Fig. 9.8 that the approximate envelopes $a^{(0)}$ and $a^{(1)}$, corresponding to the LPT, are close to the envelope of the exact solution of system (9.32). This confirms the dominating contribution of the exciting impulse in the system dynamics.

Fig. 9.8 Exact and approximate solutions: The *solid line* depicts the relative displacement $u(\tau_0)$—the numerical solution of (9.32); the *dashed* and *dotted lines* correspond, respectively, to the approximate slow envelopes $a^{(0)}$ and $a^{(1)}$

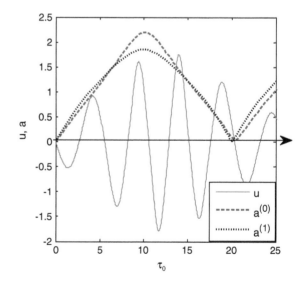

References

Kolovsky, M.Z.: Nonlinear Dynamics of Active and Passive Systems of Vibration Protection. Springer, Berlin (1999)

Kovaleva, A., Manevitch, L.I.: Resonance energy transport and exchange in oscillator arrays. Phys. Rev. E **88**, 022904 (2013)

Manevitch, L.I., Gourdon, E., Lamarque, C.H.: Towards the design of an optimal energetic sink in a strongly inhomogeneous two-degree-of-freedom system. ASME J. Appl. Mech. **74**, 1078–1086 (2007)

Meirovitch, L.: Fundamentals of Vibrations. McGraw Hill, New York (2001)

Raghavan, S., Smerzi, A., Fantoni, S., Shenoy, S.R.: Coherent oscillations between two weakly coupled Bose-Einstein condensates: Josephson effects, π-oscillations, and macroscopic quantum self-trapping. Phys. Rev. A **59**, 620–633 (1999)

Vakakis, A.F., Gendelman, O., Bergman, L.A., McFarland, D.M., Kerschen, G., Lee, Y.S.: Nonlinear Targeted Energy Transfer in Mechanical and Structural Systems. Springer, Berlin (2008)

Vaurigaud, B., Manevitch, L.I., Lamarque, C.-H.: Passive control of aeroelastic instability in a long span bridge model prone to coupled flutter using targeted energy transfer. J. Sound Vib. **330**, 2580–2595 (2011)

Chapter 10
Nonlinear Energy Channeling in the 2D, Locally Resonant, Systems

10.1 Unit Cell Model: High Energy Pulsations

Passive control of the acoustic wave propagation in metamaterials is a subject of broad scientific and practical interest in various aspects of applied sciences and engineering (Nicolaou and Motter 2012; Grima and Garuana-Gauci 2012; Deymier 2013; Boechler et al. 2011). Of late, dynamics of metamaterials has driven a considerable attention for their highly tunable, material properties (e.g., tailored band gaps) giving rise to quite an intriguing wave phenomena, e.g., cloaking, sound focusing, and more. The novel acoustic structure, namely locally resonant meta-materials (LRSM), has been first introduced by Liu et al. (2000). The main advantage of LRSM over the typical acoustically absorptive metamaterials is in their ability to form the low-frequency band gaps.

One of the rapidly developing strategies of passive and semi-active control of vibrations and waves in various acoustical structures is based on the attachment of essentially nonlinear elements (e.g., nonlinear energy sinks (NES)) in the externally loaded structure. As it was shown in many theoretical and experimental works, this inclusion of essentially nonlinear elements may invoke the well-known dynamical phenomenon (Vakakis et al. 2008) of unidirectional, broadband, passive energy transfer (TET) (see also Chap. 9). This phenomenon is characterized by the formation of strong energy exchanges between the weakly nonlinear substructure and the essentially nonlinear attachment. As it was noted in Chap. 9, the regime of strong energy transfer (a nonlinear beating phenomenon which is described adequately with using the LPT concept) is characterized by the recurrent, near complete energy exchanges between the several coupled physical systems. Basically, these regimes are caused by the internal resonant interactions leading, in particular, to the peculiar behavior which entails the non-stationary, recurrent, spatially localized energy bursts emerging on the different fragments of the complex system. Needless to say that in many real physical processes, emergence of highly non-stationary regimes is of primary importance. Apparently, the existing

© Springer Nature Singapore Pte Ltd. 2018   245
L.I. Manevitch et al., *Nonstationary Resonant Dynamics of Oscillatory Chains and Nanostructures*, Foundations of Engineering Mechanics,
DOI 10.1007/978-981-10-4666-7_10

methodology which is appropriate for studying the stationary regimes and NNMs is quite misleading when applied to the strongly non-stationary physical processes characterized by a major energy transport between the different elements of the coupled systems. Spontaneous formation of these highly non-stationary processes can also be a result of some global bifurcation undergone by the spatially localized regimes. Recently a novel type of NES was introduced based on an eccentric rotator, inertially coupled to a primary structure with the freedom to oscillate or rotate in a horizontal plane (Gendelman et al. 2012; Sigalov et al. 2012a, b). In the same study by Sigalov et al. (2012b), authors have shown that nonlinear inertial coupling between a linear oscillator and an eccentric rotator can lead to very interesting interchanges between regular and chaotic dynamical behavior.

In the present section, we perform a thorough theoretical study of highly non-linear, non-stationary regimes emerging in the 2D, locally resonant unit cell model incorporating internal rotator in the limit of high energy excitations. The main focus of the present paper is the analytical investigation of the emergence and bifurcations of highly non-stationary regimes, manifested by the recurrent, bidirectional energy transport (i.e., energy wandering between the axial and the lateral vibrations of the unit cell model) in the limit of high energy excitations. Special emphasis of the present study is devoted to the analytical investigation of the mechanism of spontaneous transition from the regime of the unidirectional energy locking (entrapment) to the recurrent, complete energy channeling (i.e., high energy pulsations) between the axial and the lateral vibrations of the 2D, unit cell model.

10.1.1   The Model

System under consideration is the 2D, unit cell, oscillatory model comprising a locally resonant, outer element (i.e., single mass element containing internal rotator) subject to the two-dimensional, nonlinear local potential. Scheme of the model under consideration is illustrated in Fig. 10.1.

Fig. 10.1 Scheme of the 2D, locally resonant, single cell model

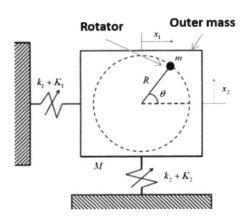

For our model, we assume the following: (1) the motion of the system is in-plane, (2) gravity is not taken into account, and (3) linear, viscous dissipation is assumed for the internal rotator.

It is easy to show that the underlying Hamiltonian system is defined by the Lagrangian:

$$L = \frac{1}{2}(M+m)\sum_{i=1}^{2}\dot{x}_i^2 + m\dot{\theta}R\left((\dot{x}_2\cos\theta - \dot{x}_1\sin\theta) + \frac{\dot{\theta}R}{2}\right)$$
$$- \sum_{i=1}^{2}\left(\frac{1}{2}k_i x_i^2 + \frac{1}{4}K_i x_i^4\right), \quad i = 1,2 \tag{10.1}$$

To account for the dissipation, introduced in the model, we use a regular Rayleigh function,

$$D = \frac{1}{2}c\,\dot{\theta}^2 \tag{10.2}$$

The governing equations of motion are easily derived from (10.1) and (10.2) using Lagrangian formalism,

$$(M+m)\ddot{x}_1 - mR\left(\ddot{\theta}\sin\theta + \dot{\theta}^2\cos\theta\right) + k_1 x_1 + K_1 x_1^3 = 0$$
$$(M+m)\ddot{x}_2 + mR\left(\ddot{\theta}\cos\theta - \dot{\theta}^2\sin\theta\right) + k_2 x_2 + K_2 x_2^3 = 0 \tag{10.3}$$
$$mR\ddot{x}_2\cos\theta - mR\ddot{x}_1\sin\theta + mR^2\ddot{\theta} = 0$$

Here M stands for the mass of the outer element, m is the mass of the internal rotator, R is the radius of the internal rotatory path (see Fig. 10.1), k_1, k_2—linear stiffness coefficients of the horizontal and vertical springs applied on the outer element, respectively, K_1, K_2—nonlinear stiffness coefficients, x_1, x_2—axial and lateral deflections of the outer mass element, θ—angular deflection of the internal rotator (see Fig. 10.1), and c—linear damping coefficient of the internal rotator. To bring system (10.3) into its non-dimensional form, we perform a regular system rescaling:

$$\tau = \omega_s t; \quad x_t = R\xi_i; \quad \varepsilon = \frac{m}{M+m}; \quad \eta = \frac{c}{mR^2\,\omega_s};$$
$$\kappa_i = \frac{k_i}{\omega_s^2(M+m)}; \quad \varepsilon\alpha_i = \frac{K_i R^2}{\omega_s^2(M+m)} \tag{10.4}$$

Introducing (10.4) into (10.3), one arrives at the non-dimensional set of the governing equations of motion,

$$
\begin{aligned}
&\xi_1'' - \varepsilon\left(\theta'' \sin\theta + \theta'^2 \cos\theta\right) + \kappa_1\xi_1 + \varepsilon\,\alpha_1\,\xi_1^3 = 0 \\
&\xi_2'' + \varepsilon\left(\theta'' \cos\theta - \theta'^2 \sin\theta\right) + \kappa_2\,\xi_2 + \varepsilon\,\alpha_2\,\xi_2^3 = 0 \qquad (10.5) \\
&\xi_2'' \cos\theta - \xi_1'' \sin\theta + \theta'' + \eta\theta' = 0
\end{aligned}
$$

In the present work, we aim at studying the limit of high energy excitations leading to the resonant, recurrent, and complete energy transport between the axial and the lateral vibrations of the outer element as well as the spontaneous, unidirectional energy localization.

To satisfy the conditions for the most fundamental resonance case (i.e., 1:1 resonance between the axial and the lateral deflections of the outer mass), we assume the perfect symmetry conditions on the system under consideration, i.e., $k_1 = k_2 = k, \kappa_1 = \kappa_2 = \kappa, \alpha_1 = \alpha_2 = \alpha$.

Accounting for the aforementioned symmetry conditions and choosing $\omega_s = \sqrt{k/(M+m)}$ yield the following symmetric system,

$$
\begin{aligned}
&\xi_1'' + \xi_1 + \varepsilon\,\alpha\,\xi_1^3 = \varepsilon\left(\theta'' \sin\theta + \theta'^2 \cos\theta\right) \\
&\xi_1'' + \xi_2 + \varepsilon\,\alpha\,\xi_2^3 = -\varepsilon\left(\theta'' \cos\theta - \theta'^2 \sin\theta\right) \qquad (10.6) \\
&\theta'' + \eta\theta' = \xi_1'' \sin\theta - \xi_2'' \cos\theta
\end{aligned}
$$

At this point, we note that ε parameter which is according to (10.4) has a physical meaning of mass ratio (i.e., ratio between the mass of the internal rotator and the total system mass) is formally declared as a small system parameter ($0 < \varepsilon \ll 1$). As is evident from (10.6), ε—scales the magnitude of nonlinear terms of the elastic forces applied on the outer element as well as the strength of the coupling between the rotator and the outer mass. In the present study, we focus on the analysis of the two peculiar non-stationary states.

Bidirectional, recurrent, energy channeling—corresponds to the complete, recurrent energy transport from axial to lateral vibrations of the outer element (controlled by the motion of the internal rotator) being initially excited strictly in the axial (or lateral) direction.

Unidirectional, energy locking—corresponds to the permanent, unidirectional energy localization in the outer element being initially excited strictly in the axial (or lateral) direction.

In the following sections, we perform an extensive analytical and numerical study of the resonant mechanisms leading to the aforementioned phenomena of *bidirectional, recurrent, energy channeling* and *unidirectional, energy locking* and their peculiar bifurcations leading to the spontaneous transitions from one state to another. As it will become clear from the further analysis, despite the complexity of the system under consideration, these mechanisms can be fully predicted by the effective reduction of the global system dynamics on the slow invariant manifold (SIM).

10.1.2 Analytical Study

In the present section, we devise a singular, multi-scale, analytical procedure enabling the reduction of the global system dynamics on the 1:1:1 resonance manifold of (10.6). Special emphasis is given to the analytical description of the mechanism of formation and annihilation of the resonant energy transport between the axial and the lateral deflections of the outer element. Local and global bifurcation analysis of the system dynamics reduced onto the 1:1:1 resonance manifold is considered.

10.1.2.1 Derivation of the Slow Flow System in the Vicinity of 1:1:1 Resonance Manifold

The asymptotic analysis of the dynamics in the vicinity of 1:1:1 resonance manifold is performed by the method of complexification–averaging (C–A). This method relies on complexification of the equations of motion, slow–fast partitioning of the transient dynamics, and, finally, averaging in terms of the dominant frequency of the fast components (Vakakis et al. 2008).

Following the C–A procedure, we introduce the complex coordinates corresponding to the axial and lateral deflections of the outer mass,

$$\psi_k = \xi_k' + i\xi_k, \quad k = 1,2 \tag{10.7}$$

Substitution of (10.7) into (10.6) yields the following set of equations of motion (EOM) given in the complex form:

$$\psi_1' - i\psi_1 + i\frac{\varepsilon\alpha}{8}(\psi_1 - \psi_1^*)^3 = \varepsilon(\theta'' \sin\theta + \theta'^2 \cos\theta)$$

$$\psi_2' - i\psi_2 + i\frac{\varepsilon\alpha}{8}(\psi_2 - \psi_2^*)^3 = -\varepsilon(\theta'' \cos\theta - \theta'^2 \sin\theta) \tag{10.8}$$

$$\theta'' + \eta\theta' = \left(\psi_1' - \frac{i}{2}(\psi_1 + \psi_1^*)\right)\sin\theta - \left(\psi_2' - \frac{i}{2}(\psi_2 + \psi_2^*)\right)\cos\theta$$

As it has been mentioned above, in the present study, we concentrate solely on the resonant interaction (i.e., 1:1:1 resonance) between the axial and lateral motions of the outer mass with the purely rotational motion of the internal device. To this end, it is rather natural to introduce the following internal resonance relations among the vibrational and rotational DOFs of the system

$$\psi_k = \varphi_k(t)\exp(it), \quad k = 1,2; \quad \theta = t + \beta(t) \tag{10.9}$$

where $\varphi_1(t), \varphi_2(t)$ are the slowly varying complex amplitudes of the axial and lateral deflections ($|d\varphi_k(t)/dt| \ll 1$), respectively, modulating the fast oscillations. It is also worth noting that the first linear term of the rotational coordinate (θ) defined in (10.9), stands for the constant, resonant rotation of the internal device (rotator), while the second term (β) corresponds to the slow phase modulation of the rotator response in the vicinity of the fundamental resonance. Following the C–A procedure, we substitute (10.9) into (10.8) and average out the fast components with respect to the dominant resonant frequency, yielding the following averaged system

$$\varphi_k' = \varepsilon \left[i\frac{3\alpha}{8}|\varphi_k|^2\varphi_k + \frac{(-i)^k}{2}\left(\beta'' + i(1+\beta')^2\right)\exp\left(i\beta(\tau)\right) \right], \quad k = 1,2$$

$$\beta'' + \eta\beta' = -\frac{1}{4}\left[\begin{array}{l} \left(\varphi_1^* \exp\left(i\beta(\tau)\right) + \varphi_1 \exp\left(-i\beta(\tau)\right)\right) - \\ -i\left(\varphi_2^* \exp\left(i\beta(\tau)\right) - \varphi_2 \exp\left(-i\beta(\tau)\right)\right) \end{array} \right] - \eta$$

(10.10)

As is clear from (10.10), the first two modulation variables (φ_1, φ_2) are slowly varying, while the third modulation variable (β) is the fast one. This property of the modulated system under consideration makes it a perfect candidate for the singular multi-scale analysis, thus allowing for the additional slow–fast scale decomposition and order reduction.

Following the commonly applied approach of the slow–fast system decomposition (see, e.g., Vakakis et al. 2008; Sigalov et al. 2012a), we proceed with the straightforward multi-scale expansion with respect to the formal small system parameter ε in the form,

$$\frac{d(\cdot)}{dt} = \frac{\partial(\cdot)}{\partial\tau_1} + \varepsilon\frac{\partial(\cdot)}{\partial\tau_2}, \quad \varphi_k(\tau) = \varphi_{k0}(\tau_1, \tau_2) + O(\varepsilon),$$

$$\beta(\tau) = \beta_0(\tau_1, \tau_2) + O(\varepsilon)$$

(10.11)

Substituting (10.11) into (10.10) and expanding with respect to the like powers of ε yields in the leading-order ($O(\varepsilon^0)$),

$$\partial\varphi_{k0}/\partial\tau_1 = 0, \quad k = 1,2 \Rightarrow \varphi_{k0} = \varphi_{k0}(\tau_2)$$

(10.12)

Accounting for (10.11) and (10.12) in the third equation of (10.10) gives

$$\frac{\partial^2\beta_0}{\partial\tau_1^2} + \eta\frac{\partial\beta_0}{\partial\tau_1} = -\frac{1}{4}\left[\begin{array}{l} \left(\varphi_{10}^*(\tau_2) \exp\left(i\beta_0\right) + \varphi_{10}(\tau_2) \exp\left(-i\beta_0\right)\right) \\ -i\left(\varphi_{20}^*(\tau_2) \exp\left(i\beta_0\right) - \varphi_{20}(\tau_2) \exp\left(-i\beta_0\right)\right) \end{array} \right] - \eta$$

(10.13)

Equation (10.13) depicts the fast evolution of the modulated phase of the rotator (β), while ($\varphi_{k0}, k = 1,2$) are the slowly varying functions which evolve with respect to the slow timescale τ_2. In the present work, we aim at studying the slow evolution of the averaged flow. To this end, we proceed with the next order of the multi-scale expansion $O(\varepsilon^1)$ applied on the first two equations of (10.10),

$$\frac{\partial \varphi_{k0}}{\partial \tau_2} = i\frac{3\alpha}{8}|\varphi_{k0}|^2 \varphi_{k0} + \frac{(-i)^k}{2}\left[\frac{\partial^2 \beta_0}{\partial \tau_1^2} + i\left(1 + \frac{\partial \beta_0}{\partial \tau_1}\right)^2\right] \exp(i\beta_0), \quad k = 1,2 \quad (10.14)$$

The derived set of Eqs. (10.13) and (10.14) depicts the multi-scaled evolution of the averaged flow of (10.6). Clearly, Eqs. (10.13) and (10.14) contain the fast and the slow evolving components of the averaged flow. To obtain the slow evolution of the latter, we seek for the stationary points of (10.13) and (10.14) with respect to the fast timescale (τ_1). Thus, requiring the nullification of all the time derivatives with respect to the fast timescale yields

$$\frac{\partial \beta_0(\tau_1, \tau_2)}{\partial \tau_1} = \frac{\partial^2 \beta_0(\tau_1, \tau_2)}{\partial \tau_1^2} = 0 \Rightarrow \beta_0 = B(\tau_2) \quad (10.15)$$

Plugging (10.15) into (10.13) and (10.14) leads to the following set of the slowly evolving averaged flow

$$\frac{\partial \varphi_{k0}}{\partial \tau_2} - i\frac{3\alpha}{8}\varphi_{k0}|\varphi_{k0}|^2 = \frac{(-i)^{k-1}}{2}\exp(iB), \quad k = 1,2$$

$$4\eta + \left[\begin{array}{c}\left(\varphi_{10}^*(\tau_2)\exp(iB(\tau_2)) + \varphi_{10}(\tau_2)\exp(-iB(\tau_2))\right) \\ -i\left(\varphi_{20}^*(\tau_2)\exp(iB(\tau_2)) - \varphi_{20}(\tau_2)\exp(-iB(\tau_2))\right)\end{array}\right] = 0 \quad (10.16)$$

We note that the last equation of (10.16) establishes a special algebraic relation between the slowly varying phase of the rotator (B) and slow evolution of the modulation coordinates ($\varphi_{k0}, k = 1,2$). This algebraic equation defines the SIM of the full averaged flow (10.10), whereas system (10.16) depicts the evolution of the averaged flow as a whole on the SIM. This evolution approximates the dynamics of the full system (10.6) in the vicinity of the 1:1:1 resonance surface.

10.1.2.2   Intrinsic Dynamics on a Slow Invariant Manifold (SIM)

In the present subsection, we analyze the global system dynamics on the SIM for both the non-dissipative ($\eta = 0$) and dissipative cases ($\eta \neq 0$). The ultimate goal of the present section is the analytical description of the mechanism of formation of the aforementioned regimes of bidirectional energy channeling as well as the unidirectional energy locking.

B1. Non-Dissipative Case ($\eta = 0$)

It is rather natural to start the analysis of the dynamics on the SIM from the simplest case of the linear elastic foundation.

Linear elastic foundation $(\alpha = 0)$

In the case of a linear elastic foundation and zero dissipation, the slow system (10.16) reads

$$\frac{\partial \varphi_{10}}{\partial \tau_2} = \frac{\exp(iB)}{2}; \quad \frac{\partial \varphi_{20}}{\partial \tau_2} = -\frac{i \exp(iB)}{2}$$

$$\begin{bmatrix} \left(\varphi_{10}(\tau_2) \exp(-iB(\tau_2)) + \varphi_{10}^*(\tau_2) \exp(iB(\tau_2))\right) \\ + i\left(\varphi_{20}(\tau_2) \exp(-iB(\tau_2)) - \varphi_{20}^*(\tau_2) \exp(iB(\tau_2))\right) \end{bmatrix} = 0 \tag{10.17}$$

Interestingly enough, system (10.17) can be solved exactly. To show that we introduce the new complex variable

$$Z = \varphi_{10} + i\varphi_{20} = |Z| \exp(iv) \tag{10.18}$$

Thus, after some trivial algebraic manipulations on (10.17) brings it into the more simplified form which can be expressed in terms of the new variable Z and the slow modulated phase of the rotator B:

$$\partial Z / \partial \tau_2 = \exp(iB), \quad Z^* \exp(iB) + Z \exp(-iB) = 0 \tag{10.19}$$

Introducing the first equation of (10.19) into the second one yields:

$$\frac{d\left(|Z|^2\right)}{d\tau_2} = 0 \Rightarrow |Z| = \text{const} \tag{10.20}$$

and

$$\cos(B(\tau_2) - v(\tau_2)) = 0 \Rightarrow B(\tau_2) = v(\tau_2) + \pi(m+1/2), \quad m = 1, 2 \tag{10.21}$$

Using (10.21), we solve the first equation of (10.19):

$$v(\tau_2) = (-1)^m |Z|^{-1} \tau_2 + v_0, \quad |Z| \neq 0 \tag{10.22}$$

Plugging (10.22) into (10.21) we get the explicit expression for $B(\tau_2)$:

$$B(\tau_2) = (-1)^m |Z|^{-1} \tau_2 + v_0, \quad |Z| \neq 0, \quad m = 1, 2 \tag{10.23}$$

where $v_0 = v_0 + \pi(1/2 + m)$. It can be easily shown that combining (10.19) with (10.17) and (10.18) and accounting for (10.22), one arrives at the explicit solutions for φ_{10} and φ_{20}, reading

$$\varphi_{k0}(\tau_2) = i^k(-1)^m\frac{|Z|}{2}\exp\left(i\left((-1)^m|Z|^{-1}\tau_2 + v_0\right)\right) + C_{k0}, \quad k = 1, 2 \quad (10.24)$$

As is clear from the results of the stability analysis brought in Appendix, the solutions are stable for m—odd and unstable for m—even. The derived solutions depict the resonant, recurrent energy transport (beating) between the axial and the lateral deflections of the outer element induced by the motion of the internal rotator. Here we note that the response given by (10.24) is essentially nonlinear because of the strong dependence of the modulation frequency on the amplitude of the response $|Z|$. Moreover, it is worthwhile emphasizing that the depth of these modulated response regimes is controlled by the constants of C_{k0} and $|Z|$. Thus, for some particular choice of $|Z|$ and C_2 (i.e., $|Z| = 2|C_2|$), the full, recurrent energy channeling from the axial to the lateral vibrations can be achieved.

Another interesting case corresponds to the stationary solution of (10.17), i.e., constant energy distribution between the axial and lateral vibrations of the outer mass satisfying the following

$$|\varphi_1(\tau_2)| = |\varphi_2(\tau_2)| = \text{const}, \quad \angle\varphi_1(\tau_2) - \angle\varphi_2(\tau_2) = (1/2 + m)\pi, \quad m = 1, 2$$
$$(10.25)$$

Thus, in the case of the linear elastic foundation, the only possible mode of energy transport between the axial and lateral vibrations of the outer mass can be characterized by the harmonically (e.g., weakly and strongly) modulated orbits given by (10.24). This result has a very important physical implication. Indeed, by eliminating the nonlinear term in the local potential ($\alpha = 0$), we rule out the possibility of the unidirectional energy localization. This point will become clear in the following subsection where the effect of nonlinearity introduced in the local potential is discussed.

Nonlinear elastic foundation ($\alpha > 0$)

Accounting for the nonlinear elastic foundation in the slow system (10.16), one has,

$$\frac{\partial\varphi_{k0}}{\partial\tau_2} - i\frac{3\alpha}{8}\varphi_{k0}|\varphi_{k0}|^2 = \frac{(-i)^{k-1}}{2}\exp(iB), \quad k = 1, 2$$
$$\begin{bmatrix} i\left(\varphi_{20}(\tau_2)\exp(-iB) - \varphi_{20}^*(\tau_2)\exp(iB)\right) \\ + \left(\varphi_{10}(\tau_2)\exp(-iB) + \varphi_{10}^*(\tau_2)\exp(iB)\right) \end{bmatrix} = 0 \quad (10.26)$$

It can be easily shown that system (10.26) possesses the two integrals of motion. The first integral reads,

$$|\varphi_{10}|^2 + |\varphi_{20}|^2 = N^2 \quad (10.27)$$

Using (10.27), it is convenient to introduce angular coordinates:

$$\varphi_{10}(\tau_2) = N \cos \Theta (\tau_2) \exp (i\delta_1(\tau_2)), \quad \varphi_{20}(\tau_2) = N \sin \Theta(\tau_2) \exp (i\delta_2(\tau_2)) \tag{10.28}$$

Substituting (10.28) into the last equation of (10.26) yields

$$\left[\begin{array}{l} (\cos \Theta - i \sin \Theta \exp (i\Delta)) \exp (i(B - \delta_1)) + \\ (\cos \Theta + i \sin \Theta \exp (-i\Delta)) \exp (-i(B - \delta_1)) \end{array} \right] = 0 \tag{10.29}$$

where $\Delta = \delta_1 - \delta_2$. After some simple algebraic manipulations with (10.29), one arrives at the following important relation,

$$\cos (B(\tau_2) - \delta_1(\tau_2) - \kappa(\tau_2)) = 0 \Rightarrow B(\tau_2)$$
$$= \pi(1/2 + m) + \delta_1(\tau_2) + \kappa(\tau_2), \quad m = 0, 1 \tag{10.30}$$

where $\kappa = \tan^{-1}\left(\frac{\sin \Theta \cos \Delta}{\cos \Theta + \sin \Theta \sin \Delta}\right)$. As it can be inferred from the results of the stability analysis brought in Appendix, the branch of the SIM corresponding to m—even is unstable while that of m—odd is stable. Using the angular representation of (10.28) along with (10.29) allows for the reduction of (10.26) to the following planar system:

$$\Theta' = (-1)^m \frac{\cos \Delta}{2N\sqrt{1 + \sin \Delta \sin 2\Theta}}$$
$$\Delta' = 2\sigma N^2 \cos 2\Theta - \frac{(-1)^m}{N}\left(\frac{\cos 2\Theta \sin \Delta}{\sin 2\Theta\sqrt{1 + \sin \Delta \sin 2\Theta}}\right) \tag{10.31}$$

where $\sigma = \frac{3\alpha N^3}{16}$. As is evident from (10.31), the local and global bifurcations of the two-dimensional flow on the phase plane (Θ, Δ) are solely governed by a single system parameter σ. It can be also shown that system (10.31) possesses an additional integral of motion,

$$H = \sigma(\cos^4 \Theta + \sin^4 \Theta) + (-1)^m \sqrt{1 + \sin 2\Theta \sin \Delta} \tag{10.32}$$

It is important to note that system (10.32) defines the two independent planar flows for the two distinct values of m (i.e., m—even, m—odd) and henceforth each case should be analyzed separately. We show below that variation of the parameter σ leads to the peculiar local and global bifurcations. Further study of the system dynamics will be fully concentrated on the analysis of (10.31) for both cases, i.e., m—even and m—odd. Let us proceed with finding the fixed points of (10.31). Apparently, fixed points of (10.31) admit the following set of algebraic equations (by setting $\Theta' = \Delta' = 0$):

$$\cos \Delta = 0, \ \cos 2\Theta \left(2\sigma \ \sin 2\Theta \sqrt{1 + \sin 2\Theta \ \sin \Delta} - (-1)^m \sin \Delta \right) = 0 \quad (10.33)$$

System (10.33) possesses the two distinct sets of stationary points:

$$(1): \quad \Delta_0^{(1)} = \pi(1/2 + l), \quad \Theta_0^{(1)} = \pi/2(1/2 + k)$$

$$(2): \quad \Delta_0^{(2)} = \pi(1/2 + l), \quad \sin 2\Theta_0^{(2)} \sqrt{1 + (-1)^l \sin 2\Theta_0^{(2)}} - \frac{(-1)^{m+l}}{2} \mu = 0$$

$$(10.34)$$

where $\mu = \sigma^{-1}$. It is worth noting that due to the periodicity of the right-hand side of (10.31) (π periodicity in Θ and 2π periodicity in Δ), we restrict the analysis of fixed points to the range of $\Delta \in [0, 2\pi]$, $\Theta \in [0, \pi]$. The first set of the stationary points corresponds to the periodic motions in both axial and lateral directions with equal amplitudes having the phase difference of $\Delta = \frac{\pi}{2}$ and $\Delta = \frac{3\pi}{2}$. The second set of solutions corresponds to the localized nonlinear normal modes, where the amplitude of the periodic response in, e.g., axial direction is significantly higher than that of the lateral one. In fact, finding the explicit solutions of the second set of (10.34) is a formidable task and beyond the scope of the current paper. However, as it was already mentioned above, the global system dynamics is fully governed by a single parameter σ. Thus, in Fig. 10.2 (Upper Panel), we plot the solutions of (10.34) versus the variation of the bifurcation parameter $\mu = \sigma^{-1}$. Unlike the fixed points given by the first set of (10.34), the localized solutions given by the second set of (10.34) depend on $\mu = \sigma^{-1}$ and their branches bifurcate from $(\Theta_0 = \pi/4)$.

As is clear from the results of the bifurcation diagram of Fig. 10.2a (Upper Panel) corresponding to the stable branch of SIM (m—even), the two branches emanating from $(\Theta_0 = \pi/4, \Delta_0 = 3\pi/2)$ at $\mu_{cr} = 0$ correspond to the localized solutions. Evidently enough, stationary solutions undergo a classical, subcritical, pitchfork bifurcation. Red lines of the bifurcation diagram correspond to the unstable solutions, while the solid lines correspond to the stable ones.

Interestingly enough, for the case of the unstable branch of the SIM (m—even), the bifurcation diagram shown in Fig. 10.2b (Upper Panel) is qualitatively different. Thus, the localized stationary solutions emanating from $(\Theta_0 = \pi/4, \Delta_0 = \pi/2)$ undergo a supercritical pitchfork bifurcation in contrast to the subcritical bifurcation obtained in the previous case. Moreover, the critical bifurcation values differ between the two cases (i.e., m—even and m—odd). In Fig. 10.2 (Lower Panel) and Fig. 10.3, we illustrate the phase portraits of (10.32) for the two distinct values of m, namely, $m = 0$ and $m = 1$, respectively. Here we note that the phase portraits are plotted in the range $\Delta \in [0, 2\pi]$, $\Theta \in [0, \pi/2]$. As we have already noted above, the main goal of the present study is to analyze the mechanism of formation of a highly non-stationary regime of recurrent energy channeling between the axial and lateral vibrations.

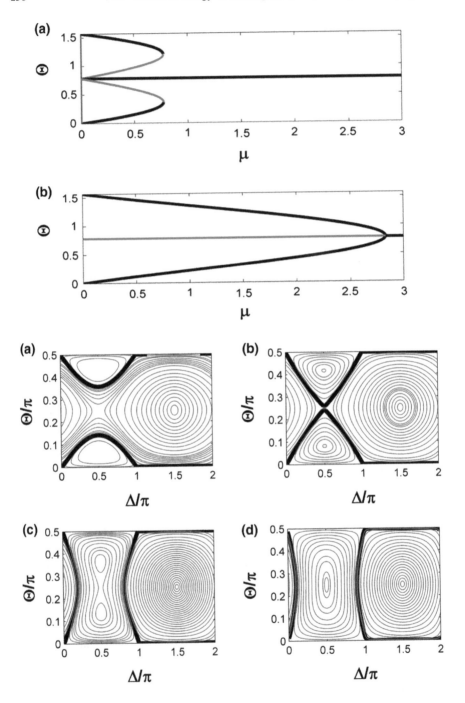

◀**Fig. 10.2**   (*Upper Panel*) Solutions of (10.34) versus the variation of μ **a** m—odd (stable branch of SIM) **b** m—even (unstable branch of SIM). Stable branches of solution are denoted by the *black solid line*, and unstable branches of solutions are denoted by the *red solid line*. (*Lower Panel*) Phase portraits of (10.31) for unstable branch for **a** $\mu = 0.91$ **b** $\mu = 1.21$ **c** $\mu = 2$ and **d** $\mu = 3.33$ $\mu = 0.55$ LPTs are denoted by the *black bold solid line*

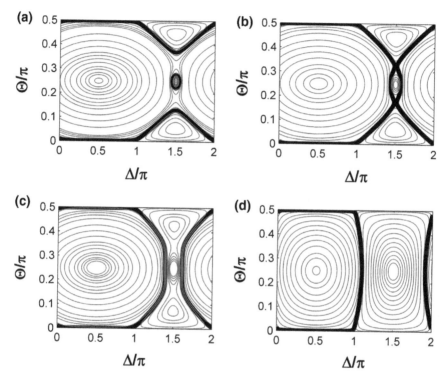

Fig. 10.3   Phase portraits of (10.31) for stable branch for **a** $\mu = 0.55$, **b** $\mu = 0.593$, **c** $\mu = 0.667$, and **d** $\mu = 3.33$. LPTs are denoted by the *black bold solid line*

In terms of the slow flow model (10.31), the aforementioned regime of recurrent energy channeling corresponds to a special orbit that departs from $\Theta = 0$ and reaches the value of $\Theta = \pi/2$. As it can be deduced from (10.28), these two conditions ensure a complete energy exchange between the axial and lateral oscillations of the outer element. This special type of trajectory is referred to in the literature as a limiting phase trajectory (Manevitch et al. 2011; Manevitch and Smirnov 2010). In Fig. 10.2a (Lower Panel), one can clearly see the existence of a special orbit satisfying the first condition (i.e., the orbit departing from the point where $\Theta = 0$).

This trajectory is denoted by a bold solid line. Apparently this orbit cannot lead to a complete energy exchange between the axial and lateral deflections of the outer mass as the trajectory by itself does not reach the value of $\Theta = \pi/2$ (which is a necessary condition for the complete energy transfer).

In the present study, we refer to this kind of the phase trajectory as LPT of the first kind. However, increasing the value of μ above a certain critical value, one observes a coalescence of the LPT of the first kind with a separatrix of a saddle point (Fig. 10.2b Lower Panel). This leads to a global bifurcation, resulting in the formation of the limiting phase trajectory of a qualitatively different type (Fig. 10.2c, d Lower Panel). In the present study, we will refer to this trajectory as a LPT of the second kind.

This type of response corresponds to the bidirectional, recurrent energy channeling, while that of the first kind corresponds to the regime of unidirectional energy locking in a single direction. As it will become evident from an inspection of the results of Fig. 10.3 and further analysis, the case of $m = 1$ is special and reveals a very interesting global bifurcation undergone by the LPT which is different from the commonly considered cases (Manevitch and Smirnov 2010; Manevitch et al. 2011). Thus, unlike the previous case (i.e., m—even) in Fig. 10.3a–c, we observe the formation of the additional pair of fixed points which is clearly a result of a subcritical pitchfork bifurcation of NNMs [Fig. 10.2a (Upper Panel)]. Obviously enough, the LPT illustrated in Fig. 10.3a is of the first kind. However, choosing the value of μ above a certain threshold, one observes a completely different scenario concerning the evolution of LPT of the first kind (Fig. 10.3a). Clearly, except the regular trajectory of LPT emanating from $\Theta = 0$, there is a formation of an additional branch of the LPT encircling a stable center. In the present work, we will refer to it as the LPT bubble. At some critical value of μ, the separatrix coalesces with the bubble as well as the lower and upper branches of LPT which results in the reconnection of LPTs and transition to the LPT of the second kind (Fig. 10.3b–d). To derive the analytical predictions for the formation of the LPT of the second kind for both cases (i.e., m—even and m—odd), we resort to the second integral given in (10.32). As is clear from the discussion above, substituting $\Theta = 0$ into (10.32) yields the exact constant value of $H(\Theta, \Delta)$ corresponding to the branches of the limiting phase trajectories,

$$H(\Theta = 0, \Delta) = \sigma + (-1)^m, \quad m = 0, 1 \tag{10.35}$$

Finding a critical value of σ corresponding to the reconnection of LPTs is quite a trivial task in the case of m—even (i.e., unstable branch of the SIM). Indeed, requiring the passage of the LPT trajectory through a saddle point ($\Delta = \pi/2, \Theta = \pi/4$), one arrives at the condition on the critical value of σ, yielding

$$2\left(\sqrt{2} - 1\right) = \sigma_{cr\_even} = \mu_{cr\_even}^{-1} \tag{10.36.a}$$

Analytical derivation of the critical value of σ corresponding to the formation of LPT of the second kind is quite a non-trivial task for the case of m—odd (i.e., stable branch of the SIM). Indeed as is clear from the phase portraits of Fig. 10.3, the global bifurcation of the LPT (i.e., the second transition) occurs when it passes through a saddle point $(\Theta_0^{(2)} = \Theta_{0\_sad}^{(2)}, \Delta_0^{(2)} = 3\pi/2)$ given by the second set of (10.34). This condition on the critical value of σ in the case of m—odd can be formulated implicitly as follows,

$$\frac{2\left(1 - \sqrt{2}\right) + \sqrt{1 - \sin\left(2\Theta_{0\_sad}^{(2)}\left(\sigma_{cr\_odd}\right)\right)}}{\cos^4\left(\Theta_{0\_sad}^{(2)}\left(\sigma_{cr\_odd}\right)\right) + \sin^4\left(\Theta_{0\_sad}^{(2)}\left(\sigma_{cr\_odd}\right)\right)} = \sigma_{cr\_odd} = \mu_{cr\_odd}^{-1} \quad (10.36.b)$$

where $\Theta_{0\_sad}^{(2)} = \Theta_{0\_sad}^{(2)}(\sigma_{cr\_odd})$ given implicitly by (10.34). Obviously enough, derivation of the explicit solution of $\Theta_{0\_sad}^{(2)}(\sigma_{cr\_odd})$ is a formidable task leading to a rather cumbersome expression which is beyond the scope of the present study.

B2. *Dissipative Case* $(\eta \neq 0)$

In the present and the following subsections, we analyze the dissipative flow of the averaged system on the SIM.

Linear elastic foundation $(\alpha = 0)$

Accounting for a dissipative term and neglecting the nonlinearity in (10.16) yield the following set of equations corresponding to the dissipative slow system evolution,

$$\frac{\partial \varphi_{10}}{\partial \tau_2} = \frac{\exp{(iB)}}{2}; \quad \frac{\partial \varphi_{20}}{\partial \tau_2} = -\frac{i \exp{(iB)}}{2}$$

$$\left[\begin{array}{l} i\left(\varphi_{20}\left(\tau_2\right) \exp{(-iB)} - \varphi_{20}^*\left(\tau_2\right) \exp{(iB)}\right) \\ + \left(\varphi_{10}\left(\tau_2\right) \exp{(-iB)} + \varphi_{10}^*\left(\tau_2\right) \exp{(iB)}\right) \end{array}\right] = -4\eta \quad (10.37)$$

Similarly to the non-dissipative case of the linear elastic foundation, explicit solution of system (10.37) can be derived. Thus, arguing as above, we use a new complex variable given by (10.18) and rewrite (10.37) in the following form:

$$\partial Z/\partial \tau_2 = \exp{(iB)}, \quad Z^* \exp{(iB)} + Z \exp{(-iB)} = -4\eta \quad (10.38)$$

Introducing the first equation of (10.38) into the second one yields:

$$d|Z|^2/d\tau_2 = -4\eta \Rightarrow |Z(\tau_2)| = \sqrt{C - 4\eta\tau_2} \quad (10.39)$$

Plugging (10.39) into (10.18) and accounting for (10.38) yield the explicit solution for Z,

$$Z(\tau_2) = \sqrt{C - 4\eta\tau_2} \exp{(i\upsilon)} \quad (10.40)$$

where the phase v is given by,

$$v = \mp \left(\frac{1}{2\eta} \sqrt{C - 4\eta\tau_2 - 4\eta^2} - \tan^{-1} \left(\frac{1}{2} \sqrt{\frac{C - 4\eta\tau_2 - 4\eta^2}{\eta}} \right) \right) + v_0 \qquad (10.41)$$

Substituting (10.40) and (10.41) into (10.38), one obtains the explicit solution for B,

$$
\begin{aligned}
B(\tau_2) &= \cos^{-1} \left(\frac{2\eta}{\sqrt{z_0^2 - 4\eta\tau_2}} \right) \mp \frac{\sqrt{z_0^2 - 4\eta\tau_2 - 4\eta^2}}{2\eta} \\
&\pm \tan^{-1} \left(\frac{1}{2} \sqrt{\frac{z_0^2 - 4\eta\tau_2 - 4\eta^2}{\eta}} \right) + v_0
\end{aligned}
\qquad (10.42)
$$

where $C = z_0^2 = |Z(0)|^2$ and $v_0 = \arg(Z(0))$. Clearly, expressions for φ_{10} and φ_{20} can now be given in the following integral forms,

$$\varphi_{10} = \frac{1}{2} \int_0^{\tau_2} \exp(iB(\xi)) d\xi + C_{10};$$

$$\varphi_{20} = -\frac{i}{2} \int_0^{\tau_2} \exp(iB(\xi)) d\xi + C_{20}$$

(10.43)

where C_{10} and C_{20} are the complex constants determined by the initial conditions. Again, the derived set of solutions is stable for m—odd and unstable for m—even. The obtained solutions depict the resonant energy exchanges between the axial and the lateral motions accompanied with the constantly decaying amplitude of the response (due to the presence of a dissipative term). In Sect. 10.1.3, we demonstrate numerically the comparison of the dissipative flow reduced on the SIM with the true response of the original system (10.6) given to the similar initial conditions.

Nonlinear elastic foundation $(\alpha > 0)$

Assuming a nonzero dissipation and accounting for the nonlinear term of the local potential, the slow system (10.16) reads

$$
\frac{\partial \varphi_{k0}}{\partial \tau_2} - \frac{3i\alpha}{8} \varphi_{k0} |\varphi_{k0}|^2 = (-i)^{k-1} \frac{\exp(iB)}{2}, \quad k = 1, 2
$$
$$
\left[\begin{array}{l} i \left(\varphi_{20}(\tau_2) \exp(-iB) - \varphi_{20}^*(\tau_2) \exp(iB) \right) \\ + \left(\varphi_{10}(\tau_2) \exp(-iB) + \varphi_{10}^*(\tau_2) \exp(iB) \right) \end{array} \right] + 4\eta = 0
$$

(10.44)

Obviously enough, system (10.44) cannot be solved exactly. However, assuming the low dissipation rate, we would like to demonstrate the peculiar mechanism of spontaneous breakdown of energy localization in the axial direction resulting in the formation of the recurrent, bidirectional energy channeling. Thus, applying the initial excitation in the axial direction and assuming the initial energy level to be above the localization threshold, one would expect that low-rate energy decay will bring the flow to the critical energy threshold resulting in the spontaneous energy delocalization. Here it is worthwhile noting that the passage through the localization threshold is certainly a necessary but not a sufficient condition for the emergence of recurrent energy channeling.

To emphasize this point better, we resort again to the Hamiltonian case, considered in the previous subsection. Revisiting the results of the phase plane diagrams obtained for both stable and unstable branches of the SIM, we note that the mechanism of formation of the limiting phase trajectory of the second kind, governing the bidirectional, recurrent energy channeling qualitatively differs between the two branches of the SIM (i.e., stable and unstable). Importantly, at the bifurcation point (i.e., coalescence of LPT with the separatrix), the central fixed point $(\Theta = \pi/4, \Delta = \pi/2)$ (see Fig. 10.2a–c) (Right Panel)) of the unstable branch of SIM is a saddle point, while the central fixed point $(\Theta = \pi/4, \Delta = 3\pi/2)$ (See Fig. 10.3) of the stable branch of SIM turns out to be a center. Thus, in presence of dissipation, the center of the stable branch of SIM $(\Theta = \pi/4, \Delta = 3\pi/2)$ becomes an attractor which prevents from the phase trajectory emanating from the initially localized state (similarly to the limiting phase trajectory of the underlying Hamiltonian system departing from $\Theta = 0$) to reach the vicinity of $\Theta = \pi/2$ (complete energy transport to the lateral vibrations). In other words, the dissipative flow on the stable branch of SIM will not show the expected transition from localization to transport as the phase trajectory escaping from the initially localized state is being attracted by the central fixed point $(\Theta = \pi/4, \Delta = 3\pi/2)$.

Interestingly enough, the flow on the unstable branch of the SIM shows a qualitatively different behavior. Indeed, same type of the phase trajectory emanating from the localized state can reach the vicinity of $\Theta = \pi/2$ being repelled from the central saddle point $(\Theta = \pi/4, \Delta = \pi/2)$. The dissipative flow on the stable and unstable branches of the SIM is shown in Fig. 10.4a, b, respectively. Initial stage of the response (before reaching the transition threshold) corresponding to the localized state is designated with the blue solid line, while the intermediate stage of the response (right after the passage of the transition threshold) is designated with the red one. Here by the intermediate stage, we refer to some finite period of the response starting at the moment when the phase trajectory reaches the transition threshold and before its escape from the SIM designating the breakdown of 1:1:1 resonance. Here we note in passing that in the case of the dissipative flow, the escape from the assumed 1:1:1 resonance is rather obvious due to the constant energy decay bringing the flow to the critical energy threshold below which system cannot sustain the full resonant rotations of the internal device.

As is clear from the results of the diagrams shown in Fig. 10.4a (stable branch of SIM), right after the passage of the transition threshold by the phase trajectory

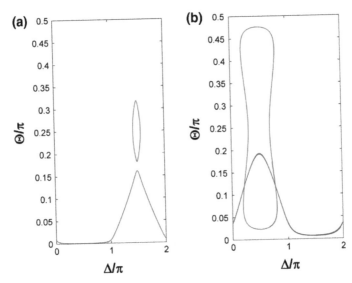

Fig. 10.4 Phase portraits, **a** stable branch of SIM, **b** unstable branch of SIM. The *blue solid line* corresponds to the initial stage of the response, and the *red solid lines* correspond to the intermediate stage

(starting at the localized state), it is being attracted by the stable focus ($\Theta = \pi/4, \Delta = 3\pi/2$) and therefore stays in its vicinity. However, phase trajectory of the unstable branch (Fig. 10.4b) (emanating from the similar localized state) after passing the threshold starts wandering in the vicinity of the two local states, i.e., $\Theta = 0$ (pure axial vibrations) and $\Theta = \pi/2$ (pure lateral vibrations). These recurrent transitions between the two local states clearly signify the formation of the bidirectional, almost complete energy channeling. To depict analytically the mechanism governing the damped transition from localized state to the recurrent, energy channeling, we resort to the results of the underlying Hamiltonian system. Thus, as is shown in the previous subsection, transition from energy localization to the recurrent energy transport occurs at some particular threshold. As it was argued above the mechanism of damped transitions from unidirectional localization to the recurrent, bidirectional energy channeling can be expected only for the unstable branch of SIM.

To show this, let us define the critical value of N (of the unstable branch) above which energy is localized in one direction and consequently below which the recurrent, bidirectional energy channeling is possible,

$$N_{cr\_even} = \left(\frac{16}{3\alpha\mu_{cr\_even}}\right)^{1/3} \qquad (10.45)$$

here μ_{cr_even} is defined implicitly in (10.36.a). Performing some basic algebraic manipulations on (10.44) and accounting for (10.27), one can show that the following holds,

$$\frac{dN^2}{d\tau_2} = -2\eta \Rightarrow N(\tau_2) = \sqrt{N_0^2 - 2\eta\tau_2} \qquad (10.46)$$

where $N^2(\tau_2) = |\varphi_{10}(\tau_2)|^2 + |\varphi_{20}(\tau_2)|^2$.

Starting at $N_0 > N_{cr\_odd}$ (localization) and assuming the adiabatically slow energy dissipation, one would expect for a spontaneous transition from the unidirectional energy channeling to the bidirectional one in the vicinity of threshold (i.e., $N = N_{cr\_odd}$). The time interval required for the transition can be easily assessed from (10.46) and yields the following,

$$T_{anal\_cr} = (2\eta)^{-1}\left(N_0^2 - N_{cr\_even}^2\right) \qquad (10.47)$$

Let us illustrate the mechanism by simulating the slow averaged flow given in (10.44). To this end, we choose the initial conditions satisfying $|\varphi_{10}(0)|^2 + |\varphi_{20}(0)|^2 > N_{cr\_odd}^2$ for the fixed values of the parameters of nonlinearity α and dissipation η. In Fig. 10.5a, we illustrate an example of the damped transition from energy localization to the recurrent energy transport. As is evident from the results of Fig. 10.5a, the transient energy localization in the axial direction breaks down spontaneously and is followed by the recurrent energy fluctuations between the axial and the lateral vibrations.

In Fig. 10.5b, c, we illustrate the phase portraits of the underlying Hamiltonian system ($\eta = 0$) corresponding to the two distinct energy levels (these energy levels satisfy $N_1 > N_{cr}$ and $N_2 < N_{cr}$, they are denoted on Fig. 10.5a with the red lines) traversed by the dissipative system (10.44). Clearly enough, simulation of the slow system (10.44) fully confirms the theoretical prediction. In the following section (B-4), we confirm the preservation of this intriguing effect of spontaneous, spatial energy delocalization in the full model (10.6).

10.1.3   Numerical Verifications

In the present section, we perform numerical verifications of the validity of a theoretical model for the four distinct cases considered above.

B-1 *Linear elastic foundation* $(\alpha = 0)$, *no dissipation* $(\eta = 0)$

We start numerical verifications from a comparison of the analytical solution derived for the slow flow reduced on the 1:1:1 resonance manifold (10.24) with the time histories of the response computed for the full model (10.6) (Fig. 10.6).

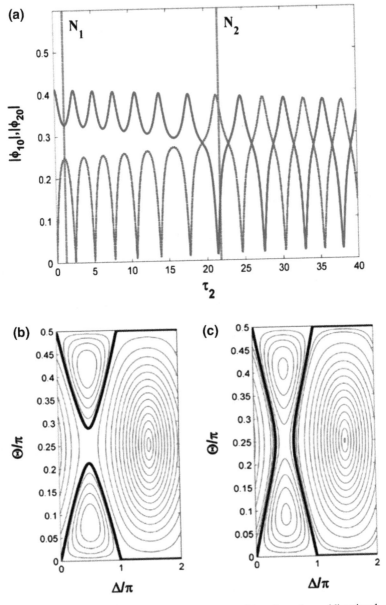

Fig. 10.5 a Time histories illustrating the damped transition from the unidirectional energy localization in the axial direction to the recurrent energy transport in the dissipative slow model (10.44), **b** and **c** phase portraits for unstable branch corresponding to the two distinct states of the corresponding Hamiltonian system ($\eta = 0$) at $\tau_2 = 1.25$ and $\tau_2 = 21.8$, respectively. The *black bold solid lines* denote the LPTs for both time instants

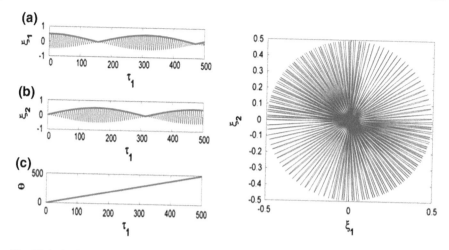

Fig. 10.6 (*Left Panel*) Time histories of the full unit cell model, **a** axial deflection, **b** lateral deflection, **c** rotator angular coordinate. The *blue solid line* corresponds to the true system response of (10.6), and the *red solid line* denotes the analytical solution (10.24). (*Right Panel*) Lissajous curves corresponding to the numerical simulations of fully energy channeling. Initial conditions: $\xi_1(0) = 0.5, \theta'(0) = 1, \xi_1'(0) = \xi_2(0) = \xi_2'(0) = \theta(0) = 0$. System parameters: $\varepsilon = 0.01$

As is clear from the analytical solution given by (10.24), the depth of modulation ranging from zero modulation (i.e., stationary response) and up to the complete energy transport is fully controlled by the choice of the initial conditions on the resonance manifold. In Fig. 10.6 (Left Panel) and (Right Panel), we illustrate the time histories and the Lissajous curves of the response corresponding to complete, bidirectional energy channeling. As it has been already explained in the theoretical part, the case of the linear elastic foundation is special. That is the motion on the 1:1:1 resonance manifold can only contain harmonically modulated orbits with recurrent, uniform, periodic energy exchange between the axial and the lateral vibrations of the outer mass. In other words, no unidirectional energy localization is possible in that case. Evidently enough results of numerical simulations of the full model are in a very good agreement with the analytical solution.

B-2 *Linear elastic foundation* $(\alpha = 0)$, *nonzero dissipation* $(\eta \neq 0)$

In the present subsection, we confirm the validity of the analytical solution derived for the slow flow model (linear elastic foundation) in the presence of the internal dissipation. Similarly to the previous subsection, we start with a comparison of the analytical solution derived for the slow dissipative flow reduced on the 1:1:1 resonance manifold (10.37) with the time histories of the response computed for the full model (10.6) (Fig. 10.7).

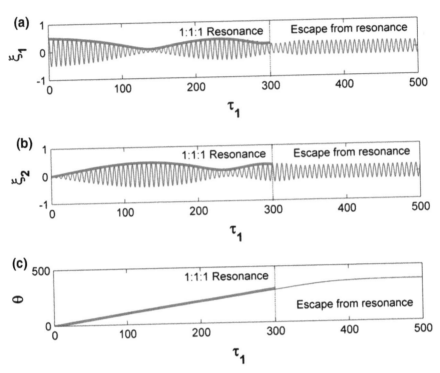

Fig. 10.7 Time histories of the full unit cell model, **a** axial deflection, **b** lateral deflection, **c** rotator angular coordinate. The *blue solid line* corresponds to the true system response of (10.6), and the *red solid line* denotes the analytical solution of (10.37). Initial conditions: $\xi_1(0) = 0.5, \theta'(0) = 1, \xi_1'(0) = \xi_2(0) = \xi_2'(0) = \theta(0) = 0$, system parameters: $\varepsilon = 0.01$, $\eta = 0.02$

As is clear from the results of Fig. 10.7, in the dissipative case, the transient evolution of the system occurs in the vicinity of the resonance (1:1:1) manifold which is followed by a spontaneous escape from resonance at approximately $\tau_1 \approx 300$. Obviously enough the proposed approximation becomes invalid far from the (1:1:1) resonant surface and thus fails to predict the system evolution in the case of the spontaneous escape from resonance.

Again, results of the numerical simulation of the full system (10.6) are in the very good agreement with the analytical model in the transient regime in the vicinity of 1:1:1 resonance surface.

B-3 Nonlinear elastic foundation ($\alpha > 0$), zero dissipation ($\eta = 0$)

In the present section, we perform the numerical verifications of the validity of the SIM analysis corresponding to the conservative case of nonlinear elastic foundation. Let us start the discussion with the comparison of the recurrent energy channeling predicted by the reduced, slow flow model with the results of numerical simulations of the full model (10.6).

The analytically predicted transition threshold (m—odd) from unidirectional energy locking to the recurrent energy channeling is compared with the numerically computed threshold ($\mu_{cr\_odd} = 0.5882$, $\mu_{cr\_odd\_num} = 0.5536$). In Fig. 10.8 (Upper Panel), we plot the time histories of the response of the unit cell model subject to the two distinct initial excitation levels, i.e., above and below the critical threshold N_{cr_odd}. As is evident from the results of Fig. 10.8, above the critical value of N_{cr_odd} [Fig. 10.8b, d (Upper Panel)], energy is localized in the axial direction; however, below the critical value of N_{cr_odd} [Fig. 10.8a, c (Upper Panel)], energy gets recurrently channeled between both the axial and the lateral vibrations of the outer element. In Fig. 10.8 (Lower Panel), we plot the Lissajous curves, corresponding to the planar motion of the outer element for both regimes, i.e., recurrent energy channeling (Fig. 10.8a—Lower Panel) as well as the unidirectional, permanent energy locking (Fig. 10.8b—Lower Panel). It is worth noting that analytical predictions are confirmed by the numerical simulations of the full model.

B-4 *Nonlinear elastic foundation* ($\alpha > 0$), *nonzero dissipation* ($\eta \neq 0$)

The second dissipative case considered in the theoretical part which requires numerical confirmation corresponds to the mechanism of the damped transition from the unidirectional energy localization in the axial vibrations to the recurrent energy transport between the axial and lateral vibrations of the outer element. The first comparison test corresponds to the slow flow model (10.44). Thus, starting with initial excitation (in the axial direction) on the unstable branch of the SIM above the critical energy threshold—we plot the response of the dissipative, slow flow (10.44) (Fig. 10.9).

In Fig. 10.9a, we plot the slow evolution of N given by (10.46). The horizontal, red, solid line of Fig. 10.9a corresponds to the transition value of $N = N_{cr\_even} = 0.4016$ predicted by the analytical model. To confirm the persistence of the mechanism of damped transition also in the original model, we plot the time history of the response of (10.6) which corresponds to the unstable branch of the SIM (Fig. 10.9b–d).

As is evident from the results of Fig. 10.9b–d, transient response of the outer element is localized initially in the axial direction; however, when reaching the theoretically predicted critical threshold ($N \sim N_{cr\_even}$) (see Fig. 10.9a) at ($\tau_1 \approx T_{anal\_cr} = 851$), the localized state breaks down resulting in the formation of a new regime of the almost complete, bidirectional, recurrent energy channeling.

It is worth emphasizing that the theoretically predicted, damped transition illustrated in Fig. 10.9b–d is quite in agreement with the predictions of the theoretical model. Another interesting result which comes out of the numerical simulations of the present section is the persistence of the mechanism dictated by the damped resonant transition occurring on the <u>unstable</u> branch of the SIM. On one hand it is rather natural to expect that starting on the unstable branch of the SIM the phase trajectory of the full model escapes from it and therefore the phenomenon of

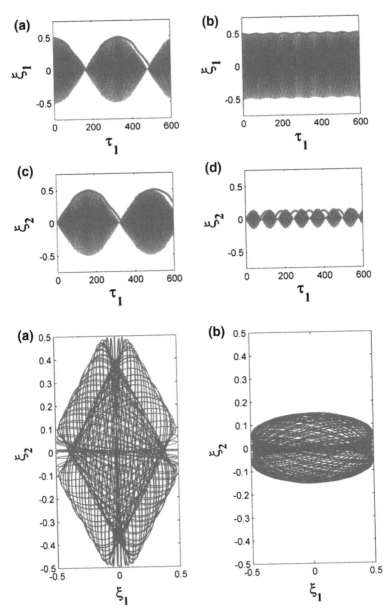

Fig. 10.8 (*Upper Panel*) Time histories of the modulated and localized response of the unit cell model. **a, c** Recurrent energy transport between the axial and the lateral vibrations [**a** deflection of the outer element in the axial direction and **c** deflection of the outer element in the lateral direction]. **b, d** Permanent energy localization in the axial direction [**b** deflection of the outer element in the axial direction and **d** deflection of the outer element in the lateral direction]. *Thin solid line* denotes the true system response (10.6) while the *bold solid line* corresponds to the slow flow model (10.26). System parameters: $\varepsilon = 0.01$, $\eta = 0$, $N = 0.5$, **a, c** $\alpha = 69.15$ **b, d** $\alpha = 106.67$ (*Lower Panel*) Lissajous curves corresponding to the **a** recurrent energy transport and **b** energy localization in the axial direction

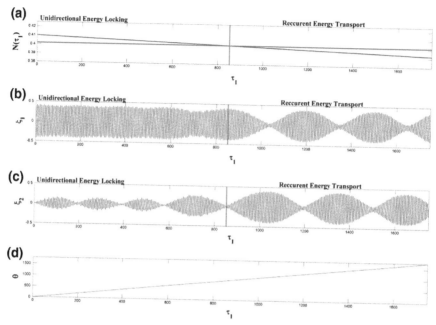

Fig. 10.9 Damped transition from the unidirectional energy localization in the axial direction to the recurrent energy transport **a** slow evolution of N given by (10.44), (**b, c, d**). Time histories of the full system (10.6) illustrating the damped transition from the unidirectional energy localization in the axial direction to the recurrent energy transport in the original model (10.6). Initial conditions: $\xi_1(0) = 0.41, \theta'(0) = 1, \xi'_1(0) = \xi_2(0) = \xi'_2(0) = 0, \theta(0) = \pi/2$. System parameters: $\alpha = 68.21, \eta = 0.0004, \varepsilon = 0.01$

damped transition (predicted by the slow flow model) would not persist in the true model. However, results of the numerous numerical tests clearly show that for some sufficiently long, transient period, phase trajectory of the full dissipative flow can stay sufficiently long in the vicinity of the unstable branch of the resonance manifold and fully confirms the theoretically predicted scenario of the damped transition.

10.1.4   Concluding Remarks

Dynamic response of the locally resonant, unit cell model to high energy excitation is considered analytically and numerically. Special analytical treatment based on the singular multi-scale analysis is developed. The basic question of the existence and bifurcations of highly non-stationary regime of massive energy transport as well as the regime of unidirectional energy locking is addressed via the reduction of the global flow on the SIM in the vicinity of the fundamental resonance manifold (1:1:1). In the present study, we have shown that below a certain energy threshold, regime of a complete, recurrent energy channeling between the axial and the lateral vibrations is possible; however, above that threshold energy channeling terminates leading to the unidirectional energy locking. A spontaneous transition from localized state to the recurrent bidirectional energy transport has been obtained for the dissipative case. The aforementioned transition thresholds from energy localization to energy transport as well as the local and global bifurcations undergone by the highly nonlinear, pulsating response regimes are predicted analytically in the frameworks of the LPT concept. Numerical simulations fully confirm the analytical predictions concerning the structure of the response regimes and reveal some of their peculiar local and global bifurcations.

10.2   Unit Cell Model: Low Energy Excitation Regimes

Of late, significant attention has been given to unidirectional devices that pass acoustic energy in only one direction. Present study is devoted to the analysis of two-dimensional, nonlinear energy transport emerging in the unit cell model subject to the low energy initial excitation.

In contrast to the high energy excitation limit, the presently considered limit of low energy excitations reveals the emergence of quite intriguing, highly nonlinear, transient regimes of unidirectional energy channeling manifested by the partial and complete, unidirectional energy flow from axial to lateral vibrations. Here we demonstrate that the phenomena of recurrent energy channeling and energy locking persist in the low energy limit as well. The ultimate goal of the present study is the analytical investigation of the intrinsic mechanisms governing the aforementioned phenomena of energy channeling in the low energy limit. This limit can be also characterized by the absence of the resonant interactions between the internal rotator and the motion of the outer element. To this end, we devise a special analytical procedure based on a regular multi-scale expansion constructed for the asymptotic limit of low energy excitations. Special emphasis is given to the analysis of nonlinear phenomenon of complete, unidirectional energy transport from axial to lateral vibrations.

10.2.1   Numerical Evidence of the Unidirectional Energy Channeling

In the present section, we illustrate numerically the existence of some peculiar response regimes manifested by partial or complete, unidirectional energy flow from axial to lateral vibrations. In scope of the present work, we refer to this phenomenon as to the (partial or complete) energy channeling depending on the amount of energy transfer.

Here it is important to note that initial excitation applied on the outer mass is assumed solely in the axial direction. Interestingly enough, the portion of the unidirectional energy transport from axial direction to the lateral one is fully controlled by the initial tuning (i.e., initial angle) of the rotator. In the following subsection, we perform a multi-scale analysis revealing the intrinsic mechanisms of this peculiar type of energy transport.

To illustrate the phenomena of partial as well as a complete energy redirection (from axial to lateral), we performed the two distinct numerical tests. In both tests, we apply the same initial impulse on the outer mass (in the axial direction) for the different tunings of the internal rotator. In Figs. 10.10 (Upper Panel) and 10.11 (Upper Panel), we illustrate the time histories of the response of the outer element recorded in the axial and the lateral directions as well as the time histories of the response of the internal rotator for the cases of complete and partial energy channeling, respectively.

Lissajous curves corresponding to the motion of the outer mass ($\xi_2(t)$ vs. $\xi_1(t)$) are brought in Figs. 10.10 (Lower Panel) and 10.11 (Lower Panel) for the cases of complete and partial energy channeling, respectively. Figures 10.10a (Lower Panel) and 10.11a (Lower Panel) correspond to the transient phase of the response (initial 5% of the total running time), while Figs. 10.10b (Lower Panel) and 10.11b (Lower Panel) correspond to the steady-state one (the final 5% of the total running time).

Results of Figs. 10.10 and 10.11 clearly show that based on the initial tuning of the rotator, the total system energy initially imported in the axial direction can be channeled either partially or completely to the lateral vibrations. In the following subsection, we show analytically that the ratio of energy partition between the axial and the lateral vibrations is fully controlled by the initial tuning of the rotator. Moreover, all the mechanisms bidirectional and unidirectional energy channeling can be completely described and predicted using a regular, multi-scale, asymptotic procedure.

10.2.2 Theoretical Study

To describe analytically the mechanism of partial and complete energy channeling as well as finding the necessary conditions for its existence, we perform an effective reduction of the full system dynamics into a neighborhood of a 1:1 resonance manifold. Assuming the 1:1 resonant interaction between the axial and the lateral vibrations of the outer element, we introduce complex variables in the following form,

$$\psi_1 = \xi_1' + i\,\xi_1$$
$$\psi_2 = \xi_2' + i\,\xi_2 \tag{10.48}$$

Substitution of (10.48) into (10.47) yields the following set of equations of motion (EOM) represented in the complex form:

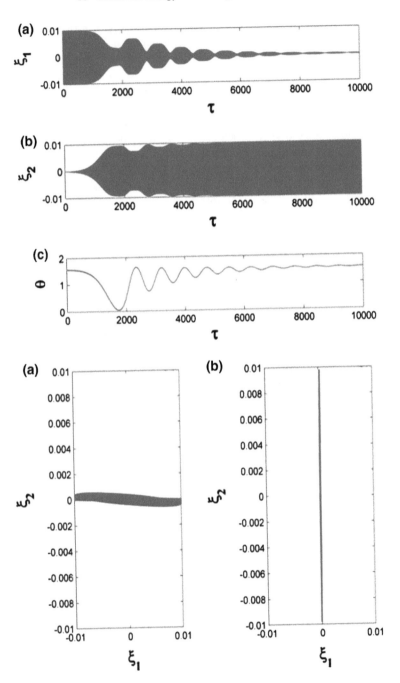

◀**Fig. 10.10** (*Upper Panel*) Time histories of the response of a single cell model (10.6) (complete energy redirection), **a** axial deflection, **b** lateral deflection, **c** rotator angle. System parameters: $\varepsilon = 0.01, \mu = 0.15, \xi_1(0) = 0.01, \theta(0) = 1.5708, \xi_1'(0) = \xi_2(0) = \xi_2'(0) = \theta'(0) = 0$ (*Lower Panel*) Lissajous curve (partial energy redirection) corresponding to the numerical test of Fig. 10.3, **a** transient response (initial 5% of the total running time) and **b** steady-state response (the final 5% of the total running time)

$$\psi_1' - i\psi_1 + i\frac{\varepsilon}{8}\,\alpha\left(\psi_1 - \psi_1^*\right)^3 = \varepsilon\left(\theta'' \sin\theta + \theta'^2 \cos\theta\right)$$

$$\psi_2' - i\psi_2 + i\frac{\varepsilon}{8}\,\alpha\left(\psi_2 - \psi_2^*\right)^3 = -\varepsilon\left(\theta'' \cos\theta - \theta'^2 \sin\theta\right) \tag{10.49}$$

$$\left(\psi_2' - i\frac{\psi_2 + \psi_2^*}{2}\right)\cos\theta - \left(\psi_1' - i\frac{\psi_1 + \psi_1^*}{2}\right)\sin\theta + \theta'' = -\varepsilon\,\mu\,\theta'$$

To analyze the dynamics of (10.49) in the limit of low energy excitations, we use the regular, multi-scale asymptotic expansion in the form:

$$\frac{d(\cdot)}{dt} = \frac{\partial(\cdot)}{\partial\tau_0} + \varepsilon\,\frac{\partial(\cdot)}{\partial\tau_1} + O\left(\varepsilon^2\right),$$
$$\psi_k(\tau) = \varepsilon\,\psi_{k0}\left(\tau_0, \tau_1\right) + \varepsilon^2\,\psi_{k1}\left(\tau_0, \tau_1\right) + O\left(\varepsilon^3\right), \quad k = 1, 2 \tag{10.50}$$
$$\theta(\tau) = \theta_0\left(\tau_0, \tau_1\right) + \varepsilon\,\theta_1\left(\tau_0, \tau_1\right) + O\left(\varepsilon^2\right)$$

Introducing (10.50) into (10.49) and expanding with respect to the like powers of ε yield, in zeroth order $\left(O\left(\varepsilon^0\right)\right)$,

$$\frac{\partial^2\theta_0}{\partial\tau_0^2} = 0 \tag{10.51}$$

Solution of (10.51) yields,

$$\theta_0(\tau_0, \tau_1) = A_1(\tau_1)\tau_0 + A_2(\tau_1) \tag{10.52}$$

Apparently, angular motion of the rotator given in (10.52) is a rapidly growing, linear function. However, in scope of the present analysis, we are seeking for the slow modulation equation depicting the slow evolution of the amplitudes and phases. To this end, we set $A_1(\tau_1)$ to be an identically zero function $(A_1(\tau_1) \equiv 0)$. Proceeding further with a multi-scale expansion yields, in the first order of approximation $\left(O\left(\varepsilon^1\right)\right)$,

$$\frac{\partial\psi_{k0}}{\partial\tau_0} - i\psi_{k0} = 0, \; k = 1, 2$$
$$\frac{\partial^2\theta_1}{\partial\tau_0^2} = -\frac{i}{2}\left(\left(\psi_{20} - \psi_{20}^*\right)\cos\theta_0 - \left(\psi_{10} - \psi_{10}^*\right)\sin\theta_0\right) \tag{10.53}$$

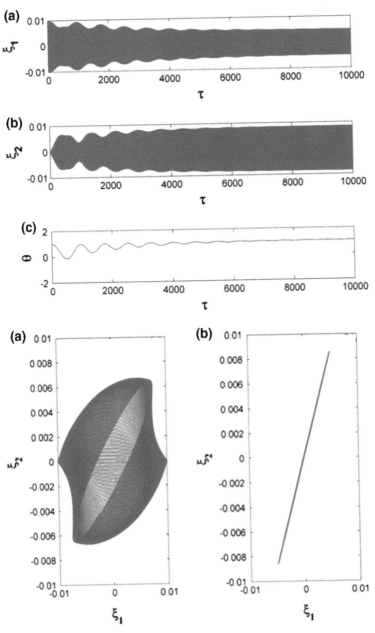

Fig. 10.11 (*Upper Panel*) Time histories of the response of a single cell model (10.6) (partial energy redirection), **a** axial deflection, **b** lateral deflection, **c** rotator angle. System parameters: $\varepsilon = 0.01$, $\mu = 0.15$, $\xi_1(0) = 0.01$, $\theta(0) = 1.048$, $\xi_1'(0) = \xi_2(0) = \xi_2'(0) = \theta'(0) = 0$. (*Lower Panel*) Lissajous curve (partial energy redirection) corresponding to the numerical test of Fig. 10.3a transient response (initial 5% of the total running time) and **b** steady-state response (the final 5% of the total running time)

Solution of the first equation of (10.53) yields

$$\psi_{k0} = \varphi_{k0}(\tau_1)\, e^{i\tau_0}, k = 1, 2 \tag{10.54}$$

Substituting (10.54) into the second equation of (10.53) and solving for θ_1 yield

$$\begin{aligned}
\theta_1 = -\frac{i}{2}\Big(&\big(\varphi_{20}(\tau_1)\, e^{i\tau_0} - \varphi_{20}^*(\tau_1)\, e^{-i\tau_0}\big)\cos\theta_0 \\
&-\big(\varphi_{10}(\tau_1)\, e^{i\tau_0} - \varphi_{10}^*(\tau_1)\, e^{-i\tau_0}\big)\sin\theta_0\Big)
\end{aligned} \tag{10.55}$$

Proceeding to the next order approximation $(O(\varepsilon^2))$, one obtains,

$$\frac{\partial\psi_{11}}{\partial\tau_0} - i\psi_{11} = -\frac{\partial\psi_{10}}{\partial\tau_1} + \frac{i}{4}(1-\cos 2\theta_0)(\psi_{10} - \psi_{10}^*) + \frac{i}{4}(\psi_{20} - \psi_{20}^*)\sin 2\theta_0$$

$$\frac{\partial\psi_{21}}{\partial\tau_0} - i\psi_{21} = -\frac{\partial\psi_{20}}{\partial\tau_1} + \frac{i}{4}(1+\cos 2\theta_0)(\psi_{20} - \psi_{20}^*) + \frac{i}{4}(\psi_{10} - \psi_{10}^*)\sin 2\theta_0$$

$$\frac{\partial^2\theta_0}{\partial\tau_1^2} = -\mu\frac{\partial\theta_0}{\partial\tau_1} - \frac{1}{4}\left(\begin{array}{l}\big(|\varphi_{10}(\tau_1)|^2 - |\varphi_{20}(\tau_1)|^2\big)\sin 2\theta_0 \\ -\big(\varphi_{20}^*(\tau_1)\varphi_{10}(\tau_1) + \varphi_{20}(\tau_1)\,\varphi_{10}^*(\tau_1)\big)\cos 2\theta_0\end{array}\right) \tag{10.56}$$

Next, accounting for (10.54), we get rid of the secular terms in the first two equations of (10.56) deriving the two slow flow equations corresponding to the slow modulation of the amplitude and phase of axial and lateral coordinates of the outer element

$$\begin{aligned}
\varphi_{10}'(\tau_1) &= \frac{i}{4}\varphi_{10}(\tau_1)(1-\cos 2\theta_0) - \frac{i}{4}\varphi_{20}(\tau_1)\sin 2\theta_0 \\
\varphi_{20}'(\tau_1) &= \frac{i}{4}\varphi_{20}(\tau_1)(1+\cos 2\theta_0) - \frac{i}{4}\varphi_{10}(\tau_1)\sin 2\theta_0
\end{aligned} \tag{10.57}$$

It is easy to see that combination of the third equation of (10.56) with system (10.57) constitutes a slow flow model describing a slow modulation of amplitude and phases of axial and lateral coordinates as well as the angle of the rotator. The slow flow model under consideration reads

$$\begin{aligned}
\varphi_{10}'(\tau_1) &= \frac{i}{4}\varphi_{10}(\tau_1)(1-\cos 2\theta_0) - \frac{i}{4}\varphi_{20}(\tau_1)\sin 2\theta_0 \\
\varphi_{20}'(\tau_1) &= \frac{i}{4}\varphi_{20}(\tau_1)(1+\cos 2\theta_0) - \frac{i}{4}\varphi_{10}(\tau_1)\sin 2\theta_0 \\
\frac{\partial^2\theta_0}{\partial\tau_1^2} &= -\mu\frac{\partial\theta_0}{\partial\tau_1} - \frac{1}{4}\left(\begin{array}{l}\big(|\varphi_{10}(\tau_1)|^2 - |\varphi_{20}(\tau_1)|^2\big)\sin 2\theta_0 \\ -\big(\varphi_{20}^*(\tau_1)\varphi_{10}(\tau_1) + \varphi_{20}(\tau_1)\,\varphi_{10}^*(\tau_1)\big)\cos 2\theta_0\end{array}\right)
\end{aligned} \tag{10.58}$$

Indeed, despite the presence of a dissipative term, system (10.58) possesses an integral of motion which is usually referred to in the literature as an "occupation number." This integral of motion reads:

$$|\varphi_{10}(\tau_1)|^2 + |\varphi_{20}(\tau_1)|^2 = N^2 \tag{10.59}$$

The form of the first integral of motion allows for a very convenient change of coordinates, leading to an angular representation.

$$\begin{aligned} \varphi_{10}(\tau_1) &= N \cos \Theta(\tau_1) e^{i\delta_1(\tau_1)} \\ \varphi_{20}(\tau_1) &= N \sin \Theta(\tau_1) e^{i\delta_2(\tau_1)} \end{aligned} \tag{10.60}$$

Substituting (10.60) back into (10.58) and performing some trivial algebraic manipulations yield:

$$\begin{aligned} \frac{\partial \Theta}{\partial \tau_1} &= \frac{1}{4} \sin \Delta \sin 2\theta_0 \\ \frac{\partial \Delta}{\partial \tau_1} &= -\frac{1}{2} [\cos 2\theta_0 - \cos \Delta \sin 2\theta_0 \cot 2\Theta] \\ \frac{\partial^2 \theta_0}{\partial \tau_1^2} &= -\mu \frac{\partial \theta_0}{\partial \tau_1} - \frac{N^2}{4} (\cos 2\Theta \sin 2\theta_0 - \cos \Delta \sin 2\Theta \cos 2\theta_0) \end{aligned} \tag{10.61}$$

where $\Delta = \delta_1 - \delta_2$. It is also worthwhile emphasizing that by deriving system (10.61), we successfully reduced the dimension of the phase space from six to four. The underlying conservative system of (10.61) reads

$$\begin{aligned} \frac{\partial \Theta}{\partial \tau_1} &= \frac{1}{4} \sin \Delta \sin 2\theta_0 \\ \frac{\partial \Delta}{\partial \tau_1} &= -\frac{1}{2} [\cos 2\theta_0 - \cos \Delta \sin 2\theta_0 \cot 2\Theta] \\ \frac{\partial^2 \theta_0}{\partial \tau_1^2} &= -\frac{N^2}{4} (\cos 2\Theta \sin 2\theta_0 - \cos \Delta \sin 2\Theta \cos 2\theta_0) \end{aligned} \tag{10.62}$$

System (10.62) possesses the following two integrals,

$$\begin{aligned} L &= \frac{1}{2} N^2 \sin \Delta \sin 2\Theta + \frac{\partial \theta_0}{\partial \tau_1} \\ H &= -\cos 2\Theta \cos 2\theta_0 - \cos \Delta \sin 2\Theta \sin 2\theta_0 + 4N^{-2} \theta_0'^2 \end{aligned} \tag{10.63}$$

Using the first conserved quantity of (10.63), the four-dimensional slow flow model (10.62) can be reduced to the three-dimensional phase space, yielding

$$\frac{\partial \Theta}{\partial \tau_1} = \frac{1}{4} \sin \Delta \sin 2\theta_0$$

$$\frac{\partial \Delta}{\partial \tau_1} = -\frac{1}{2} [\cos 2\theta_0 - \cos \Delta \sin 2\theta_0 \cot 2\Theta] \qquad (10.64)$$

$$\frac{\partial \theta_0}{\partial \tau_1} = L - \frac{1}{2} N^2 \sin \Delta \sin 2\Theta$$

We start the analytical treatment of (10.64) from the analysis of the stationary points corresponding the stationary regimes, i.e., nonlinear normal modes (NNMs) of the full model (47). Thus, seeking for the fixed points of the reduced slow flow model (10.64), we equate all the time derivatives to zero $(\partial \Theta/\partial \tau_1 = \partial \Delta/\partial \tau_1 = \partial \theta_0/\partial \tau_1 = 0)$. This leads to the following set of nonlinear equations

$$\sin \Delta \sin 2\theta_0 = 0$$

$$\cos 2\theta_0 - \cos \Delta \sin 2\theta_0 \cot 2\Theta = 0 \qquad (10.65)$$

$$2L - N^2 \sin \Delta \sin 2\Theta = 0$$

It is easy to show that system (10.65) has a solution only in the case of zero angular momentum, i.e., $L = 0$. Thus, assuming $(L = 0)$ and solving (10.65) one arrives at the two sets of stationary points

$$\Delta = \pi n, \quad \theta_0 = (-1)^n \Theta + \frac{\pi m}{2}, \quad n, m \in Z, m--\text{odd}$$

$$\Delta = \pi n, \quad \theta_0 = (-1)^n \Theta + \frac{\pi m}{2}, \quad n, k \in Z, m--\text{even} \qquad (10.66)$$

Before proceeding with the stability analysis of the fixed points (10.66), it is important to make a note that the first set of fixed points cannot be attributed to any possible normal mode of the full system. Obviously enough, internal rotator cannot be stationary (as required by the stationary points analysis) as long as its orientation is perpendicular toward the direction of the linear motion of the outer mass. Thus, the first set of stationary points provided by the slow flow model (10.64) requires farther clarification. To understand the origin for the emergence of this "strange" set of stationary points, we bring both physical and mathematical reasoning. Thus, in the limit of small-amplitude vibrations of the outer mass (considered in the present study), the leading-order approximation does not account for the effect of the inertial excitation applied on the internal rotator by the outer mass. Indeed, as is clear from the devised asymptotical procedure, due to the assumed limit of small energy vibrations of the outer element, the effect of the inertial excitation enters into consideration only in the next order approximation. Thus, to arrive at the final resolution concerning the essence of the first ("strange") set of stationary points, we resort to the next order approximation derived in (10.55). Substituting the first and the second sets of stationary solutions (10.66) into (10.55) yields

$$\theta_1(\tau_0, \tau_1) = -N(-1)^k \sin(\delta_1(\tau_1) + \tau_0) \quad \text{(First set)}$$
$$\theta_1(\tau_0, \tau_1) = 0 \qquad\qquad\qquad\qquad \text{(Second set)} \tag{10.67}$$

Clearly, the first set of fixed points of (10.66) corresponds to the low amplitude $(O(\varepsilon))$ vibrations of the rotator, while the second one corresponds to the perfectly periodic motion of the outer element with internal rotator being permanently oriented toward the direction of motion of the outer element. Thus, the second set of fixed points is the only one which can be related to the real, physical periodic motion of the full system, i.e., nonlinear normal modes (NNMs) manifested by the linear, periodic motion of the external element oscillating in the direction parallel to the internal orientation of the rotator.

Before proceeding with the stability analysis of each set, it is important to clarify once again that the second set of stationary points of (10.66) corresponds to the perfectly periodic motion of the full system under consideration (10.47) where the relative phase between the axial and the lateral motions is an integer multiple of π. Interestingly enough the angular coordinate Θ which according to its definition (10.60) governs the ratio of energy partition between the axial and the lateral vibrations can assume arbitrary values. In fact this is an important result also from the viewpoint of practical applications meaning that internal rotator can be tuned such that it enables an arbitrary energy partition between the axial and the lateral periodic motion given that the response is stable. Let us proceed with the linear stability analysis of the stationary points corresponding to the second set of (10.66). To this end, we assume small deviations $(\widetilde{\Theta}, \widetilde{\Delta}, \widetilde{\theta}_0)$ in the vicinity of stationary points:

$$\Delta = \Delta_S + \widetilde{\Delta}, \quad \theta_0 = \theta_{0S} + \widetilde{\theta}_0, \quad \Theta = \Theta_S + \widetilde{\Theta} \tag{10.68}$$

where $\Delta_S, \theta_{0S}, \Theta_S$ stand for the stationary points. Substituting (10.68) into (10.64) and performing a trivial linearization, one arrives at the following linear system,

$$\begin{pmatrix} \widetilde{\Theta}' \\ \widetilde{\Delta}' \\ \widetilde{\theta}'_0 \end{pmatrix} = \begin{bmatrix} 0 & \dfrac{(-1)^n \sin(2\theta_{0S})}{4} & 0 \\ \dfrac{(-1)^{n+1}}{\sin(2\theta_{0S})} & 0 & \dfrac{1}{\sin(2\theta_{0S})} \\ 0 & \dfrac{(-1)^{m+1}}{2} N^2 \sin(2\theta_{0S}) & 0 \end{bmatrix} \begin{pmatrix} \widetilde{\Theta} \\ \widetilde{\Delta} \\ \widetilde{\theta}_0 \end{pmatrix} \tag{10.69}$$

The characteristic polynomial of the Jacobian matrix of (10.69) reads

$$\lambda \left(\lambda^2 + \frac{1}{4}\left(2(-1)^m N^2 + 1\right) \right) = 0 \tag{10.70}$$

It is easy to show that the second set of stationary points is neutrally stable for m—even and an arbitrary value of N. Here we also note that stability (or instability) of the stationary points corresponding to the second set is independent of the values of the stationary points (i.e., θ_{0S} and Θ_S). It is worth noting that the general analysis of non-stationary regimes (e.g., weak and strong energy pulsations between the horizontal and vertical vibrations of the outer element) remains a rather complex problem. However, as we show below for a certain choice of the system parameters, one can completely describe and predict analytically the emergence of the intriguing phenomenon of "bidirectional energy channeling." In the same study, this special response has been defined as a complete, recurrent energy transport from axial to lateral vibrations of the outer element (controlled by the motion of the internal rotator) being initially excited in the axial direction. To demonstrate the persistence of similar phenomena for the low energy limit (this limit is also manifested by the lack of strong resonant interactions between the internal rotator and the outer mass), we resort again to the analysis of the slow flow model (10.64) under assumption of zero angular momentum $(L = 0)$. Before proceeding with the analysis, we would like to make a note concerning this choice of the system parameters, i.e., zero angular momentum. In fact $(L = 0)$ corresponds to the very important physical case where internal rotator is initially at rest. This special type of initial condition has a very special practical implication in the design of the locally resonant absorptive metamaterials and sonic structures, where at the initial state of the system, all the internal devices (e.g., internal resonators) are at rest. Thus, for instance, in various physical and mechanical models, vibration absorbers tuned to protect the initially excited primary structure are assumed to be initially at rest. Our system under consideration reads

$$\frac{\partial \Theta}{\partial \tau_1} = \frac{1}{4} \sin \Delta \sin 2\theta_0$$

$$\frac{\partial \Delta}{\partial \tau_1} = -\frac{1}{2} [\cos 2\theta_0 - \cos \Delta \sin 2\theta_0 \cot 2\Theta] \qquad (10.71)$$

$$\frac{\partial \theta_0}{\partial \tau_1} = -\frac{1}{2} N^2 \sin \Delta \sin 2\Theta$$

It is easy to see that system (10.71) possesses an additional conserved quantity C, given by

$$C = 2N^2 \cos 2\Theta + \cos 2\theta_0 \qquad (10.72)$$

In fact (10.72) defines a projection of the entire phase space of (10.30) onto the (θ_0, Θ) plane. Four distinct projections (for the different values of N) are illustrated in Fig. 10.12. To understand better the mechanism of bidirectional energy channeling, it is convenient to define the two basic states for the projections under consideration. The first state is defined as $\Theta \equiv 0$ which corresponds to the complete energy localization in the axial direction, while $\Theta \equiv \pi/2$ is the second state corresponding to the complete energy localization in the lateral direction. As is clear from the projections illustrated in Fig. 10.12, there exist three types of projection curves: (1) curves that depart from a certain state (e.g., $\Theta \equiv 0$ or $\Theta \equiv \pi/2$) and return to the same state without piercing the other one, (2) curves interconnecting both states [i.e., ($\Theta \equiv 0$) with ($\Theta \equiv \pi/2$)], (3) curves which do not pierce any of the two states, and (4) curves which separate the two distinct families of projection curves. For the sake of brevity, we define the first type of curves as *returning,* the second type as *channeling,* the third type as *disconnecting,* and the last one as *separating.* Obviously enough, the aforementioned regime of bidirectional, complete energy channeling emerging between the axial and lateral vibrations corresponds to the channeling curves of the projected phase space. In the subplots (a) and (b) of Fig. 10.12, one can clearly see the special regions containing the channeling curves. These are the regions of the special interest which will be referred to as "channeling regions." As it is also evident from the results of Fig. 10.12, as we increase the value of N, the "channeling regions" gradually shrink and at a certain critical value of $N = N_{CR}$, (Fig. 10.12c) completely disappears. Increasing the value of N further ($N > N_{CR}$) brings to a global change in the topology of the projection. Thus, the whole family of the channeling curves vanishes giving rise to the formation of the third type of projection curves, namely the disconnecting curves.

The first goal of the present subsection is in finding the threshold value of $N = N_{CR}$ corresponding to the destruction of the "channeling regions" as well as calculating their zones of existence for each state in the case of ($N \le N_{CR}$). To this end, we use the conserved quantity given in (10.72) and define

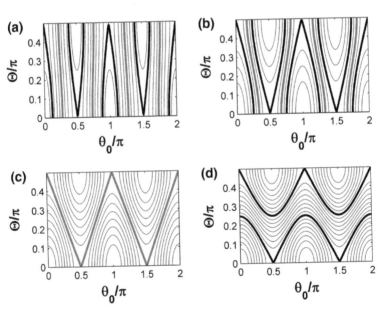

Fig. 10.12 Projection of the phase space of (10.64) onto the (θ_0, Θ) plane for the different values of N: **a** $N = 0.25$, **b** $N = 0.5$, **c** $N = 1/\sqrt{2}$, and **d** $N = 1$. Channeling path corresponding to the complete energy transport is denoted with the *bold lines*, separatrices are denoted with the *bold solid lines*, and regular paths are denoted with the *thin solid lines*. System parameters: $L = 0$

$$\theta_0^- = \theta_0(\Theta = 0), \ \theta_0^+ = \theta_0(\Theta = \pi/2) \tag{10.73}$$

here $\theta_0^\pm$ stands for the values of the rotator angle corresponding to the first and the second states.

Thus, for the first and the second states, Eq. (10.72) reads

$$\begin{aligned} \Theta = 0: \quad & C_- = 2N^2 + \cos 2\theta_0^- \\ \Theta = \frac{\pi}{2}: \quad & C_+ = -2N^2 + \cos 2\theta_0^+ \end{aligned} \tag{10.74}$$

Obviously enough, "channeling curves" should satisfy

$$C_- = C_+ = C \tag{10.75}$$

Equation (10.75) yields the very important relations which will be used below,

$$\begin{aligned} \cos\left(2\theta_0^+\right) + \cos\left(2\theta_0^-\right) &= 2C \\ \cos\left(2\theta_0^+\right) - \cos\left(2\theta_0^-\right) &= 4N^2 \end{aligned} \tag{10.76}$$

The critical value of $N = N_{CR}$ can be easily derived from the solvability conditions of the second equation of (10.76) yielding

$$N_{CR} = 2^{-1/2} \tag{10.77}$$

Using the solvability conditions of (10.74) and accounting for (10.75) yield the conditions for the channeling interval,

$$\left|C - 2N^2\right| \leq 1 \quad \text{and} \quad \left|C + 2N^2\right| \leq 1 \tag{10.78}$$

From (10.74) to (10.78), one infers that channeling curves satisfy

$$2N^2 - 1 \leq C \leq 1 - 2N^2, \quad N \leq 2^{-1/2} \tag{10.79}$$

Thus, the intervals of the "channeling regions" of both states can be readily derived from (10.74)

$$\begin{aligned} \Theta = 0: \quad & \theta_{OL}^- = \frac{1}{2}\arccos\left(1 - 4N^2\right) + \pi(n-1), \ \theta_{OR}^- = \pi/2 + \pi n, \ N < N_{CR} \\ \Theta = \frac{\pi}{2}: \quad & \theta_{OR}^- = \pi n, \ \theta_{OL}^+ = \frac{1}{2}\arccos\left(4N^2 - 1\right) + \pi n, \ N < N_{CR} \end{aligned}$$

$$\tag{10.80}$$

Here $\theta_{OL}^-, \theta_{OR}^-$ stand for the left and the right boundaries of the "channels" of the lower state ($\Theta = 0$), whereas $\theta_{OL}^+, \theta_{OR}^+$ stand for the left and the right boundaries of the "channels" of the upper one ($\Theta = \pi/2$). As it was already mentioned above, the diagrams of Fig. 10.12 show the projection of the entire phase space of (10.71) onto the plane. Therefore, presence of the "channeling regions" on the projection plane (θ_0, Θ) does not necessarily mean that each trajectory of the full phase space of (10.71) emanating from the lower or upper states and following one of the channeling curves will reach the second state. There might be also a case where following one of the "channeling curves," the true trajectory of the entire phase space turns back and arrives at the original state through the same path without piercing the upper one. This also means that the main condition for bidirectional, complete energy transport is not satisfied. Thus, the final aim of the present subsection would be to single out the special "channeling curves" corresponding to the pure energy transport.

To derive the additional conditions on the phase trajectories corresponding to the complete energy channeling, we use the second integral of motion of (10.63). Arguing exactly as above, we compute the values of Hamiltonian for both states

$$H(\Theta = 0) = H^- = -\cos 2\theta_0^-, H(\Theta = \pi/2) = H^+ = \cos 2\theta_0^+ \qquad (10.81)$$

Obviously, each phase trajectory corresponding to the complete energy channeling must satisfy

$$H^- = H^+ = h \Rightarrow \cos(2\theta_0^+) + \cos(2\theta_0^-) = 0 \qquad (10.82)$$

Thus, accounting for (10.82) in the first equation of (10.76) singles out a projection curve belonging to the "channeling region" and satisfying $(C = 0)$. This special projection curve is given by

$$2N^2 \cos 2\Theta + \cos 2\theta_0 = 0 \qquad (10.83)$$

In fact Eq. (10.83) defines a special path corresponding to the complete energy transport between the axial and the lateral vibrations of the outer element and is denoted with the red bold color on the diagrams of Fig. 10.12. Apparently, there exists a unique phase trajectory corresponding to the aforementioned bidirectional energy channeling regimes which can be defined implicitly by the Hamiltonian. Thus, from (10.83) one has,

$$h = H^- = \cos 2\theta_0^- = -2N^2 \qquad (10.84)$$

Substituting (10.84) into the expression for the Hamiltonian and setting $L = 0$ yields,

$$2N^2 - \cos 2\Theta \cos 2\theta_0 - \cos \Delta \sin 2\Theta \sin 2\theta_0 + N^2 \sin^2 \Delta \sin^2 2\Theta = 0 \qquad (10.85)$$

Thus, (10.85) together with (10.83) defines the unique phase trajectory corresponding to the complete, recurrent energy transport. In Fig. 10.13, we plot the time histories of the response of the slow flow model (10.64) corresponding to the regime of complete recurrent energy channeling.

Here we mention in passing that the derived, special phase trajectory constitutes the three-dimensional analog of the LPT introduced in the Part 1. As it was mentioned several times, the LPT was initially defined (Manevitch L.I. et al.) as a special phase trajectory corresponding to the complete recurrent energy transport emerging in the systems of two weakly coupled anharmonic oscillators (ranging from linear and up to the purely nonlinear oscillators). The analytical approximation of this unique higher dimensional LPT defined by (10.83) and (10.85) deserves a separate study and is beyond the scope of the present book. Clearly, in the dissipative case, angular momentum is a time varying quantity, reading

$$L(\tau_1) = \frac{\partial \theta_0}{\partial \tau_1} + \frac{1}{2} N^2 \sin \Delta \sin 2\Theta \qquad (10.86)$$

Differentiating (10.86) with respect to a slow timescale yields,

$$\frac{dL}{d\tau_1} = \frac{1}{2} N^2 \left(\frac{\partial \Delta}{\partial \tau_1} \cos \Delta \sin 2\Theta + 2\frac{\partial \Theta}{\partial \tau_1} \sin \Delta \cos 2\Theta \right) + \frac{\partial^2 \theta_0}{\partial \tau_1^2} \qquad (10.87)$$

Accounting for (10.61) in (10.87) and performing some simple algebraic manipulations, one can show that (10.87) reduces to the following simple form

$$\frac{dL}{d\tau_1} = -\mu \frac{\partial \theta_0}{\partial \tau_1} \qquad (10.88)$$

Integrating (10.88) once with respect to τ_1 allows one to find the following useful relation,

$$L = C - \mu \theta_0 \qquad (10.89)$$

where $C = L(0) + \mu \theta_0(0)$. Thus, accounting for (10.89), the dissipative slow flow system can be reformulated as,

$$\frac{\partial \Theta}{\partial \tau_1} = \frac{1}{4} \sin \Delta \sin 2\theta_0$$
$$\frac{\partial \Delta}{\partial \tau_1} = -\frac{1}{2} [\cos 2\theta_0 - \cos \Delta \sin 2\theta_0 \cot 2\Theta] \qquad (10.90)$$
$$\frac{\partial \theta_0}{\partial \tau_1} = C - \mu \theta_0 - \frac{1}{2} N^2 \sin \Delta \sin 2\Theta$$

As it was explained above, seeking for the stationary points of the reduced slow flow model (10.90), we set all the time derivatives to zero

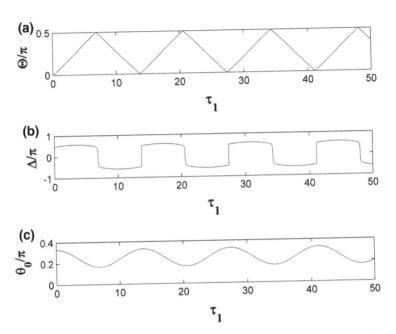

Fig. 10.13 Time histories of the response of a reduced slow flow model (10.71) corresponding to the complete recurrent energy channeling, **a** normalized angular coordinate (Θ) describing the instantaneous energy partition between the axial and lateral vibrations of the outer mass, **b** normalized relative phase (Δ) between the axial and lateral vibrations of the outer mass, and **c** normalized internal rotator angle (θ_0). System parameters: $N = 0.5$, $\Theta(0) = 0$, $\Delta(0) = \pi/2$, $\theta_0 = \pi/3$

($\partial\Theta/\partial\tau_1 = \partial\Delta/\partial\tau_1 = \partial\theta_0/\partial\tau_1 = 0$). This leads to the following set of nonlinear, algebraic equations,

$$\sin \Delta \sin 2\theta_0 = 0$$
$$\cos 2\theta_0 - \cos \Delta \sin 2\theta_0 \cot 2\Theta = 0 \qquad (10.91)$$
$$C - \mu\,\theta_0 - \frac{1}{2}N^2 \sin \Delta \sin 2\Theta = 0$$

Similarly to the conservative case, it is easy to show that system (10.91) possesses the following set of fixed points:

$$\Delta = \pi n, \quad \theta_0 = \frac{C}{\mu}, \quad \Theta = (-1)^n \theta_0 + \frac{\pi m}{2} \qquad (10.92)$$

Let us proceed with the linear stability analysis of the stationary points of (10.92). To this end, we assume again small deviations ($\tilde{\Theta}, \tilde{\Delta}, \tilde{\theta}_0$) in the vicinity of stationary points:

$$\Delta = \Delta_S + \tilde{\Delta}, \ \theta_0 = \theta_{0S} + \tilde{\theta}_0, \ \Theta = \Theta_S + \tilde{\Theta} \tag{10.93}$$

where $\Delta_S, \theta_{0S}, \Theta_S$ stand for the stationary points. Substituting (10.93) into (10.90) and performing a trivial linearization in the vicinity of the stationary points of the set (10.92), one arrives at the following linear system,

$$
\begin{pmatrix} \tilde{\Theta}' \\ \tilde{\Delta}' \\ \tilde{\theta}_0' \end{pmatrix} =
\begin{bmatrix}
0 & \frac{(-1)^n \sin(2\theta_{0S})}{4} & 0 \\
\frac{(-1)^{n+1}}{\sin(2\theta_{0S})} & 0 & \frac{1}{\sin(2\theta_{0S})} \\
0 & \frac{(-1)^{m+1}}{2} N^2 \sin(2\theta_{0S}) & -\mu
\end{bmatrix}
\begin{pmatrix} \tilde{\Theta} \\ \tilde{\Delta} \\ \tilde{\theta}_0 \end{pmatrix} \tag{10.94}
$$

The characteristic polynomial of the Jacobian matrix of (10.94), reads

$$\lambda^3 + \mu \lambda^2 + \frac{1}{4}\left(2(-1)^m N^2 + 1\right)\lambda + \frac{\mu}{4} = 0 \tag{10.95}$$

It is easy to show (e.g., Routh–Hurwitz stability criterion) that the set of fixed points given in (10.92) is stable for (m—even) which as it has been pointed out above corresponds to the true NNMs of the full model.

In Section III, we demonstrated the intriguing nonlinear phenomenon of unidirectional, partial (Fig. 10.11), and complete (Fig. 10.10) energy channeling from axial to lateral vibrations of the outer element. This section focuses on the analytical description of the mechanism governing this nonlinear phenomenon. Authors believe that the analysis of the present subsection provides very efficient and simple theoretical tools which have a potential to facilitate the design of optimally tuned locally resonant mechanical metamaterials and acoustic structures with the unique dynamical properties enabling the efficient, 2D wave redirection and wave arrest. Thus, based on the results of the previous subsection, we formulate a simple control strategy based on the initial tuning of the internal rotator leading to the partial and complete, unidirectional energy flow. To this end, we define the initial state of the rotator as $\theta_0^I = \theta_0(0)$ and the final state of the rotator as $\theta_0^F = \theta_0(\tau_1 \to \infty)$. At this point, it would be rather convenient to define the map from the initial state of the system to the final one. Thus, using (10.92) and the results of the stability analysis, one obtains the following map,

$$\Delta^F = \pi n, \quad \theta_0^F = \frac{L(0) + \mu\theta_0^I}{\mu}, \quad \Theta^F = (-1)^n \frac{L(0) + \mu\theta_0^I}{\mu} + \pi k \tag{10.96}$$

where $\{\Theta^F, \Delta^F, \theta_0^F\}$ define the final (steady) state of the system. Clearly, results of (10.96) can be used for initial tuning of the internal rotator such that at the final state, system settles down at the required orientation where the external mass performs the one-directional, steady-state oscillations (at a certain orientation angle prescribed by the initial orientation of the rotator). In the same final state, internal rotator is stationary and is oriented toward the direction of the 2D (linear) motion of the outer mass. Thus, to achieve the complete unidirectional energy channeling from axial to lateral vibrations as is shown in Fig. 10.10 (Upper Panel), let us consider the following initial state $\Delta^I = \pi n$, $\theta_0^I = \frac{\pi\mu/2 - l}{\mu}$, $\Theta^I = 0, L(0) = l$. Here we note that the values of Θ^I and Δ^I are assumed to be known and correspond to the initial excitation of the outer mass. In the considered case, we assume the initial excitation of the outer mass applied strictly in the axial direction $\Theta^I = 0$ (e.g., initial deflection). To achieve the complete energy transport from axial to lateral vibrations for the given initial state, one should require that $\Theta^F = \pi/2$. Plugging the required value of Θ^F into the last equation of (10.96), one arrives at the initial value of the rotator angle $(\theta_0^I = \frac{\pi\mu/2 - l}{\mu} + \pi k)$ that assures the required complete energy transport. We note that in the special case of $L(0) = 0$, the initial value of the rotator corresponding to the complete energy transport is independent of the dissipation rate. In summary, we would like to stress that using (10.96) one can tune the rotator for the arbitrary ratio of energy partition including the two limiting cases, i.e., zero energy channeling ($\Theta^F = 0$) and complete energy channeling ($\Theta^F = \frac{\pi}{2}$). In the following section, we confirm the results of analytical predictions for partial and complete energy channeling with the numerical simulations of the full and reduced models.

10.2.3 Numerical Verifications

In the present section, we perform numerical verifications of the validity of a theoretical model devised in the previous section. In Fig. 10.14, we plot the time histories of a simple periodic response (NNM) exhibited by the full model (47) (assuming zero dissipation) and compare it with the predictions of stationary analysis applied on the slow flow model (10.64). As is clear from the results of Fig. 10.14, the analytical prediction of the stationary response derived from the slow flow model depicts fairly well the response of the original one, and above all, the correspondence of analytical and numerical models is very good. To verify the occurrence of complete and incomplete recurrent energy channeling emerging in the underlying Hamiltonian system, we performed the two additional numerical runs. In Fig. 10.15a, we plot the response of the full (10.47) and the slow (10.71) models. The time histories corresponding to the predicted regime of complete ($N < N_{CR} = 2^{-1/2}$) recurrent energy channeling (i.e., "*limiting phase trajectory*") of the slow flow model (10.71) are compared with these of the full model (47) (Fig. 10.15, Upper Panel).

The time histories corresponding to the regimes of mild energy transport predicted by the analytical model (i.e., destruction of "*Channeling Regions*" for $N > N_{CR} = 2^{-1/2}$) are illustrated in (Fig. 10.15, Lower Panel). The results of numerical simulations presented in Fig. 10.15 confirm the analytically predicted transition in the response regimes from bidirectional energy channeling to the

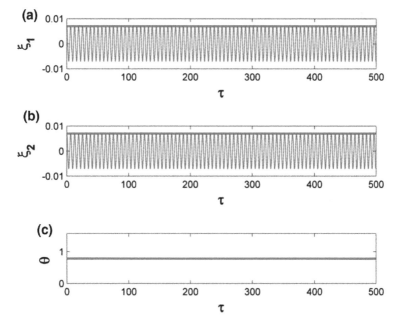

Fig. 10.14 Time histories of the response of a single cell model (10.47) (stationary response regime), **a** axial deflection, **b** lateral deflection, **c** rotator angle. The *blue solid line* corresponds to the response of the full model while the *red solid line* is related to the response of the reduced slow flow model (10.47). System parameters: $\varepsilon = 0.01$, $\xi_1(0) = \xi_2(0) = 0.0071$, $\theta(0) = \pi/4$, $\xi_1'(0) = \xi_2'(0) = \theta'(0) = 0$

"unidirectional energy locking" manifested by the mild energy exchange between the axial and lateral vibrations of the outer element and significant energy localization in the axial direction. As it has been proved analytically in the previous section, this transition from bidirectional energy transport to the unidirectional energy locking is fully governed by the value of N. Let us proceed with the comparison of the time histories of the true system response (10.47) with that of a dissipative, reduced, slow flow model (10.90) for the two distinct cases, namely complete energy channeling (Fig. 10.16, Upper Panel) as well as a partial energy channeling (Fig. 10.16, Lower Panel). As is clear from the results of Fig. 10.16, the derived slow flow model depicts fairly well the response of the full model and above all faithfully predicts the emergence of both phenomena of complete as well as the partial energy channeling. Evidently enough numerical simulations of the full model (10.47) are in the very good correspondence with the analytical predictions derived from the analysis of the reduced one (10.90).

In summary, we would like to stress once again that the results of Figs. 10.14, 10.15 and 10.16 fully confirm the validity of the analytical model as well as the analytical predictions of the intrinsic dynamical mechanisms of partial and complete, unidirectional energy channeling. Importantly, numerical simulations of the present section confirm the main result of the analytical study showing that in the final (steady) state of the response, energy distribution between the axial and the lateral vibrations of the outer mass is completely controlled by the initial state of the internal rotator.

10.2.4   Concluding Remarks

In the present work, we considered the dynamics of a unit cell model comprising an outer mass incorporating internal rotator and mounted on the 2D, nonlinear elastic foundation in the limit of low energy excitations. Special analytical treatment based on the regular multi-scale expansion in the low energy limit is developed. Analysis of the derived slow flow model reveals the peculiar mechanisms of the bidirectional recurrent energy channeling for the Hamiltonian case as well as the (partial or complete) unidirectional energy channeling in the dissipative one. Surprisingly enough, the derived slow flow system is fully integrable. This property of the system enabled to find certain symmetries of the system and significantly reduce its complexity for the certain cases of fundamental and practical importance. Thus, in the case of zero initial angular momentum ($L = 0$), authors derived the global analytical predictions of the mechanism of formation and bifurcations of highly non-stationary regime of complete, recurrent energy transport (bidirectional energy channeling) established between the axial and the lateral vibrations of the outer mass. Here we would like to note that the case of ($L = 0$) has a very important practical implication as it corresponds to the case of zero initial (angular) velocity of the internal rotator. This type of initial condition is commonly assumed in the problems dealing with vibration and shock absorption, targeted energy transport, design of absorptive metamaterials, and seismic protection devices. In all these cases, the LPT concept plays a key role in the understanding and description of strong non-stationary dynamics. Another interesting phenomena revealed and analyzed in the present study correspond to the two-dimensional, unidirectional partial, and complete energy channeling. Using the asymptotical model derived from the regular multi-scale expansion in the limit of low energy excitations, we describe the entire mechanism governing the partial and complete unidirectional energy channeling. Moreover, the derived analytical tools enable the prediction of the final states of permanent energy localization. The devised analytical prediction can be further exploited for the design of smart acoustic metamaterials with the unique dynamical properties enabling the 2D energy redirection and efficient spatial wave arrest. Results of the analytical model and numerical simulations are found to be in a very good correspondence.

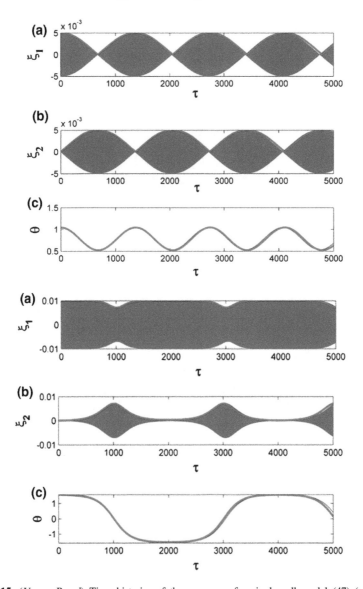

Fig. 10.15 (*Upper Panel*) Time histories of the response of a single cell model (47) (complete recurrent energy channeling), **a** axial deflection, **b** lateral deflection, **c** rotator angle. The *blue solid line* corresponds to the response of the full model, while the *red solid line* is related to the response of the reduced slow flow model (10.71). System parameters: $\varepsilon = 0.01$, $\xi_1(0) = 0.005$, $\theta(0) = 1.0472, \xi_2(0) = \xi_1'(0) = \xi_2'(0) = \theta'(0) = 0$. (*Lower Panel*) Time histories of the response of a single cell model (10.47) (incomplete recurrent energy channeling), **a** axial deflection, **b** lateral deflection, **c** rotator angle. The *blue solid line* corresponds to the response of full model, and *red solid line* is related to the response of the reduced slow flow model (10.71). System parameters: $\varepsilon = 0.01$, $\xi_1(0) = 0.01$, $\theta(0) = 1.5708$, $\xi_2(0) = \xi_1'(0) = \xi_2'(0) = \theta'(0) = 0$

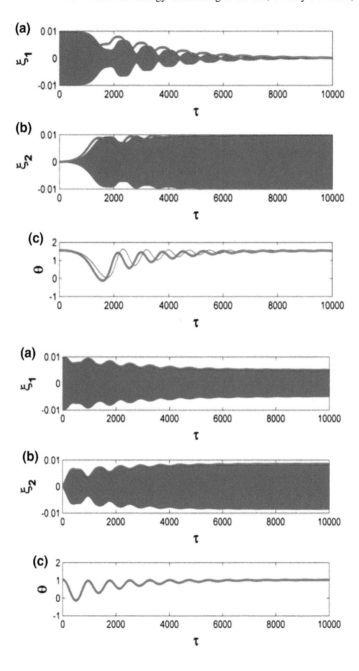

◄**Fig. 10.16** (*Upper Panel*) Time histories of the response of a single cell model (10.47) (complete energy redirection), **a** axial deflection, **b** lateral deflection, **c** rotator angle. The *blue solid line* corresponds to the response of full model, and *red solid line* is related to the response of the reduced slow flow model (10.90). System parameters: $\varepsilon = 0.01, \mu = 0.15, \xi_1(0) = 0.01$, $\theta(0) = 1.5708, \xi_1'(0) = \xi_2(0) = \xi_2'(0) = \theta'(0) = 0$. (*Lower Panel*) Time histories of the response of a single cell model (10.47) (partial energy redirection), **a** axial deflection, **b** lateral deflection, **c** rotator angle. The *blue solid line* corresponds to the response of full model, and *red solid line* is related to the response of the reduced slow flow model (10.90). System parameters: $\varepsilon = 0.01, \mu = 0.15, \xi_1(0) = 0.01, \theta(0) = 1.048, \xi_1'(0) = \xi_2(0) = \xi_2'(0) = \theta'(0) = 0$

Appendix

Stability of both branches of the SIM can be directly inferred from the Eq. (10.13),

$$\frac{\partial^2 \beta}{\partial \tau_1^2} + \eta \frac{\partial \beta}{\partial \tau_1} + \eta = -\frac{1}{4} \left[\begin{array}{c} \left(\varphi_{10}^*(\tau_2) \exp(i\beta) + \varphi_{10}(\tau_2) \exp(-i\beta)\right) \\ -i\left(\varphi_{20}^*(\tau_2) \exp(i\beta) - \varphi_{20}(\tau_2) \exp(-i\beta)\right) \end{array} \right] \quad (10.97)$$

Introducing the new coordinates,

$$Z = \varphi_{10} + i\varphi_{20} = |Z| \exp(i\upsilon), \quad \psi(\tau_1) = \beta(\tau_1, \tau_2) - \upsilon(\tau_2) \quad (10.98)$$

we rewrite (10.97) in the more compact form,

$$\frac{\partial^2 \psi}{\partial \tau_1^2} + \eta \frac{\partial \psi}{\partial \tau_1} + \frac{|Z(\tau_2)|}{2} \cos(\psi) = -\eta \quad (10.99)$$

Equation (10.99) is the well-known equation of the damped mathematical pendulum driven by the constant external torque. The fixed points (ψ_m) of (10.99) satisfy

$$\psi_m = \arccos\left(\frac{2\eta}{|Z(\tau_2)|}\right) + \pi m, \quad \eta \leq |Z(\tau_2)|/2, \quad m \in Z \quad (10.100)$$

These fixed points define the branches of the SIM which correspond to the state of the exact 1:1:1 resonance. Using the straightforward, linear stability analysis of (10.99), one can easily show that m—odd corresponds to the stable branch of SIM while m—even to the unstable one.

References

Boechler, N., Theocharis, G., Daraio, C.: Bifurcation-based acoustic switching and rectification. Nat. Mater **10**, 665–668 (2011)

Deymier P.A.: Acoustic Metamaterials and Phononic Crystals. Springer, Berlin (2013)

Gendelman, O.V., Sigalov, G., Manevitch, L.I., Mane, M., Vakakis, A.F., Bergman, L.A.: Dynamics of an eccentric rotational nonlinear energy sink. J. Appl. Mech **79**(1), 011012 (2012)

Grima, J.N., Caruana-Gauci, R.: Mechanical metamaterials: materials that push back. Nature Mater **11**, 6–565 (2012)

Liu, Z., Zhang, X., Mao, Y., Zhu, Y.Y., Yang, Z., Chan, C.T., Sheng, P.: Locally resonant sonic materials. Science **289**, 1734 (2000)

Manevitch, L.I., Kovaleva, A.S., Shepelev, D.S.: Non-smooth approximations of the limiting phase trajectories for the Duffing oscillator near 1:1 resonance. Phys. D—Nonlinear Phenom **240**(1), 1–12 (2011)

Manevitch L.I., Smirnov V.V.: Limiting phase trajectories and the origin of energy localization in nonlinear oscillatory chains. Phys. Rev. E **82**(3) (2010).

Nicolaou, Z.G., Motter, A.E.: Mechanical metamaterials with negative compressibility transitions. Nature Mater **11**, 13–608 (2012)

Sigalov, G., Gendelman, O.V., AL-Shudeifat, M.A., Manevitch, L.I., Vakakis, A.F., Bergman, L.A.: Resonance captures and targeted energy transfers in an inertially-coupled rotational nonlinear energy sink. Nonlinear Dyn **69**, 1693–1704 (2012a)

Sigalov, G., Gendelman, O.V., AL-Shudeifat, M.A., Manevitch, L.I., Vakakis, A.F., Bergman, L.A.: Alternation of regular and chaotic dynamics in a simple two-degree-of-freedom system with nonlinear inertial coupling. Chaos **22**, 013118 (2012b)

Vakakis, A.F., Gendelman, O.V., Kerschen, G., Bergman, L.A., McFarland, D.M., Lee, Y.S.: Nonlinear Targeted Energy Transfer in Mechanical and Structural Systems I. Springer, Berlin and New York (2008)

Wiggins, S.: Introduction to Nonlinear Dynamics and Chaos. Springer, New York (1990)

Chapter 11
Nonlinear Targeted Energy Transfer and Macroscopic Analogue of the Quantum Landau-Zener Effect in Coupled Granular Chains

11.1 Introduction

Resonance is the main mechanism for energy propagation in spatially periodic linear/nonlinear systems. For the case of two weakly coupled identical Hamiltonian oscillators in resonance, any amount of energy imparted to one of the oscillators gets transferred back and forth between these oscillators with a frequency proportional to the coupling. On the other hand, localization of energy is also possible in such systems provided that resonance is broken. This phenomenon is trivial for linear systems through the use of structural disorder. However, for nonlinear systems, it has been shown that the inherent nonlinearity plays a "double game" (Kopidakis and Aubry 1999, 2000a, b; Morgante et al. 2002; Johansson et al. 2002; Aubry et al. 2001) when one attempts to localize energy in one of the oscillators, in the sense that nonlinearity can either restore the resonance among the oscillators or maintain the resonance breakup.

In nonlinear spatially periodic systems, localized nonlinear modes can be formed. These modes are referred as intrinsic localized modes or discrete breathers—DBs (Campbell and Peyrard 1990; Scott 1999). Focusing and transport of energy through discrete breathers can find application to a broad range of physical, chemical, and biological systems. For example, in chemical systems the concept of DB has been introduced in studies of the vibrational states of molecules (Scott 1999), whereas in physics it has been applied in studies of nonlinear lattice problems (Flach and Willis 1998; Aubry 1997; Kopidakis et al. 2001a, b). Various other systems have also experienced DBs, such as waveguide arrays (Eisenberg et al. 1998), low-dimensional crystals (Swanson et al. 1999), antiferromagnetic spin lattices (Sato and Sievers 2004), underdamped Josephson-junction arrays (Trias et al. 2000) and Josephson-junction ladders (Binder et al. 2000), charge-transfer solids, photonic structures and micromechanical oscillator arrays (Campbell et al. 2004), α-helices (Elder et al. 2004), nonlinear networks proteins (Juanico et al. 2007), and α-uranium systems (Manley et al. 2006).

© Springer Nature Singapore Pte Ltd. 2018
L.I. Manevitch et al., *Nonstationary Resonant Dynamics of Oscillatory Chains and Nanostructures*, Foundations of Engineering Mechanics,
DOI 10.1007/978-981-10-4666-7_11

The formation of DBs in ordered granular media is, however, a relatively new area of study. The dynamics of granular media has been the subject of considerable attention among the engineers and scientists. One of the prime reasons lies in their highly tunable dynamical properties which can, indeed, range from being weakly nonlinear and smooth (when these media are highly pre-compressed) to being strongly nonlinear and non-smooth (when there is weak or absence of pre-compression) (Nesterenko 2001). In fact, uncompressed ordered granular media have been characterized as "*sonic vacua*" (Nesterenko 2001) since they possess zero speed of sound (in the classical sense) due to complete lack of linearized acoustics. In spite of this, it was shown that this type of ordered granular media possesses interesting nonlinear dynamics in the form of solitary waves (Nesterenko 2001), nonlinear traveling waves (Starosvetsky and Vakakis 2010), nonlinear normal modes (Jayaprakash et al. 2011a, b), and countable infinities of resonance and anti-resonance phenomena (Jayaprakash et al. 2011a, b:2). Some recent studies related to DBs in granular crystals include metastable breathers in one-dimensional acoustic vacua (Sen and Krishna Mohan 2009); DBs in compressed granular chains at interfaces between diatomic chains and monoatomic chains (Hoogeboom et al. 2010); and wave localization phenomena (Job et al. 2009) and localized breathing modes (Theocharis et al. 2009) in granular chains with mass defects.

The majority of published works on ordered granular media concern one-dimensional media. Starosvetsky et al. (2012) analyzed the dynamics of two weakly coupled, strongly nonlinear granular chains with no pre-compression, and identified three different mechanisms for complete and recurrent energy exchanges. These mechanisms involved the excitation of nonlinear beat phenomena involving spatially periodic traveling waves, standing localized breathers, or propagating localized breathers. Rather, in the present subsection we focus on targeted energy transfer from an excited to an absorbing chain, and thus on localization of energy in a non-excited initially granular chain instead of beating phenomena leading to recurring energy exchanges. In a more general context, this work aims to be a first step toward studying nonlinear irreversible energy transfer and passive wave redirection in weakly coupled, highly discontinuous and strongly nonlinear ordered granular media. Irreversible energy transfer can be regarded as *targeted energy transfer—TET*, which has gained much attention in the recent literature (Kosevich et al. 2007, 2008, 2009; Kopidakis et al. 2001a, b; Maniadis and Aubry 2005; Manevitch 1999). However, unlike previous studies of weakly coupled oscillatory chains, the dynamical systems considered herein incorporate both non-smooth effects due to possible separations between neighboring beads (granules), as well as strongly nonlinear interparticle Hertzian interactions. We show that these systems exhibit very rich and complex dynamics that, however, can be accurately captured by analytical approximations. Specifically, we will demonstrate that any particular amount of energy propagating as a discrete breather in one of the interacting granular chains can be almost completely and irreversibly transferred to another weakly coupled chain.

11.2 System Description

The system shown in Fig. 11.1 consists of two semi-infinite weakly coupled, uncompressed ordered homogeneous granular chains mounted on linear elastic foundations and coupled by weak linear stiffnesses. Each chain consists of a number of identical linearly elastic spherical granular beads, which are in touch with one another, so their Hertzian interactions are essentially nonlinear (i.e., nonlinearizable); moreover, in the absence of compression bead separations may occur leading to collisions and providing an additional source of strong nonlinearity. Here, we denote by k_1 and k_2 the stiffness coefficients of the linear elastic foundations and of the linear coupling elements, respectively, and assume weak coupling by imposing the condition $k_1 \gg k_2$. In addition, we assume that the beads of both chains are constrained to move in the horizontal direction only, and that no dissipative forces exist in the system.

Assuming Hertzian contact law interaction between beads, the kinetic and potential energies of the two semi-infinite granular chains are defined as follows:

$$T = \frac{1}{2}m \sum_{n=1}^{\infty} (\dot{x}_n)^2 + \frac{1}{2}m \sum_{n=1}^{\infty} (\dot{y}_n)^2$$

$$U = \frac{2}{5}\frac{E(2R)^{1/2}}{3(1-v^2)} \sum_{n=1}^{\infty} \left[(x_n - x_{n+1})_+^{5/2} \right] + \frac{2}{5}\frac{E(2R)^{1/2}}{3(1-v^2)} \sum_{n=1}^{\infty} \left[(y_n - y_{n+1})_+^{5/2} \right]$$

$$+ \frac{1}{2}k_1 \sum_{n=1}^{\infty} (x_n)^2 + \frac{1}{2}k_1 \sum_{n=1}^{\infty} (y_n)^2 + \frac{1}{2}k_2 \sum_{n=1}^{\infty} (x_n - y_n)^2$$

$$(11.1)$$

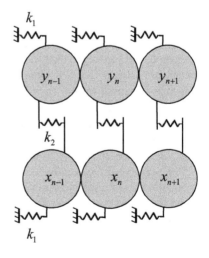

Fig. 11.1 Schematic of the weakly coupled homogeneous granular chains on elastic foundations

where the variables x_n and y_n denote the displacements of the n-th beads for the lower and upper chains, respectively, m denotes the mass of each spherical bead, E its elastic modulus, R its radius, and v the Poisson's ratio of the linearly elastic material of the beads. We assume that all the beads have the same material properties. The subscript (+) in (11.1) indicates that only nonnegative values in the parentheses should be taken into account, with zero values being assigned otherwise; this accounts for possible separations between beads that may occur in the absence of compression. Then, the equations of motion for this system can be expressed in following non-dimensionless form:

$$
\ddot{x}_n + x_n + 2\varepsilon\lambda\left(x_n - y_n\right) = \alpha\varepsilon^{3/4}\left[\left(x_{n-1} - x_n\right)_+^{3/2} - \left(x_n - x_{n+1}\right)_+^{3/2}\right]
$$
$$
\ddot{y}_n + y_n + 2\varepsilon\lambda\left(y_n - x_n\right) = \alpha\varepsilon^{3/4}\left[\left(y_{n-1} - y_n\right)_+^{3/2} - \left(y_n - y_{n+1}\right)_+^{3/2}\right] \qquad (11.2)
$$
$$
n = 1, 2, 3, \ldots
$$

In (11.2), α is the normalized stiffness coefficient of the nonlinear Hertzian interaction between the beads of each chain, λ the normalized parameter scaling the linear coupling between chains and $0 < \varepsilon \ll 1$ the small scaling parameter of the problem. We assume that an impulsive excitation is applied to the first bead of the lower chain which is referred to as the "*excited chain*", whereas the upper chain is initially at rest and is designated as the "*absorbing chain*" . A vectorized fourth-order Runge-Kutta time integration scheme is used to numerically calculate the dynamics of system (11.2).

11.3 Recurrent Energy Exchange Phenomena in the System of Coupled Granular Chains

As a first step, we apply an initial impulse to the first bead of the excited chain and numerically compute the bead responses of the two chains. In Fig. 11.2, we plot the relative displacement profile of each bead both in the excited and absorbing chains. The results clearly depict that *nearly complete but reversible energy exchange* is occurring between the two chains. Interestingly enough, this repetitive energy exchange is caused by the excitation of nonlinear beat phenomena, whereby initially all input energy is localized in the excited chain, but with progressing time nearly all of this energy gets transferred to the absorbing chain, as energy is almost completely "drained" from the excited chain. At a later phase of the dynamics, the energy gets transferred back to the excited chain, after which this energy exchange cycle repeats itself recurrently. We note that this repetitive energy exchange occurs in the absence of dissipative forces in the system. Now, we would like to analytically study these almost complete and recurrent energy exchanges, which are similar to nonlinear beats in weakly coupled FPU chains described in the Sect. 11.5 with using the LPT concept. In the following analysis, we focus only in the case of

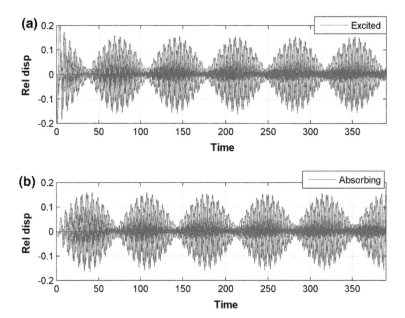

Fig. 11.2 Numerical simulation of the dynamics of (11.2): **a** Relative displacements $(x_n - x_{n+1})$ and **b** $(y_n - y_{n+1})$ of the first 100 beads of the two chains for $\varepsilon = 0.05$, $\alpha = 10$, $\lambda = 1.05$, and initial conditions $x_n = y_n = \dot{x}_n = \dot{y}_n = 0$, $n = 1, 2, \ldots$ except $x_1 = \varepsilon^{1/2}V_0$, $V_0 = 1$

1:1 internal resonance, i.e., when the oscillations of the beads of the two chains possess nearly identical frequencies. To carry the analysis, we will employ the complexification-averaging (CX-A) technique (Manevitch 1999) and then applied to various problems related to passive nonlinear targeted energy transfer (Vakakis et al. 2008).

To this end, we rewrite Eq. (11.2) as follows:

$$\ddot{x}_n + x_n = \alpha\varepsilon^{3/4}\left[(x_{n-1} - x_n)_+^{3/2} - (x_n - x_{n+1})_+^{3/2}\right] + 2\varepsilon\lambda y_n$$
$$\ddot{y}_n + y_n = \alpha\varepsilon^{3/4}\left[(y_{n-1} - y_n)_+^{3/2} - (y_n - y_{n+1})_+^{3/2}\right] + 2\varepsilon\lambda x_n \qquad (11.3)$$

Introduce the condition of 1:1 internal resonance, and apply the CX-A methodology (Manevitch 1999; Vakakis et al. 2008). To this end, we introduce the following new complex variables:

$$\dot{x}_n + ix_n \equiv \psi_n^x, \quad \dot{y}_n + iy_n \equiv \psi_n^y, \quad \dot{x}_n - ix_n \equiv \psi_n^{x*}, \quad \dot{y}_n - iy_n \equiv \psi_n^{y*} \qquad (11.4)$$

where asterisk denotes complex conjugate and $i = (-1)^{1/2}$. By relations (11.4), we implicitly assume that a condition of 1:1 resonance exists between the bead oscillations of the two chains and that all beads oscillate with normalized frequency equal to unity. The transformation of variables $(\dot{x}_n, x_n) \to \psi_n^x$ corresponds the transfer to a convenient complex representation of the phase plane of the i-th particle in terms of varying amplitude and phase. In physics and engineering, this representation is known as phasor (phase vector) possessing the information (in a single complex variable) about the amplitude and phase variation of the oscillatory response rather than keeping separately the displacement and velocity. Thus, the filed ψ_n^x can also be viewed as a rotating coordinate system (polar coordinate system) when $(\dot{x}_n, x_n)$ as a stationary frame. These complex conjugate linear combinations of displacement and velocities can be visually presented as vectors of equal length rotating in opposite directions.

From (11.4), we can represent x_n, $\dot{x}_n$, $\ddot{x}_n$, y, $\dot{y}_n$ and $\ddot{y}_n$ as follows:

$$x_n = \frac{1}{2i}\left(\psi_n^x - \psi_n^{x*}\right), \quad y_n = \frac{1}{2i}\left(\psi_n^y - \psi_n^{y*}\right), \quad \dot{x}_n = \frac{1}{2}\left(\psi_n^x + \psi_n^{x*}\right), \quad \dot{y}_n = \frac{1}{2}\left(\psi_n^y + \psi_n^{y*}\right),$$

$$\ddot{x}_n = \frac{d\psi_n^x}{dt} - \frac{i}{2}\left(\psi_n^x + \psi_n^{x*}\right), \quad \ddot{y}_n = \frac{d\psi_n^y}{dt} - \frac{i}{2}\left(\psi_n^y + \psi_n^{y*}\right)$$

$$(11.5)$$

Substituting (11.4) and (11.5) into (11.3) and performing simplifications, we obtain the following system of first-order complex equations, which is still exact and completely equivalent to (11.2):

$$\frac{d\psi_n^x}{dt} - i\psi_n^x = \alpha\varepsilon^{3/4}\left\{\left[\frac{\left(\psi_{n-1}^x - \psi_{n-1}^{x*}\right)}{2i} - \frac{\left(\psi_n^x - \psi_n^{x*}\right)}{2i}\right]_+^{3/2}\right.$$
$$\left. - \left[\frac{\left(\psi_n^x - \psi_n^{x*}\right)}{2i} - \frac{\left(\psi_{n+1}^x - \psi_{n+1}^{x*}\right)}{2i}\right]_+^{3/2}\right\} - i\varepsilon\lambda\left(\psi_n^y - \psi_n^{y*}\right)$$

$$\frac{d\psi_n^y}{dt} - i\psi_n^y = \alpha\varepsilon^{3/4}\left\{\left[\frac{\left(\psi_{n-1}^y - \psi_{n-1}^{y*}\right)}{2i} - \frac{\left(\psi_n^y - \psi_n^{y*}\right)}{2i}\right]_+^{3/2}\right.$$
$$\left. - \left[\frac{\left(\psi_n^y - \psi_n^{y*}\right)}{2i} - \frac{\left(\psi_{n+1}^y - \psi_{n+1}^{y*}\right)}{2i}\right]_+^{3/2}\right\} - i\varepsilon\lambda\left(\psi_n^x - \psi_n^{x*}\right)$$

$$(11.6)$$

Now, we introduce the multiple scales method to obtain the approximate solution of (11.6) under condition of 1:1 resonance. To leading order, we assume the existence of two timescales in the dynamics, namely a *fast timescale* $\tau_0 = t$ and a *slow timescale*, $\tau_1 = \varepsilon\tau_0$. The fast timescale is the characteristic timescale of oscillations governed by the linear foundations of the two chains, whereas the slow timescale is the characteristic timescale governing the dynamics of nonlinear energy exchanges between the two chains. Rescaling the complex variables in (11.6) and expanding in asymptotic series as follows:

$$\psi_n^x(t) = \varepsilon^{1/2} \{ \psi_{n0}^x(\tau_0, \tau_1, \ldots) + \varepsilon \psi_{n1}^x(\tau_0, \tau_1, \ldots) + \ldots \}$$
$$\psi_n^y(t) = \varepsilon^{1/2} \{ \psi_{n0}^y(\tau_0, \tau_1, \ldots) + \varepsilon \psi_{n1}^y(\tau_0, \tau_1, \ldots) + \ldots \}$$
$$\frac{d}{dt} = \frac{\partial}{\partial \tau_0} + \varepsilon \frac{\partial}{\partial \tau_1} + \ldots$$

(11.7)

Substituting into (11.6) and matching terms multiplying different powers of the small parameter ε, we obtain a hierarchy of subproblems governing the dynamics at different approximations.

Considering $O(\varepsilon^{1/2})$ terms, we derive the following leading-order approximation:

$$\frac{\partial \psi_{n0}^x}{\partial \tau_0} - i \psi_{n0}^x = 0 \Rightarrow \psi_{n0}^x = \varphi_{no}^x(\tau_1) \exp(i\tau_0)$$
$$\frac{\partial \psi_{n0}^y}{\partial \tau_0} - i \psi_{n0}^y = 0 \Rightarrow \psi_{n0}^y = \varphi_{no}^y(\tau_1) \exp(i\tau_0)$$

(11.8)

Viewed in a different context, (11.8) represents a slow/fast partition of the dynamics of (11.6) or (11.2), with the complex exponential term $\exp(i\tau_0)$ representing the *fast* oscillating parts of the responses of the beads of the two chains (at the common fast frequency unity), and $\varphi_{no}^{x,y}(\tau_1)$ the *slow* envelopes (modulations) of the fast oscillations.

Proceeding to the $O(\varepsilon^{3/2})$ approximation, we derive the following system, which yields the following averaged slow flow system:

$$i \frac{\partial \varphi_{n0}^x}{\partial \tau_1} = \tilde{\alpha} \left(F_{(n-1)0}^x \left| F_{(n-1)0}^x \right|^{1/2} - F_{(n)0}^x \left| F_{(n)0}^x \right|^{1/2} \right) + \tilde{\lambda} \varphi_{n0}^y$$
$$i \frac{\partial \varphi_{n0}^y}{\partial \tau_1} = \tilde{\alpha} \left(F_{(n-1)0}^y \left| F_{(n-1)0}^y \right|^{1/2} - F_{(n)0}^y \left| F_{(n)0}^y \right|^{1/2} \right) + \tilde{\lambda} \varphi_{n0}^x$$

(11.9)

where

$$\tilde{\alpha} = \alpha * \gamma_1 / 2\pi, \quad \tilde{\lambda} = \lambda \gamma_2$$
$$\gamma_1 = \oint_{2\pi} \{ \cos \Phi_n^x \}_+^{3/2} \exp(-i\Phi_n^x) \, d\Phi_n^x = (2\alpha/\pi) \oint_{2\pi} \{ \cos \Phi_n^x \}_+^{5/2} d\Phi_n^x$$
$$\gamma_2 = \oint_{2\pi} \{ \cos(\Phi_n^x - \theta_n^x) \}_+^{3/2} \exp[-i(\Phi_n^x - \theta_n^x)] d(\Phi_n^x - \theta_n^x)$$

The above equation is the slow flow modulation equation governing the nearly complete but recurrent energy exchange between the excited and absorbing granular chains mounted on elastic foundations. The derivation of formula (11.9) is detailed in the appendix. If we plot the response of the coupled slow flow Eq. (11.9) for the same system parameters and initial condition as in Fig. 11.2, we get the responses depicted in Fig. 11.3. These results clearly show that the averaged slow flow (11.9) accurately captures the energy exchange phenomenon between the granular chains that is initiated when an initial impulse is applied to the excited chain.

To study the response of the averaged slow flow model (11.9) in space, in Fig. 11.4 we plot the spatial profile of the relative displacements of the two chains at three different (normalized) time instants (the same parameters and initial condition of Figs. 11.2 and 11.3 were employed). Referring to Fig. 11.4, we note that at time instant $t_1 = 419$, nearly the entire energy of the oscillation is localized in the excited chain whereas the response of the absorbing chain is almost zero. At a later instant of time $t_2 = 453$, however, the reverse is observed, since almost all of the energy is transferred to the absorbing chain leaving an insignificant amount of energy in the excited chain. Furthermore, at an even later time instant $t_3 = 488$, the energy of the oscillation is transferred back to the excited chain. Hence, we can clearly observe the recurrent nonlinear energy exchange (beat) phenomenon in the weakly coupled granular chains. Moreover, we note that during the nonlinear beats the energy travels toward the far field on the right, and that the energy localization occurs in the span of 4–5 beads. The residual response observed in the tail of the

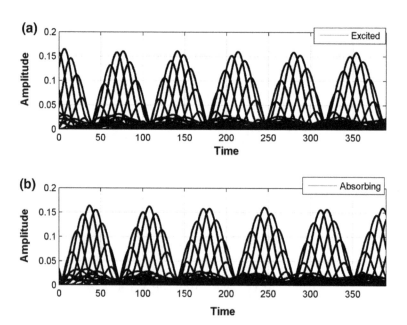

Fig. 11.3 Numerical simulation of the theoretically predicted averaged slow flow model (11.9) for the same parameters and initial conditions as in Fig. 11.2; the plots show the slow envelopes of the relative responses of the first 100 beads in the excited and absorbing chains

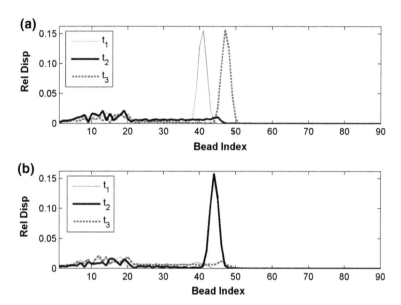

Fig. 11.4 Spatial waveforms of the responses predicted by the averaged slow flow model (9):
a Excited chain and **b** absorbing chain; three snapshots of the spatial waveforms are shown with
$t_3 > t_2 > t_1$

propagating pulse (i.e., from beads 1–35 in Fig. 11.4) is due to partial scattering of
the propagating primary pulse when it encounters the beads of the excited and
absorbing chains, resulting in the formation of "oscillating tails" in its wake.

11.4   Nonlinear Targeted Energy Transfer and Energy Exchange: Analysis

From the results of Fig. 11.4, it is clearly observed that complete and reversible
energy exchange among the two granular chains is possible through the excitation
of nonlinear beat phenomena realized under the condition of 1:1 resonance. Now,
the natural question to address is whether it is possible to completely localize the
energy in the absorbing chain in a *one-way irreversible fashion* so that energy does
not "spread back" to the excited chain. To address this issue, we will need to extend
the analysis of the previous section. From the exact governing Eqs. (11.2) or (11.6),
and the averaged slow flow model (11.9), it is clear that the dynamics depends on
two parameters, namely the stiffness of the elastic foundation k_1, and the stiffness of
coupling parameter λ. Below, we will show that, indeed, it is possible to almost
completely localize the impulsive energy in the absorbing granular chain through
suitable tuning of either one of these two system parameters. To this end, two
distinct mechanisms will be used: (a) energy localization through complete

decoupling, and (b) Landau-Zener tunneling in space by suitable stratification of the elastic foundation of the granular chains. We examine each of these two mechanisms separately.

11.4.1   Localization of Energy by Complete Decoupling

Referring to the snapshots of Fig. 11.4, we note that at the particular time instant $t = t_2$, energy localization in the absorbing chain reaches a maximum (occurring at approximately bead 43), whereas at the same time instant the response of the excited chain is almost negligible, consisting of low-amplitude oscillating tails due to pulse scattering. Hence, it is reasonable to assume that if we decouple the two chains starting at the 43rd beads of the excited and absorbing chains (cf. Fig. 11.5), we will be able to interrupt further development of the nonlinear beat phenomenon and energy will be permanently localized in the absorbing chain. Hence, irreversible energy transfer and localization in the absorbing chain can be achieved by introducing decoupling at the appropriate phase of the nonlinear beat.

   This decoupling was performed for the impulsively excited system discussed in the previous section, and the resulting responses are depicted in Fig. 11.6. For this series of simulations, the system parameters and initial impulse of the excited chain were identical to the simulations depicted in Fig. 11.2. From the results of Fig. 11.6, it is clear that if we completely decouple the two chains after the 43rd beads of the excited and absorbing chains (when the response in the absorbing chain reaches a maximum), then we are able to localize most of the energy in the absorbing chain in a permanent, irreversible fashion. We observe that before decoupling is applied, complete and recurring energy exchanges occur between the two chains as discussed in the previous section. However, once the decoupling is applied at the appropriate phase of the nonlinear beat (i.e., after the 43rd beads of the two chains), the energy of the oscillation is almost completely localized to the

Fig. 11.5 Schematic of the system of granular chains with partial coupling

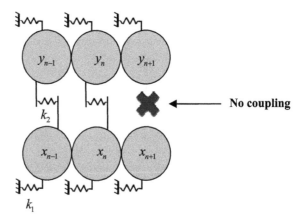

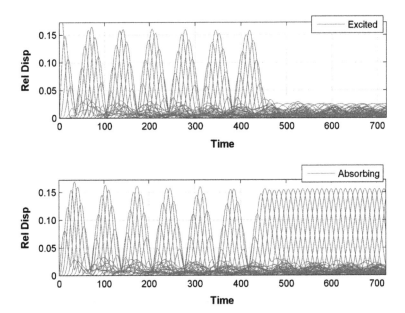

Fig. 11.6 Numerical simulation of theoretically predicted averaged slow flow model (11.9) for the parameters and initial condition as shown in Fig. 11.2 after complete decoupling starting at the 43rd beads; the plots show the slow envelopes of the relative responses of the first 100 beads of the chains

absorbing chain. The same localization phenomenon can be observed if we plot the spatial waveform of the propagating pulse as shown in Fig. 11.7.

11.5   Targeted Energy Transfer Through the Landau-Zener Tunneling Effect in Space

The results of previous section confirmed that it is possible to localize most of the impulsive energy initiated in the excited chain to the absorbing chain when we completely decouple the two chains at an appropriate phase of the recurring nonlinear beat coupling. Given that this is not always practical, in this section we will demonstrate an alternative mechanism for irreversible targeted energy transfer from the excited to the absorbing chain, based on inducing the macroscopic analogue of quantum Landau-Zener tunneling (LZT) effect (Razavy 2003; Zener 1932) in space. Indeed, instead of complete decoupling between the two chains, we will consider spatial variation (stratification) of the elastic foundations of the granular chains. Such a macroscopic analogue of LZT (in time, however) was reported in (Manevitch et al. 2011) considering a system of two weakly coupled pendula, with the length of one of the pendula varying with time. LZT is a dynamical transition

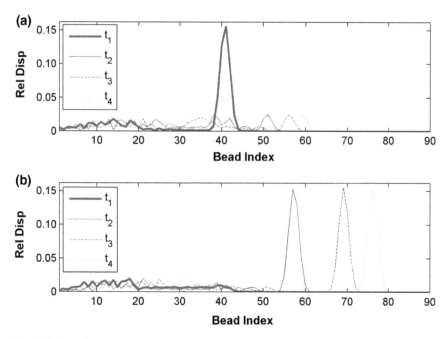

Fig. 11.7 Spatial waveforms of the responses predicted by the averaged slow flow model (9) with complete decoupling starting from bead 43: **a** Excited chain and **b** absorbing chain; four snapshots of the spatial waveforms are shown with $t_4 > t_3 > t_2 > t_1$

where a quantum system tunnels across an energy gap between two anti-crossed energy levels (states) (Razavy 2003; Zener 1932). Quantum LZT was observed in semiconductor superlattices for electrons (Rosam et al. 2003), in optical lattices for ultracold atoms (Anderson and Kasevich 1998) and in Bose-Einstein condensates (Kovaleva et al. 2011). The common feature of the aforementioned applications of *nonadiabatic LZT* is the irreversible (and almost unidirectional) exchange of energy between two states caused by external forcing or perturbation. This form of energy exchange is especially suitable for oscillating systems, such as the weakly coupled granular chain system considered herein, where impulsively induced vibration energy originally localized in one state (the excited chain) could be irreversibly transferred to an alternative state (the absorbing chain). It turns out that such a classical system, governed by equations similar to those of a quantum system (Manevitch et al. 2011; Kosevich et al. 2010; Kovaleva et al. 2011), can in fact be realized if the analysis is carried out in the spatial (instead of the temporal) domain.

From a practical point of view, the proposed alternative method of stratification is more feasible in implementation in material designs compared to complete decoupling. For simplicity, in the following analysis we keep the elastic foundation of the absorbing chain uniform while spatially varying (decreasing) the elastic foundation of the excited chain from bead to bead. It follows that the normalized governing equations of motion of such a stratified system are given by:

$$\ddot{x}_n + x_n + 2\varepsilon\lambda\,(x_n - y_n) - 2\varepsilon\gamma_n x_n = \alpha\varepsilon^{3/4}\left[(x_{n-1} - x_n)_+^{3/2} - (x_n - x_{n+1})_+^{3/2}\right]$$

$$\ddot{y}_n + y_n + 2\varepsilon\lambda\,(y_n - x_n) = \alpha\varepsilon^{3/4}\left[(y_{n-1} - y_n)_+^{3/2} - (y_n - y_{n+1})_+^{3/2}\right]$$

$$n = 1, 2, \ldots$$

$$(11.10)$$

In comparison with the original dynamical system (11.2), system (11.10) possesses one more parameter, namely the coefficient γ_n, $n = 1, 2, \ldots$ denoting the detuning parameter of the elastic foundation of the excited chain. This parameter varies in space and will play an important role for realizing passive targeted energy and nearly irreversible energy localization in the absorbing chain. Moreover, an initial impulse of intensity $\sqrt{\varepsilon}V_0$ is applied to the first bead of the excited chain at $t = 0$, with the system being initially at rest. In this study, we are interested in construction of a simplified reduced-order model being able to correctly predict and explain the mechanism of a near complete targeted energy transport from the excited chain to the absorbing one. To this extent, we resort to the previously introduced concept of "effective particle" introduced to model momentum transfer in homogeneous granular chains. To this end, taking separately the sum of the responses of all beads of the excited and absorbing chains in (11.10), we obtain the following:

$$\sum \{\ddot{x}_n + x_n + 2\varepsilon\lambda\,(x_n - y_n) - 2\varepsilon\gamma_n x_n\} = \sum \left\{\alpha\varepsilon^{3/4}\left[(x_{n-1} - x_n)_+^{3/2} - (x_n - x_{n+1})_+^{3/2}\right]\right\}$$

$$\sum \{\ddot{y}_n + y_n + 2\varepsilon\lambda(y_n - x_n)\} = \sum \left\{\alpha\varepsilon^{3/4}\left[(y_{n-1} - y_n)_+^{3/2} - (y_n - y_{n+1})_+^{3/2}\right]\right\}$$

$$(11.11)$$

where each of the summations is with respect to n, where $n = 1, 2, \ldots$. At this point, we define the coordinates of the two effective particles of system (11.10) as follows:

$$u_1(t) \equiv \sum x_n(t), \quad u_2(t) \equiv \sum y_n(t) \tag{11.12}$$

which upon substituting into (11) yields the system of coupled effective particles,

$$\ddot{u}_1(t) + G(t) + 2\varepsilon\lambda\,[u_1(t) - u_2(t)] = 0$$

$$\ddot{u}_2(t) + u_2(t) + 2\varepsilon\lambda\,[u_2(t) - u_1(t)] = 0$$

$$(11.13)$$

where

$$G(t) \equiv u_1(t) - 2\varepsilon \sum \gamma_n x_n(t) \tag{11.14}$$

In order to simplify further the analysis, we now make the important assumption that due to a compacton-like localization of the propagating disturbances considered in the coupled granular chains, the collective effect of the spatial variation of the stiffness of elastic foundation of the excited chain can be modeled as if it is a single oscillator with a slowly, monotonically varying (decreasing) stiffness. Hence, we introduce the following approximation:

$$G(t) = u_1(t) - 2\varepsilon \sum \gamma_n x_n(t) \cong [1 - 2\varepsilon\xi(t_1)]u_1(t) \tag{11.15}$$

which enables us to approximate the time-dependent term $G(t)$ in terms of the coordinate of the effective particle of the excited chain. To achieve this, we replaced the *discrete spatial* variation of the elastic foundation of the excited chain γ_n, by the *continuous temporal* variation $\xi(t_1)$. Then, the system of effective particles (11.13) reduces to the following approximate form:

$$\ddot{u}_1(t) + u_1(t) + 2\varepsilon\lambda \left[u_1(t) - u_2(t)\right] - 2\varepsilon\xi(t_1)u_1(t) = 0$$
$$\ddot{u}_2(t) + u_2(t) + 2\varepsilon\lambda \left[u_2(t) - u_1(t)\right] = 0 \tag{11.16}$$

The function $\xi(t_1)$ is chosen to model the stiffness variation in time and depends on the new timescale t_1. The quite natural question to be asked is how the time and space representations of the variation (decrease) of the elastic foundation of the excited granular chain will correlate in (11.16). The answer can be found by considering elements from the theory of discrete breathers propagating in granular media with stratified potentials. Starosvetsky et al. (2012) were able to formulate an analytical approximation for the propagation of breather solutions in this type of media. Using this result, we can estimate the time it takes for the propagating signal (the localized breather) to propagate the distance between two beads of the granular lattice. It is precisely this time estimate the characteristic time step that is needed for the construction of the function $\xi(t_1)$. In Fig. 11.8, we depict the correlation between the spatial and temporal dependencies of the stiffness variation that we realize for such an analysis. Here, the discrete reduction of the stiffness parameter in space (i.e., as the pulse propagates between two neighboring beads of the excited granular chain) can be approximated by a continuous function of the same stiffness parameter in time (e.g., the time-varying stiffness of the effective particle of the excited chain).

In summary, using the concept of effective particles, we attempt to depict the intrinsic dynamics of the system (11.10) of coupled granular chain by a reduced-order model (11.16) consisting *two weakly coupled linear oscillators with*

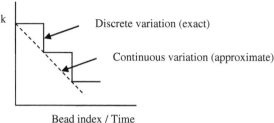

Fig. 11.8 Correlation between the discrete (exact) and continuous (approximate) spatial and temporal dependencies of the stiffness variation of the excited granular chain

a time-dependent parameter. In turn, the reduced-order model can support a macroscopic analogue of the linear Landau-Zener tunneling in time. That is, by appropriate design of the stiffness detuning parameter $2\varepsilon\xi(\tau_1)$, we will be able to achieve targeted and irreversible energy transfer from the excited to the absorbing effective particle (oscillator) in (11.16); in turn, by inversing the previous transformations, we will demonstrate irreversible targeted energy transfer in space in the full model (11.10); in effect, this will amount to *passive pulse redirection* in the system of coupled granular chains.

The validity of the reduced-order model (11.16) with its numerical verifications is presented in the last section of this work. In the remainder of this section, we analyze the dynamics of the reduced model with an applied initial impulse to the excited u_1-oscillator (effective particle) and with the u_2-oscillator designated as the absorbing oscillator (effective particle). Introducing the notation $\tau_0 \equiv t$ in (11.16), we express this system in the form:

$$\frac{d^2 u_1}{d\tau_0^2} + u_1 + 2\varepsilon\lambda(u_1 - u_2) - 2\varepsilon\xi(\tau_1)u_1 = 0$$

$$\frac{d^2 u_2}{d\tau_0^2} + u_2 + 2\varepsilon\lambda(u_2 - u_1) = 0$$

(11.17)

In Eq. (11.17), the variables u_1 and u_2 denote the normalized responses of the excited and absorbing oscillators, respectively, and λ defines the strength of the linear coupling between the two oscillators. The coefficient, $\xi(\tau_1) = 2\beta^2\tau_1 - \sigma$, defines the detuning modulation parameter, with β^2 characterizing the rate of resonance crossing and σ defining the strength of the foundation stiffness of the excited oscillator in its initial stage of detuning at the timescale τ_1.

To formulate an asymptotic analysis, the solution of the second equation in (11.17) is expressed in the following integral form:

$$u_2(\tau_0) = u_2(0)\cos(\omega\tau_0) + \frac{\dot{u}_2(0)}{\omega}\sin(\omega\tau_0) + \frac{1}{\omega}\int_0^{\tau_0} 2\varepsilon\lambda u_1(s)\sin[\omega(\tau_0 - s)]\mathrm{d}s$$

$$= 2\varepsilon\lambda\omega^{-1}\int_0^{\tau_0} u_1(s)\sin[\omega(\tau_0 - s)]\mathrm{d}s$$

$$(11.18)$$

with the following, initial conditions $u_1(0) = 0$, $\dot{u}_1(0) = \sqrt{\varepsilon}V_0$ and $u_2(0) = 0$, $\dot{u}_2(0) = 0$ are assumed, and $\omega = (1 + 2\varepsilon\lambda)^{1/2}$. Now, substituting (11.18) into the first of Eq. (11.17), we derive the following linear integro-differential equation with time-dependent coefficient for u_1:

$$\frac{\mathrm{d}^2 u_1}{\mathrm{d}\tau_0^2} + \omega u_1 - 2\varepsilon\xi(\tau_1)u_1 = 4\varepsilon^2\lambda^2\omega^{-1}\int_0^{\tau_0} u_1(s)\sin[\omega(\tau_0 - s)]\mathrm{d}s \qquad (11.19)$$

Hence, we are able to reduce the problem to a single integro-differential equation for u_1.

The asymptotic analysis of (11.19) has been carried out with the help of CX-A technique. To do so, we again introduce a complex conjugate variable pair,

$$\psi = \dot{u}_1 + iu_1, \quad \psi^* = \dot{u}_1 - iu_1 \qquad (11.20)$$

where again asterisk represents complex conjugate. From (11.20), we can represent u_1 and $\dot{u}_1$ in terms of the complex variable ψ, with initial condition $\psi(0) = \dot{u}_1(0) + iu_1(0) = \sqrt{\varepsilon}V_0$. Expressing (11.19) in terms of ψ, we obtain the complex ordinary differential equation:

$$\frac{\mathrm{d}\psi}{\mathrm{d}\tau_0} - i\omega\psi + i\varepsilon\xi(\tau_1)(\psi - \psi^*) = -2i\varepsilon^2\omega^{-1}\lambda^2\int_0^{\tau_0}[\psi(s,\varepsilon) - \psi^*(s,\varepsilon)]\sin[\omega(\tau_0 - s)]\mathrm{d}s$$

$$(11.21)$$

Now, we analyze the asymptotic approximation of (11.21) employing the method of multiple scales, by introducing the fast timescale τ_0 and the slow timescale $\tau_0 = \varepsilon\tau_1$. Applying the asymptotic analysis, we express the dependent complex variable as $\psi(\tau_0, \varepsilon) = \sqrt{\varepsilon}[\psi_0(\tau_0, \tau_1) + \varepsilon\psi_1(\tau_0, \tau_1) + \ldots]$, which upon substitution into (11.21) and expression of the time derivatives by the chain rule with respect to the two timescales yields an hierarchy of problems at the different orders of approximation.

Considering the leading-order $O(\varepsilon^{1/2})$ problem, we obtain the following solution, expressed as a fast oscillation modulated by the slowly varying envelope $\varphi_0(\tau_1, \varepsilon)$:

$$\frac{\partial \psi_0}{\partial \tau_0} - i\omega \psi_0 = 0 \quad \Rightarrow \quad \psi_0 = \varphi_0(\tau_1, \varepsilon) \exp(i\omega \tau_0) \tag{11.22}$$

This envelope is evaluated by considering the $O(\varepsilon^{3/2})$ subproblem,

$$\frac{\partial \psi_1}{\partial \tau_0} + \frac{\partial \psi_0}{\partial \tau_1} - i\omega \psi_1 + i\xi(\tau_1)(\psi_0 - \psi_0^*)$$

$$- 2i\varepsilon\omega^{-1}\lambda^2 \int\limits_0^{\tau_0} [\psi_0(s, \varepsilon) - \psi_0^*(s, \varepsilon)] \sin[\omega(\tau_0 - s)]ds \tag{11.23}$$

which, taking into account (11.22), is expressed as follows:

$$\frac{\partial \varphi_0}{\partial \tau_1} + i\omega \varphi_0 + \frac{\partial \varphi_1}{\partial \tau_0} \exp(-i\omega \tau_0) - i\omega \varphi_1 \exp(-i\omega \tau_0) + i\xi(\tau_1)[\varphi_0 - \varphi_0^* \exp(-2i\omega \tau_0)]$$

$$= -\omega^{-1}\lambda^2 \left[\int\limits_0^{\varepsilon\tau_0} \varphi_0(\varepsilon\tau_0, \varepsilon)[1 - \exp\{-2i\omega_2(\tau_0 - s)\}]ds \right.$$

$$\left. + \exp(-2i\omega_2\tau_0) \int\limits_0^{\varepsilon\tau_0} \varphi_0^*(\varepsilon\tau_0, \varepsilon)[1 - \exp\{-2i\omega_2(\tau_0 - s)\}]ds \right] \tag{11.24}$$

Finally, suppressing secular terms in (11.24), we obtain the following slow flow equation, i.e., the complex modulation equation governing the envelope $\varphi_0(\tau_1, \varepsilon)$ of the $O(\varepsilon^{1/2})$ approximation (11.22):

$$\frac{d\varphi_0}{d\tau_1} + i[\omega + \xi(\tau_1)]\varphi_0 = -\omega^{-1}\lambda^2 \left[\int\limits_0^{\varepsilon\tau_0} \varphi_0(\varepsilon\tau_0, \varepsilon)ds \right] = -\omega^{-1}\lambda^2 \int\limits_0^{\tau_1} \varphi_0(r, \varepsilon)dr \Rightarrow$$

$$\frac{d\varphi_0}{d\tau_1} + i(\omega + 2\beta^2\tau_1 - \sigma)\varphi_0 = -\omega^{-1}\lambda^2 \int\limits_0^{\tau_1} \varphi_0(r, \varepsilon)dr \tag{11.25}$$

Alternatively, the slow flow (11.25) can be rewritten as the following complex second-order differential equation:

$$\frac{d^2\varphi_0}{d\tau_1^2} + i\left(\omega + 2\beta^2\tau_1 - \sigma\right)\frac{d\varphi_0}{d\tau_1} + \left(\omega^{-1}\lambda^2 + 2i\beta^2\right)\varphi_0 = 0 \qquad (11.26)$$

which represents the slow flow of the excited oscillator. To compare the slow flow Eq. (11.26) with the full model (11.17), first we numerically integrate the full model with the prescribed initial conditions (corresponding to an initial impulse of intensity $\sqrt{\varepsilon}V_0$ applied to the excited chain), and also simulate (11.26) with the corresponding initial conditions $\varphi_0(0) = \sqrt{\varepsilon}V_0$, and $\frac{d\varphi_0(0)}{d\tau_1} = -i(\omega - \sigma)\sqrt{\varepsilon}V_0$. We note that (11.26) provides an approximation of the slow flow of the excited oscillator (effective particle); later, we show how based on this result we can compute the slow flow of the absorbing oscillator.

In Fig. 11.9, we depict the responses of the oscillators of the system of effective particles (11.17) with parameters $\beta = 0.53$, $\sigma = 1.35$, $\varepsilon = 0.05$, and $\lambda = 1.05$, where the excited oscillator is forced by an initial impulse of intensity $\sqrt{\varepsilon}V_0$, with $V_0 = 1$. From this result, it is clear that as time progress, the response of the excited oscillator decreases, and simultaneously the response of the absorbing oscillator increases; hence, irreversible passive targeted energy transfer occurs following the LZT effect. Indeed, initially the two effective particles are in 1:1 resonance (as evidenced by their nearly equal oscillation frequencies and the clear nonlinear beat developing at earlier times), but with increasing time and the gradual decrease of the elastic foundation of the excited oscillator there occurs escape from the regime

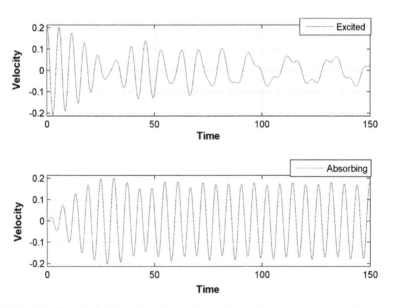

Fig. 11.9 LZT effect: Velocities of excited and absorbing effective particles based on the full model (11.17) for an initial impulse applied to the excited oscillator and a gradually decreasing elastic foundation for the excited oscillator

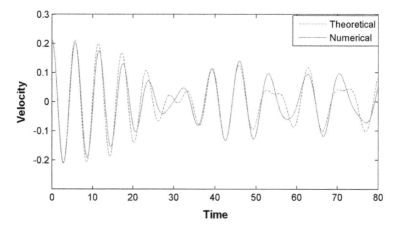

Fig. 11.10 Comparison between the exact and slow flow models for the excited effective particle

of 1:1 resonance. As a result, the nonlinear beating phenomenon is gradually interrupted and the energy transferred to the absorbing oscillator is localized permanently to that oscillator. To get a comparison of this result with the previous theoretical slow flow model, we numerically integrated (11.26) and computed an analytical approximation of the response of the excited effective particle through relations (11.20) and (11.22). In Fig. 11.10, we compare the exact numerical response of the excited oscillator depicted in Fig. 11.9 with the theoretically obtained response based on the slow flow model (11.26). Good agreement is noted between the exact and the analytical model which is observed, especially at early times where the major part of the impulsive energy gets transferred from the excited to the absorbing chain in an irreversible fashion. This clearly demonstrates that the theoretical model can fairly capture the exact nonlinear transient dynamics and the governing LZT effect.

Indeed, based on the previous analysis further insight can be gained for the LZT energy transfer in the initial stage of the dynamics. Based on the previous asymptotic analysis and the transformations, we derive the following leading-order approximation for the excited effective particle (oscillator):

$$
\begin{aligned}
u_{10}(\tau_0, \tau_1) &\approx -\frac{i\sqrt{\varepsilon}}{2}\left[\varphi_0(\tau_1)\exp(i\omega\tau_0) - \varphi_0^*(\tau_1)\exp(-i\omega\tau_0)\right] \\
&= \sqrt{\varepsilon}|\varphi_0(\tau_1)|\sin[\omega\tau_0 + \alpha(\tau_1)] \\
\dot{u}_{10}(\tau_0, \tau_1) &\approx \frac{\sqrt{\varepsilon}}{2}\left[\varphi_0(\tau_1)\exp(i\omega\tau_0) + \varphi_0^*(\tau_1)\exp(-i\omega\tau_0)\right] \\
&= \sqrt{\varepsilon}|\varphi_0(\tau_1)|\cos[\omega\tau_0 + \alpha(\tau_1)] \\
\alpha(\tau_1) &= \arg\varphi_0(\tau_1)
\end{aligned}
\tag{11.27}
$$

Hence, we can express the equation for the slowly varying partial energy of the excited oscillator as follows:

$$e_{10}(\tau_1) \approx \frac{1}{2}\left[\langle u_{10}^2 \rangle + \langle \dot{u}_{10}^2 \rangle\right] = \frac{\varepsilon}{2}|\varphi_0(\tau_1)|^2 \tag{11.28}$$

However, for small values of the slow time τ_1 (i.e., in the initial phase of the LZT effect during which passive targeted energy transfer occurs), the slow flow envelope can be approximated directly from the slow flow (11.26) by taking into account the initial conditions (Kovaleva et al. 2011).

To get similar analytical approximations for the absorbing oscillator (effective particle), we need to extend the previous slow flow analysis. To this end, from the second of Eq. (11.17), we can approximate the governing equation for the absorbing oscillator as

$$\frac{d^2 u_2}{d\tau_0^2} + \omega u_2 = 2\varepsilon\lambda u_{10} + O(\varepsilon^2) \tag{11.29}$$

with initial conditions $u_2(0) = 0$, $\dot{u}_2(0) = 0$. Again we apply the CX-A method, by introducing the complex variable $y = \dot{u}_2 + iu_2$, and transforming (11.29) to the following complex ordinary differential equation:

$$\frac{dy}{d\tau_0} - i\omega y = -i\varepsilon\lambda[\psi - \psi^*] \tag{11.30}$$

Again, we use a similar multiple scales asymptotic analysis and represent the complex variable as $y(\tau_0, \tau_1) = \sqrt{\varepsilon}[y_0(\tau_0, \tau_1) + \varepsilon y_1(\tau_0, \tau_1) + \ldots]$. The leading-order approximation is then evaluated as:

$$y_0 = \eta_0(\tau_1)\exp(i\omega\tau_0) \tag{11.31}$$

where the slowly varying complex envelope is governed by the following slow flow equation for the absorbing oscillator:

$$\frac{\partial \eta_0}{\partial \tau_1} = -i\lambda\varphi_0(\tau_1) \tag{11.32}$$

For sufficiently small values of the slow time τ_1, the solution of (11.32) for zero initial conditions can be approximated as $\eta_0(\tau_1) \approx -i\lambda\sqrt{\varepsilon}V_0\tau_1$, yielding the following leading-order approximation for the response of the absorbing oscillator:

$$u_{20}(\tau_0, \tau_1) \approx -\frac{i\sqrt{\varepsilon}}{2}\left[\eta_0(\tau_1)\exp(i\omega\tau_0) - \eta_0^*(\tau_1)\exp(-i\omega\tau_0)\right]$$
$$= \sqrt{\varepsilon}|\eta_0(\tau_1)|\sin[\omega\tau_0 + \delta(\tau_1)]$$
$$\dot{u}_{20}(\tau_0, \tau_1) \approx \frac{\sqrt{\varepsilon}}{2}\left[\eta_0(\tau_1)\exp(i\omega\tau_0) + \eta_0^*(\tau_1)\exp(-i\omega\tau_0)\right] \qquad (11.33)$$
$$= \sqrt{\varepsilon}|\eta_0(\tau_1)|\cos[\omega\tau_0 + \delta(\tau_1)]$$
$$\delta(\tau_1) = \arg\eta_0(\tau_1)$$

In Fig. 11.11, we compare the exact response for the absorbing effective particle computed by numerically integrating system (11.17) with the theoretical approximation (11.33). At the earlier stage of the dynamics, there is a close correspondence between the two models; however, with progressing time the theoretical response of the absorbing oscillator deviates from the exact response. Two important factors play an important role for this deviation. First, in the asymptotic analysis of the absorbing oscillator, we are only considering the leading-order term. Moreover, the slow flow approximation of the absorbing oscillator is dependent on the corresponding approximation of the excited oscillator, which, however, in itself is approximate. Based on the approximation (11.33), we can estimate the instantaneous energy of the absorbing effective particle as follows:

$$e_{20}(\tau_1) = \frac{1}{2}\left[\langle u_{20}^2\rangle + \langle \dot{u}_{20}^2\rangle\right] \approx \frac{\varepsilon}{2}|\eta_0(\tau_1)|^2 \qquad (11.34)$$

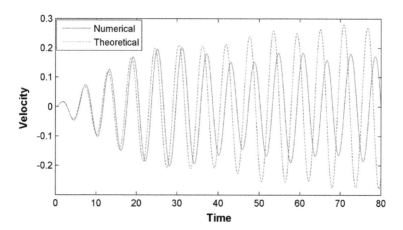

Fig. 11.11 Comparison between the exact and slow flow models for the absorbing effective particle

which leads to the following early time approximation:

$$e_{20}(\tau_1) \approx \frac{\varepsilon}{2}(\lambda V_0 \tau_1)^2 \tag{11.35}$$

Interestingly enough, using the analytical approximations (11.26), (11.34), and (11.35), we can estimate the slow time instant $\tau = \tau^*$ (Kovaleva et al. 2011) at which equipartition of energy occurs, i.e., when $e_{10}(\tau^*) = e_{20}(\tau^*)$. This time instant is found to be the following:

$$\tau^* = \sqrt{\frac{1}{\lambda^2 - (\omega - \sigma)^2}} \quad \text{(Energy equipartition)} \tag{11.36}$$

In the next section, we will compare the estimate (11.36) to the exact value derived by considering the instant of energy equipartition from direct numerical simulations of the original Eq. (11.10).

11.5.1   Nonlinear Targeted Energy Transfer and Irreversible Energy Exchange: Simulation

A series of numerical simulations for the discrete system of coupled granular chains (11.10) was performed in order to verify the theoretical predictions that were based on the model of effective particles (11.17). As shown in the previous section, in order to realize the LZT effect in space (leading to passive targeted energy transfer from the excited to the absorbing chain), it is necessary to gradually decrease the stiffness coefficient of the elastic foundation of the excited chain in space, while keeping the elastic foundation of the absorbing chain and the coupling stiffness uniform. Based on the previous theoretical predictions, for a decrease of the elastic foundation of the excited chain at a rate of $2\beta^2 \tau_1 = 2\varepsilon\beta^2 \tau_0$, the peak-to-peak time delay for the propagating breather is $\tau_0 \approx 8.88$. Hence, the required gradual spatial reduction rate of the elastic foundation of the excited chain should be approximately 22% over the leading 4 beads of the excited chain.

The results of the numerical simulation of the system (11.10) for parameters $\varepsilon = 0.05$, $\alpha = 10$, $\lambda = 1.05$ and impulsive excitation $\sqrt{\varepsilon}V_0$, with $V_0 = 1$, applied to the excited chain are shown in Fig. 11.12, where we plot the velocity profile of each bead (starting from bead 6) both for the excited and absorbing chain. According to the previous discussion, the elastic foundation of the excited chain was reduced by the 22% over the leading four beads; this was performed by setting $\gamma_1 = 0$, $\gamma_2 = 0.247$, $\gamma_3 = 0.494$, $\gamma_4 = 0.741$, $\gamma_n = 0.988$ for $n \geq 5$ in (11.10). We can clearly observe the targeted energy transfer to the absorbing chain due to the LZT effect. It follows that these results numerically confirm the theoretical prediction that appropriate spatial variation of the foundation of one of the two chains leads to

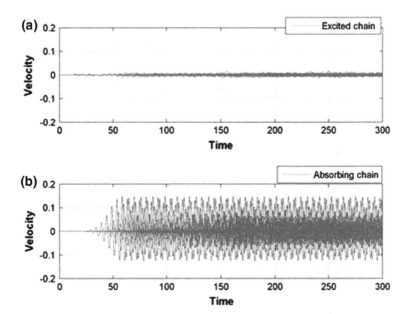

Fig. 11.12 Velocity profiles of the first 100 beads of **a** the excited and **b** the absorbing granular chain (starting from the 6th beads) of system (11.10) after gradual reduction by 22% the elastic foundation of the excited chain

passive targeted energy transfer, or pulse redirection from the excited to the absorbing chain. In Fig. 11.13, we present the same response in a space-time diagram. In that figure, the contour plots of total instantaneous energies of each of the beads of the two granular chains are depicted, with bright shades corresponding to high, and dark shades to low energy levels. From these results, the passive energy redirection from the excited to the absorbing chain is clear. One can also observe that there are regions of oscillating tails in the trail of the propagating primary pulse in the absorbing chain appearing as traces of secondary propagating waves.

In Fig. 11.14, we provide the time series of the total instantaneous energies of the excited and absorbing chains for the same response. We note that as time progresses, the total energy in the excited chain decreases and at the same time the total energy in the absorbing chain increases. At normalized time $\tau^* \approx 20$, there occurs energy equipartition between the two chains, compared to the theoretically predicted value of $\tau^* \approx 19.88$ using relation (11.36). After that time instant, most of the input energy remains localized in the absorbing chain, while the part of total energy that retains in the excited chain is small. It is interesting to note that passive energy transfer occurs at a rather fast scale in the site of the leading four beads of the excited chain, where the elastic foundation is gradually reduced. Furthermore, in Fig. 11.15, we compare the response of the slow flow envelope obtained from (11.26) to (11.27) with the direct numerical simulation of the exact equations of

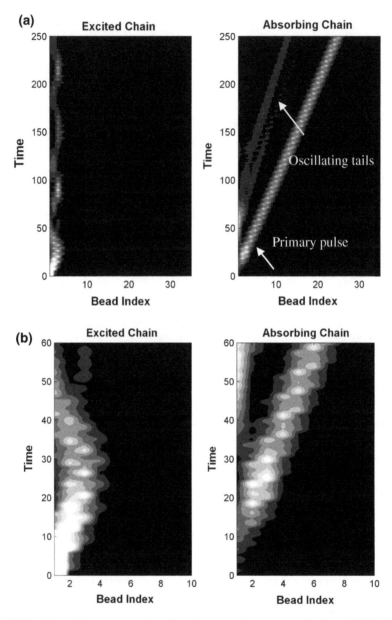

Fig. 11.13 **a** Space-time representation of passive targeted energy transfer due to LZT effect in the system of granular chains (11.10); **b** Early time details of the plots presented in (**a**)

motion of the system of coupled granular chains (11.10); by asterisks, we denote the maximum velocity of each bead realized during the propagation of the breather solution in the excited chain.

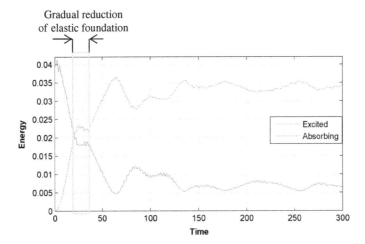

Fig. 11.14 Temporal dependence of the total instantaneous energy in the excited and absorbing chains

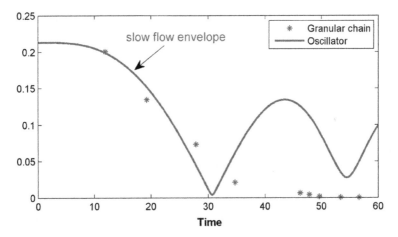

Fig. 11.15 Comparison between the slow flow prediction (26) with the direct numerical simulation of the exact system (11.10) for the excited chain; *asterisk* denotes the maximum velocity of each bead (except for the first bead)

From this comparison, we note that the theoretically predicted slow flow envelope fairly captures the envelope of the exact breather solution in the critical initial stage of strong passive targeted energy transfer. The reason for the discrepancy between the theoretical prediction and the exact numerical solution is due to the fact that the theoretical prediction takes into account only the leading-order approximation of the asymptotic analysis. It is well known that such a leading-order approximation can accurately model the transient response at early times, so the correlation between theory and numeric can be further improved over longer time intervals by adding higher-order terms in the asymptotic analysis of the excited effective particle (oscillator).

Finally, to clearly visualize the LZT phenomenon in space caused by the stratification of the elastic foundation of the excited chain, instead of applying the stratification from the second bead, we stratify the elastic foundation of the excited chain after the 24th bead where due to the nonlinear beat phenomenon the response in the excited chain attains a maximum and the response of the absorbing chain a minimum (the same system and forcing parameters as in the previous simulations depicted in Figs. 11.12, 11.13, 11.14, 11.15 were used for this simulation). The result is shown in Fig. 11.16 for a reduction rate of 22% of the elastic foundation of the excited chain over four beads; for this simulation, the following values for the detuning stiffness coefficients of the excited chain were used in (11.10):

$$\gamma_p = 0, \quad p = 1, \ldots, 24, \quad \gamma_{25} = 0.247, \quad \gamma_{26} = 0.494, \quad \gamma_{27} = 0.741,$$
$$\gamma_n = 0.988, \quad n \geq 28$$

We note that before the stratification is applied, nearly complete and recurring exchange of energy between the two chains occurs. However, once the stratification is applied, the LZT effect takes place, and the energy is passively redirected to the absorbing chain and remains localized there (Fig. 11.16). We note the sudden and fast targeted energy transfer to the absorbing chain, indicating the feasibility of the proposed energy redirection mechanism in practical material designs.

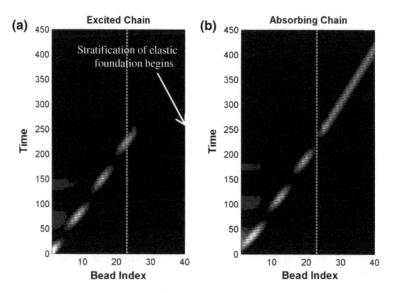

Fig. 11.16 Space-time representation of passive targeted energy transfer due to LZT effect in the system of granular chains (11.10): **a** Excited chain, **b** absorbing chain, with a spatial decrease rate by 22% of the elastic foundation of the excited granular chain starting at the 24th bead

11.6 Conclusions

In this work, we have discussed nonlinear dynamical mechanisms governing the irreversible energy transfers in weakly coupled granular networks leading to passive wave redirection. In particular, we considered two homogeneous granular chains mounted on linear elastic foundations and coupled by weak linear stiffnesses and showed both theoretically and numerically that efficient targeted energy transfer is possible from the directly excited chain to the absorbing one by two different dynamical mechanisms. The first mechanism is based on interrupting the coupling stiffness between the two chains at an appropriate phase of the developing nonlinear beat phenomenon, i.e., at the phase when maximum energy exchange occurs between chains. The second mechanism relies on the realization of a macroscopic analogue of the Landau-Zener tunneling quantum effect in space. This is achieved by appropriately varying the stiffness of the elastic foundations of the coupled chains so that conditions of resonance energy transfer are developed in order to transfer energy from the excited to the absorbing chain, and then escape from resonance occurs in order to confine (localize) the transmitted energy to the absorbing chain without possibility for back scattering.

The analysis presented in this work fully reveals (at least to leading order) the nonlinear dynamics that governs these two distinct energy transfer mechanisms, and paves the way for predictive designs for practical implementations of these mechanisms in material systems with embedded granular media with passive energy redirection properties.

Appendix

In this appendix, we develop the approximate expression of the slow flow modulation equation shown in Eq. (11.9). Rewriting Eq. (11.8), which shows the slow/fast partition of the dynamics of the system:

$$
\begin{aligned}
\frac{\partial \psi_{n0}^x}{\partial \tau_0} - i\psi_{n0}^x = 0 &\Rightarrow \psi_{n0}^x = \varphi_{no}^x(\tau_1)\exp(i\tau_0) \\
\frac{\partial \psi_{n0}^y}{\partial \tau_0} - i\psi_{n0}^y = 0 &\Rightarrow \psi_{n0}^y = \varphi_{no}^y(\tau_1)\exp(i\tau_0)
\end{aligned}
\tag{11.37}
$$

Proceeding to the $O(\varepsilon^{3/2})$ approximation, we derive the following system:

$$\frac{\partial \psi_{n0}^x}{\partial \tau_1} + \frac{\partial \psi_{n1}^x}{\partial \tau_0} - i\psi_{n1}^x = \alpha \left\{ \frac{\psi_{(n-1)0}^x - \psi_{(n-1)0}^{x*} - \psi_{(n)0}^x + \psi_{(n)0}^{x*}}{2i} \right\}_+^{3/2}$$

$$- \alpha \left\{ \frac{\psi_{(n)0}^x - \psi_{(n)0}^{x*} - \psi_{(n+1)0}^x + \psi_{(n+1)0}^{x*}}{2i} \right\}_+^{3/2} - i\lambda \left[\psi_{(n)0}^y - \psi_{(n)0}^{y*} \right]$$

$$\frac{\partial \psi_{n0}^y}{\partial \tau_1} + \frac{\partial \psi_{n1}^y}{\partial \tau_0} - i\psi_{n1}^y = \alpha \left\{ \frac{\psi_{(n-1)0}^y - \psi_{(n-1)0}^{y*} - \psi_{(n)0}^y + \psi_{(n)0}^{y*}}{2i} \right\}_+^{3/2}$$

$$- \alpha \left\{ \frac{\psi_{(n)0}^y - \psi_{(n)0}^{y*} - \psi_{(n+1)0}^y + \psi_{(n+1)0}^{y*}}{2i} \right\}_+^{3/2} - i\lambda \left[\psi_{(n)0}^x - \psi_{(n)0}^{x*} \right]$$

$$(11.38)$$

Introducing (11.37) into (11.38) yields the following:

$$\frac{\partial \left[\varphi_{n0}^x \exp(i\tau_0) \right]}{\partial \tau_1} + \frac{\partial \psi_{n1}^x}{\partial \tau_0} - i\psi_{n1}^x$$

$$= \alpha \left\{ \frac{\varphi_{(n-1)0}^x \exp(i\tau_0) - \varphi_{(n-1)0}^{x*} \exp(-i\tau_0) - \varphi_{(n)0}^x \exp(i\tau_0) + \varphi_{(n)0}^{x*} \exp(-i\tau_0)}{2i} \right\}_+^{3/2}$$

$$- \alpha \left\{ \frac{\varphi_{(n)0}^x \exp(i\tau_0) - \varphi_{(n)0}^{x*} \exp(-i\tau_0) - \varphi_{(n+1)0}^x \exp(i\tau_0) + \varphi_{(n+1)0}^{x*} \exp(-i\tau_0)}{2i} \right\}_+^{3/2}$$

$$- i\lambda \left[\phi_{n0}^y \exp(i\tau_0) - \phi_{n0}^{y*} \exp(-i\tau_0) \right]$$

$$\frac{\partial \left[\varphi_{n0}^y \exp(i\tau_0) \right]}{\partial \tau_1} + \frac{\partial \left[\psi_{n1}^y \right]}{\partial \tau_0} - i\psi_{n1}^y$$

$$= \alpha \left\{ \frac{\varphi_{(n-1)0}^y \exp(i\tau_0) - \varphi_{(n-1)0}^{y*} \exp(-i\tau_0) - \varphi_{(n)0}^y \exp(i\tau_0) + \varphi_{(n)0}^{y*} \exp(-i\tau_0)}{2i} \right\}_+^{3/2}$$

$$- \alpha \left\{ \frac{\varphi_{(n)0}^y \exp(i\tau_0) - \varphi_{(n)0}^{y*} \exp(-i\tau_0) - \varphi_{(n+1)0}^y \exp(i\tau_0) + \varphi_{(n+1)0}^{y*} \exp(-i\tau_0)}{2i} \right\}_+^{3/2}$$

$$- i\lambda \left[\phi_{n0}^x \exp(i\tau_0) - \phi_{n0}^{x*} \exp(-i\tau_0) \right]$$

$$(11.39)$$

Upon imposing solvability conditions in (11.39), yields the following *slow flow*, i.e., the system of modulation equations in the slow timescale governing the (slow) evolutions of the complex envelopes in (11.37):

$$\frac{\partial \varphi_{n0}^x}{\partial \tau_1} = \alpha \left\{ \frac{\varphi_{(n-1)0}^x \exp(i\tau_0) - \varphi_{(n-1)0}^{x*} \exp(-i\tau_0) - \varphi_{(n)0}^x \exp(i\tau_0) + \varphi_{(n)0}^{x*} \exp(-i\tau_0)}{2i} \right\}_+^{3/2}$$

$$\times \exp(-i\tau_0)$$

$$- \alpha \left\{ \frac{\varphi_{(n)0}^x \exp(i\tau_0) - \varphi_{(n)0}^{x*} \exp(-i\tau_0) - \varphi_{(n+1)0}^x \exp(i\tau_0) + \varphi_{(n+1)0}^{x*} \exp(-i\tau_0)}{2i} \right\}_+^{3/2}$$

$$\times \exp(-i\tau_0) - i\lambda \left[\phi_{n0}^y \exp(i\tau_0) - \phi_{n0}^{y*} \exp(-i\tau_0) \right] \exp(-i\tau_0)$$

$$\frac{\partial \varphi_{n0}^y}{\partial \tau_1} = a \left\{ \frac{\varphi_{(n-1)0}^y \exp(i\tau_0) - \varphi_{(n-1)0}^{y*} \exp(-i\tau_0) - \varphi_{(n)0}^y \exp(i\tau_0) + \varphi_{(n)0}^{y*} \exp(-i\tau_0)}{2i} \right\}_+^{3/2}$$

$$\times \exp(-i\tau_0)$$

$$- \alpha \left\{ \frac{\varphi_{(n)0}^y \exp(i\tau_0) - \varphi_{(n)0}^{y*} \exp(-i\tau_0) - \varphi_{(n+1)0}^y \exp(i\tau_0) + \varphi_{(n+1)0}^{y*} \exp(-i\tau_0)}{2i} \right\}_+^{3/2}$$

$$\times \exp(-i\tau_0) - i\lambda \left[\phi_{n0}^x \exp(i\tau_0) - \phi_{n0}^{x*} \exp(-i\tau_0) \right] \exp(-i\tau_0)$$

$$(11.40)$$

To evaluate the non-smooth terms in (11.40), we follow (Starosvetsky et al. 2012) and introduce the Fourier expansions:

$$F_{(n-1)0}^x \equiv \varphi_{(n-1)0}^x - \varphi_{n0}^x = \left| F_{(n-1)0}^x \right| \exp\left(i\theta_{n-1}^x\right), \quad F_{(n)0}^x \equiv \varphi_{n0}^x - \varphi_{(n+1)0}^x = \left| F_{(n)0}^x \right| \exp\left(i\theta_n^x\right)$$

$$F_{(n-1)0}^y \equiv \varphi_{(n-1)0}^y - \varphi_{n0}^y = \left| F_{(n-1)0}^y \right| \exp\left(i\theta_{n-1}^y\right), \quad F_{(n)0}^y \equiv \varphi_{n0}^y - \varphi_{(n+1)0}^y = \left| F_{(n)0}^y \right| \exp\left(i\theta_n^y\right)$$

$$(11.41)$$

which when substituted into the non-smooth terms on the right-hand sides of the slow flow (11.40) lead to the following expressions:

$$\left\{\frac{F^x_{(n-1)0}\exp(i\tau_0) - F^{x*}_{(n-1)0}\exp(-i\tau_0)}{2i}\right\}^{3/2}_+ = \left|F^x_{(n-1)0}\right|^{3/2}\left\{\sin(\tau_0 + \theta^x_{n-1})\right\}^{3/2}_+$$

$$= \left|F^x_{(n-1)0}\right|^{3/2}\left\{\cos\Phi^x_{n-1}\right\}^{3/2}_+$$

$$\left\{\frac{F^x_{(n)0}\exp(i\tau_0) - F^{x*}_{(n)0}\exp(-i\tau_0)}{2j}i\right\}^{3/2}_+ = \left|F^x_{(n)0}\right|^{3/2}\left\{\sin(\tau_0 + \theta^x_n)\right\}^{3/2}_+$$

$$= \left|F^x_{(n)0}\right|^{3/2}\left\{\cos\Phi^x_n\right\}^{3/2}_+$$

$$\left\{\frac{F^y_{(n-1)0}\exp(i\tau_0) - F^{y*}_{(n-1)0}\exp(-i\tau_0)}{2i}\right\}^{3/2}_+ = \left|F^y_{(n-1)0}\right|^{3/2}\left\{\sin(\tau_0 + \theta^y_{n-1})\right\}^{3/2}_+$$

$$= \left|F^y_{(n-1)0}\right|^{3/2}\left\{\cos\Phi^y_{n-1}\right\}^{3/2}_+$$

$$\left\{\frac{F^y_{(n)0}\exp(i\tau_0) - F^{y*}_{(n)0}\exp(-i\tau_0)}{2i}\right\}^{3/2}_+ = \left|F^y_{(n)0}\right|^{3/2}\left\{\sin(\tau_0 + \theta^y_n)\right\}^{3/2}_+$$

$$= \left|F^y_{(n)0}\right|^{3/2}\left\{\cos\Phi^y_n\right\}^{3/2}_+$$

$$(11.42)$$

where $\Phi^{x,y}_k + \frac{\pi}{2} = \tau_0 + \theta^{x,y}_k$. Substituting (11.42) into (11.40) and performing averaging with respect to the fast timescale τ_0 yields the following *averaged slow flow* equations:

$$\frac{\partial\varphi^x_{n0}}{\partial\tau_1} = (\alpha/2\pi)\left\{\left|F^x_{(n-1)0}\right|^{3/2}\exp\left[i(\theta^x_{n-1} - \pi/2)\right]\oint_{2\pi}\left\{\cos\Phi^x_{n-1}\right\}^{3/2}_+\exp(-i\Phi^x_{n-1})\,d\Phi^x_{n-1}\right.$$

$$\left. - \left|F^x_{(n)0}\right|^{3/2}\exp\left[i(\theta^x_n - \pi/2)\right]\oint_{2\pi}\left\{\cos\Phi^x_n\right\}^{3/2}_+\exp(-i\Phi^x_n)\,d\Phi^x_n\right\}$$

$$+ (\lambda/\pi)\exp(-i\pi/2)\oint_{2\pi}\Phi^y_{n0}\left\{\cos(\Phi^y_n - \theta^y_n)\right\}\exp\left\{-i(\Phi^y_n - \theta^y_n)\right\}\,d(\Phi^y_n - \theta^y_n)$$

$$\frac{\partial\varphi^y_{n0}}{\partial\tau_1} = (\alpha/2\pi)\left\{\left|F^y_{(n-1)0}\right|^{3/2}\exp\left[i(\theta^y_{n-1} - \pi/2)\right]\oint_{2\pi}\left\{\cos\Phi^y_{n-1}\right\}^{3/2}_+\exp(-i\Phi^y_{n-1})\,d\Phi^y_{n-1}\right.$$

$$\left. - \left|F^y_{(n)0}\right|^{3/2}\exp\left[i(\theta^y_n - \pi/2)\right]\oint_{2\pi}\left\{\cos\Phi^y_n\right\}^{3/2}_+\exp(-i\Phi^y_n)\,d\Phi^y_n\right\}$$

$$+ (\lambda/\pi)\exp(-i\pi/2)\oint_{2\pi}\Phi^x_{n0}\left\{\cos(\Phi^x_n - \theta^x_n)\right\}\exp\left\{-i(\Phi^x_n - \theta^x_n)\right\}\,d(\Phi^x_n - \theta^x_n)$$

$$(11.43)$$

Using Fourier expansion, retaining only the leading-order harmonics, and rearranging terms yields the following averaged slow flow system:

$$
\begin{aligned}
i\frac{\partial \varphi_{n0}^x}{\partial \tau_1} &= \tilde{\alpha}\left(F_{(n-1)0}^x \left| F_{(n-1)0}^x \right|^{1/2} - F_{(n)0}^x \left| F_{(n)0}^x \right|^{1/2} \right) + \tilde{\lambda}\varphi_{n0}^y \\
i\frac{\partial \varphi_{n0}^y}{\partial \tau_1} &= \tilde{\alpha}\left(F_{(n-1)0}^y \left| F_{(n-1)0}^y \right|^{1/2} - F_{(n)0}^y \left| F_{(n)0}^y \right|^{1/2} \right) + \tilde{\lambda}\varphi_{n0}^x
\end{aligned}
\tag{11.44}
$$

References

Anderson, B.P., Kasevich, M.A.: Macroscopic quantum interference from atomic tunnel arrays. Science **282**, 1686–1689 (1998)

Aubry, S.: Breathers in nonlinear lattices: existence, linear stability and quantization. Physica D **103**, 201–250 (1997)

Aubry, S., Kopidakis, S., Morgante, A.M., Tsironis, G.P.: Analytic conditions for targeted energy transfer between nonlinear oscillators for discrete breathers. Phys. B **296**, 222–236 (2001)

Binder, P., Abraimov, D., Ustinov, A.V., Flach, S., Zolotaryuk, Y.: Observation of breathers in Josephson ladders. Phys. Rev. Lett. **84**, 745–748 (2000)

Campbell, D.K., Peyrard, M.: In: Campbell, D.K. (ed.) CHAOS-Soviet American Perspectives on Nonlinear Science. American Institute of Physics, New York (1990)

Campbell, D.K., Flach, S., Kivshar, Y.S.: Localizing energy through nonlinearity and discreteness. Phys. Today **57**, 43–49 (2004)

Edler, J., Pfister, R., Pouthier, V., Falvo, C., Hamm, P.: Direct observation of self-trapped vibrational states in α-helices. Phys. Rev. Lett. **93**, 106405 (2004)

Eisenberg, H.S., Silberberg, Y., Morandotti, R., Boyd, A.R., Aitchison, J.S.: Discrete spatial optical solitons in waveguide arrays. Phys. Rev. Lett. **81**, 3383–3386 (1998)

Flach, S., Willis, C.R.: Discrete breathers. Phys. Rep. **295**, 181–264 (1998)

Hoogeboom, C., Theocharis, G., Kevrekidis, P.G.: Discrete breathers at the interface between a diatomic and a monoatomic granular chain. Phys. Rev. E **82**, 061303 (2010)

Jayaprakash, K.R., Starosvetsky, Y., Vakakis, A.F.: New family of solitary waves in granular dimer chains with no precompression. Phys. Rev. E **83**, 036606 (2011a)

Jayaprakash, K.R., Starosvetsky, Y., Vakakis, A.F., Peeters, M., Kerschen, G.: Nonlinear normal modes and band zones in granular chains with no pre-compression. Nonlinear Dyn. **63**, 359–385 (2011b)

Job, S., Santibanez, F., Tapia, F., Melo, F.: Wave localization in strongly nonlinear Hertzian chains with mass defect. Phys. Rev. E **80**, 025602(R) (2009)

Johansson, M., Morgante, A.M., Aubry, S., Kopidakis, G.: Eur. Phys. J. B **29**, 279283 (2002)

Juanico, B., Sanejouand, Y.-H., Piazza, F., De Los Rios, P.: Discrete breathers in nonlinear network models of proteins. Phys. Rev. Lett. **99**, 238104 (2007)

Kopidakis, G., Aubry, S.: Intraband discrete breathers in disordered nonlinear systems. I. Delocalization. Physica D **130**, 155–186 (1999)

Kopidakis, G., Aubry, S.: Discrete breathers and delocalization in nonlinear disordered systems. Phys. Rev. Lett. **84**, 3236–3239 (2000a)

Kopidakis, G., Aubry, S.: Intraband discrete breathers in disordered nonlinear systems. II. Localization. Physica D **139**, 247–275 (2000b)

Kopidakis, G., Aubry, S., Tsironis, G.P.: Targeted energy transfer through discrete breathers in nonlinear systems. Phys. Rev. Lett. **87**, 165501 (2001a)

Kopidakis, G., Aubry, S., Tsironis, G.P.: Targeted energy transfer through discrete breathers in nonlinear systems. Phys. Rev. Lett. **87**, 165501 (2001b)

Kosevich, Y.A., Manevitch, L.I., Savin, A.V.: Energy transfer in coupled nonlinear phononic waveguides: transition from wandering breather to nonlinear self-trapping. J. Phys.: Conf. Ser. **92**, 012093 (2007)

Kosevich, Y.A., Manevitch, L.I., Savin, A.V.: Wandering breathers and self-trapping in weakly coupled nonlinear chains: classical counterpart of macroscopic tunneling quantum dynamics. Phys. Rev. E **77**, 046603 (2008)

Kosevich, Y.A., Manevitch, L.I., Savin, A.V.: Energy transfer in weakly coupled nonlinear oscillator chains: Transition from a wandering breather to nonlinear self-trapping. J. Sound. Vib. **322**, 524–531 (2009)

Kosevich, Y.A., Manevitch, L.I., Manevitch, E.L.: Vibrational analogue of nonadiabatic Landau–Zener tunneling and a possibility for the creation of a new type of energy trap. Phys. Usp. **53**, 1281–1286 (2010)

Kovaleva, A., Manevitch, L.I., Kosevich, Y.A.: Fresnel integrals and irreversible energy transfer in an oscillatory system with time-dependent parameters. Phys. Rev. E **83**, 026602 (2011)

Manevitch, L.I., Kosevich, Y.A., Mane, M., Sigalov, G., Bergman, L.A., Vakakis, A.F.: Towards a new type of energy trap: classical analog of quantum Landau-Zener tunneling. Int. J. Non Linear Mech. 46, 247–252 (2011). doi:10.1016/ijnonlinmec.20.08.01010

Manevitch, L.I.: Complex representation of dynamics of coupled nonlinear oscillators. In Mathematical Models of Non-Linear Excitations: Transfer, Dynamics, and Control in Condensed Systems and Other Media, pp. 269–300. Kluwer Academic, Plenum Publishers, New York (1999)

Maniadis, P., Aubry, S.: Targeted energy transfer by Fermi resonance. Physica D **202**, 200–217 (2005)

Manley, M.E., Yethiraj, M., Sinn, H., Volz, H.M., Alatas, A., Lashley, J.C., Hults, W.L., Lander, G.H., Smith, J.L.: Formation of a new dynamical mode in α-Uranium observed by inelastic X-ray and neutron scattering. Phys. Rev. Lett. **96**, 125501 (2006)

Morgante, A.A., Johansson, M., Kopidakis, G., Aubry, S.: Physica D. **162**, 53 (2002)

Nesterenko, V.F.: Dynamics of Heterogeneous Materials. Springer, Berlin, New York (2001)

Razavy, M.: Quantum Theory of Tunneling. World Scientific, Singapore (2003)

Rosam, B., Leo, K., Gluck, M., Keck, F., Korsch, H., Zimmer, F., Kohler, K.: Lifetime of Wannier-Stark states in semiconductor superlattices under strong Zener tunneling to above-barrier bands. Phys. Rev. B **68**, 125301 (2003)

Sato, M., Sievers, A.J.: Direct observation of the discrete character of intrinsic localized modes in an antiferromagnet. Nature **432**, 486–488 (2004)

Scott, A.: Nonlinear Science: Emergence and Dynamics of Coherent Structures. Oxford University Press, New York (1999)

Sen, S., Krishna Mohan, T.R.: Dynamics of metastable breathers in nonlinear chains in acoustic vacuum. Phys. Rev. E **79**, 036603 (2009)

Sias, C., Zenesini, A., Lignier, H., Wimberger, S., Ciampini, D., Morsch, O., Arimondo, E.: Resonantly enhanced tunneling of Bose-Einstein condensates in periodic potentials. Phys. Rev. Lett. **98**, 120403 (2007)

Sievers, A.J., Takeno, S.: Intrinsic localized modes in anharmonic crystals. Phys. Rev. Lett. **61**, 970–973 (1988)

Starosvetsky, Y., Vakakis, A.F.: Traveling waves and localized modes in one-dimensional homogeneous granular chains with no pre-compression. Phys. Rev. E **82**, 026603 (2010)

Starosvetsky, Y., Jayaprakash, K.R., Vakakis, A.F., Manevitch, L.I.: Effective particles and classification of periodic orbits of homogeneous granular chains with no pre-compression. Phys. Rev. E (in press)

Starosvetsky, Y., Hasan, M.A., Vakakis, A.F., Manevitch, L.I.: Strongly nonlinear beat phenomena and energy exchanges in weakly coupled granular chains on elastic foundations. SIAM J. Appl. Math. **72**(1), 337–361 (2012)

Swanson, B.I., Brozik, J.A., Love, S.P., Strouse, G.F., Shreve, A.P., Bishop, A.R., Wang, W.-Z., Salkola, M.I.: Observation of intrinsically localized modes in a discrete low-dimensional material. Phys. Rev. Lett. **82**, 3288–3291 (1999)

Theocharis, G., Kavousanakis, M., Kevrekidis, P.G., Daraio, C., Porter, M.A., Kevrekidis, I.G.: Localized breathing modes in granular crystals with defects. Phys. Rev. E **80**, 066601 (2009)

Trias, E., Mazo, J.J., Orlando, T.P.: Discrete breathers in nonlinear lattices: experimental detection in a Josephson array. Phys. Rev. Lett. **84**, 741–744 (2000)

Vakakis, A.F., Gendelman, O., Bergman, L.A., McFarland, D.M., Kerschen, G., Lee, Y.S.: Nonlinear targeted energy transfer in mechanical and structural systems. Springer, Berlin, New York (2008)

Zener, C.: Non-adiabatic crossing of energy levels. Proc. R. Soc. Lond. A Math. Phys. Sci. **137**, 696–702 (1932)

Chapter 12
Forced Pendulum

Harmonically forced pendulum is one of the basic models of nonlinear dynamics which has numerous applications in different fields of physics and mechanics (Baker and Blackburn 2005; Sagdeev et al. 1988; Scott 2003; Arnold et al. 2006). There are two main directions in the study of this model. First of them can be denoted as application of general mathematical perturbation theory in which integrable conservative system is a generating model (Sagdeev et al. 1988; Nayfeh and Mook 2004; Bogolubov and Mitropolsky 1961; Hale 1963; Chirikov and Zaslavsky 1972, Neishtadt 1975; Neishtadt and Vasiliev 2005). The obtained results relate mainly to the quasi-linear approach (small amplitude) and reflect the common features of nonlinear oscillator: finiteness of the resonance amplitude and possibility of abrupt its changes due to instability of a stationary state. The second direction was developed by physicists and mechanicians and deals with analytical description of the stationary states (in the quasi-linear approximation also), analysis of their stability, and numerical study of nonstationary dynamics (Nayfeh and Mook 2004; Manevitch and Manevitch 2005; Manevitch and Musienko 2009). The concept of limiting phase trajectories allows to find an efficient analytical description of highly nonstationary resonance dynamics in which the oscillator (pendulum) takes off a maximum possible (at given conditions) energy from its source (periodic field). The goal of this section is to remove the restrictions on the amplitude of pendulum oscillations. For this, we use a semi-inverse approach in combination with the LPT concept. In such a case, the analytical description of qualitative transitions in both stationary and highly nonstationary dynamics can be found.

© Springer Nature Singapore Pte Ltd. 2018   327
L.I. Manevitch et al., *Nonstationary Resonant Dynamics of Oscillatory Chains and Nanostructures*, Foundations of Engineering Mechanics,
DOI 10.1007/978-981-10-4666-7_12

12.1   The Model

We discuss the undamped dynamics of a pendulum excited by a harmonic excitation and undergoing unidirectional motion. Corresponding equation of motion is well-known

$$\frac{d^2q}{dt^2} + \sin q = f \sin \Omega t \tag{12.1}$$

where q is the angular coordinate of the pendulum, f and Ω are the harmonic forcing amplitude and frequency, respectively.

By introducing the complex amplitude of the pendulum oscillations as

$$\psi = \frac{1}{\sqrt{2}} \left(\frac{1}{\sqrt{\omega}} \frac{dq}{dt} + i\sqrt{\omega} q \right) \tag{12.2}$$

$$q = \frac{-i}{\sqrt{2\omega}} (\psi - \psi^*); \quad \frac{dq}{dt} = \sqrt{\frac{\omega}{2}} (\psi + \psi^*)$$

one can rewrite Eq. (12.1) as follows:

$$i\frac{d\psi}{dt} + \frac{\omega}{2}(\psi + \psi^*) + \frac{1}{2\omega}\sum_{k=0}^{\infty} \frac{1}{(2k+1)!} \left(\frac{1}{2\omega}\right)^k (\psi - \psi^*)^{2k+1}$$
$$= \frac{f}{2\sqrt{2\omega}} \left(e^{i\Omega t} - e^{-i\Omega t} \right) \tag{12.3}$$

To find the stationary solution of Eq. (12.3), one should assume that $\omega = \Omega$ and

$$\psi = \sqrt{X} e^{i\Omega t} \tag{12.4}$$

where X = const is the amplitude.

Substituting solution (12.4) into Eq. (12.3) and multiplying the result on $\exp(i\Omega t)$, one can obtain after integration over period $2\pi/\Omega$.

$$\frac{\Omega}{2}\sqrt{X} + \frac{1}{\sqrt{2\Omega}} J_1 \left(\sqrt{\frac{2X}{\Omega}} \right) = \frac{f}{2\sqrt{2\Omega}} \tag{12.5}$$

With taking into account the relation

$$X = \frac{\omega}{2} Q^2 \tag{12.6}$$

between the moduls of complex function and amplitude, the forcing frequency can be expressed via the amplitude of stationary oscillations as follows:

$$\Omega = \sqrt{\frac{1}{Q}(2J_1(Q) - f)} \tag{12.7}$$

The amplitude–frequency relationship (12.7) for different forces in range (-0.2, 0.2) is shown in Fig. 12.1.

Figure 12.1 shows that all branches under backbone (black) curve possess two stationary states—stable and unstable ones. The condition

$$\frac{d\Omega}{dQ} = 0 \tag{12.8}$$

determines the high boundary of the frequency range, where three stationary states occur.

However, if we interest ourself in the nonstationary oscillations of the driven pendulum, one should consider a solution

$$\psi = \varphi e^{i\omega t} \tag{12.9}$$

where φ is a slowly changing function and $\omega = \Omega - s$ ($s \ll \Omega$).

Substituting solution (12.9) into Eq. (12.3) and assuming that the detuning parameter s is small enough, one can consider φ as a function of the slow time $\tau = st$. The latter is supposed to be a new variable that is independent on the "fast" time t.

Then, multiplying Eq. (12.3) by the $e^{-i\omega t}$ and integrating it with respect to the "fast" time t, the condition, which provides excluding the resonance (secular) terms, is obtained as:

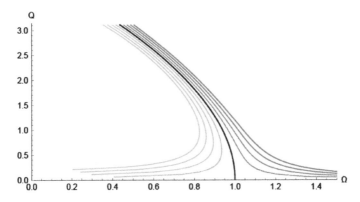

Fig. 12.1 Driving frequency–amplitude relationship for different values of the force: $f = -0.2, ...,$ 0.2 (colors—from *red* to *green*). *Black curve* shows the backbone frequency of free oscillations ($f = 0$)

$$is\frac{\partial\varphi}{\partial\tau} - \frac{\omega}{2} + \frac{1}{\sqrt{2\omega}}J_1\left(\sqrt{\frac{2}{\omega}}|\varphi|\right)\frac{\varphi}{|\varphi|} = \frac{f}{\sqrt{2\omega}}e^{i\tau} \qquad (12.10)$$

It is easy to check that the function $\varphi = \sqrt{X}e^{i\tau}$ is the solution of Eq. (12.10) if the frequency ω satisfies the relation

$$s - \frac{\omega}{2} + \frac{1}{\sqrt{2\omega}}J_1\left(\sqrt{\frac{2X}{\omega}}\right) = \frac{f}{\sqrt{2X\omega}} \qquad (12.11)$$

This equation permits to derive a very simple expression for the frequency of pendulum oscillations as a function of their amplitude:

$$\omega = -s + \sqrt{\frac{1}{Q}(2J_1(Q) - f) + s^2} \qquad (12.12)$$

One can see that the limit $s \to 0$ leads to expression (12.7).

12.2   Nonstationary Dynamics and Dynamical Transitions

The nonstationary dynamics of the forced pendulum can be tackled by introducing the phase $\delta(\tau)$ and the amplitude $a = \sqrt{\frac{\omega}{2}q}$, such that the slowly varying function φ can now be expressed as $\varphi = a\,e^{-i\delta}$. Having introduced the phase shift $\Delta = \tau - \delta$, the equation of motion (12.10) can be written as

$$sa\frac{\partial\Delta}{\partial\tau} - \left(s + \frac{\omega}{2}\right)a + \frac{1}{\sqrt{2\omega}}J_1\left(\sqrt{\frac{2X}{\omega}}a\right) = \frac{f}{\sqrt{2\omega}}\cos\Delta \qquad (12.13)$$

$$sa\frac{\partial a}{\partial\tau} = \frac{f}{\sqrt{2\omega}}\sin\Delta \qquad (12.14)$$

The corresponding integral of motion reads

$$H = \frac{s}{2}\left[\left(s + \frac{\omega}{2}\right)a^2 + J_0\left(\sqrt{\frac{2}{\omega}}a\right) - 1 + \frac{af}{\sqrt{2\omega}}\cos\Delta\right] \qquad (12.15)$$

In order to study the nonstationary processes, one should consider the phase portrait of the system (12.15) (Fig. 12.2a–d). There are three control parameters in the considered system: detuning s, forcing f, and frequency ω. They are coupled

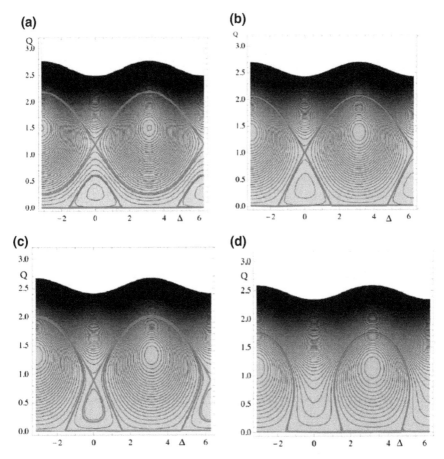

Fig. 12.2 Evolution of the $\Delta - Q$ phase portrait for $f = 0.06$ for increasing frequency ω. **a** Before the first transition, $\omega = 0.79$; **b** after the first transition, $\omega = 0.8106$; **c** at the second transition, $\omega = 0.82$; **d** after the second transition, $\omega = 0.85$

with the stationary oscillation amplitude Q by relation (12.7). Let us fix detuning s and the force amplitude f and will vary the pendulum frequency ω.

Figure 12.2a shows the phase portrait in terms of amplitude a and phase shift Δ, which is typical for the "low" frequency region. Three stationary points associate with stable and unstable brunches in Fig. 12.1. It is important that the attraction areas of the stable states are demarcated by two specific trajectories. The first one is the separatrix that passes through the unstable state and surrounds the large-amplitude stationary point.

The initial conditions with zero amplitudes (without dependence on the initial phase shift) lead to the motion along the closed trajectory, bounding the attraction area of the stable stationary points at $\Delta = 0$ (red curve in Fig. 12.2a). The principal difference of this trajectory from the separatrix is the finite time of its passing. Due to this, such trajectory is the most distant from the stationary points, and we refer it as the limiting phase trajectory (LPT). The LPT corresponds to the most intensive energy taking off by the pendulum from the energy source (at given initial conditions), see also Chap. 6. All other trajectories require nonzero initial conditions. Increasing the frequency ω is accompanied by enlarging of areas inside both the LPT and the separatrix. This growth is finished when the LPT coincides with the separatrix (Fig. 12.2b). At this moment, the separatrix is becoming homoclinic and it makes up two branches of LPT, one of them surrounds the stable state with phase shift $\Delta = 0$ and another one envelopes the stationary point with $\Delta = \pi$. The latter turns out to be essentially larger than the first one. Therefore, the amplitudes of nonstationary oscillations increase stepwise, and the energy flow from the energy source to the pendulum, respectively, grows. The further rise of the frequency ω leads to weak decreasing of LPT as well as the separatrix up to the annihilation of latter at the frequency, the value of which is determined by equation (12.8) (see Fig. 12.2c, d).

In order to estimate the threshold of the stepwise changing of oscillation amplitude, one should note that the value of Hamiltonian (12.15) turns out to be equal to zero at the bifurcation point. Thus, solving the equation

$$H(Q, \omega)|\Delta = 0 = 0$$

jointly with Eq. (12.11) with respect to ω and Q at fixed values f and s, one can calculate the threshold frequency as a function of detuning s and forcing f. Figure 12.3 shows the threshold values of ω at detuning $s = 0.1$ as the function of forcing f (red curves). The bifurcation value of ω that leads to the annihilation of unstable stationary point is represented in Fig. 12.3 by blue curves.

12.3   Poincaré Sections and Onset of Chaotic Motion

In this section, a numerical validation of the dynamic regimes exhibited by the forced pendulum is proposed by resorting to Poincaré sections obtained from direct integration of the starting equation of motion (12.1). Besides confirming the two dynamical transitions discussed in Sect. 12.3, the Poincaré sections allow to

Fig. 12.3 Analytic dynamical transitions thresholds on the $(f - \omega)$ plane for $s = 0.1$. First (*blue*) and second (*red*) thresholds, analytical (*solid*), numerical (*dashed*), and points corresponding to the phase portraits shown in Fig. 12.2 and Poincaré sections shown in Fig. 12.4

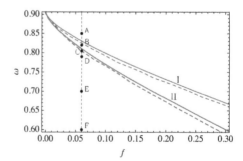

identify the onset of non-regular pendulum response and its connection with the LPTs and the dynamic separatrix. Having fixed the forcing amplitude and detuning, the dynamic regimes evolution is described for varying forcing frequency. For $s = 0.1$, in Figs. 12.4 and 12.5, Poincaré sections are shown for $f = 0.06$ and $f = 0.08$, respectively. In more detail, in Fig. 12.4a, the scenario at $\omega = 0.85$ corresponding to point A in Fig. 12.3, whose phase plane is shown in Fig. 12.2a, is depicted. It is characterized by the presence of, the in-phase NNM and the LPT (red curve) encircles it. Next, in Fig. 12.4b, the Poincaré sections at $\omega = 0.82$, corresponding to point B in Fig. 12.3, is shown. In this case the newborn out-of-phase NNM and unstable hyperbolic point can be seen together with the LPT (red curve) encircling the in-phase NNM and the heteroclinic separatrix (blue curve). As the value of ω is

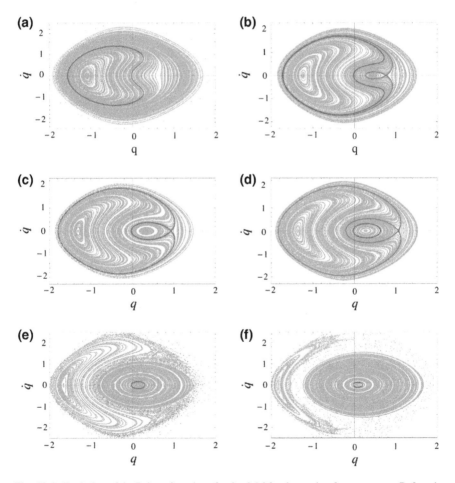

Fig. 12.4 Evolution of the Poincaré sections for $f = 0.06$ for decreasing frequency ω. **a** Before the first transition, $\omega = 0.85$; **b** after the first transition, $\omega = 0.82$; **c** at the second transition, $\omega = 0.8047$; **d** after the second transition, $\omega = 0.79$; **e** $\omega = 0.7$; **f** $\omega = 0.6$

lowered to 0.8047, the second transition, corresponding to point C in Fig. 12.3 and phase plane in Fig. 12.3c, occurs; the LPT (red curve) and the heteroclinic separatrix coalesce implying the most intense energy exchange between the source of excitation and the pendulum. At last, after the second transition, the LPT (red curve) localization is clearly seen in Fig. 12.4d—in which, for $\omega = 0.79$, consistently with the phase plane shown in Fig. 12.2d, the LPT amplitude undergoes a significant reduction entailing a weak energy exchange with the harmonic forcing. For lower values of ω, the regular motion region, so far characterizing the whole phase plane, splits into two regions, separated by chaotic sea and surrounding the two stationary points (see Fig. 12.4e, f). The Poincaré sections reported in Fig. 12.5, corresponding to the case with $f = 0.08$, show the analogous qualitative evolution of the previous case with

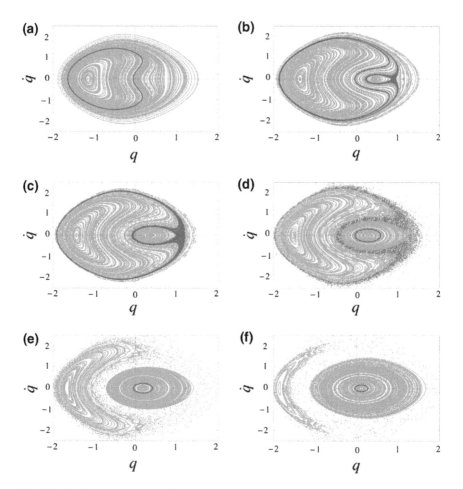

Fig. 12.5 Evolution of the Poincaré sections for $f = 0.08$ for decreasing frequency ω. **a** Before the first transition, $\omega = 0.85$; **b** after the first transition, $\omega = 0.8$; **c** at the second transition, $\omega = 0.7834$; **d** after the second transition, $\omega = 0.75$; **e** $\omega = 0.7$; **f** $\omega = 0.6$

$f = 0.06$. The main difference stems from the onset of separatrix chaos that now occurs for higher values of ω, in between the two dynamic transitions (see Fig. 12.5b). As shown in Fig. 12.5c, at the second transition such chaotization involves the LPT as well. Afterward, as ω decreases, a trend similar to the case for $f = 0.06$ can be observed. It is seen that at relatively weak intensity of forcing (Fig. 12.3), one can observe a regular behavior which corresponds exactly to analytical predictions. Increase of this intensity leads to manifestation of chaotic behavior (Figs. 12.4 and 12.5). Two scenarios of chaotization are possible. This phenomenon can indeed appear after both dynamical transitions (Fig. 12.4) and may be identified as breaking of regular separatrix (Fig. 12.5e). The second scenario is realized at $f = 0.08$ (Fig. 12.5b–f), and its manifestation occurs after stationary transition (Fig. 12.4b) and becomes especially clear at the nonstationary transition (Fig. 12.4c).

References

Arnold, V.I., Kozlov, V.V., Neishtadt, A.I.: Mathematical Aspects of Classical and Celestial Mechanics, 3rd ed. Encyclopaedia of Mathematical Sciences, vol. 3. Springer, Berlin (2006)

Baker, G.I., Blackburn, J.A.: The Pendulum. A Case Study in Physics. Oxford University Press, New York (2005)

Bogoliubov, N.N., Mitropolskii, Y.A.: Asymptotic Methods in the Theory of Nonlinear Oscillations. Gordon and Breach, New York (1961)

Chirikov, B.V., Zaslavsky, G.M.: Stochastic instability of nonlinear oscillations. Sov. Phys. Uspekhi **14**, 549 (1972)

Hale, J.K.: Oscillations in Nonlinear Systems. McGraw-Hill, New York (1963)

Manevitch, A.I., Manevitch, L.I.: The Mechanics of Nonlinear Systems with Internal Resonances. Imperial College Press, London (2005)

Manevitch, L.I., Musienko, A.I.: Limiting phase trajectories and energy exchange between anharmonic oscillator and external force. Nonlinear Dyn. **58**, 633 (2009)

Nayfeh, A.H., Mook, D.T.: Nonlinear Oscillations. Wiley-VCH Verlag GmbH and Co., Germany (2004)

Neishtadt, A.I.: Passage through a separatrix in a resonance problem with a slowly-varying parameter. J. Appl. Math. Mech. **39**(4), 594 (1975)

Neishtadt, A.I., Vasiliev, A.A.: Capture into resonance in dynamics of a classical hydrogen atom in an oscillating electric field. Phys. Rev. E **71**, 056623 (2005)

Sagdeev, R.Z., Usikov, D.A., Zaslavsky, G.M.: Nonlinear Physics: From the Pendulum to Turbulence and Chaos. Harwood Academic Publishers, New York (1988)

Scott, A.: Nonlinear Sciense: Emergence and Dynamics of Coherent Sructures. Oxford University Press, New York (2003)

Chapter 13
Classical Analog of Linear and Quasi-Linear Quantum Tunneling

In this part of the report, we develop an analytical framework to investigate irreversible energy transfer in a system of two *unforced* weakly coupled oscillators with slowly time-varying frequencies. In the system under consideration, one of the oscillators is initially excited by an initial impulse, while the second one is initially at rest. As shown in the previous sections, these initial conditions provide motion along the LPT with a maximum possible energy transfer from the excited oscillator to the second one.

The analysis developed in this section gives special attention to an analogy between energy transfer in a system of classical oscillators and the quantum Landau–Zener tunneling. Due to its generality, the Landau–Zener scenario has been applied to numerous problems in various contexts, such as laser physics (Sahakyan et al. 2010), semiconductor super-lattices (Rosam et al. 2003), tunneling of optical (Trompeter et al. 2006) or acoustic waves (Sanchis-Alepuz et al. 2007; de Lima et al. 2010), and quantum information processing (Saito et al. 2006), to name just a few examples. Although a passage between two energy levels is an intrinsic feature of all above-mentioned processes, manifestation of a direct analogy between energy transfer in a classical system with time-varying parameters and quantum Landau–Zener tunneling is a recent development (Manevitch et al. 2011). It was shown that the equations of the adiabatic passage through resonance in a system of two weakly coupled linear oscillators with a slowly varying frequency detuning are identical to the equations of the Landau–Zener tunneling problem. This conclusion may be treated as an extension of a previously found analogy between *adiabatic* quantum tunneling and energy exchange in weakly coupled oscillators with constant parameters demonstrated in Chap. 2.

It is well known that an analytical solution for the problem of transient tunneling is prohibitively difficult even in the linear case (Zener 1932). The purpose of this section is to derive an explicit asymptotic approximation, which depicts irreversible energy transfer on the LPT. In addition, the obtained asymptotic solution provides a simple and accurate prediction of tunneling over a finite time interval, unlike a

© Springer Nature Singapore Pte Ltd. 2018

L.I. Manevitch et al., *Nonstationary Resonant Dynamics of Oscillatory Chains and Nanostructures*, Foundations of Engineering Mechanics, DOI 10.1007/978-981-10-4666-7_13

common approach considering fixed initial conditions at $t \to -\infty$ and a stationary solution at $t \to \infty$.

An approximate solution of the quasi-linear problem is found with the help of an iteration procedure, wherein the linear solution is chosen as an initial approximation. Correctness of the constructed approximations is confirmed by numerical simulations.

13.1   Two Weakly Coupled Linear Oscillators

In this section, we study energy transport in a system of two weakly coupled linear oscillators; the first oscillator of mass m_1 and linear stiffness c_1 is excited by an initial impulse V; the coupled oscillator of mass m_2 and time-dependent linear stiffness $C_2(t) = c_2 - (k_1 - k_2 t)$ is initially at rest; the oscillators are connected by a linear coupling of stiffness c_{12}. The absolute displacements and velocities of the oscillators are denoted by u_i and $V_i = du_i/dt$, $i = 1, 2$. We will demonstrate that the second oscillator with time-dependent frequency acts as an energy sink and ensures a visible reduction of oscillations of the excited mass. The dynamics of the system is described by the equations

$$
\begin{aligned}
m_1 \frac{d^2 u_1}{dt^2} + c_1 u_1 + c_{12}(u_1 - u_2) &= 0, \\
m_2 \frac{d^2 u_2}{dt^2} + C_2(t) u_2 + c_{12}(u_2 - u_1) &= 0,
\end{aligned}
\tag{13.1}
$$

with initial conditions $u_1 = u_2 = 0$; $V_1 = V$ and $V_2 = 0$ at $t = 0$. We recall that these conditions determine the limiting phase trajectory (LPT) corresponding to motion with maximum energy transfer from the first to the excited oscillator to the second one being initially at rest.

Quasi-resonance interactions between the oscillators imply that $(c_1/m_1)^{1/2} = (c_2/m_2)^{1/2} = \omega$. Assuming weak coupling, we define the small parameter of the problem by the equality $2\varepsilon = c_{12}/c_2 \ll 1$. Then, considering the rescaled dimensionless parameters

$$
c_{12}/c_r = 2\varepsilon\lambda_r, \, r = 1, 2; \quad \lambda_2 = 1, \quad k_1/c_2 = 2\varepsilon\sigma, \quad k_2/(c_2\omega) = 2\varepsilon^2\beta^2
\tag{13.2}
$$

and introducing the dimensionless timescales $\tau_0 = \omega t$, $\tau_1 = \varepsilon\tau_0$, Eq. (13.1) are rewritten as

$$
\begin{aligned}
\frac{d^2 u_1}{d\tau_0^2} + u_1 + 2\varepsilon\lambda_1(u_1 - u_2) &= 0, \\
\frac{d^2 u_2}{d\tau_0^2} + u_2 + 2\varepsilon\lambda_2(u_2 - u_1) - 2\varepsilon\zeta(\tau_1)u_2 &= 0,
\end{aligned}
\tag{13.3}
$$

where $\zeta(\tau_1) = \sigma - 2\beta^2\tau_1$; initial conditions at $\tau_0 = 0$ are given by $u_1 = u_2 = 0$; $v_1 = V/\omega = V_0$, $v_2 = 0$, $v_i = du_i/d\tau_0$. It is important to note that system (13.3) may be considered as resonant only in a finite time interval wherein $|\zeta(\tau_1)| \sim 1$. In this case, the value of $\varepsilon\zeta(\tau_1)$ is small, and instant frequencies of the overall system remain close.

In analogy to the previous sections, approximate solutions of system (13.3) are sought with the help of the multiple scales techniques. First, we introduce the complex-valued amplitudes η and φ by formulas

$$\eta(\tau_0, \varepsilon) = (v_1 - iu_1)e^{-i\omega_\varepsilon\tau_0}, \ \varphi(\tau_0, \varepsilon) = (v_2 + iu_2)e^{-i\omega_\varepsilon\tau_0},$$

where $\omega_\varepsilon = (1 + 2\varepsilon\lambda_1)^{1/2}$. In the next step, the functions η and φ are constructed in the form of the asymptotic series

$$\begin{aligned}\eta(\tau_0, \varepsilon) &= \eta_0(\tau_1) + \varepsilon\eta_1(\tau_0, \tau_1) + \varepsilon^2\ldots \\ \varphi(\tau_0, \varepsilon) &= \varphi_0(\tau_1) + \varepsilon\varphi_1(\tau_0, \tau_1) + \varepsilon^2\ldots\end{aligned} \tag{13.4}$$

Reproducing standard arguments, we derive the following first-order equations for the slow complex amplitudes η_0 and φ_0

$$\begin{aligned}\frac{d\eta_0}{d\tau_1} &= -i\lambda_1\varphi_0(\tau_1), \eta_0(0) = V_0, \\ \frac{d\varphi_0}{d\tau_1} - i\Omega(\tau_1)\varphi_0 &= -i\lambda_2\eta_0.\end{aligned} \tag{13.5}$$

where $\Omega(\tau_1) = \rho + 2\beta^2\tau_1$, $\rho = \lambda_2 - \lambda_1 - \sigma$ (see (Kosevich et al. 2010; Kovaleva et al. 2011; Manevich et al. 2011) for more details).

Once the envelopes $\eta_0(\tau_1)$ and $\varphi_0(\tau_1)$ are found, the leading-order approximations to the solutions u_1 and u_2 are given by

$$\begin{aligned}u_{10}(\tau_0, \varepsilon) &= |\eta_0(\tau_1)| \sin(\omega_\varepsilon\tau_0 + \delta(\tau_1)), \delta(\tau_1) = \arg(\eta_0(\tau_1)), \\ u_{20}(\tau_0, \tau_1) &= |\varphi_0(\tau_1)| \sin(\omega_\varepsilon\tau_0 + \alpha(\tau_1)), \alpha(\tau_1) = \arg(\varphi_0(\tau_1)).\end{aligned} \tag{13.6}$$

Partial energy of the oscillators on the LPT is approximately expressed as

$$\begin{aligned}e_{10}(\tau_1) &= \frac{1}{2}(<u_{10}^2> + <v_{10}^2>) = \frac{1}{2}|\eta_0(\tau_1)|^2. \\ e_{20}(\tau_1) &= \frac{1}{2}(<u_{20}^2> + <v_{20}^2>) = \frac{1}{2}|\varphi_0(\tau_1)|^2,\end{aligned} \tag{13.7}$$

where $<\cdot>$ denotes the averaging over the "fast" period $T = 2\pi/\omega_\varepsilon$. Note that expressions (13.7) ignore the residual terms of $O(\varepsilon)$ associated with potential energy of weak coupling and small slow change of linear stiffness.

We obtain from (13.5) to (13.7) that for small τ_1, the following approximations are valid:

$$\eta_0(\tau_1) = V_0(1 - \frac{1}{2}\lambda_1\lambda_2\tau_1^2), \quad e_{10}(\tau_1) = \frac{1}{2}V_0^2(1 - \lambda_1\lambda_2\tau_1^2).$$
$$\varphi_0(\tau_1) \approx -i\lambda 2 V_0\tau_1, \quad e_{20}(\tau_1) \approx \frac{1}{2}(\lambda_2 V_0\tau_1)^2. \tag{13.8}$$

It now follows from (6.39) that in the initial time interval, the energy of the excited oscillator is decreasing while the energy of the second oscillator is increasing. An instant τ_1^* at which $e_{10}(\tau_1^*) = e_{20}(\tau_1^*)$ is defined by the equality $(\lambda_2 V_0\tau_1)^2 = V_0^2(1 - \lambda_1\lambda_2\tau_1^2)$,

$$\tau_1^* = \frac{1}{\sqrt{\lambda_2(\lambda_1 + \lambda_2)}} \tag{13.9}$$

It is obvious that τ_1^* decreases with an increase in coupling. This conclusion agrees with the experimental results presented in (Manevitch et al. 2011; Kosevich et al. 2010).

Finally, we note that Eq. (13.5) are equivalent to the second-order differential equation

$$\frac{d^2\varphi_0}{d\tau_1^2} - i\Omega(\tau_1)\frac{d\varphi_0}{d\tau_1} + (\lambda_1\lambda_2 - i\beta^2)\varphi_0 = 0 \tag{13.10}$$

with initial conditions $\varphi_0 = 0$, $d\varphi_0/d\tau_1 = -i\lambda_2 V_0$ at $\tau_1 = 0$. The equivalence of Eq. (13.10) to the equation of the Landau–Zener transient tunneling problem [(Landau 1932; Zener 1932) is discussed in (Kovaleva et al. 2011).

13.2 Approximate Analysis of Energy Transfer in the Linear System

It is well known that an analytical solution for the Landau–Zener problem of transient tunneling dynamics is prohibitively complicated even in the linear case (Zener 1932). However, it follows from (13.10) that asymptotic solutions can be greatly simplified if $2\beta^2 \gg \lambda_1\lambda_2$ (Kovaleva et al. 2011).

For brevity, we let $\lambda_1 = \lambda_2 = \lambda$. Under these assumptions, Eq. (13.10) is approximated as

$$\frac{d^2\tilde{\varphi}_0}{d\tau_1^2} - i\Omega_0(\tau_1)\frac{d\tilde{\varphi}_0}{d\tau_1} - i\beta^2\tilde{\varphi}_0 = 0, \tag{13.11}$$

where $\Omega_0(\tau_1) = (\rho_0 + 2\beta^2\tau_1)$, $\rho_0 = -\sigma$. We thus have

$$\tilde{\varphi}_0(\tau_1) = -i\lambda V_0 I(\tau_1), I(\tau_1) = \frac{1}{\beta}e^{iB(\tau_1)}F(\tau_1, \theta_0)e^{i\theta_0^2}, \qquad (13.12)$$

where $B(s) = \rho_0 s + (\beta s)^2 = (\beta s + \theta_0)^2$, $\theta_0 = -\sigma/2\beta$, and

$$F(\tau_1, \theta_0) = \int\limits_{\theta_0}^{\beta\tau_1 + \theta_0} e^{-ih^2} dh = [C(\beta\tau_1 + \theta_0) - C(\theta_0)]$$
$$- i[S(\beta\tau_1 + \theta_0) - S(\theta_0)].$$

$C(x)$ and $S(x)$ are the cos- and sin-Fresnel integrals. Once the envelope $\tilde{\varphi}_0(\tau_1)$ is found, the envelope $\tilde{\eta}_0(\tau_1)$ is expressed as

$$\tilde{\eta}_0(\tau_1) = V_0 - i\lambda \int\limits_0^{\tau_1} \tilde{\varphi}_0(s)ds. \qquad (13.13)$$

An approximate solution of system (13.3) is now given by

$$\tilde{u}_1(\tau_0, \varepsilon) = |\tilde{\eta}_0(\tau_1)| \sin(\omega_\varepsilon\tau_0 + \delta(\tau_1)), \delta(\tau_1) = \arg(\tilde{\eta}_0(\tau_1)),$$
$$\tilde{u}_2(\tau_0, \varepsilon) = |\tilde{\varphi}_0(\tau_1)| \sin(\omega_\varepsilon\tau_0 + \tilde{\alpha}(\tau_1)), \tilde{\alpha}(\tau_1) = \arg(\tilde{\varphi}_0(\tau_1)) \qquad (13.14)$$

We evaluate the amplitudes of oscillations in two limiting cases:

1. If $\beta\tau_1 \ll \sqrt{2}$, it follows from (13.15) to (13.18) that

$$|\varphi_0(\tau_1)| \approx \lambda V_0\tau_1 \qquad (13.15)$$

2. If $\beta\tau_1 \ll \sqrt{2}$, then it follows from the properties of the Fresnel integrals (Gradshtein and Ryzhik 2000) that

$$\tilde{\varphi}_0(\tau_1) \to \bar{\varphi}_0 = -i\frac{\lambda V_0}{\beta}\left\{\left[\sqrt{\frac{\pi}{8}} - C(\theta_0)\right] - i\left[\sqrt{\frac{\pi}{8}} - S(\theta_0)\right]\right\}$$
$$\|\bar{\varphi}_0\| = \frac{\lambda V_0}{\beta}\left\{\left[\sqrt{\frac{\pi}{8}} - C(\theta_0)\right]^2 + \left[\sqrt{\frac{\pi}{8}} - S(\theta_0)\right]^2\right\}^{1/2}, \quad \text{as } \tau \to \infty. \qquad (13.16)$$

Expressions (6.47) imply that at large times, the second oscillator (the energy sink) exhibits quasi-stationary oscillations with constant amplitude $|\bar{\varphi}_0|$ and energy $\bar{e}_{20} = 1/2|\bar{\varphi}_0|^2$. This illustrates almost irreversible energy transfer from the initially excited oscillator to the sink being initially at rest.

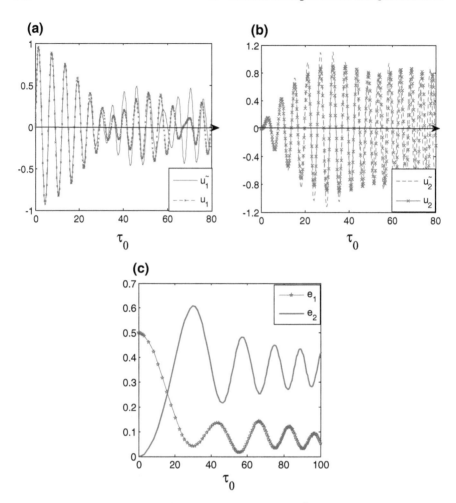

Fig. 13.1 Exact (numerical) and approximate solutions $u_{1,2}$ and $\tilde{u}_{1,2}$, respectively [plots (**a**), (**b**)]; energy of the oscillator e_1 and the sink e_2 [plot (**c**)]

We now compare exact (numerical) solutions of the initial systems (13.3) with approximations (13.14). The parameters of the system are given by

$$\varepsilon = 0.05; \quad V_0 = 1; \quad \varepsilon\sigma = 0.1125; \quad \varepsilon\beta = 0.05; \quad \lambda = 1.$$

In Fig. 13.1, one can observe close proximity of the exact solutions and their approximations for each oscillator separately (Fig. 3.51a, b). Almost irreversible energy transfer manifested as a decrease in energy of the first oscillator with a simultaneously growing energy of the second oscillator is shown in Fig. 13.1c.

13.3 Classical Analog of Quasi-Linear Quantum Tunneling

In this section, we expose asymptotic equivalence between the slow passage through resonance in two weakly coupled quasi-linear oscillators and quasi-linear Landau–Zener tunneling, thereby extending the previously found mathematical analogy to quasi-linear systems.

The system considered consists of two weakly coupled oscillators. A linear oscillator with constant parameters is excited by an initial impulse, while a coupled nonlinear oscillator with time-dependent stiffness is initially at rest. We will demonstrate that the nonlinear oscillator acts as an energy sink and ensures a visible reduction of the amplitude of oscillations of the excited mass.

The nonlinear equations of motion are similar to (13.1) but include an additional nonlinearity

$$
\begin{aligned}
m_1 \frac{d^2 u_1}{dt^2} + c_1 u_1 + c_{12}(u_1 - u_2) &= 0, \\
m_2 \frac{d^2 u_2}{dt^2} + C_2(t) u_2 + c_{12}(u_2 - u_1) + k_3 u_2^3 &= 0,
\end{aligned}
\tag{13.17}
$$

Coefficients in Eq. (6.48) coincide with coefficients of Eq. (13.1); k_3 denotes nonlinearity of the sink.

As in Sect. 13.1, we define the small parameter $2\varepsilon = c_{12}/c_2 \ll 1$ and coefficients (6.33) together with dimensionless nonlinearity $k_3/c_2 = 8\varepsilon\alpha$. Then, introducing the fast and slow timescales $\tau_0 = \omega t$ and $\tau_1 = \varepsilon\tau_0$, the original system (6.48) is rewritten as

$$
\begin{aligned}
\frac{d^2 u_1}{d\tau_0^2} + u_1 + 2\varepsilon\lambda_1(u_1 - u_2) &= 0, \\
\frac{d^2 u_2}{d\tau_0^2} + u_2 + 8\varepsilon\alpha u_2^3 + 2\varepsilon\lambda_2(u_2 - u_1) - 2\varepsilon\zeta(\tau_1)u_2 &= 0,
\end{aligned}
\tag{13.18}
$$

with detuning $\zeta(\tau_1) = \sigma - 2\beta^2\tau_1$ and initial conditions $u_1 = u_2 = 0$; $v_1 = V_0, v_2 = 0$. We recall that these initial conditions determine the LPT of system (13.18) and provide maximum possible energy transfer from the excited nonlinear oscillator to the linear sink being initially at rest.

As in the linear case, solutions of system (13.18) are sought by means of the change of variables $\eta(\tau_0,\varepsilon) = (v_1 + iu_1)e^{-i\omega_s\tau_0}$, $\varphi(\tau_0,\varepsilon) = (v_2 + iu_2)e^{-i\omega_s\tau_0}$, where the complex amplitudes η and φ are constructed in the form of the multiple scales expansions (13.4) with slowly varying main terms $\eta_0(\tau_1)$, $\varphi_0(\tau_1)$. The series of obvious transformations (Kovaleva and Manevitch 2012) yield the following equations for the slow envelopes $\eta_0(\tau_1)$ and $\varphi_0(\tau_1)$:

$$\frac{d\eta_0}{d\tau_1} = -i\lambda_1\varphi_0(\tau_1), \eta_0(0) = V_0,$$

$$\frac{d\varphi_0}{d\tau_1} = i\Omega(\tau_1)\varphi_0 - i\lambda_2\eta_0 + 3i\alpha\varphi_0|\varphi_0|^2, \varphi_0(0) = 0. \tag{13.19}$$

where $\Omega(\tau_1) = \rho + 2\beta^2\tau_1$ and $\rho = \lambda_2 - \lambda_1 - \sigma$. Once the solutions $\varphi_0(\tau_1)$ and $\eta_0(\tau_1)$ are found, the leading-order approximations u_{10}, u_{20} to the solutions u_1, u_2 are calculated by (6.37).

Equation (6.50), which are central to our investigation, are equivalent to the nonlinear Landau–Zener equations and have a broad scope of physical applications, from the study of Bose–Einstein condensate in a time-varying double-well trap to optics, laser physics, etc., (see, e.g., Khomeriki 2011; Ishkhanyan et al. 2009; Itin and Watanabe 2007; Liu et al. 2002; Nakamura 2002; Sahakyan et al. 2010; Trimborn et al. 2010, and references therein). Equivalence of the mathematical description implies that a classical system of two weakly coupled oscillators may be treated as an adequate model of a wide variety of complicated physical processes.

For further analysis, it is convenient to rewrite (13.19) in the form

$$\frac{d\eta_0}{d\tau_1} = -i\lambda_1\varphi_0(\tau_1), \eta_0(0) = V_0,$$

$$\frac{d\varphi_0}{d\tau_1} - i\Omega(\tau_1)\varphi_0 - 3i\alpha\varphi_0|\varphi_0|^2 = -i\lambda_2 V_0 - \lambda_1\lambda_2 \int_0^{\tau_1} \varphi_0(r)dr, \tag{13.20}$$

$$\varphi_0(0) = 0$$

The solution of the coupled problem (13.20) can be significantly simplified if the integral term in the second equation is negligible in the main approximation. Using the arguments of Sect. 13.2, one can show that this assumption holds under the above-mentioned condition $\lambda_1\lambda_2 \ll 2\beta^2$. Using this condition, we develop a relevant iteration procedure for a quasi-linear system [see (Kovaleva and Manevitch 2012) for details and discussion].

For brevity, we consider a symmetric system with $m_1 = m_2 = m$, $c_1 = c_2 = c$ and weak coupling $c_{12}/c = 2\varepsilon\lambda_0$. The initial iteration $\eta_0^{(0)}$, $\varphi_0^{(0)}$ is defined as a solution of the linear equations:

$$\frac{d\eta_0^{(0)}}{d\tau_1} = -i\lambda_0\varphi_0^{(0)}, \eta_0^{(0)}(0) = V_0,$$

$$\frac{d\varphi_0^{(0)}}{d\tau_1} - i\Omega_0\varphi_0^{(0)} = -i\lambda_0 V_0, \varphi_0^{(0)}(0) = 0, \tag{13.21}$$

where $\Omega_0(\tau_1) = -\sigma + 2\beta^2\tau_1$. The initial linear iteration determines the shape of the solution, while successive iterations improve the accuracy of approximations.

The first iteration $\varphi_0^{(1)}$ satisfies the linearized equation that approximately considers the effect of weak nonlinearity:

$$\frac{d\varphi_0^{(1)}}{d\tau_1} + i\Omega_0(\tau_1)\varphi_0^{(1)} - 3i\varepsilon\alpha|\varphi_0^{(0)}|\varphi_0^{(1)} = -i\lambda_0 V_0, \quad \varphi_0^{(1)} = 0. \tag{13.22}$$

It is easy to verify that the initial iteration (13.21) approximately depicts the slow envelopes of the linear system (13.3), while the first iteration (13.22) corresponds to the truncated system similar to (13.23). In both cases, we ignore the effect of slow changes of the envelope $\eta_0(\tau_1)$ on the evolution of the envelope $\varphi_0(\tau_1)$. The improved iteration $\varphi_0^{(2)}$ takes into account both nonlinearity and slow changes of the envelope $\eta_0(\tau_1)$:

$$\frac{d\eta_0^{(2)}}{d\tau_1} = -i\lambda_0\varphi^{(0)}(\tau_1), \quad \eta_0(0) = V_0,$$

$$\frac{d\varphi_0^{(2)}}{d\tau_1} + i\Omega_0(\tau_1)\varphi_0^{(2)} - 3\alpha|\varphi_0^{(0)}|^2\varphi_0^{(2)} \tag{13.23}$$

$$= -i\lambda_0 V_0 - \lambda_0^2 \int_0^{\tau_1} \varphi_0^{(0)}(s)ds, \quad \varphi_0^{(2)}(0) = 0.$$

Note that the solution of the generic problem (13.21) is expressed through the Fresnel integral; linear Eqs. (13.22) and (13.23) also can be solved analytically (Kovaleva and Manevitch 2012). In this section, we compare the exact (numerical) solution of (13.18) with parameters

$$\varepsilon = 0.05, \quad \lambda_0 = 1, \quad V_0 = 1, \quad \sigma = 2.25, \quad 2\beta^2 = 2.25; \quad \alpha = 0.25.$$

with approximations calculated by formulas (13.16), in which the amplitudes $|\eta_0|$, $|\varphi_0|$ are changed to their iterations $\left|\eta_0^{(j)}\right|$, $\left|\varphi_0^{(j)}\right|$, $j = 0, 1$. A simple calculation proves that $\lambda_0^2 < 2\beta^2$, $k = 3\alpha/4\lambda_0 = 0.1875$. Therefore, the hypotheses of weak coupling and weak nonlinearity hold.

Figure 13.2 depicts the energy of the oscillator (1) and the trap (2) in the systems with coefficients $k = 0.1875$. Energy of the nonlinear system is calculated by formulas $e_1 = 1/2|\eta_0|^2$ and $e_2 = 1/2|\varphi_0|^2$ with η_0 and φ_0 satisfying Landau–Zener Eq. (13.19) rescaled to the fast timescale τ_0; approximate values are calculated by formulas $e_1^{(0)} = 1/2|\eta_0^{(0)}|^2$ and $e_2^{(0)} = 1/2|\varphi_0^{(0)}|^2$. It is easy to conclude that an increase in nonlinearity renders a difference between the nonlinear and linear dynamics more pronounced.

Finally, we note that Eq. (13.2), as well as approximations (13.21)–(13.23) are valid for both classical and quantum problems. The equivalence of the mathematical descriptions implies that a classical system of weakly coupled oscillators

Fig. 13.2 Energy of the
excited oscillator (*1*) and the
trap (*2*) for the nonlinear
system with parameter
$k = 0.1875$ (*solid lines*);
linear approximations are
depicted with *dashed lines*

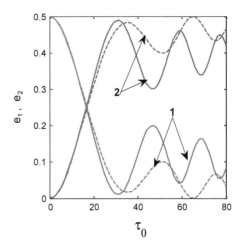

may serve as a simple but adequate model of complicated physical processes. Moreover, the mathematical equivalence, in principle, enables the substitution of mechanical modeling for complicated and costly quantum experiments (Kosevich et al. 2010; Manevitch et al. 2011).

13.4   Moderately and Strongly Nonlinear Adiabatic Tunneling

This section explores we examine irreversible energy transfer along the LPTs for moderately and strongly nonlinear regimes in a system with slowly time-varying parameters. We demonstrate the equivalence of the equations for the slow passage through resonance in the classical system and the equations of nonlinear LZ tunneling. It is noted here that the well-known LZ tunneling is typical for moderately nonlinear regimes, while the system with strongly nonlinear behavior can exhibit both the transition from the energy localization to intense energy exchange (with a large inductive period) as well as the rapid passage through the separatrix. It is important to underline that the revealed mathematical equivalence suggests a unified approach to the study of such physically different processes as energy transfer in classical oscillatory systems under slow driving and nonlinear quantum LZ tunneling.

13.4.1   Moderately Nonlinear Regimes

We consider adiabatic tunneling in a system of two coupled oscillators similar to (13.17), namely

$$\frac{d^2 u_1}{d\tau_0^2} + u_1 + 2\varepsilon(u_1 - u2) + 8\varepsilon\alpha u_1^3 = 0,$$

$$\frac{d^2 u_2}{d\tau_0^2} + (1 + 2\varepsilon g(\tau_2))u_2 + 2\varepsilon(u_2 - u_1) + 8\varepsilon\alpha u_2^3 = 0,$$ (13.24)

where $g(\tau_2) = g_0 + g_1\tau_2$, $\tau_2 = \varepsilon^2\tau_0$, $\tau_1 = \varepsilon\tau_0$. It is well known that energy transfer between coupled oscillators with adiabatically changed parameters is divided into two stages, with each characterizing by adiabatic invariance but separated by an abrupt jump at a moment of tunneling. Breaking of adiabaticity under slow driving has been intensively studied over last decades using various approximations [see, e.g., (Itin and Törmä 2009, 2010; Itin and Watanabe 2007; Khomeriki 2011; Sahakyan et al. 2010)]. We investigate slow transient processes before and after tunneling. The asymptotic analysis accounts for the fact that in the first interval of motion most part of energy is localized on the excited oscillator while the residual energy of the coupled oscillator is small enough. After tunneling, localization of energy takes place on the coupled oscillator, but energy of the initially excited oscillator becomes small. An introduction of the small parameter characterizing a relative energy level allows an explicit asymptotic solution.

As in Sect. 13.1, we introduce the change of variables $v_j + iu_j = Y_j e^{i\tau_0}$, $j = 0,1$, and then obtain that $Y_j = \varphi_j^{(0)} + \varepsilon\varphi_j^{(0)} + O(\varepsilon^2)$, where $\varphi_1^{(0)} = ae^{i\tau_1}$, $\varphi_2^{(0)} = be^{i\tau_1}$. The equations for the slow complex envelopes a and b are given by

$$\frac{da}{d\tau_1} + ib - 3i\alpha|a|^2 a = 0,$$

$$\frac{db}{d\tau_1} + ia - 3i\alpha|b|^2 b - 2ig(\tau_2)b = 0,$$ (13.25)

with initial conditions $a(0) = 1$, $b(0) = 0$ corresponding to the initial unit impulse applied to the first oscillators. It is easy to prove that the conservation law $|a|^2 + |b|^2 = 1$ holds true for system (13.25) despite its non-stationarity. We recall that given initial conditions correspond to the LPT of the system, thus providing maximum irreversible energy transfer from the excited oscillator to the attached one. Also, we note that Eq. (13.25) are identical to the nonlinear analog of the LZ equations of quantum tunneling (see, e.g., [94, 191, 209, 213] for further details).

The change of variables $a = \cos\theta\, e^{i\delta_1}$, $b = \sin\theta\, e^{i\delta_2}$, $\Delta = \delta_1 - \delta_1$ reduces (13.25) to the form

$$\frac{d\theta}{d\tau_1} = \sin\Delta,$$

$$\sin 2\theta \frac{d\Delta}{d\tau_1} = 2(\cos\Delta + 2k\sin 2\theta)\cos 2\theta - 2g(\tau_2)\sin 2\theta,$$ (13.26)

with initial conditions $\theta(0) = 0$, $\Delta(0) = \pi/2$. Figure 13.3 depicts the plots of the energy $2e_1 = |a|^2$, $2e_2 = |b|^2$ in the interval $0 \leq \tau_1 \leq 1000$ for system (13.26) with parameters $k = 0.65$, $g_0 = -0.5$, and $\varepsilon g_1 = 0.001$.

It is observed that the dynamics of each oscillator is close to motion along the LPT of the time-invariant system in the initial interval of motion, whereas the change of detuning is negligible (Fig. 13.3a), but afterward the change of energy becomes evident and the occurrence of tunneling at $T^* \approx 585$ becomes evident (Fig. 13.3b). The value of detuning $g = g_0 + \varepsilon g_1 T^* = 0.085$ at the instant of tunneling is close to critical detuning $g^* \approx 0.083$ calculated in Sect. 1.1 for a time-invariant system. Figure 13.3c represents the phase portrait in the plane $(\theta, V = \sin\Delta)$, which also demonstrates an instant jump between energy levels.

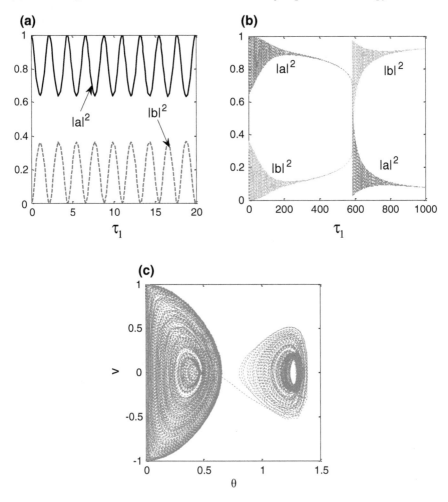

Fig. 13.3 Energy of the oscillators: **a** motion along the unperturbed LPT with maximum energy exchange between the oscillators in the initial interval of motion; **b** an instant jump between the energy levels; **c** the phase portrait of system (13.26)

Now, we briefly analyze the dynamical behavior in the interval $S_1 : 0 \leq \tau_1 < T^*$ before tunneling. We employ the change of variables

$$\Psi = |b|e^{i\Delta} = |\sin\theta|e^{i\Delta} \qquad (13.27)$$

which reduces (6.57) to a complex-valued equation

$$\frac{d\Psi}{d\tau_1} = 4i\left[\frac{1 - 1/2(\Psi^2 + 2|\Psi|^2)}{\sqrt{1 - |\Psi|^2}} + 2\omega(\tau_2)\Psi - 8k\Psi|\Psi|^2\right], \quad \Psi(0) = 0, \quad (13.28)$$

where $\omega(\tau_2) = 2k - g(\tau_2) = \omega_0 - g_1\tau_2$, $\omega_0 = 2k - g_0$. It is important to note that the frequency $\omega(\tau_2)$ directly depends on the coefficient k, thereby reflecting the effect of nonlinearity even in the linear approximation of Eq. (13.28).

Since $|b| = |\Psi| \ll 1$ in S_1, we introduce the rescaled variable $\psi = \varepsilon^{-1/2}\Psi$ and the new timescale $s = \varepsilon^{-1/2}\tau_1$, as well as rescaled coefficients $8k = \varepsilon^{-1/2}\kappa$, $2\omega_0 = \varepsilon^{-1/2}w_0$, where κ and w_0 are of $O(1)$. Finally, we denote $\varepsilon^{3/2}g_1 = \beta^2/4$ (keeping in mind that $\beta \ll 1$). When we substitute the rescaled variables and coefficients into (13.28) and ignore the terms of orders higher than ε, we obtain the equation

$$\frac{d\psi}{ds} = 4i\left[w(s)\psi + 1 - \frac{1}{2}\varepsilon(\psi^2 + |\psi|^2) - \kappa\varepsilon|\psi|^2\psi\right], \quad \psi(0) = 0, \qquad (13.29)$$

where $w(s) = w_0 - \beta^2 s/2$. Thus, we get a quasi-linear equation, and the earlier developed iteration procedure can be employed to construct an approximate solution. The initial iteration $\psi_0(s)$ is chosen as a solution of the linear equation

$$\frac{d\psi_0}{ds} = 4i[w(s)\psi_0 + 1], \quad \psi_0(0) = 0,$$
$$\psi_0(s) = 4i\beta^{-1}e^{i(\phi_0(s)-\alpha^2)}\Phi_0(s), \qquad (13.30)$$

where $\phi_0(s) = w_0s - (\beta s)^2, \alpha = w_0/\beta$, and

$$\Phi_0(s) = \int_{-\alpha}^{\beta s-\alpha} e^{ih^2}\,dh = [C(\beta s - \alpha) + C(\alpha)] + i[S(\beta s - \alpha) + S(\alpha)],$$

with $C(h)$ and $S(h)$ being the cos- and sin-Fresnel integrals, respectively. Once ψ_0 is found, the main approximation to the function $|b|$ is calculated as $|b_0| = \varepsilon^{1/2}|\psi_0|$. Note that the parameter w_0 depends on the coefficient of nonlinearity k, and thus, the behavior of the solution $\psi_0(s)$ is conditioned by nonlinearity.

In the next step, the first iteration ψ_1 is found from the linearized equation

$$
\frac{d\psi_1}{ds} = 4i\left\{ [w(s) - \varepsilon\kappa|\psi_0(s)|^2]\psi_1 - \varepsilon\left(\frac{1}{2}\psi_0^2(s) + |\psi_0(s)|^2\right) + 1\right\},
$$

$$
\psi_1(0) = 0,
$$

$$
\psi_1(s) = \psi_0(s) + 4ie^{i\phi_1(s)}\Phi_1(s),
$$

(13.31)

where

$$
\phi_1(s) = \phi_0(s) - \varepsilon\kappa\int_0^s |\psi_0(z)|^2 dz,
$$

$$
\Phi_1(s) = \int_0^s e^{-i\phi_1(z)} dz - \varepsilon\int_0^s e^{-i\phi_1(z)} R(z) dz,
$$

$$
R(z) = \frac{1}{2}\psi_0^2(z) + |\psi_0(z)|^2.
$$

Once the solution ψ_1 is derived, the first iteration to the function $|b|$ is calculated as $|b_1| = \varepsilon^{1/2}|\psi_1|$. The exact solution $|b(\tau_1)|^2$ and the iterations $|b_0(\tau_1)|^2$ and $|b_1(\tau_1)|^2$ are presented in Fig. 13.4a. Figure 13.4a clearly indicates that in the first half of the interval S_1 the maximal divergence between the exact solution and its linear approximation is less than 15%. The increased divergence in the second part of S_1 is due to the different behavior of the exact and approximate solutions near the point of transition, while the exact solution demonstrates a sudden increase at an instant of tunneling, the linear approximation tends to a certain limit.

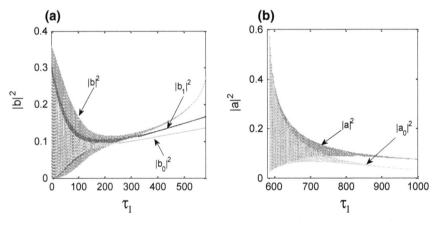

Fig. 13.4 Exact and approximate solutions before and after tunneling: **a** the exact solution $|b|^2$ and the iterations $|b_0|^2$ and $|b_1|^2$ in the interval S_1; **b** the exact solution $|a|^2$ and the iteration $|a_0|^2$ in the interval S_2

The dynamical behavior in the interval $S_2 : \tau_1 > T$ after tunneling is studied in a similar way. Given $|a| \ll 1$, the variable analogous to (13.27) is introduced

$$Z = -|a|e^{i\Delta} = -|\cos\theta|e^{i\Delta} \tag{13.32}$$

Transformations of the same sort that led to Eq. (13.29) yield the following equation:

$$\frac{dZ}{d\tau_1} = 4i\left[2\omega_1(\tau_2)Z - 1 + \frac{1}{2}(Z^2 + |Z|^2) - 8\kappa Z|Z|^2\right],$$

$$Z(T) = p_0, \tag{13.33}$$

where $\omega_1(\tau_2) = 2k + g(\tau_2) = \omega_{10} + g_1\tau_2$, $\omega_{10} = 2k + g_0$. As shown in Sect. 3.1, initial condition for Eq. (13.33) at $\tau_1 = T$ should be defined from the condition of the coalescence of the stable and unstable states. This means that the quantity $\theta(T) = \theta_T$ determines the initial condition $|a(T)| = |\cos\theta_T| = p_0$.

Given $|Z| \ll 1$ in S_2, we introduce the following transformations of the variables and the parameters: $Z = \varepsilon^{1/2}z$, $\tau_1 = \varepsilon^{1/2}s$, $8k = \varepsilon^{-1/2}\kappa$, $2\omega_{10} = \varepsilon^{-1/2}w_{10}$. Then, we denote $\varepsilon^{3/2}g_1 = \beta^2/4$, $\varepsilon^{-1/2}T = s_0$, $\varepsilon^{1/2}p_0 = p_{10}$. Substituting the rescaled quantities into (13.33) and ignoring the higher-order terms, we obtain the quasi-linear equation similar to (13.29)

$$\frac{dz}{ds} = 4i\left[w_1(s)z - 1 + \frac{1}{2}\varepsilon(z^2 + 2|z|^2) - \varepsilon\kappa z|z|^2\right],$$

$$z(s_0) = p_{10}, \tag{13.34}$$

with the adiabatically increasing parameter $w_1(s) = w_{10} + \beta^2 s/2$. The initial iteration $z_0(s)$ is given by the following expressions

$$\frac{dz_0}{ds} = 4i[w_1(s)z_0 - 1], \quad z_0(s_0) = p_{10},$$

$$z_0(s) = (p_1 - 4i\Theta(s))e^{i\delta(s)}, \tag{13.35}$$

where $\delta(s) = 4[w_{10}s + (\beta s)^2]$, $\Theta(s) = e^{i\alpha_1^2}F_1(s)/2\beta$, and

$$F_1(s) = \int_{\alpha_1}^{2\beta s + \alpha_1} e^{-ih^2}\,dh = [C(2\beta s + \alpha_1) - C(\alpha_1)]$$

$$- i[S(2\beta s + \alpha_1) - S(\alpha_1)].$$

Applying the inverse rescaling, we get the initial iteration $|a_0| = \varepsilon^{1/2}|z_0|$. As shown in Fig. 13.4(b), the plot of $|a_0(\tau_1)|^2$ is close to $|a(\tau_1)|^2$. Finally, calculating the stationary state $|\bar{a}_0| = \lim|a_0(\tau_1)|$ as $\tau_1 \to \infty$, we obtain

$$|\bar{a}_0| = \varepsilon^{1/2}\bar{z}_0 = |p_0 + 1/\omega_{10}|. \tag{13.36}$$

The resulting non-zero value of $|\bar{a}_0|$ implies the existence of the residual energy depending on the initial condition p_0. Given $k = 0.65$, $g_0 = -0.5$, we obtain $|\bar{a}_0| = 0.445$. It follows from (13.36) that in the interval $S_2 : \tau_1 > T$ the quantity e_2 tends to the limiting value $\bar{e}_2 = |\bar{b}_0|^2/2 = (1 - |\bar{a}_0|^2)/2 = 0.401$ as $\tau_1 \to \infty$. The quantity $\bar{e}_2$ characterizes the amount of energy transferred from the excited oscillator with initial energy $e_{10} = 1/2$ to the coupled oscillator being initially at rest.

13.5   Strongly Nonlinear Regimes

The analysis of strongly nonlinear dynamics of a time-invariant system demonstrates that the change of detuning g may entail a transition from weak to strong energy exchange. In this section, we show that this conclusion remains valid for adiabatic strongly nonlinear tunneling.

Figure 13.5 demonstrates the phase portrait and the transient evolution of the angle θ for system (13.36) with parameters $k = 0.9$, $g_0 = -0.25$; $\varepsilon g_1 = 0.001$.

The phenomenon of energy localization with relatively small oscillations around the slowly varying steady state is observed in the initial interval $S_1 : 0 \leq \tau_1 < T^*$; then, in the interval $S_2 : \tau_1 > T^*$, energy localization changes to intense exchange. Note that tunneling occurs at $T^* = 580$ that corresponds to $g = 0.33$.

The adiabatic convergence to the transition point at the first stage of motion can be examined in the same way as in Sect. 13.5. In the current section, we briefly analyze the dynamical behavior in the interval S_2, where the variations of $\theta(\tau_1)$ are large enough, and the asymptotic approach of Sect. 3.1 is inapplicable. We note that the local minima $\theta^-(\tau_1)$ and maxima $\theta^+(\tau_1)$ of $\theta(\tau_1)$ lie on the slowly varying envelops $Q^-(\tau_2)$ and $Q^+(\tau_2)$, respectively, (Fig. 13.5c). Given that at the initial moment the quantities $|Q^-|$ and $|\pi/2 - Q^+|$ are small enough (Fig. 13.5c), the initial approximation can be chosen as $\theta_0^- = Q_0^- = 0$, $\theta_0^+ = Q_0^+ = \pi/2$. This yields the first approximation

$$Q_1^{\pm}(\tau_1) = \theta_0^{\pm} - \varepsilon g_1(\tau_1 - T). \tag{13.37}$$

[see (Manevitch and Kovaleva 2013) for more details]. Figure 13.5c demonstrates a good agreement of approximations (13.37) with the precise (numerical) solution in the interval S_2.

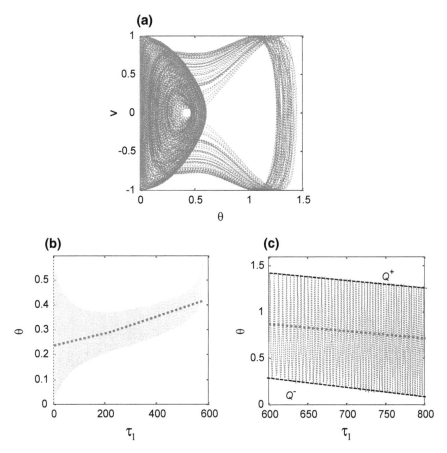

Fig. 13.5 Transient processes in the interval $0 \leq 1 \leq 900$: **a** the phase portrait of the system; **b** the plot of θ (τ_1) before tunneling; **c** the plot of θ (τ_1) after tunneling; *dotted lines* in plots **b**, **c** depict the backbone curves

References

Gradshteyn, I.S., Ryzhik, I.M.: Tables of Integrals, Series, and Products, 6th edn. Academic Press, San Diego, CA (2000)

Ishkhanyan, A., Joulakian B., Suominen, K.-A.: Variational ansatz for the nonlinear Landau–Zener problem for cold atom association. J. Phys. B. **42**, 221002–221006 (2009)

Itin, A.P., Törmä, P.: Dynamics of a many-particle Landau-Zener model: inverse sweep. Phys. Rev. A. **79**, 055602(1–3) (2009)

Itin, A.P., Törmä, P.: Dynamics of quantum phase transitions in Dicke and Lipkin-Meshkov-Glick models, arXiv:0901.4778v1 (2010)

Itin, A.P., Watanabe, S.: Universality in nonadiabatic behavior of classical actions in nonlinear models with separatrix crossings. Phys. Rev. E. **76**, 026218(1–16) (2007)

Khomeriki, R.: Multiple Landau-Zener tunnelling in two weakly coupled waveguide arrays. Euro. Phys. J. D: Atom. Mole. Opt. Phys **61**, 193–197 (2011)

Kosevich, YuA, Manevitch, L.I., Manevitch, E.L.: Vibrational analogue of nonadiabatic Landau-Zener tunneling and a possibility for the creation of a new type of energy traps. Phys. Uspekhi **53**, 1281–1287 (2010)

Kovaleva, A., Manevitch L.I.: Classical analog of quasilinear Landau-Zener tunneling. Phys. Rev. E. **85**, 016202(1–8) (2012)

Kovaleva, A., Manevitch, L.I., Kosevich, Y.A.: Fresnel integrals and irreversible energy transfer in an oscillatory system with time-dependent parameters. Phys. Rev. E **83**, 026602(1–12) (2011)

Landau, L.D.: Zur Theorie der Energieubertragung. II, Phys. Z. Sowjetunion **2**, 46–51 (1932)

de Lima Jr. M.M., Kosevich, Y.A., Santos, P.V., Cantarero, A.: Surface acoustic Bloch oscillations, the Wannier-Stark ladder, and Landau-Zener tunneling in a solid. Phys. Rev. Lett. **104**, 165502(1–4) (2010)

Liu, J., Fu, L., Ou, B.-Y., Chen, S.-G., Choi, D.-I., Wu, B., Niu, Q.: Theory of nonlinear Landau-Zener tunneling. Phys. Rev. A **66**, 023404(1–7) (2002)

Manevitch, L.I., Kovaleva, A.: Nonlinear energy transfer in classical and quantum systems. Phys. Rev. E **87**, 022904(1–12) (2013)

Manevitch, L.I., Kosevich, Y.A., Mane, M., Sigalov, G., Bergman, L.A., Vakakis, A.F.: Towards a new type of energy trap: classical analog of quantum Landau-Zener tunneling. Int. J. Nonlinear Mech **46**, 247–252 (2011)

Nakamura, H.: Nonadiabatic Transitions: Concepts, Basic Theories and Applications. World Scientific, Singapore (2002)

Rosam, B., Leo, K., Glück, M., Keck, F., Korsch, H.J., Zimmer, F., Köhler, K.: Lifetime of Wannier-Stark states in semiconductor superlattices under strong Zener tunneling to above-barrier bands. Phys. Rev. B **68**, 125301(1–7) (2003)

Sahakyan, N., Azizbekyan, H., Ishkhanyan, H., Sokhoyan, R., Ishkhanyan, A.: Weak coupling regime of the Landau-Zener transition for association of an atomic Bose-Einstein condensate. Laser Phys **20**, 291–297 (2010)

Saito, K., Wubs, M., Kohler, S., Hanggi, P., Kayanuma, Y.: Quantum state preparation in circuit QED via Landau-Zener tunneling. Europhys. Lett **76**, 22–28 (2006)

Sanchis-Alepuz, H., Kosevich, Y.A., Sanchez-Dehesa, J.: Acoustic analogue of Bloch oscillations and Zener tunneling in ultrasonic superlattices. Phys. Rev. Lett. **98**, 134301(1–4) (2007)

Trimborn, F., Witthaut, D., Kegel, V., Korsch, H.J.: Nonlinear Landau–Zener tunneling in quantum phase space. New J. Phys. **12**, 05310(1–20) (2010)

Trompeter, H., Pertsch, T., Lederer, F., Michaelis, D., Streppel, U., Bräuer, A., Peschel, U.: Visual observation of Zener tunneling. Phys. Rev. Lett. **96**, 023901(1–4) (2006)

Zener, C.: Non-adiabatic crossing of energy levels. Proc. R. Soc. London A **137**, 696–702 (1932)

Chapter 14
Strongly Nonlinear Lattices

14.1 The Large-Amplitude Oscillations in the Discrete Finite Frenkel–Kontorova Model

The majority of physical, mechanical and engineering problems dealing with nonlinear processes cannot be solved by the direct methods. The main goal of the researchers is the construction of a simplified model, which can be solved by an approximate method without loosing the physical content. The asymptotic methods, in particular, the averaging ones, based on using a small parameter, are the well-developed tools, which are effectively applied in the nonlinear dynamics (Bogolubov and Mitropolski 1961; Kuehn 2015; Sanders and Verhulst 2007). Moreover, there is a popular jest that a nonlinear problem is unsolvable, if a small parameter cannot be found. In this context, we would like to separate the problems with an external small parameter (ESP) connected with a small perturbation of the solvable system, and the problems, the small parameter of which is an internal one (Internal Small Parameter—ISP). Besides, in the latter, an ISP can be a priori unclear (i.e., ISP is hidden). The most evident example of the system with ESP can be found in the celestial mechanics. It is a small perturbation of the stationary orbit of a planet by a remote heavenly body.

The example of the system with the ISP can be found in the theory of nonlinear oscillations. Let us consider two weakly coupled pendula. The small parameter is determined by the coupling rigidity, and it is the internal parameter because the stationary dynamics of the system has to be considered in the terms of normal modes, but not of the pendula. The small coupling has to be accounted in the main asymptotic approximation. The resonant interaction of the nonlinear normal modes (NNMs) of the well-known Fermi–Pasta–Ulam nonlinear lattice gives an example of the hidden ISP (Manevitch and Smirnov 2010a, b). Really, considering the slow evolution of the interacting NNMs, we have to remove the fast motion and to

© Springer Nature Singapore Pte Ltd. 2018

L.I. Manevitch et al., *Nonstationary Resonant Dynamics of Oscillatory Chains and Nanostructures*, Foundations of Engineering Mechanics,
DOI 10.1007/978-981-10-4666-7_14

analyze the envelopes of NNMs. Such an approach leads to the multi-scale expansion, the small parameter of which turns out to be the NNMs' frequency splitting. This parameter depends on the length of the lattice and on the wave number.

In this section, we would like to use the effective method of the solution of nonlinear problems which do not content any small parameter in their initial formulation. This method is based on the averaging procedure, the validity of which is proven in the process of solution finding. The source of the ISP as well as its value is dictated by the solution obtained and is verified by the comparison with the numerical simulation data. Omitting the general discussion of the method, we will demonstrate its efficiency in the application to both stationary and non-stationary dynamical processes.

The model

Let us consider the periodic system of N coupled pendula. This system has a wide application in physics modeling an array of the particles in the field of the local (on-site) potential (Braun and Kivshar 2004). The best known example of such systems is the Frenkel–Kontorova model describing a dislocation or interstitial atom in the crystal lattice (Frenkel and Kontorova 1938). Replacing the pendula by the nonlinear oscillators with two-well on-site potential, we obtain the Klein–Gordon model, which is used for the description of the displacement disorder in the crystals. In the linear approximation, these models lead to equations of the same type (linear discrete Klein–Gordon equations), the properties of which in the continual limit are well studied. However, in relatively small systems, the discreteness plays an important role, and the traditional approach consists in the application of the nonlinear normal mode (NNMs) concept.

Let us write the Hamiltonian of the system with length N in the dimensionless form:

$$H = \sum_{j=1}^{N} \left(\frac{p_j^2}{2} + \frac{\beta}{2} \left(q_{j+1} - q_j \right)^2 + 1 - \cos q_j \right) \tag{14.1}$$

Under periodic boundary conditions: $q_{N+1} = q_1$. The corresponding equation of motion

$$\frac{d^2 q_j}{dt^2} - \beta \Delta_2 q_j + \sin q_j = 0$$
$$\Delta_2 q_j = \left(q_{j+1} - 2q_j + q_{j-1} \right) \tag{14.2}$$

may be represented in the terms of complex variables:

$$\Psi_j = \frac{1}{\sqrt{2}} \left(\frac{1}{\sqrt{\omega}} \frac{dq_j}{dt} + i\sqrt{\omega} q_j \right) \tag{14.3}$$

where ω is a frequency of the oscillations, which will be later defined.

$$i\frac{d\Psi_j}{dt} + \frac{\omega}{2}\left(\Psi_j + \Psi_j^*\right) - \frac{\beta}{2\omega}\Delta_2\left(\Psi_j - \Psi_j^*\right)$$
$$+ \frac{1}{2\omega}\sum_{k=0}^{\infty}\frac{1}{(2k+1)!}\left(\frac{1}{2\omega}\right)^k\left(\Psi_j - \Psi_j^*\right)^{2k+1} = 0 \qquad (14.4)$$

Let us represent the solution of Eq. (14.4) in the form:

$$\Psi_j = \psi_j e^{i\omega t} \qquad (14.5)$$

where ψ_j is the amplitude.

One can show that function (14.5) is the solution of the resonant equation, which is obtained by averaging of Eq. (14.4) over the period $2\pi/\omega$, if the relation

$$\frac{\beta}{2\omega}\Delta_2\psi_j - \frac{\omega}{2}\psi_j + \frac{1}{\sqrt{2\omega}}J_1\left(\sqrt{\frac{2}{\omega}}|\psi_j|\right)\frac{\psi_j}{|\psi_j|} = 0 \qquad (14.6)$$

is satisfied. (Here, J_1 is the Bessel function of the first kind.)

Equation (14.6) has the sense of dispersion relation, if the function ψ_j is represented as follows:

$$\psi_j = \sqrt{X}e^{i\kappa j} \qquad (14.7)$$

where $X = $ const and with $\kappa = 2\pi k/N$ and $k = 0, 1, 2, ..., N - 1$.

Now, one can understand the origin of the frequency ω. Really, one can see that functions (14.5) and (14.7) describe the nonlinear normal mode with the wave number κ. Then, the parameter ω is its frequency.

According to definition (14.3) of the function Ψ, the amplitude X is expressed via the amplitude of the displacement Q:

$$X = \frac{\omega}{2}Q^2 \qquad (14.8)$$

Taking into account this relation and Eq. (14.7), one can rewrite dispersion relation (14.6) as follows:

$$\omega^2 = \frac{2}{Q}J_1(Q) + 4\beta\sin^2\frac{\kappa}{2} \qquad (14.9)$$

If the wave number is zero, $\kappa = 0$, Eq. (14.9) describes the amplitude dependence of the uniform (gap) mode, when all pendula oscillate in phase:

$$\omega = \sqrt{\frac{2}{Q}J_1(Q)} \qquad (14.10)$$

In such a case, this frequency has to correspond to oscillation frequency of single pendulum, the exact value of which is well known:

$$\omega_p = \frac{\pi}{2K\left(\sin\frac{Q}{2}\right)}, \qquad (14.11)$$

where K is the complete elliptic integral of first kind. The comparison of these frequencies is shown in Fig. 14.1.

One can see that the frequencies (14.10) and (14.11) are well accorded, excluding a vicinity of the limiting oscillation amplitude $= \pi$.

To understand the discrepancy, one can assume that Eq. (14.6) is the stationary version of more general equation:

$$i\frac{\partial \psi_j}{\partial \tau} + \frac{\beta}{2\omega}\Delta_2\psi_j - \frac{\omega}{2}\psi_j + \frac{1}{\sqrt{2\omega}}J_1\left(\sqrt{\frac{2}{\omega}}|\psi_j|\right)\frac{\psi_j}{|\psi_j|} = 0 \qquad (14.12)$$

Equation (14.12) can be obtained under assumption that a specific timescale for the variation of ψ_j is much more than the period of modes $2\pi/\omega$. In spite of that it is not clearly appeared, the small parameter is the ratio of mode period to the envelope variation time. The frequency of pendulum oscillations converges to zero in the

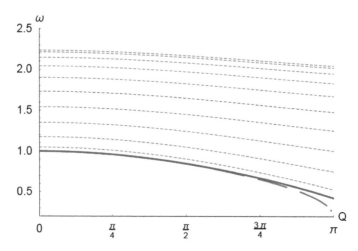

Fig. 14.1 NNMs frequencies calculated according to dispersion ratio (14.9) for the chain with 20 particles and the wave numbers $k = 1, \ldots, 10$. *Solid blue* and *long-dashed red curves* correspond to the uniform (zone-bounding) mode frequencies, calculated according to Eqs. (14.11) and (14.10), respectively. The coupling parameter is $\beta = 1.0$

vicinity of the limiting oscillation amplitude $Q = \pi$, and the separation of the times turns out to be invalid.

Equation (14.12) describes a slow variation of the envelope function ψ_j that results from the nonlinear interaction with surrounding modes. Therefore, one can use it for analyzing the resonant mode interaction.

Equation (14.12) correspond to the Hamilton function in the form:

$$H_a = \sum_{j=1}^{N} \frac{1}{2}\left[1 - \omega|\psi_j|^2 - J_0\left(\sqrt{\frac{2}{\omega}}|\psi_j| \right) + \frac{\beta}{\omega}|\psi_{j+1} - \psi_j|^2 \right] \qquad (14.13)$$

The additional integral of motion is the "occupation" number:

$$X = \frac{1}{N}\sum_{j=1}^{N} |\psi_j|^2 \qquad (14.14)$$

In such a case, the key small parameter is a difference between the mode frequencies. Let us consider the non-stationary dynamics of the chain following from the interaction of the low-frequency modes with the wave numbers $\kappa_0 = 0$ and $\kappa_1 = 2\pi/N$. It was early shown (Smirnov and Manevitch 2011) that the combination of these modes leads to dividing the system on two domains, inside of which the particles move almost coherently. These domains we will designate as the "coherent domains." It is convenient to introduce the domain coordinates:

$$\chi_1 = \frac{1}{\sqrt{2N}}\sum_{j=1}^{N}\left[1 + (\cos \kappa_1 j + \sin \kappa_j) \right]\psi_j$$

$$\chi_2 = \frac{1}{\sqrt{2N}}\sum_{j=1}^{N}\left[1 - (\cos \kappa_1 j + \sin \kappa_j) \right]\psi_j \qquad (14.15)$$

(It is useful to note that the domain coordinates (14.15) are transformed into coordinates of particles, if the chain length $N = 2$.) The inverse transformation to the complex amplitude of the particles is written as follows:

$$\psi_j = \frac{1}{\sqrt{2N}}\left[(\chi_1 + \chi_2) + (\chi_1 - \chi_2)(\cos \kappa_1 j + \sin \kappa_1 j) \right]. \qquad (14.16)$$

One can see that the amplitude vectors $(\chi_1, \chi_2) = (1, 0)$ and $(\chi_1, \chi_2) = (0, 1)$ correspond to the maximum of pendulum displacements in the one and second domains, respectively.

After substituting the expression (14.16) into Eq. (14.13), the equations of motion for the domain coordinates in the terms of complex amplitudes may be obtained immediately by variation of the Hamiltonian with respect to the domain coordinates χ_1, χ_2. However, the resulting equations are lengthy enough and do not allow to analyze the chain dynamics clearly. Therefore, one should simplify the procedure. First of all, one can see that transformation (14.16) preserves the integral of occupation number (14.14):

$$X = |\chi_1|^2 + |\chi_2|^2 \tag{14.17}$$

With using the integral (14.17), the dimension of the system's phase space may be reduced by introducing the relative amplitudes of the domain coordinates χ_j (Manevitch and Smirnov 2010a, b):

$$\chi_1 = \sqrt{X} \cos \theta e^{i\delta_1}, \quad \chi_2 = \sqrt{X} \sin \theta e^{i\delta_2}$$

In such a case, the parameter X specifies the total excitation of the system, while the "angle" θ reflects the relative excitation of the coherent domains. In fact, the energy of the system does not depend on the absolute values of the phases δ_1 and δ_2, but it is the function of their difference $\Delta = \delta_1 - \delta_2$ only:

$$H(\theta, \Delta) = -\frac{\omega}{2} X + \frac{\beta X}{\omega} \sin^2 \frac{\kappa_1}{2} (1 - \cos \Delta \sin 2\theta) - \sum_{j=1}^{N} J_0(\xi_j)$$

$$\xi_j = \sqrt{\frac{2X}{\omega N}} \sqrt{1 + \cos 2\theta (\cos \kappa_1 j + \sin \kappa_1 j) + \sin \kappa_1 j \cos \kappa_1 j (1 - \cos \Delta \sin 2\theta)}$$

$$\tag{14.18}$$

The first term in Hamiltonian (14.18) corresponds to the kinetic part of the oscillation energy. The second term describes the linear splitting of the NNMs frequencies. And the third term is associated with the nonlinear interaction of the NNMs.

Hamiltonian (14.18) allows to analyze the phase portrait of the system and to define the bifurcations of the phase trajectories under different excitation level that is specified by the excitation number X.

This procedure was well discussed for the coupled nonlinear oscillators (Manevitch 2007; Manevitch and Romeo 2015) as well as for the nonlinear chains (Manevitch and Smirnov 2010a, b, 2011) in the small-amplitude approximation. However, the dispersion relation (14.9) and the asymptotic Eq. (14.12) are not restricted by any assumptions with respect to smallness of the oscillation amplitudes. In such a case, one can find the bifurcations at any given amplitude as a function of the chain parameters.

One can show that the stationary points ($\theta = \pi/4$, $\Delta = 0$) and ($\theta = \pi/4$, $\Delta = \pi$) of system (14.18) correspond to the NNMs, while the values of θ, which are equal to 0 and $\pi/2$, relate to the coherent domains χ_1 and χ_2, respectively. The "domain states" are the parts of the trajectory, which divides the attraction areas of the NNMs (it is named the limiting phase trajectory—LPT). It is known (Smirnov and Manevitch 2011) that the uniform (zone-bounding) mode can lose its stability at the certain conditions, if the nonlinearity is soft. In such a case, two new stationary points, which relate to the weakly localized states, appeared. If the oscillation amplitudes increase or the coupling parameter can be decreased, the separatrix passing through the unstable NNM expands and it can reach the domain states $\theta = 0$ and $\theta = \pi$. At this moment, any paths from one domain state to another one turn out to be forbidden. Thus, the energy, initially concentrated in one of the domains, becomes to be captured in it.

Taking into account all mentioned above, it is easy to formulate the bifurcation conditions. The lost of stability occurs when Hamiltonian (14.18) plateaus near the lowest NNM:

$$\frac{\partial^2 H(\theta, \Delta)}{\partial \theta^2}\Bigg|_{(\theta=\frac{\pi}{4}, \Delta=0)} = 0 \tag{14.19}$$

Solving Eq. (14.19) with respect to coupling parameter β, one can obtain the instability threshold:

$$\beta_{\text{ins}} = \frac{J_2(Q)}{2 \sin^2 \frac{\kappa_1}{2}} \tag{14.20}$$

The global bifurcation occurs when the energy of unstable stationary point $(\theta = \pi/4, \Delta = 0)$ becomes equal to the energy of "domain states" $(\theta = \pi/2, \Delta = \pi/2)$ and $(\theta = 0, \Delta = \pi/2)$. Under this condition, the solution of respective equation leads to the localization threshold as follows:

$$\beta_{\text{loc}} = 2 \frac{\frac{1}{N} \sum_{j=1}^{N} J_0\left(\frac{Q}{\sqrt{2}} f_j\right) - J_0(Q)}{Q^2 \sin^2 \frac{\kappa_1}{2}} \tag{14.21}$$

$$f_j = 1 + \cos \kappa_1 j + \sin \kappa_1 j$$

The first fact of worth is that instability threshold (14.20) and localization one (14.21) are in inverse proportion to the squared sin $(\kappa_1/2)$. Taking into account that $\kappa_1 \sim 1/N$, one can conclude that the crucial values of coupling grow while the chain length increases. However, it is clear that the real parameter, which determines the resonant conditions, is a value of the gap between zone-bounding and the first non-uniform modes. This value is defined by the "effective coupling constant" that is the production $\beta \sin^2 (\pi/N)$ (Smirnov and Manevitch 2011). Such conclusion coincides with that the production $\beta_{\text{ins}} \sim \sin^2 \kappa_1/2$ does not depend on the length of the chain and is equal to ε_{ins} for the pair of coupled pendula (Manevitch and Romeo 2015). One should notice that threshold (14.21) coincides exactly with the values that were obtained in the work (Manevitch and Romeo 2015), if $N = 2$. Figure 14.2a shows the "effective" threshold values for the instability and localization bifurcations.

The evolution of the phase trajectories of the system can be conveniently analyzed in the Poincare map for the domain coordinates (Fig. 14.2b–d). Figure 14.2b

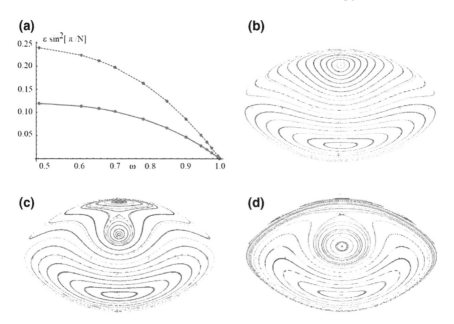

Fig. 14.2 **a** Effective thresholds (14.20), (14.21) (*blue dashed* and *red solid curves*, respectively) versus oscillation frequency ω for the chain with 16 pendula. **b–d** Poincare maps for the chain with 16 pendula at oscillation amplitude $Q = 3\pi/4$ and different coupling parameters: **b** $\beta > \beta_{\text{ins}}$, **c** $\beta = \beta_{\text{loc}}$, **d** $\beta < \beta_{\text{loc}}$

shows the Poincare map for the chain with 16 coupled pendula and the coupling parameter $\beta > \beta_{\text{ins}}$. It is well seen that two stationary points correspond to the modes with wave numbers $\kappa = 0$ and $\kappa = 2\pi/N$. Figure 14.2c, d shows the Poincare map at the localization threshold and below of it. In Fig. 14.2c, one can see that the separatrix, passing through the unstable stationary point, surrounds two new stable stationary points, which were created as a result of instability of the lower mode with the wave number $\kappa = 0$. In contrast, the separatrix in Fig. 14.2d encircles the stationary point corresponding to mode with wave number $\kappa = 2\pi/N$, and no path from one localization state to another one exists.

14.2 Large-Amplitude Nonlinear Normal Modes of the Discrete Sine-Lattices

The peculiarity of the model condered in the previous section is the presence of the periodic on-site potential, while the inter-particle interaction can be described by potentials with nonlinearities of different types. Periodic inter-atomic potentials arise, in particular, while dealing with magnetic systems, unzipping the DNA molecule and oscillations of the flexible crystalline polymers (see, e.g., Takeno and Homma 1986;

Takeno and Peyrard 1996, 1997, where the existence of the highly localized soliton-like solution has been proved in the framework of the sine-lattice model).

The lucky star of the Frenkel–Kontorova model is the existence of the integrable continuum limit of the respective equation of motion (sine-Gordon equation). Due to full integrability of the latter, its spectra of the nonlinear periodic and localized excitations have been studied in detail (Novikov et al. 1984). The continuum limit of the discrete model with the nonlinear periodic inter-atomic interaction leads to the same Sine-Gordon equation with understandable restrictions on the wavelengths (accounting the discreteness effects in the framework of approach introduced by Rosenau (1986) results in the improvement of the long-wavelength approximation only).

In this section, we study the NNMs of discrete lattice with nonlinear on-site and inter-site interactions in the wide range of the oscillation amplitudes and wavelengths. We use the semi-inverse asymptotic approach, which was successfully implemented in the previous section to the chain of coupled pendula.

The model

We consider a finite chain of coupled particles with periodic on-site and inter-particle potentials; each of them is described by a harmonic function with, generally speaking, different periods. This system will be referred to as sine-lattice (SL), in contrast to the classic FK system. Keeping in mind the coincidence of the mathematical descriptions, we will discuss all results in terms of coupled pendulums. The energy of such system may be written as follows:

$$H = \sum_{j=1}^{N} \left[\frac{1}{2} \left(\frac{dq_j}{dt} \right)^2 + \frac{\beta}{\alpha^2} \left(1 - \cos\left(\alpha(q_{j+1} - q_j) \right) \right) + (1 - \cos q_j) \right] \quad (14.22)$$

where qj is the deviation of the jth pendulum, while β and α are the parameters, which specify the rigidity and the period of inter-pendulum coupling. We use the periodic boundary conditions as the most appropriate for the analysis of the chain dynamics; i.e., we assume that $q_{N+1} = q_1$ and $q_0 = q_N$.

The respective equations of motion can be written as follows:

$$\frac{d^2 q_j}{dt^2} - \frac{\beta}{\alpha} \left(\sin\left(\alpha(q_{j+1} - q_j) \right) - \sin\left(\alpha(q_j - q_{j-1}) \right) \right) + \sin q_j = 0 \quad (14.23)$$

Introducing the complex variables given by

$$\Psi_j = \frac{1}{\sqrt{2}} \left(\frac{1}{\sqrt{\omega}} \frac{dq_j}{dt} + i\sqrt{\omega} q_j \right)$$

$$q_j = -\frac{i}{\sqrt{2\omega}} (\Psi_j - \Psi_j^*); \quad \frac{dq_j}{dt} = \sqrt{\frac{\omega}{2}} \left(\Psi_j + \Psi_j^* \right) \quad (14.24)$$

and substituting them in (14.23), one can rewrite Eq. (14.23) as:

$$i\frac{d\Psi_j}{dt} + \frac{\omega}{2}\left(\Psi_j + \Psi_j^*\right) + \frac{1}{2\omega}\sum_{k=0}^{\infty}\frac{1}{(2k+1)!}\left(\frac{1}{2\omega}\right)^k\left[(\Psi_j - \Psi_j^*)^{2k+1}\right.$$
$$-\beta\alpha^{2k}\left(((\Psi_{j+1} - \Psi_{j+1}^*) - (\Psi_j - \Psi_j^*))^{2k+1}\right. \tag{14.25}$$
$$\left.\left.-\left((\Psi_j - \Psi_j^*) - (\Psi_{j-1} - \Psi_{j-1}^*)\right)^{2k+1}\right] = 0$$

where the nonlinear terms in Eq. (14.23) are represented as the series of their arguments.

The semi-inverse resonance approach to the dynamic analysis without any restrictions on the oscillation amplitudes assumes that the considered system admits two timescales (fast and slow). In the framework of this approach, corresponding small parameter as well as the frequency ω can be not presented in starting equations of motion (14.23) and they have to be determined later. In order to demonstrate these sentences, one can start from the stationary solution of Eq. (14.25):

$$\Psi_j = \phi_j e^{i\omega t}, \tag{14.26}$$

where ϕ_j = const. Inserting solution (14.26) into Eq. (14.25) with further multiplying the latter on the factor $\exp(-i\omega t)$ and integrating over the period $2\pi/\omega$ leads to the transcendental equations for the envelope function ϕ_j:

$$-\frac{\omega}{2}\varphi_j - \frac{\beta}{\alpha\sqrt{2\omega}}\left[J_1\left(\alpha\sqrt{\frac{2}{\omega}}|\varphi_{j+1} - \varphi_j|\right)\frac{\varphi_{j+1} - \varphi_j}{|\varphi_{j+1} - \varphi_j|}\right.$$
$$\left.-J_1\left(\alpha\sqrt{\frac{2}{\omega}}|\varphi_j - \varphi_{j-1}|\right)\frac{\varphi_j - \varphi_{j-1}}{|\varphi_j - \varphi_{j-1}|}\right] + \frac{1}{\sqrt{2\omega}}J_1\left(\sqrt{\frac{2}{\omega}}|\varphi_j|\right)\frac{\varphi_j}{|\varphi_j|} = 0 \tag{14.27}$$

where J_1 is the Bessel function of the first order.

It is easy to see that the procedure used above is somewhat similar to the harmonic balance method, which is widely used in the analysis of nonlinear oscillations (Mickens 2010).

In spite of the complexity of Eq. (14.27), one can directly check that the simple expression

$$\varphi_j = \sqrt{X}e^{-i\kappa j} \tag{14.28}$$

with the wave number $\kappa = 2\pi k/N$ (k is an integer, $k \leq N/2$) satisfies it, if the frequency ω is the solution of the equation

$$-\frac{\omega}{2} - \frac{2\beta}{\alpha\sqrt{2\omega X}} J_1\left(2\alpha\sqrt{\frac{2X}{\omega}}\sin\frac{\kappa}{2}\right)\sin\frac{\kappa}{2} + \frac{1}{\sqrt{2\omega}} J_1\left(\sqrt{\frac{2X}{\omega}}\right) = 0 \qquad (14.29)$$

The latter equation is strongly simplified, if we use the relationship between the modulus of complex function X and the amplitude of oscillations Q, which results from definition (14.24) of complex variable Ψ:

$$X = \frac{\omega}{2} Q^2$$

Taking into account, this relationship leads to the expression for the NNMs' frequency of the oscillations with the given amplitude Q:

$$\omega^2 = \frac{2}{Q}\left(2\frac{\beta}{\alpha} J_1\left(2\alpha Q \sin\frac{\kappa}{2}\right)\sin\frac{\kappa}{2} + J_1(Q)\right) \qquad (14.30)$$

Prior to the analysis of eigenfrequencies (14.30), one should test the limiting case that corresponds to the oscillations of a single pendulum. Really, if the coupling parameter $\beta = 0$, Hamiltonian (14.22) describes a set of independent pendulums, the oscillation frequency of which depends on the amplitude Q. In such a case, frequency (14.30) has the form:

$$\omega = \sqrt{\frac{2}{Q} J_1(Q)} \qquad (14.31)$$

Equation (14.30) can be compared with the exact oscillation frequency of pendulum:

$$\omega_e = \frac{\pi}{2K\left(\sin\left(\frac{Q}{2}\right)\right)} \qquad (14.32)$$

where K is the complete elliptic integral of the first kind.

One can see from Fig. 14.1 that the agreement for all amplitudes is excellent up to $Q \le 3\pi/4$ and turns out to be good enough even for $Q = 9\pi/10$.

In order to understand the origin of the frequencies divergence, one should notice that Eq. (14.27) is the limiting stationary case of more commonly equation, which describes a slow evolution of the envelope function φ_j. The latter results from the interaction of the NNMs with close frequencies (it is an analogue of the beating phenomenon in the linearized system). The specific time of this evolution is determined by the relative difference of the NNMs' frequencies. Until these values are small enough, the timescales may be separated well. However, when the oscillation amplitude Q approaches its maximum value π, the frequency rapidly diminishes and the difference mentioned above turns out to be non-small. So, the small parameter, which determines the timescale separation, is related with the gap between NNMs frequencies. We will demonstrate using this parameter at the analysis of the stability of the NNMs.

Equation (14.30) describes the NNM "zone" structure, i.e., the dispersion ratio for the SL model at the arbitrary oscillation amplitude (excluding the vicinity of the "rotation limit" $Q = \pi$).

One should note that because model (14.22) leads to the discrete FK chain in the "long-wavelength" limit $(\alpha(q_{j+1} - q_j) \ll 1)$, Eq. (14.30) has to describe the respective spectrum. Actually, considering the wave number κ as a small value, one can expand the Bessel function as a power series. The first term of Eq. (14.30) becomes $2\alpha Q \sin \kappa/2$, and the eigenfrequency is reduced as follows:

$$\omega^2 = \left(\frac{2}{Q} J_1(Q) + 4\beta \sin^2 \frac{\kappa}{2} \right) \tag{14.33}$$

Figure 14.1 (in the previous section) shows the zone structure for the FK chain with 20 particles under periodic boundary conditions.

The low-frequency mode, which bounds the zone, corresponds to the uniform oscillations of the chain or to the oscillations of the single pendulum (14.31), while the high-frequency bounding mode corresponds to the out-of-phase pendulum oscillations ("π"-mode). One can see that frequency (14.33) is the monotonically increasing function of the wave number κ, but the difference $\Delta\omega^2 = \omega^2(\kappa = \pi) - \omega^2(\kappa = 0)$ does not depend on the amplitude of oscillations.

Figure 14.3 shows zone structure for the harmonically coupled pendulums. The comparison of Figs. 14.1 and 14.3 shows a cardinal distinction between them. Firstly, the width of the SL zone depends on the amplitude of oscillations. It is more important that the dispersion relation at a fixed amplitude $Q \geq \pi/2$ (the threshold value depends on the parameters α and β) is a non-monotonic function of the wave number. As a result, the frequency of the zone-bounding π-mode turns out to be smaller than the frequency of the uniform mode for large Q. In such a case, the multiple resonances occur in the vicinity of right edge of the spectrum, and their existence is defined by non-monotonic character of the dispersion relation rather than by the length of the chain. Figure 14.4 shows the dispersion relation for SL chain with 20 pendulums and the oscillation amplitude $Q = \pi/10$ in comparison with the same for $Q = 9\pi/10$.

The structure of the NNMs zone for the SL chain has been checked by the direct numerical integration of Eq. (14.23).

Figure 14.5 allows comparing the positions of the zone-bounding modes and some intermediate ones. One can see that the mutual positions of the modes correspond to the dispersion relations that are shown in Fig. 14.4.

Due to non-monotonic behavior of the dispersion relations at the large amplitudes (Fig. 14.4), a multitude of the resonances can exist for both the NNMs with the nearby wave numbers and for the modes, the wave numbers of which differ significantly. Moreover, an almost flat dispersion relation can be obtained for the certain combination of the lattice parameters such as α and β. In such a case, the

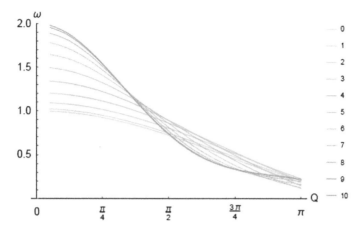

Fig. 14.3 The zone structure for the SL chain with 20 pendulums under periodic boundary conditions. *Light blue* and *light brown curves* correspond to the zone-bounding uniform and π-modes, respectively. The coupling parameters are $\beta = 0.25$, $\alpha = 1.2$. The *numbers* in the figure legend show the mode's number

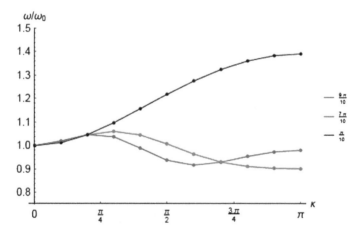

Fig. 14.4 The comparison of the dispersion relations for the SL chain with 20 pendulums at the different oscillation amplitudes: *Black, red*, and *blue points* correspond to the amplitudes $Q = \pi/10$, $Q = 7\pi/10$, and $Q = 9\pi/10$, respectively. The relative frequencies ω/ω_0 (ω_0 is the frequency of the uniform modes) are shown. The potential parameters are as follows: $\beta = 0.25$, $\alpha = 1.2$

thermoconductivity of the system decreases essentially due to the effective phonon scattering (Han et al. 2016).

An initial excitation of general nature a priori contains some combination of the modes with different amplitudes. In contrast to a linear system, where any interaction between normal modes is absent, the NNMs in the essentially nonlinear systems can interact efficiently if the resonant conditions occur (Chirikov and

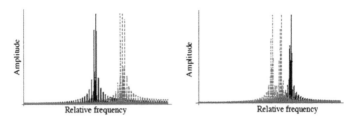

Fig. 14.5 Fourier spectra of the oscillations of SL chain with 8 pendulums. Left panel shows the frequencies of the normal mode with $\kappa = 0$, $3\pi/4$, π (*solid black, dot-dashed red,* and *dashed blue lines,* respectively) at the amplitude $Q = \pi/10$. *Right panel* shows the same as left one at the amplitude $Q = 9\pi/10$. The potential parameters are as follows: $\beta = 0.25$, $\alpha = 1.2$

Zaslavsky 1972). Therefore, the problem of internal resonances is very important for study of the dynamics of the sine-lattice.

Figure 14.6 show the "map" of resonantly interacting modes for two amplitudes of the oscillations: $Q = 3\pi/10$ and $Q = 7\pi/10$. The panels (a–f) differ in the partial amplitudes of interacting modes Q_1 and Q_2 at the constant sum of them: $Q = Q_1 + Q_2$.

One can see that for a small amplitude Q, the resonantly interacting modes have the close numbers: k_1, k_2 for any ratios of the partial amplitudes of the modes [Q_1 and Q_2 (see Fig. 14.6a)]. In contrast to that, the oscillations with a large amplitude (Fig. 14.6c) do not contain the resonantly interacting modes for small values Q_2 and include the multitude of resonant modes in the short wavelength domain of the spectrum. Figure 14.6 shows the number of resonances as a function of the ratio Q_2/Q_1 for three values of the oscillation amplitude: $Q = 3/\pi/10$, $\pi/2$, $7\pi/10$ (Fig. 14.7).

There are at least two reasons for the importance of the NNMs interactions. As it was shown early, the resonant interaction of the NNMs leads to the localization effect (the capture of the energy of oscillations in some domain of the chain) (Manevitch and Smirnov 2010a, b: 2, 2011). The necessary condition of such localization is the instability of one of interacting modes (Dauxois and Peyrard 1993).

The second reason arises from the occurrence of the chaotic regimes of the oscillations under conditions of internal nonlinear resonances (Chirikov and Zaslavsky 1972; Sagdeev et al. 1988; Neishtadt 1975; Arnold et al. 2006). As it was mentioned above, the continuum approximation of Eq. (14.23) leads to the well-known sine-Gordon equation, which corresponds to the integrable system and does not allow any chaotic behavior. In contrast with latter, the considered system is not an integrable one. Taking into account the multitude of internal resonances, one should expect the existence of the chaotic trajectories in the phase space of the system. The chaotic trajectories arise inside the stochastic layer nearby the separatrix passing via the unstable singularity, which is formed when one of resonantly interacting modes losses its stability (Zaslavsky 1998).

Therefore, the stability of the NNMs is one of the key problems of the dynamics of the system under consideration.

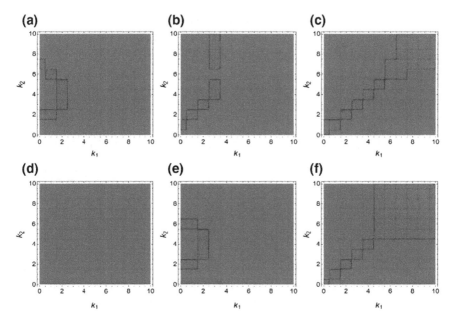

Fig. 14.6 Maps of resonances for two values of oscillation amplitudes: $Q = \pi/2$ (**a–c**) and $Q = 7\pi/10$ (**d–f**). k_1 and k_2 are the modes' numbers. The panels (**a–c**) and (**d–e**) differ in the partial amplitudes of the modes Q_1 and Q_2. The *red squares* correspond to the modes the frequencies of which differ not more than 5%

Fig. 14.7 The number of resonances versus partial amplitude ratio for total oscillation amplitudes $Q = 3\pi/10$, $\pi/2$, $7\pi/10$

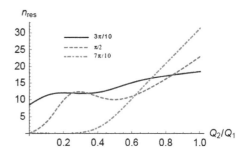

In order to analyze the stability of the NNMs, we will use the non-stationary extension of Eq. (14.25). Let us assume that the initial conditions contain alongside with considered NNM (stationary solution) a small perturbation corresponding to mode with different wave number:

$$\varphi_j = \sqrt{X}\left(1 + \varepsilon\chi_j\right)e^{i\kappa j} = \varphi_{j,0} + \varepsilon\chi_j \tag{14.34}$$

where $\varphi_{j,0}$ is the solution of stationary Eq. (14.27). In such a case, one should wait that Eq. (14.27) is not satisfied. It is easy to understand that this envelope will have a large timescale (in comparison with the period of the carrier $2\pi/\omega$) if the frequencies of the interacting modes turn out to be close. The short discussion of the time separation procedure is presented in the Appendix 1. As a result, the time-dependent equations can be written as follows:

$$
\begin{aligned}
&i\frac{\partial \varphi_j}{\partial \tau} - \frac{\omega}{2}\varphi_j - \frac{\beta}{\alpha\sqrt{2\omega}}\left[J_1\left(\alpha\sqrt{\frac{2}{\omega}}|\varphi_{j+1} - \varphi_j|\right)\frac{\varphi_{j+1} - \varphi_j}{|\varphi_{j+1} - \varphi_j|}\right.\\
&\left.- J_1\left(\alpha\sqrt{\frac{2}{\omega}}|\varphi_j - \varphi_{j-1}|\right)\frac{\varphi_j - \varphi_{j-1}}{|\varphi_j - \varphi_{j-1}|}\right] + \frac{1}{\sqrt{2\omega}}J_1\left(\sqrt{\frac{2}{\omega}}|\varphi_j|\right)\frac{\varphi_j}{|\varphi_j|} = 0
\end{aligned}
$$

$$(14.35)$$

Assuming the function (14.34) in the form

$$
\varphi_j = \sqrt{X}\left(e^{i\kappa j} + s(\tau)e^{i\nu j}\right) = \sqrt{X}\left(e^{i\kappa j} + (u(\tau) + iv(\tau))e^{i\nu j}\right) \tag{14.36}
$$

where u and v are the real functions of the slow time τ_1, one can analyze the effect of the mode with wave number ν on the NNM with wave number κ. Expanding Eq. (14.35) in the vicinity of the NNM φ_j and keeping the perturbation of first order only, one can obtain the equation for the real part of the function s as follows:

$$
\frac{\partial^2 u}{\partial \tau^2} + \Lambda u = 0 \tag{14.37}
$$

where the parameter Λ is

$$
\begin{aligned}
\Lambda = {} & \frac{1}{8\omega\sin\frac{\kappa}{2}}\left[2\frac{\beta}{\alpha}J_1\left(2\alpha Q\sin\frac{\kappa}{2}\right)(\cos\kappa - \cos\nu) - 4\beta Q J_2\left(2\alpha Q\sin\frac{\kappa}{2}\right)\sin^2\frac{\nu}{2}\sin\frac{\kappa}{2} - QJ_2(Q)\right]^2\\
& - \frac{Q^2\sin\frac{\kappa}{2}}{8\omega}\left[2\beta(\cos(\kappa - \nu) - \cos k)J_2\left(2Q\alpha\sin\frac{\kappa}{2}\right) + J_2(Q)\right]^2.
\end{aligned}
$$

$$(14.38)$$

The stability of the NNM is determined by the sign of the parameter Λ. If $\Lambda > 0$, the perturbation remains a small, but it increases if $\Lambda < 0$. The parameter Λ depends on the lattice parameters α and β as well as on the amplitude Q. Let us determine the coupling parameter β at the threshold of stability. Solving the equation

$$
\Lambda = 0
$$

with respect to β, one can obtain its critical value as follows:

$$
\beta_{\text{ins}} = \frac{\alpha Q J_2(Q)\sin\frac{\kappa}{2}}{J_1\left(2\alpha Q\sin\frac{\kappa}{2}\right)(\cos\kappa - \cos\nu) - 4\alpha Q J_2\left(2\alpha Q\sin\frac{\kappa}{2}\right)\sin^2\frac{\kappa}{2}\sin\frac{\nu}{2}\cos\left(\frac{1}{2}(\kappa - \nu)\right)}
$$

$$(14.39)$$

As it was shown (Peyrad and Bishop 1989; Manevitch and Smirnov 2010a, b, 2011) the stability of the zone-bounding mode with $\kappa = 0$ is important for the localization of the oscillations. Namely, the lost of the stability of this mode is the first step to stationary non-uniform distribution of the oscillation energy along the chain. Assuming $\kappa = 0$ and $v = 2\pi/N$, one can obtain the instability threshold for the zone-bounding mode as follows:

$$\beta_{\text{ins}} = \frac{1}{2} \frac{J_2(Q)}{\sin^2\left(\frac{\pi}{N}\right)}, \tag{14.40}$$

that correlates well with the estimation of analogous instability threshold for the Frenkel–Kontorova chain $\beta = (3Q/16\pi)^2 N^2$ in the small-amplitude limit (Smirnov and Manevitch 2011).

The "instability map" of the zone-bounding mode for the chain with $N = 20$ is shown in Fig. 14.8. One can see that the instability threshold increases significantly while the oscillation amplitude grows. It seems from the physical viewpoint that a large coupling parameter has no sense. It means that the large-amplitude uniform oscillations will be unstable in the majority of the physical systems.

We will not consider the stability of other resonantly interacting modes because due to the simpleness of Eq. (14.39) this problem can be analyzed for any physical systems.

However, the lost of stability of the zone-bounding mode does not imply the creation of the localized oscillation in the chain. There is the second bifurcation after that the processes of the energy redistribution are forbidden (Manevitch and Smirnov 2010a, b). This bifurcation occurs when the energy of the unstable mode turns out to be equal the energy of the mixed state of the stable and unstable modes. Such a state corresponds to the limiting phase trajectory (LPT), which describes the non-stationary dynamics of the chain with the extremely large energy exchange between some parts of the system. The motion of the pendula inside these parts is close to the coherent oscillations, while the dynamics of the pendula in different parts differ essentially. The process of the energy exchange between these parts (clusters or coherent domains) of the system is similar to the beating in the system of two weakly coupled oscillators, while the energy localization is analogous to the energy capture of one of them.

In order to estimate the localization threshold, one should use the Hamiltonian corresponding to Eq. (14.35):

$$H = \sum_{j=1}^{N} \left[-\frac{\omega}{2}|\varphi_j|^2 + \frac{\beta}{\alpha^2}\left(1 - J_0\left(\alpha\sqrt{\frac{2}{\omega}}|\varphi_{j+1} - \varphi_j|\right)\right) \right. $$
$$\left. + \left(1 - J_0\left(\sqrt{\frac{2}{\omega}}|\varphi_j|\right)\right) \right] \tag{14.41}$$

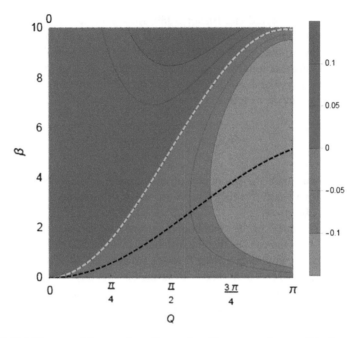

Fig. 14.8 Stability map of the zone-bounding mode with wave number $\kappa = 0$ in the coordinates (Q, β). The *blue* and *red domains* correspond to the stability and instability of the mode. The figure legend shows the specific values of the parameter Λ. The boundary between domains is marked by *bright dashed line*. The threshold of the localization is shown as the *black dashed curve* (see text). The length of chain is equal to 20 and parameter $\alpha = 1.25$

As it was mentioned above, the regimes with the energy localization occur when the NNMs with wave number $\kappa = 0$ and $\kappa = 2\pi/N$ interact resonantly. Setting energy (14.41) of the zone-bounding mode solution ($\varphi_j = \mathrm{const}$) equal to the energy of the mode mixture ($\varphi_j \sim 1 + \exp(i2\pi/N)$), one can get the localization threshold of the coupling parameter as follows:

$$\beta_{\mathrm{loc}} = \frac{\alpha^2}{4} \frac{4J_0^2\left(\frac{Q}{2}\right) - 4J_0(Q) - QJ_1(Q)}{1 - J_0\left(\alpha Q \sin\left(\frac{\pi}{N}\right)\right)} \tag{14.42}$$

The black dashed curve in Fig. 14.8 shows the value of the localization threshold (14.42) for the chain with 20 pendula and parameter $\alpha = 1.25$. One can see that the localization threshold β_{loc} grows essentially slower than the value β_{ins}; however, it reaches a large enough values for the large oscillation amplitudes $Q \geq \pi/2$. On the other side, one should notice that the small-amplitude expansion of Eq. (14.20) shows that the value $\beta \sin^2 \pi/N$ turns out to be small if the length of the chain is large enough. It correlates strongly with the assumptions about the resonance of the considered modes, because splitting between them is proportional to this value.

Figure 14.8a–c show the energy distribution along the chain for different values of the coupling parameter β—before and after threshold value β_{loc}. These figure were obtained by the direct numerical integration of the equations of motion which correspond to initial Hamiltonian (14.1) under the periodic boundary conditions. The period of the zone-bounding mode at the amplitude $Q = \pi/2$ is $T = 2\pi/\omega \sim 7.4$ time units.

Figure 14.9a shows the periodic energy redistribution along the chain before the localization ($\beta = 1.76$): The bright and dark areas change their location with the period which is essentially larger than the oscillation one. Figure 14.9b demonstrates the energy distribution at the coupling parameter $\beta = 1.75$, that is, under previous value less than one percent. One can see that the main part of the energy is localized near the center of the chain. Finally, Fig. 14.9c shows the well-localized oscillations at the coupling parameter $\beta = 1.00 \ll \beta_{\mathrm{loc}}$. One should notice that the estimation of the localization threshold with Eq. (14.42) for the used parameters gives the value $\beta = 2.044$, while it is shown from Fig. 14.9b that the localization occurs at $\beta \sim 1.755$. The difference between the numerical result and the analytical estimation is approximately 15%.

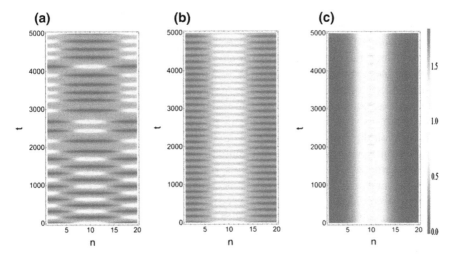

Fig. 14.9 Energy distribution along the chain with 20 pendula for three values of the coupling parameter β: 1.76 (**a**), 1.75 (**b**), and 1.50 (**c**). Parameter, $\alpha = 1.25$; amplitude of oscillation, $Q = \pi/2$. The figure legend (*right panel*) shows the energy value in the dimensionless units. n number of pendulum and t dimensionless time in the reverse gap frequency

14.3 Is Energy Localization Possible in the Conditions of Acoustic Vacuum?

In this section, we discuss results of analytical and numerical study of planar dynamics of a string with uniformly distributed discrete masses without a preliminary stretching. Each mass is also affected by grounding support with cubic characteristic (which is equivalent to transversal unstretched string). We consider the most important case of low-energy transversal dynamics. This example is especially instructive because the considered system which cannot be linearized and therefore oscillates in the conditions of acoustic vacuum is the model of efficient energy trap. Adequate analytical description of resonance non-stationary processes which correspond to intensive energy exchange between different clusters of the particles in low-frequency domain was obtained in terms of LPTs. We have revealed also in these terms the conditions of energy localization on the initially excited cluster. Analytical results are in agreement with the results of numerical simulations. It is shown that the effectiveness of the system as energy trap considered system can be strongly promoted when using grounding supports. Coerrespondinig two particle system was considered in Chap. 2.

It was mentioned in Chap. 2 that in the limit of low energy, a fixed-fixed chain of linearly coupled particles performing in-plane transverse oscillations possesses strongly nonlinear dynamics and acoustics due to geometric nonlinearity, forming a nonlinear acoustic vacuum (see also Manevitch and Vakakis 2014). This designation denotes the fact that the speed of sound as defined in the sense of classical acoustic theory is zero in that medium, so the resulting equations of motion lack any linear stiffness components. A significant feature of that system was the presence of strongly non-local terms in the governing equations of motion (in the sense that each equation directly involves all particle displacements), in spite of the fact that the physical spring-mass chain has only local (nearest-neighbor) interactions between particles. These non-local terms constitute a time-dependent "effective speed of sound" for this medium, which is completely tunable with energy. A rich structure of resonance manifolds of varying dimensions was identified in the nonlinear sonic vacuum, and 1:1 resonance interactions are studied asymptotically to prove the possibility of strong energy exchanges between nonlinear modes.

One of distinctive features of chain without grounding support was that its NNMs (Vakakis et al. 1996) could be exactly determined. Moreover, the analysis has shown that the number of NNMs in the sonic vacuum was equal to the dimensionality of the configuration space and that no NNM bifurcations were possible. In addition, the most intensive 1:1 resonance intermodal interaction was the one realized by the two NNMs with the highest wave numbers. However, the unstretched string model considered in Manevitch and Vakakis (2014) is in some sense a special case, since one of the most significant features of dynamical systems with homogeneous potentials is that the number of NNMs may exceed the number of degrees of freedom due to mode bifurcations (Rosenberg 1966). One can expect that such NNM bifurcations will also lead to drastic modification of the non-stationary resonance dynamics of the sonic vacuum described by LPTs. Thus, it is of great interest to consider an extension of the nonlinear sonic vacuum developed in Manevitch and Vakakis (2014) so that the modified system has the capacity to undergo NNM bifurcations. Such a study can provide us with the opportunity to investigate how these bifurcations can affect the non-stationary resonant dynamics corresponding to resonant energy exchange and localization.

These questions were discussed Chap. 2 and in paper devoted to unstretched string with grounding support but carrying only two discrete masses (Koroleva Kikot et al. 2015). Here, we present an extension to much more complicated system with arbitrary finite number of discrete masses.

14.3.1 The Model and Equations of Motion

Let us consider unstretched string with uniformly distributed equal masses and returning forces, proportional to cubes of deformations (see Fig. 14.10). The equations of motion are as follows:

$$m\frac{d^2 U_j}{dt^2} + T_j \cos\theta_j - T_{j+1}\cos\theta_{j+1} = 0, \quad j = 1,\ldots,N,$$
$$m\frac{d^2 V_j}{dt^2} + cV_j^3 + T_j \sin\theta_j - T_{j+1}\sin\theta_{j+1} = 0, \quad j = 1,\ldots,N,$$
(14.43)

with U_j, V_j being longitudinal and transversal displacements of jth mass, respectively; θ_j being angle between jth segment and its equilibrium position. Tensile forces are proportional to deformations and may be written as follows:

$$T_j = K\frac{1}{l}\left[(U_j - U_{j-1}) + \frac{1}{2l}(V_j - V_{j-1})^2\right],$$

with l being equilibrium length of one segment and K being stiffness coefficient.

The mechanism of non-local force formation was discussed in the paper (Manevitch and Vakakis 2014). According to this mechanism, the tensile forces in all segments are approximately equal to their mean value:

$$T = \langle T_j \rangle = \frac{1}{N+1}K\frac{1}{2l^2}\sum_{s=0}^{N}(V_{s+1} - V_s)^2$$

Introducing the "slow" timescale $\tau_0 = \varepsilon t$, where small parameter ε describes the relative smallness of transversal frequencies ($\varepsilon = a/l$, with a being an amplitude of transversal oscillations), we obtain the following equation system for transversal motion (parameter $\mu = K/Cl^3$ describes relation between contributions of string itself and grounding supports):

$$\frac{d^2 v_j}{d\tau_0^2} + \frac{1}{\mu} v_j^3 + \frac{1}{2(N+1)} \sum_{s=0}^{N} (v_{s+1} - v_s)^2 (2v_j - v_{j+1} - v_{j-1}) = 0, \; v_0 = v_{N+1} = 0,$$

$$(14.44)$$

where $V_j = \varepsilon v_j$ and v_j are normalized displacements, and $\omega_0 = \sqrt{K/ml}$ (Fig. 14.10).

14.3.2   Two-Mode Approximation

We consider 1:1 resonance on the frequency ω and rewrite the system (14.46) in the following form:

$$(\ddot{v}_j + \omega^2 v_j) + \varepsilon_1 \gamma \left(\frac{1}{\mu} v_j^3 - \omega^2 v_j + \frac{2v_j - v_{j-1} - v_{j+1}}{2(N+1)} \sum_{s=0}^{N} (v_{s+1} - v_s)^2 \right) = 0$$

$$(14.45)$$

Combination in the right-hand side should be small (since we consider a system near resonance). It is reflected by introducing the small parameter ε_1. We introduce parameter $\gamma = \varepsilon_1^{-1}$ to provide an equivalence of systems (14.44) and (14.45). We introduce complex variables as follows:

$$\psi_j = \dot{v}_j + i\omega v_j, \; \Phi_j = \psi_{j+1} - \psi_j, \quad j = 1, \ldots, N.$$

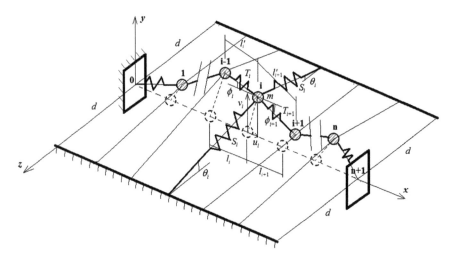

Fig. 14.10 Oscillator chain with elastic support

Then,

$$\dot{\psi}_j - i\omega\psi_j = -\varepsilon\gamma\left\{\frac{1}{\mu}\left(\frac{\psi_j - \psi_j^*}{2i\omega}\right)^3 + \frac{i\omega}{2}(\psi_j - \psi_j^*) + \frac{1}{2(N+1)}\right.$$

$$\left.\left[\left(\frac{\Phi_{j-1} - \Phi_{j-1}^*}{2i\omega}\right) - \left(\frac{\Phi_j - \Phi_j^*}{2i\omega}\right)\right]\sum_{s=0}^{N}\left(\frac{\Phi_s - \Phi_s^*}{2i\omega}\right)^2\right\}$$

Applying a procedure of multi-scale expansion, we introduce a super-slow timescale $\tau_1 = \varepsilon_1\tau_0$. Taking into account that

$$\frac{d}{d\tau_0} = \frac{\partial}{\partial\tau_0} + \varepsilon_1\frac{\partial}{\partial\tau_1} + \cdots,$$

we are looking for a solution in the following form: $\psi_j = \psi_{j0} + \varepsilon_1\psi_{j1} + \cdots$, $j = 1,\ldots,N$. We substitute this expansion into the system (14.45) and equate the terms of each order by parameter ε_1 to zero. In the first approximation, we get: ε_1, $j = 1,\ldots$, N. We substitute this expression into the equation for complex variables and consider next order of smallness. To avoid appearance of secular terms while integrating over time τ_0, coefficient before $e^{i\omega\tau_0}$ should be zero. Thus, we obtain the system which determines the "amplitude" functions $\varphi_j(\tau_1)$, $j = 1, 2$ in super-slow time τ_1:

$$\frac{\partial\varphi_{j,0}}{\partial\tau_1} - \frac{1}{2(N+1)}\frac{\gamma}{2i\omega}\left(2\varphi_{j,0}^* - \phi_{j-1,0}^* - \phi_{j+1,0}^*\right)\sum_{s=0}^{N}\frac{(\varphi_{s+1,0} - \varphi_{s,0})^2}{4\omega^2}$$

$$-\frac{1}{2(N+1)}\frac{\gamma}{2i\omega}\left(2\varphi_{j,0} - \varphi_{j-1,0} - \varphi_{j+1,0}\right)\sum_{s=0}^{N}\frac{\left|\varphi_{s+1,0} - \varphi_{s,0}\right|^2}{2\omega^2}$$

$$+\frac{\gamma}{\mu}\frac{3\varphi_{j,0}\left|\varphi_{j,0}\right|^2}{8i\omega^3} + \gamma\frac{i\omega}{2}\varphi_{j,0} = 0$$

Now, the functions $\varphi_j^m = a_m(\tau_1)e^{i\pi mj/(N+1)}$ and $\varphi_j^m = a_m(\tau_1)e^{-i\pi mj/(N+1)}$ are the exact solutions of the equation. Sine-like expression $\varphi_j^m = a_m(\tau_1)\sin(\pi mj/(N+1))$ is an exact solution only in the case without grounding supports. Also, it meets the boundary conditions (fixed ends). That is because we find a solution in two-mode approximation as a sum of two modes:

$$\varphi_j^m = A_m\sin(\pi mj/(N+1)) + A_n\sin(\pi nj/(N+1)).$$

Projecting these equations onto modes (replacing

$$\sin\frac{\pi m j}{N+1} = \frac{1}{2i}\left(e^{\frac{i\pi m j}{N+1}} - e^{-\frac{i\pi m j}{N+1}}\right)$$

getting the coefficient near $e^{\frac{i\pi m j}{N+1}}$), we obtain the following:

$$\dot{A}_m - \frac{3i\gamma}{32\mu\omega^3}\cdot\left(3A_m|A_m|^2 + 2A_k^2 A_m^* + 4A_m|A_k|^2\right)$$
$$+ \frac{i\gamma\omega_m^2}{32\omega^3}\left[3\omega_m^2|A_m|^2 A_m + 2\omega_k^2|A_k|^2 A_m + \omega_k^2 A_k^2 A_m^*\right] + \gamma\frac{i\omega}{2}A_m = 0$$
$$\dot{A}_k - \frac{3i\gamma}{32\mu\omega^3}\cdot\left(3A_k|A_k|^2 + 2A_m^2 A_k^* + 4A_k|A_m|^2\right)$$
$$+ \frac{i\gamma\omega_k^2}{32\omega^3}\left[3\omega_k^2|A_k|^2 A_k + 2\omega_m^2|A_m|^2 A_k + \omega_m^2 A_m^2 A_k^*\right] + \gamma\frac{i\omega}{2}A_k = 0$$

Here, for shortness and convenience, we denote:

$$\omega_k^2 = 4\sin^2\frac{\pi k}{2(N+1)}.$$

The obtained system is integrable because besides the integral of energy it possesses a second integral:

$$N = |A_m|^2 + |A_k|^2, \tag{14.46}$$

what can be verified directly. Due to existence of second integral, it is possible to introduce angular variables:

$$A_m = \sqrt{N}\cos\theta e^{i\delta_1}, \quad A_k = \sqrt{N}\sin\theta e^{i\delta_2}.$$

Here, θ and $\Delta = \delta_1 - \delta_2$ characterize relationship between amplitudes of two modes and phase shift between them.

In angular variables, equations are the following:

$$\sin 2\theta\dot{\Delta} = \frac{3\gamma N}{16\mu\omega^3}\left(\frac{1}{4}\sin 4\theta + \frac{1}{2}\sin 4\theta\cos 2\Delta\right)$$
$$+ \frac{\gamma N}{16\omega^3}\left(\frac{3}{2}\sin 2\theta(\omega_k^4\sin^2\theta - \omega_m^4\cos^2\theta) + \frac{\omega_m^2\omega_k^2}{4}\sin 4\theta(\cos 2\Delta + 2)\right)$$
$$\dot{\theta} = \frac{3\gamma N}{32\mu\omega^3}\sin 2\theta\sin 2\Delta + \frac{\gamma N}{32\omega^3}\omega_k^2\omega_m^2\frac{1}{2}\sin 2\theta\sin 2\Delta$$

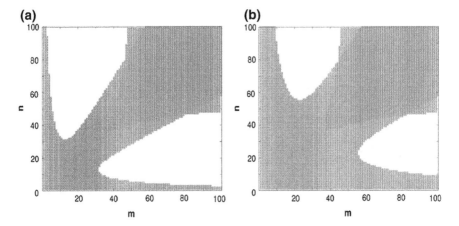

Fig. 14.11 Resonance domain for different μ: **a** $\mu = 200$, **b** $\mu = 20$. Chain consists of $N = 100$ particles

Occurence of resonance is defined by existence of stationary point of this system. From the second equation, we get: $\Delta_e = \pi/2$. Substituting in equation for θ, we get:

$$-\frac{3}{2\mu}\cos 2\theta + \left[\frac{3}{2}\left(\omega_k^4 \sin^2 \theta - \omega_m^4 \cos^2 \theta\right) - \frac{1}{2}\omega_m^2\omega_k^2 \cos 2\theta\right] = 0$$

Therefore

$$\mathrm{tg}^2\theta_e = \frac{\frac{3}{\mu} + 3\omega_m^4 - \omega_m^2\omega_k^2}{\frac{3}{\mu} + 3\omega_k^4 - \omega_m^2\omega_k^2}$$

So, condition of resonance occurence is the following (Fig. 14.11):

$$\frac{\frac{3}{\mu} + 3\omega_m^4 - \omega_m^2\omega_k^2}{\frac{3}{\mu} + 3\omega_k^4 - \omega_m^2\omega_k^2} > 0$$

14.3.3   Cluster Variables

We introduce the cluster variables as follows:

$$Y_1 = \frac{A_m + A_k}{2}, Y_2 = \frac{A_m - A_k}{2}.$$

In these variables, we come to equations:

$$\dot{Y}_1 - \frac{3i\gamma}{32\mu\omega^3}\left(9Y_1|Y_1|^2 + Y_2^2 Y_1^* + 2Y_1|Y_2|^2\right)$$

$$+ \frac{i\gamma}{64\omega^3}\left(3AY_1|Y_1|^2 + 6CY_2|Y_1|^2 + BY_2^2 Y_1^* + 3CY_1^2 Y_2^*\right.$$

$$\left. + 2BY_1|Y_2|^2 + 3CY_2|Y_2|^2\right) - \gamma\frac{i\omega}{2}Y_1 = 0$$

$$\dot{Y}_2 - \frac{3i\gamma}{32\mu\omega^3}\left(9Y_2|Y_2|^2 + Y_1^2 Y_2^* + 2Y_2|Y_1|^2\right)$$

$$+ \frac{i\gamma}{64\omega^3}\left(3AY_2|Y_2|^2 + 6CY_1|Y_2|^2 + BY_1^2 Y_2^* + 3CY_2^2 Y_1^*\right.$$

$$\left. + 2BY_2|Y_1|^2 + 3CY_1|Y_1|^2\right) - \gamma\frac{i\omega}{2}Y_2 = 0$$

Here, we introduce denotations, similar to the paper (Koroleva Kikot and Manevitch 2015):

$$A = \left(\omega_k^2 + \omega_m^2\right)^2, B = 3\omega_k^4 - 2\omega_k^2\omega_m^2 + 3\omega_m^4, C = \omega_m^4 - \omega_k^4.$$

14.3.4 Equations in Angular Variables in Cluster Variant

The obtained system is integrable because besides the integral of energy it possesses a second integral:

$$N = |Y_1|^2 + |Y_2|^2, \tag{14.47}$$

what can be verified directly. Due to existence of second integral, it is possible to introduce angular variables:

$$Y_1 = \sqrt{N}\cos\theta e^{i\delta_1}, \quad Y_2 = \sqrt{N}\sin\theta e^{i\delta_2}.$$

Here, θ and $\Delta = \delta_1 - \delta_2$ characterize relationship between amplitudes of two clusters and phase shift between them. In these variables, we obtain the following system:

$$\frac{1}{2}\sin 2\theta\dot{\Delta} = -\frac{3}{\mu}\left(-\frac{7}{4}\sin 4\theta + \frac{1}{4}\sin 4\theta\cos 2\Delta\right)$$

$$+ \frac{1}{2}\left(\frac{3A}{4}\sin 4\theta - \frac{B}{4}\sin 4\theta(\cos 2\Delta + 2) - 3C\cos 2\theta\cos\Delta\right) \tag{14.48}$$

$$\dot{\theta} = -\frac{3}{2\mu}\sin 2\theta\sin 2\Delta - \frac{1}{2}(B\sin\theta\cos\theta\sin 2\Delta + 3C\sin\Delta)$$

Here, overdot denotes differentiation with respect to normalized (for convenience) time $\tau_1^* = \frac{\gamma N}{32\omega^3}\tau_1$.

This first-order system of real equations possesses the energy integral:

$$
H = -\frac{3}{\mu}\left(\frac{9}{2}\sin^4\theta + \frac{9}{2}\cos^4\theta + \frac{1}{4}\sin^2 2\theta(\cos 2\Delta + 2)\right)
$$
$$
+ \frac{1}{2}\left(-\frac{3A}{2}(\sin^4\theta + \cos^4\theta) - 3C\sin 2\theta\cos\Delta - B\sin^2\theta\cos^2\theta(\cos 2\Delta + 2)\right),
$$

$$(14.49)$$

and hence, it is integrable. In angular variables, the stationary (equilibrium) points correspond to NNMs of initial system.

14.3.5 Phase Plane

Due to existence of the integral of motion, the simplest way of investigation is to study a topology of phase plane. Comparing phase planes for different values of parameter μ, we reveal two dynamical transitions which are reflected in the phase plane topology. The first one is caused by instability and bifurcation of the highest NNMs (Fig. 14.12).

When $\mu > \mu_{cr1}$ (as in a particular case also, when there is no grounding supports, $1/\mu = 0$), there are four critical points. When $\mu < \mu_{cr1}$, a bifurcation is observed: The point ($\theta = \pi/4$, $\Delta = 0$) (corresponding to in-phase motion of clusters) becomes unstable, and two additional equilibrium points appear. The first topological transition caused by bifurcation of considered NNM and appearance of new NNMs is a significant stage of the system evolution (in parametric space). This stage precedes to second topological transition which leads to spontaneous energy localization on initially excited cluster, when $\mu < \mu_{cr2}$ (complete energy exchange becomes impossible). It is possible to find a critical value μ_{cr2} analytically from the condition of coincidence of separatrix and LPT: $H(\pi/4,0) = H(0,\pi/2)$. Hence,

$$
\mu_{cr2} = \frac{3}{-A + 4C + B}.
$$

For two highest NNMs, $\mu_{cr2} = 1.35$ if $N = 10$.

The obtained results are confirmed by numerical integration of initial system (14.45) with initial conditions corresponding to excitation of one cluster which is formed by resonance interaction of two highest modes $\left(v_j = \sin(\pi(N-1)j/(N+1)) + \sin(\pi Nj/N+1)\right)$. When $\mu < \mu_{cr2}$, the energy localization is realized; when $\mu > \mu_{cr2}$, we observe complete energy exchange.

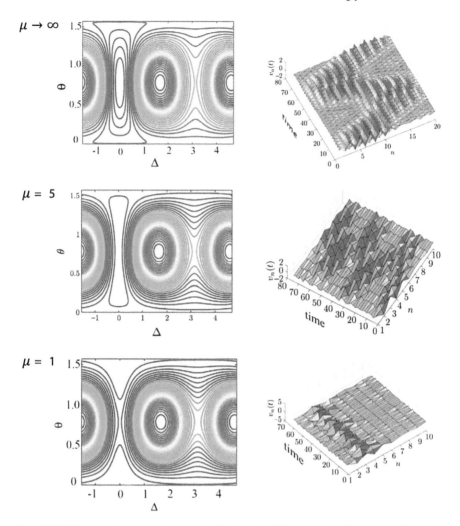

Fig. 14.12 Energy exchange and energy localization in initial variables and corresponding phase plane in angular variables

14.3.6   Analytical solution for LPT

Complete energy exchange between clusters described by LPT is a fundamental non-stationary process very important for applications. So it is important to have an analytical description of the process. Analytical description may be obtained in terms of angular variables obeying the Eq. (14.48). Since LPT is given by a condition:

$$H(\theta, \Delta) = H(0,0)$$

where energy integral is given by (14.49), we have the following equation:

$$-\frac{3}{\mu}\left(\frac{9}{2}\sin^4\theta + \frac{9}{2}\cos^4\theta + \frac{1}{4}\sin^2 2\theta(\cos 2\Delta + 2)\right)$$
$$+\frac{1}{2}\left(-\frac{3A}{2}(\sin^4\theta + \cos^4\theta) - 3C\sin 2\theta\cos\Delta\right.$$
$$\left.- B\sin^2\theta\cos^2\theta(\cos 2\Delta + 2)\right) = -\frac{93}{2\mu} - \frac{3A}{4}.$$

So, we have a following relationship between θ and Δ on LPT:

$$-\frac{3}{\mu}\left(-\frac{9}{4}\sin^2 2\theta + \frac{1}{4}\sin^2 2\theta(\cos 2\Delta + 2)\right)$$
$$+\frac{1}{2}\left(\frac{3A}{4}\sin^2 2\theta - 3C\sin 2\theta\cos\Delta - \frac{B}{4}\sin^2 2\theta(\cos 2\Delta + 2)\right) = 0$$

Either $\theta = 0$, $\theta = \pi/2$, what corresponds to straight-line parts of LPT, either

$$\cos\Delta = \frac{\frac{3}{2}C - \sqrt{\left(\frac{3}{2}C\right)^2 + \left(2\frac{3}{\mu} - \frac{B}{8} + \frac{3A}{8}\right)\left(2\frac{3}{\mu} + B\right)\sin^2 2\theta}}{\left(-\frac{3}{\mu} - \frac{B}{2}\right)\sin 2\theta}$$

Thus, we have an analytical description of LPT:

$$\theta = \frac{\pi}{2}\tau(t/a)$$

$$\Delta = -\arccos\left(\frac{\frac{3}{2}C - \sqrt{\left(\frac{3}{2}C\right)^2 + \left(\frac{6}{\mu} - \frac{B}{8} + \frac{3A}{8}\right)\left(\frac{6}{\mu} + B\right)\sin^2 \pi\tau}}{\left(-\frac{3}{\mu} - \frac{B}{2}\right)\sin \pi\tau}\right) \qquad (14.50)$$

Here, τ is sawtooth function with period $T = 2a$:

$$\tau(\tau_1) = 0.5((2/\pi)\arcsin(\sin(\pi\tau_1/a - \pi/2)) + 1),$$

$e(\tau)$ is its derivative in terms of generalized functions: $e(\tau_1) = d\tau/d\tau_1$. The period may be found from temporal equation for θ (Fig. 14.13).

$$T = 4a = 4\int_0^{\pi/2}\frac{d\theta}{\frac{3}{\mu}\sin 2\theta\sin 2\Delta + \frac{1}{2}(B\sin\theta\cos\theta\sin 2\Delta + 3C\sin\Delta)}$$

14.3.7   Poincare Sections

Since initial system (14.44) remains non-integrable even after projecting onto two-dimensional manifold, corresponding to NNM with minimal (in inverted coordinates) wave number, it is interesting to consider Poincare sections, corresponding to the system (14.48).

Let us consider a set of trajectories with fixed value of energy. The section plane is $\mathrm{Im}Y_2 = 0$. Intersection points of each trajectory with this plane are projected onto the plane $(\mathrm{Re}Y_1, \mathrm{Im}Y_1)$. LPTs are obtained by the same way when initial conditions correspond to exciting of only one cluster.

Obtained section maps are depicted on Fig. 14.14. One can see that a number of equilibrium points and LPT correspond to phase portrait obtained by asymptotical analysis in angular variables. Quite unexpected is the absence of chaotic behavior in all three domains of the parameter μ.

In particular, we reveal that for a string with arbitrary number of discrete masses in conditions of acoustic vacuum, there exists a regular regime of complete energy exchange between different domains of the string (clusters) and non-stationary energy localization on the excited cluster, alongside with NNMs and stationary energy localization. These regimes have been described analytically, and corresponding thresholds in parametric space were defined. Possibility of existence of different regimes in the same system is due to nonlinear grounding support, which enables also to widen the resonance domain. Therefore, the considered string can be used as an efficient energy sink.

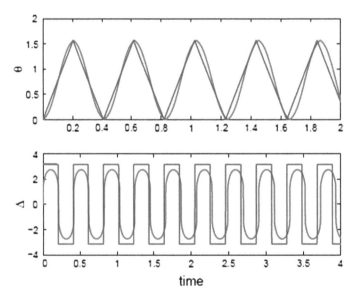

Fig. 14.13 Comparison of results of numerical solution of the system (14.48) and analytical approximation (14.50) for LPT, $\mu = 20$

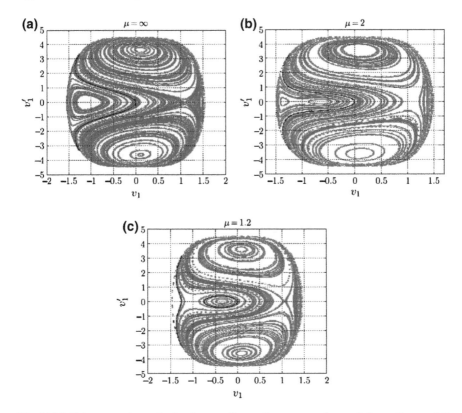

Fig. 14.14 Poincare sections **a** in the absence of grounding supports ($\mu = \infty$), **b** $\mu = 2$, **c** $\mu = 1.2$

Appendix 1: Timescale Separation

In the framework of the multi-scale method, the "fast" and "slow" times are determined by the rules:

$$\tau_0 = t; \quad \tau_1 = \varepsilon\tau_0$$

and the derivative with respect to the time are as follows:

$$\frac{d}{dt} = \frac{\partial}{\partial\tau_0} + \varepsilon\frac{\partial}{\partial\tau_0}$$

Let us rewrite Eq. (14.27) as follows:

$$\frac{\partial\varphi_j}{\partial\tau_0} + \varepsilon\frac{\partial\varphi_j}{\partial\tau_1} - \frac{\omega}{2}\varphi_j + F(\varphi_j) = 0 \tag{14.51}$$

where

$$
F(\varphi_j) = -\frac{\beta}{\alpha\sqrt{2\omega}}\left[J_1\left(\alpha\sqrt{\frac{2}{\omega}}|\varphi_{j+1}-\varphi_j|\right)\frac{\varphi_{j+1}-\varphi_j}{|\varphi_{j+1}-\varphi_j|}\right.
$$
$$
\left.-J_1\left(\alpha\sqrt{\frac{2}{\omega}}|\varphi_j-\varphi_{j-1}|\right)\frac{\varphi_j-\varphi_{j-1}}{|\varphi_j-\varphi_{j-1}|}\right]+\frac{1}{\sqrt{2\omega}}J_1\left(\sqrt{\frac{2}{\omega}}|\varphi_j|\right)\frac{\varphi_j}{|\varphi_j|}
$$

Let us consider the solution in the form $\phi_j = \phi_{j,0} + \varepsilon\chi_j$, where $\phi_{j,0}$ satisfies Eq. (14.27) and ε is a small parameter. Taking into account that $\phi_{j,0} = \text{const}$, one can write Eq. (14.51) as follows:

$$
\varepsilon\frac{\partial\chi_j}{\partial\tau_0}+\varepsilon^2\frac{\partial\chi_j}{\partial\tau_1}-\frac{\omega}{2}(\varphi_{j,0}+\varepsilon\chi_j)+F(\varphi_{j,0}+\varepsilon\chi_j)=0 \tag{14.52}
$$

One can consider this equation from the viewpoint of the order of small parameter ε:

$$
\varepsilon\frac{\partial\chi_j}{\partial\tau_0}+\varepsilon^2\frac{\partial\chi_j}{\partial\tau_1}-\frac{\omega}{2}\varphi_{j,0}+F(\varphi_{j,0})-\varepsilon\left(\frac{\omega}{2}-\left(\frac{\partial}{\partial x}F(x)\right)_{x=\varphi_{j,0}}\right)\chi_j=0 \tag{14.53}
$$

Taking into account Eq. (14.27), one can get:

$$
\varepsilon\frac{\partial\chi_j}{\partial\tau_0}+\varepsilon^2\frac{\partial\chi_j}{\partial\tau_1}-\varepsilon\left(\frac{\omega}{2}-\left(\frac{\partial}{\partial x}F(x)\right)_{x=\varphi_{j,0}}\right)\chi_j
$$
$$
=\varepsilon\frac{\partial\chi_j}{\partial\tau_0}+\varepsilon^2\frac{\partial\chi_j}{\partial\tau_1}-\varepsilon\frac{\partial}{\partial x}\left[\frac{\omega}{2}x-F(x)\right]_{x=\varphi_{j,0}}\chi_j=0 \tag{14.54}
$$

So, if the last term in Eq. (14.54) corresponds to a small value of order ε, one can separate the different orders of the small parameter ε:

$$
\varepsilon:\quad \frac{\partial\chi_j}{\partial\tau_0}=0
$$
$$
\varepsilon^2:\quad \frac{\partial\chi_j}{\partial\tau_1}-\frac{1}{\varepsilon}\frac{\partial}{\partial x}\left[\frac{\omega}{2}x-F(x)\right]_{x=\varphi_{j,0}}\chi_j=0
$$

One should notice that the term in the square brackets has to be proportional to the relative difference of the modes' frequencies $\Delta\omega/\omega$ and the small parameter ε can be estimated as follows:

$$
\varepsilon\cong\Delta\omega/\omega
$$

Because the stationary solution $\phi_{j,0}$ does not depend on the time, the nonlinear evolution equation may be written as Eq. (14.35).

Appendix 2: Projection onto Two Modes—Formulas

$$\frac{e^{\frac{ij\pi m}{N+1}} - e^{\frac{-ij\pi m}{N+1}}}{2i}\dot{A}_m + \frac{e^{\frac{ij\pi k}{N+1}} - e^{\frac{-ij\pi k}{N+1}}}{2i}\dot{A}_k$$

$$-\frac{3i\gamma}{8\mu\omega^3}\left(\frac{e^{\frac{ij\pi m}{N+1}} - e^{\frac{-ij\pi m}{N+1}}}{2i}A_m + \frac{e^{\frac{ij\pi k}{N+1}} - e^{\frac{-ij\pi k}{N+1}}}{2i}A_k\right)^2\left(\frac{e^{\frac{ij\pi m}{N+1}} - e^{\frac{-ij\pi m}{N+1}}}{2i}A_m^* + \frac{e^{\frac{ij\pi k}{N+1}} - e^{\frac{-ij\pi k}{N+1}}}{2i}A_k^*\right)$$

$$+\frac{i\gamma}{16\omega^3(N+1)}\left(2\frac{e^{\frac{ij\pi m}{N+1}} - e^{\frac{-ij\pi m}{N+1}}}{2i}A_m^* + 2\frac{e^{\frac{ij\pi k}{N+1}} - e^{\frac{-ij\pi k}{N+1}}}{2i}A_k^*\right.$$

$$-\frac{e^{\frac{i(j+1)\pi m}{N+1}} - e^{\frac{-i(j+1)\pi m}{N+1}}}{2i}A_m^* - \frac{e^{\frac{i(j+1)\pi k}{N+1}} - e^{\frac{-i(j+1)\pi k}{N+1}}}{2i}A_k^*$$

$$\left.-\frac{e^{\frac{i(j-1)\pi m}{N+1}} - e^{\frac{-i(j-1)\pi m}{N+1}}}{2i}A_m^* - \frac{e^{\frac{i(j-1)\pi k}{N+1}} - e^{\frac{-i(j-1)\pi k}{N+1}}}{2i}A_k^*\right)\sum_{s=0}^{N}\left(\varphi_{s+1,0} - \varphi_{s,0}\right)^2$$

$$+\frac{i\gamma}{8\omega^3(N+1)}\left(2\frac{e^{\frac{ij\pi m}{N+1}} - e^{\frac{-ij\pi m}{N+1}}}{2i}A_m + 2\frac{e^{\frac{ij\pi k}{N+1}} - e^{\frac{-ij\pi k}{N+1}}}{2i}A_k\right.$$

$$-\frac{e^{\frac{i(j+1)\pi m}{N+1}} - e^{\frac{-i(j+1)\pi m}{N+1}}}{2i}A_m - \frac{e^{\frac{i(j+1)\pi k}{N+1}} - e^{\frac{-i(j+1)\pi k}{N+1}}}{2i}A_k$$

$$\left.-\frac{e^{\frac{i(j-1)\pi m}{N+1}} - e^{\frac{-i(j-1)\pi m}{N+1}}}{2i}A_m - \frac{e^{\frac{i(j-1)\pi k}{N+1}} - e^{\frac{-i(j-1)\pi k}{N+1}}}{2i}A_k\right)\sum_{s=0}^{N}\left|\varphi_{s+1,0} - \varphi_{s,0}\right|^2$$

$$+\frac{\gamma}{\mu}\frac{i\omega}{2}\left(\frac{e^{\frac{ij\pi m}{N+1}} - e^{\frac{-ij\pi m}{N+1}}}{2i}A_m + \frac{e^{\frac{ij\pi k}{N+1}} - e^{\frac{-ij\pi k}{N+1}}}{2i}A_k\right) = 0$$

We calculate sums in brackets separately; they do not depend on index j.

$$\sum_{s=0}^{N}\left(\varphi_{s+1,0} - \varphi_{s,0}\right)^2 = \sum_{s=0}^{N}\left(\sin\frac{\pi(s+1)m}{N+1}A_m + \sin\frac{\pi(s+1)k}{N+1}A_k\right.$$

$$\left.- \sin\frac{\pi sm}{N+1}A_m - \sin\frac{\pi sk}{N+1}A_k\right)^2 = A_m^2(N+1)\left(1 + \cos\frac{m\pi}{N+1}\right)$$

$$+ A_k^2(N+1)\left(1 + \cos\frac{k\pi}{N+1}\right) = \frac{N+1}{2}\left(\omega_m^2 A_m^2 + \omega_k^2 A_k^2\right)$$

Analogously,

$$\sum_{s=0}^{N} |\varphi_{s+1,0} - \varphi_{s,0}|^2 = \sum_{s=0}^{N} \left| \sin\frac{\pi(s+1)m}{N+1}A_m + \sin\frac{\pi(s+1)k}{N+1}A_k \right.$$

$$\left. - \sin\frac{\pi sm}{N+1}A_m - \sin\frac{\pi sk}{N+1}A_k \right|^2 = |A_m|^2(N+1)\left(1 + \cos\frac{m\pi}{N+1}\right)$$

$$+ |A_k|^2(N+1)\left(1 + \cos\frac{k\pi}{N+1}\right) = \frac{N+1}{2}\left(\omega_m^2|A_m|^2 + \omega_k^2|A_k|^2\right)$$

Now, we want to project these equations onto modes; that is, we get the coefficient near $e^{i\pi mj/N+1}$. We obtain the following equations:

$$\dot{A}_m - \frac{3i\gamma}{32\mu\omega^3} \cdot \left(3A_m|A_m|^2 + 2A_k^2 A_m^* + 4A_m|A_k|^2\right)$$

$$+ \frac{i\gamma}{2\omega}\frac{1}{16\omega^2}2\left[|A_m|^2\omega_m^2 + |A_k|^2\omega_k^2\right]\omega_m^2 A_m$$

$$+ \frac{i\gamma}{2\omega}\frac{1}{16\omega^2}\left[A_m^2\omega_m^2 + A_k^2\omega_k^2\right]\omega_m^2 A_m^* + \gamma\frac{i\omega}{2}A_m = 0$$

$$\dot{A}_k - \frac{3i\gamma}{32\mu\omega^3} \cdot \left(3A_k|A_k|^2 + 2A_m^2 A_k^* + 4A_k|A_m|^2\right)$$

$$+ \frac{i\gamma}{2\omega}\frac{1}{16\omega^2}2\left[|A_m|^2\omega_m^2 + |A_k|^2\omega_k^2\right]\omega_k^2 A_k$$

$$+ \frac{i\gamma}{2\omega}\frac{1}{16\omega^2}\left[A_m^2\omega_m^2 + A_k^2\omega_k^2\right]\omega_k^2 A_k^* + \gamma\frac{i\omega}{2}A_k = 0$$

Simplifying these equations, we get the equation. In cluster variables,

$$\frac{i\gamma}{32\omega^3}\left(\frac{3}{2}\left(\omega_k^2 + \omega_m^2\right)^2 Y_1|Y_1|^2 + 3\left(\omega_m^4 - \omega_k^4\right)Y_2|Y_1|^2 + \frac{1}{2}\left(3\omega_k^4 - 2\omega_k^2\omega_m^2 + 3\omega_m^4\right)Y_2^2 Y_1^*\right.$$

$$+ \frac{3}{2}\left(\omega_m^4 - \omega_k^4\right)Y_1^2 Y_2^* \frac{i\gamma}{32\omega^3}\left(\frac{3}{2}\left(\omega_k^2 + \omega_m^2\right)^2 Y_1|Y_1|^2 + 3\left(\omega_m^4 - \omega_k^4\right)Y_2|Y_1|^2\right.$$

$$+ \frac{1}{2}\left(3\omega_k^4 - 2\omega_k^2\omega_m^2 + 3\omega_m^4\right)Y_2^2 Y_1^* + \frac{3}{2}\left(\omega_m^4 - \omega_k^4\right)Y_1^2 Y_2^*$$

$$\left.\left. + \left(3\omega_k^4 - 2\omega_k^2\omega_m^2 + 3\omega_m^4\right)Y_1|Y_2|^2 + \frac{3}{2}\left(\omega_m^4 - \omega_k^4\right)Y_2|Y_2|^2\right)\right) - \gamma\frac{i\omega}{2}Y_1 = 0$$

$$\dot{Y}_2 - \frac{3i\gamma}{32\mu\omega^3}\left(9Y_2|Y_2|^2 + Y_1^2 Y_2^* + 2Y_2|Y_1|^2\right)$$

$$+ \frac{i\gamma}{32\omega^3}\left(\frac{3}{2}\left(\omega_k^2 + \omega_m^2\right)^2 Y_2|Y_2|^2 + 3\left(\omega_m^4 - \omega_k^4\right)Y_1|Y_2|^2 + \frac{1}{2}\left(3\omega_k^4 - 2\omega_k^2\omega_m^2 + 3\omega_m^4\right)Y_1^2 Y_2^*\right.$$

$$\left. + \frac{3}{2}\left(\omega_m^4 - \omega_k^4\right)Y_2^2 Y_1^* + \left(3\omega_k^4 - 2\omega_k^2\omega_m^2 + 3\omega_m^4\right)Y_2|Y_1|^2 + \frac{3}{2}\left(\omega_m^4 - \omega_k^4\right)Y_1|Y_1|^2\right) - \gamma\frac{i\omega}{2}Y_2 = 0$$

References

Arnold, V.I., Kozlov, V.V., Neishtadt, A.I.: Mathematical Aspects of Classical and Celestial Mechanics, 3rd edn. Encyclopaedia mathematical sciences, vol. 3. Springer, Berlin (2006)

Bogoliubov, N.N., Mitropolskii, Y.A.: Asymptotic Methods in the Theory of Nonlinear Oscillations. Gordon and Breach, New York (1961)

Braun, O.M., Kivshar, Y.S.: The Frenkel–Kontorova Model. Springer, Berlin (2004)

Chirikov, B.V., Zaslavsky, G.M.: Sov. Phys. Uspekhi **14**, 549 (1972)

Dauxois, T., Peyrard, M.: Energy localization in nonlinear lattices. Phys. Rev. Lett **70**, 3935 (1993)

Frenkel, Y., Kontorova, T.: On the theory of plastic deformation and twining. Phys. Z. Sowietunion **13**, 1 (1938)

Han, H., Feng, L., Xiong, S., Shiga, T., Shiomi, J., Volz, S., Kosevich, Y.A.: Long-range interatomic forces can minimize heat transfer: from slowdown of longitudinal optical phonons to thermal conductivity minimum. Phys. Rev. B **94**, 054306 (2016)

Koroleva (Kikot), I., Manevitch, L.I., Vakakis, A.F.: Non-stationary resonance dynamics of a nonlinear sonic vacuum with grounding supports. J. Sound Vib. **357**, 349–364 (2015a)

Koroleva (Kikot), I., Manevitch, L.I.: Oscillatory chain with grounding support in conditions of acoustic vacuum. Rus. J. Nonlin. Dyn. **11**(3), 487–502 (2015b) (Russian)

Kuehn, C.: Multiple timescale Dynamics. Springer, New York (2015)

Manevitch, L.I.: New approach to beating phenomenon in coupled nonlinear oscillatory chains. Arch. Appl. Mech. **77**, 301 (2007)

Manevitch, L.I., Romeo, F.: Non-stationary dynamics of weakly coupled pendula. Europhys. Lett. **112**, 30005 (2015)

Manevitch, L.I., Smirnov, V.V.: Limiting phase trajectories and thermodynamics of molecular chains. Doklady Physics **55**, 324 (2010a)

Manevitch, L.I., Smirnov, V.V.: Limiting phase trajectories and the origin of energy localization in nonlinear oscillatory chains. Phys. Rev. E **82**, 036602 (2010b)

Manevitch, L.I., Vakakis, A.F.: Nonlinear oscillatory acoustic vacuum. SIAM J **74**, 1742–1762 (2014)

Mickens, R.E.: Truly Nonlinear Oscillators: An Introduction to Harmonic Balance, Parameter Expansion, Iteration, and Averaging Methods. World Scientific Publishing Co. Pte. Ltd, Singapore (2010)

Neishtadt, A.I.: Passage through a separatrix in a resonance problem with slowly varying parameter. J. Appl. Math. Mech. **39**, 594 (1975)

Novikov, S., Manakov, S., Pitaevskii, L., Zakharov, V.: Theory of Solitons: The Inverse Scattering Method, Monographs in Contemporary Mathematics, vol. 276. Springer, US (1984)

Peyrard, M., Bishop, A.R.: Statistical mechanics of a nonlinear model for DNA denaturation. Phys. Rev. Lett **62**, 2755 (1989)

Rosenau, P.: Dynamics of non-linear mass-spring chains near the continuous limit. Phys. Lett. A **118**, 222 (1986)

Rosenberg, R.M.: On nonlinear vibrations of systems with many degrees of freedom. Adv. Appl. Mech **9**, 156–243 (1966)

Sagdeev, R.Z., Usikov, D.A., Zaslavsky, G.M.: Nonlinear Physics: From the Pendulum to Turbulence and Chaos, p. 315. Harwood Academic Publishers, New York (1988)

Sanders, J.A., Verhulst, F., Murdock, J.: Averaging Methods in Nonlinear Dynamical Systems. Springer, New York (2007)

Smirnov, V.V., Manevitch, L.I.: Limiting phase trajectories and dynamic transitions in nonlinear periodic systems. Acoust. Phys. **57**(2), 271–276 (2011)

Takeno, S., Homma, S.: A sine-lattice (sine-form discrete sine-gordon) equation—one and two-Kink solution and physical models. J. Phys. Soc. Jpn. **55**, 65 (1986)

Takeno, S., Peyrard, M.: Nonlinear modes in coupled rotator models. Phys. D **92**, 140 (1996)

Takeno, S., Peyrard, M.: Nonlinear rotating modes: Green's-function solution. Phys. Rev. E **55**, 1922 (1997)

Vakakis, A.F., Manevitch, L.I., Mikhlin, YuV, Pilipchuk, V.N., Zevin, A.A.: Normal Modes and Localization in Nonlinear Systems. Wiley, New York (1996)

Zaslavsky, G.M.: Physics of Chaos in Hamiltonian Dynamics. Imperial College Press, London (1998)

Chapter 15
Nonlinear Vibrations of the Carbon Nanotubes

The previous sections concern the discrete systems, where the discreteness of the oscillation spectra is defined by the finiteness of the number of the particles forming the system under consideration. However, the discreteness of the normal mode frequencies can also exist in the systems with the distributed parameters. Such a system with a finite length is described by the set of partial differential equations; however, the oscillation spectrum is defined by the set of the waves, which satisfy the boundary conditions. Due to the finiteness of the system, the number of such waves is denumerable; therefore, the respective normal mode frequencies turn out to be separated by the finite gaps. Under these circumstances, the resonant conditions are defined by the relative value of the frequencies' difference. As it was mentioned above, such conditions can exist near the edges of the optical-type spectra, where the nonlinear coupling between the frequencies and the wave numbers occurs.

This chapter contains the analysis of the nonlinear vibrations of the single-walled carbon nanotubes (SWCNTs) in the framework of the thin elastic shell theory. The efficiency of the LPT concept in combination with the semi-inverse asymptotic method becomes apparent in finding phenomena such as the slow energy exchange and energy localization, that results from the nonlinear resonant interaction of the normal modes.

The nonlinear dynamics of the CNTs is of great interest from the viewpoint of understanding the vibrations of the nanotubes embedded into the elastic medium (Soltani et al. 2012; Mahdavi et al. 2011), the strong modes coupling in the nanotube resonators (Eichler et al. 2012; Greaney and Grossman 2007), the processes of the phonon–phonon interactions (Gambetta et al. 2006; Greaney et al. 2009; De Martino et al. 2009), and the thermal properties of the CNTs (Li et al. 2005; Savin et al. 2009; Zhang et al. 2012).

The majority of the CNT dynamics studies deal with the molecular dynamics (MD) simulation (Greaney and Grossman 2007; Savin et al. 2009; Rafiee and Moghadam 2014; Hu et al. 2012; Maruyama 2002; Srivastava et al. 2007; Chen and Kumar 2011; Pine et al. 2011) or the continuum approach (the elastic thin shell or

© Springer Nature Singapore Pte Ltd. 2018

L.I. Manevitch et al., *Nonstationary Resonant Dynamics of Oscillatory Chains and Nanostructures*, Foundations of Engineering Mechanics,
DOI 10.1007/978-981-10-4666-7_15

beam theory) (Kahn et al. 2001; Mahan 2002; Wang et al. 2004; Chico et al. 2006; Liew and Wang 2007; Shi et al. 2009a, b; Ghavanloo and Fazelzaden 2012; Soltani et al. 2011). As concerns the continuum approach, because of the extreme complexity of the phonon band structure, the dynamics of the different types of the CNT vibrations cannot be described in the framework of the unified viewpoint. In particular, the reduced nonlinear theory of the elastic thin shell under appropriate physical hypotheses is required to study the low-frequency circumferential flexure modes (Smirnov et al. 2014; Strozzi et al. 2014). To investigate analytically the nonlinear effects, which may occur in the radial breathing branch of the CNT vibration, a new model in the framework of the thin elastic shell theory is needed.

The specific feature of the optical-type vibrational branches is crowding the frequencies near the long wave edge of the spectrum. Thus, the possibility of the resonant interaction of the nonlinear normal modes (NNMs) appears.

This section contains the comparative analysis of the resonant interaction of the nonlinear normal modes in the cases of two optical-type oscillations: the circumferential flexure mode (CFM) and radial breathing mode (RBM). We consider the resonant interaction of the NNMs near the long wavelength (low-frequency) edge in the framework of the nonlinear dynamical equation for the radial component of the displacement field.

15.1   Nonlinear Optical Vibrations of Single-Walled Carbon Nanotubes

The RBMs (the circumferential wave number $n = 0$) belong to the most well-known Raman-active oscillation branch of the carbon nanotubes vibrations, and there are many studies of the RBM, which are based on the different approaches since the continuum shell theory up to the quantum ab initio methods (Ye et al. 2004; Chang 2007; Kürti et al. 2003; Chico et al. 2006; Lawler et al. 2005). The measurement of the RBM frequency allows to identify the CNT (Rao et al. 1997; Saito et al. 1998; Dresselhaus and Eklund 2000) as well as to determine the critical pressure of the structural transition (Yang et al. 2007; Lebedkin et al. 2006).

The CFMs belong to the most low-frequency optical-type modes with the gap frequency ~ 20 cm^{-1}, which are specific for the deformation of the CNTs under uniaxial loading in the direction, which is normal to the CNT axis. The respective circumferential wave number $n = 2$. The CFMs are specified by intensively changing the lateral section area without any appreciable changes in the contour length. Thus, the circumferential and shear deformations are negligible and the main variations regard to the curvatures and longitudinal deformation. The specific transversal section of the CNT during the radial breathing and circumferential flexure modes is shown in Fig. 15.1.

Fig. 15.1 The specific changes of lateral section of CNT during the radial breathing (*long-dashes*) and circumferential flexure (*dashes*) oscillations. The *solid curves* show the undistorted profile of the CNT. The *arrows* show the specific radial displacements which occur during the radial breathing (w_{RBM}) and circumferential flexure (w_{CFM}) oscillations

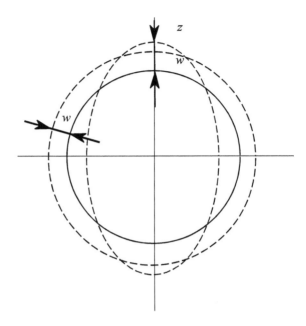

15.1.1 The Model

The key problem for successful study of CNT's nonlinear dynamics is a construction of adequate reduced model. The applicability of the thin elastic shell theory (TTES) to the mechanics of the CNTs has been discussed in the number of works (Harik et al. 2002; Harik 2002; Rafiee and Moghadam 2014; Silvestre et al. 2011; Silvestre 2012). It is noteworthy that, in contrast to macroscopic mechanics, there is no restriction in its application caused by plastic deformation. The TTES allows us to obtain an effective description of the vibrational spectrum in the framework of the linear approximation (Mahan 2002; Hu et al. 2008, 2012; Rafiee and Moghadam 2014; Gibson et al. 2007; Jiang et al. 2010). The reduced model presented below admits efficient study of both linear and nonlinear dynamics of CNTs under arbitrary boundary conditions.

As it was mentioned above, the consideration of the oscillations of different types needs the distinguished approaches, which have to take into account the specific relationships between CNT deformations. The equations of motion for the CFMs derived in this chapter are based on the hypothesis of a smallness of circumferential and shear deformations, while the equations for the RBMs have been obtained under hypothesis of the smallness of the Poisson ratio. However, all equations were obtained in the framework of the nonlinear Sanders-Koiter thin shell theory (Amabili 2008), and the asymptotic limit leads to the equations, which like the nonlinear Schrodinger equation with the specific type of the nonlinearity. The spectra of both branches for the CNTs with various aspect ratios and under different boundary conditions are compared with those obtained by the direct numerical integration of the equations of the Sanders-Koiter thin shell theory.

15.1.2 Radial Breathing Mode

In this section, we consider the radial breathing oscillations of the CNT in the framework of nonlinear Sanders-Koiter thin shell theory and their resonant interaction.

The dimensionless energy of elastic deformation of CNT elastic deformation can be written as follows:

$$
E = \iint d\xi d\varphi \frac{1}{2} \left(\varepsilon_\xi^2 + 2v\varepsilon_\xi\varepsilon_\varphi + \varepsilon_\varphi^2 + \frac{1}{2}(1-v)\varepsilon_{\xi\varphi}^2 \right.
$$
$$
\left. + \frac{1}{12}\beta^2 \left(\kappa_\xi^2 + 2v\kappa_\xi\kappa_\varphi + \kappa_\varphi^2 + \frac{1}{2}(1-v)\kappa_{\xi\varphi}^2 \right) \right)
$$
(15.1)

where ε_ξ, ε_φ, and $\varepsilon_{\xi\varphi}$ are the longitudinal, circumferential, and shear deformations; κ_ξ, κ_φ, and $\kappa_{\xi\varphi}$ are the longitudinal and circumferential curvatures, and torsion, respectively; ξ is the dimensionless (in the unit of the CNT's length L) coordinate along the CNT axis; φ is the circumferential one.

The dimensionless energy and time variables are measured in the units:

$$
E_0 = YRLh/\left(1-v^2\right) \quad \text{and} \quad t_0 = \sqrt{\frac{\rho R^2(1-v^2)}{Y}},
$$

respectively. Here, Y is the Young's modulus of graphene sheet, ρ is its mass density, v is the Poisson ratio of CNT, and R, L, and h are the CNT's radius, length, and effective thickness of the wall. Two dimensionless geometric parameters specify the CNT: the inverse aspect ratio $\alpha = R/L$ and the relative thickness of the effective shell $\beta = h/R$.

The Sanders-Koiter approximation of the defectless thin shell (Amabili 2008) allows to write the nonlinear deformations (ε) and curvatures (κ) in the following form:

$$
\begin{aligned}
\varepsilon_\xi &= \alpha\partial_\xi u + \frac{1}{2}(\alpha\partial_\xi w)^2 + \frac{1}{8}(\alpha\partial_\xi v - \partial_\varphi u)^2 \\
\varepsilon_\varphi &= \partial_\varphi v + w + \frac{1}{2}(\partial_\varphi w - v)^2 + \frac{1}{8}(\partial_\varphi u - \alpha\partial_\xi v)^2 \\
\varepsilon_{\xi\varphi} &= \partial_\varphi u + \alpha\partial_\xi v + \alpha\partial_\xi w(\partial_\varphi w - v) \\
\kappa_\xi &= -\alpha^2\partial_{\xi\xi}^2 w \\
\kappa_\varphi &= \partial_\varphi v - \partial_{\varphi\varphi}^2 w \\
\kappa_{\xi\varphi} &= -2\alpha\partial_{\xi\varphi}^2 w + \frac{3}{2}\alpha\partial_\xi v - \frac{1}{2}\partial_\varphi u
\end{aligned}
$$
(15.2)

where u, v, and w are the dimensionless (in the unit of CNT's radius R) longitudinal, tangential, and radial displacements, respectively. The symbols $\partial_\xi = \partial/\partial\xi$ and $\partial^2_{\varphi\varphi} = \partial^2/\partial\varphi^2$ correspond to the partial derivatives with respect to independent variables.

Varying energy (15.1) with respect to the variables u, v, and w, one can estimate the vibration spectrum for different values of the circumferential wave number n in the linear approximation. The frequencies of oscillations with $n = 0$, 1, and 2 are shown in Fig. 15.2 for the CNT with aspect ratio $1/\alpha = 30$ and $\beta = 0.1$, and the Poisson ratio $v = 0.19$. The unit frequency corresponds to the gap of RBM, the dimensional value of which is equal to $\omega_{\text{RBM}} = \sqrt{Y/\rho R^2(1 - v^2)}$.

Taking into account that the RBMs have not depend on the azimuthal angle φ (the respective azimuthal wave number $n = 0$) and the transversal displacement $v = 0$, one can obtain the equations of motion with taking into account the elastic deformation energy (15.1):

$$\frac{\partial^2 u}{\partial t^2} - \alpha v \frac{\partial w}{\partial \xi} - \alpha^2 \frac{\partial^2 u}{\partial \xi^2} - \alpha^3 \frac{\partial w}{\partial \xi} \frac{\partial^2 w}{\partial \xi^2} = 0$$

$$\frac{\partial^2 w}{\partial t^2} + w + \frac{\partial^2 w}{\partial \xi^2} + \alpha v \frac{\partial w}{\partial \xi} - \alpha^2 \frac{\partial}{\partial \xi}\left(vw \frac{\partial w}{\partial \xi} - \alpha \frac{\partial w}{\partial \xi} \frac{\partial u}{\partial x}\right) - \frac{3}{2}\alpha^4 \left(\frac{\partial w}{\partial \xi}\right)^2 \frac{\partial^2 w}{\partial \xi^2} = 0$$

$$(15.3)$$

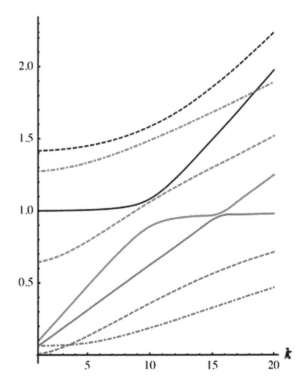

Fig. 15.2 The vibration spectrum according to the linearized Sanders-Koiter thin shell theory for the CNT with aspect ratio $L/R = 30$ at the periodic boundary conditions: *Solid curves* correspond to circumferential wave number $n = 0$; *dashed ones*, to $n = 1$; and *dot-dashed ones*, to $n = 2$. All the frequencies ω are measured in dimensionless units, and k denotes the number of longitudinal half-waves along the CNT

The dispersion curves of the linearized problem contain two branches:

$$\omega = \frac{1}{2}\left(1 \pm \left(k^2\alpha^2 - \sqrt{1 - (1 - 2v^2)2k^2\alpha^2 + k^4\alpha^4}\right)\right) \qquad (15.4)$$

the lower of which corresponds to the longitudinal acoustic modes and the higher one to the radial breathing modes (k is the longitudinal wave number). Figure 15.3 shows the frequencies of the long wavelength modes for the CNTs with different aspect ratios. In spite of that the real wave number for the periodic boundary conditions starts from the value $k = 1$, the product αk is small if the CNT is long enough.

The long wavelength limit of (15.4) shows the spectrum crowding:

$$\omega = 1 + k^2\alpha^2 v^2 + k^4\left(\alpha^4 v^2 - \alpha^4 v^4\right) \qquad (15.5)$$

The respective eigenvector

$$\{u, w\} = \left\{-\frac{-1 + k^2\alpha^2 + \sqrt{1 - 2k^2\alpha^2 + k^4\alpha^4 + 4k^2\alpha^2 v^2}}{2k\alpha v}, 1\right\} \sim \{-\alpha v k, 1\} \qquad (15.6)$$

shows the relationship between longitudinal and radial components of the displacement field.

Taking into account the expression (15.6), one can find the relation between u and w in the coordinate space as:

$$u = -\alpha v \frac{\partial w}{\partial \xi} \qquad (15.7)$$

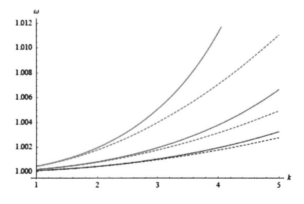

Fig. 15.3 The dimensionless RBM frequencies for the CNTs with different aspect ratios at periodic boundary conditions according to Eq. (15.4) (*solid lines*) and Eq. (15.5) (*dashed lines*). The aspect ratios are equal to 20 (*red*), 30 (*blue*), and 40 (*black*)

Using this expression, one can rewrite second equation of (15.3) as follows:

$$\frac{\partial^2 w}{\partial t^2} + w - \alpha^2 v^2 \frac{\partial^2 w}{\partial \xi^2} - \frac{\alpha^2 v}{2} \frac{\partial}{\partial \xi}\left(w \frac{\partial w}{\partial \xi}\right)$$
$$+ \alpha^4 \left(v \frac{\partial}{\partial \xi}\left(\frac{\partial w}{\partial \xi} \frac{\partial^2 w}{\partial \xi^2}\right) - \frac{3}{2}\left(\frac{\partial w}{\partial \xi}\right)^2 \frac{\partial^2 w}{\partial \xi^2}\right) = 0$$

(15.8)

To perform the asymptotic analysis of the long wavelength dynamics of the RB modes, it is convenient to rewrite Eq. (15.8) using complex variables:

$$\psi = \frac{1}{\sqrt{2}}\left(\frac{\partial w}{\partial t} + iw\right)$$

Returning to expression (15.6), one can see that the assumption of the Poisson ratio smallness $v < 1$ allows to use it as a small parameter. Using the multiple timescales, $\tau_0 = \tau$, $\tau_1 = v\tau$, $\tau_2 = v^2\tau$, etc., and expanding the function ψ into series of the small parameter v:

$$\psi = v\psi_0 + v^2\psi_1 + \dots$$

one can get the equation for the main order amplitude in the "slow" time τ_2 (see, e.g., Manevitch and Smirnov 2010a, b, c; Smirnov et al. 2016a, b for details):

$$i\frac{\partial \chi_0}{\partial \tau_2} - \frac{\alpha^2}{2}\frac{\partial^2 \chi_0}{\partial \xi^2} - \frac{3}{8}\alpha^4 \frac{\partial}{\partial \xi}\left(\left|\frac{\partial \chi_0}{\partial \xi}\right|^2 \frac{\partial \chi_0}{\partial \xi}\right) = 0,$$

(15.9)

where $\chi_0 = \chi_0(\xi, \tau_2)$ is the slow-varying envelope of the function $\psi = \chi_0(\xi, \tau_2)e^{i\tau_0}$. Equation (15.9) admits the plane wave solution:

$$\chi_0(\xi, \tau_2) = Ae^{i(\tilde{\omega}\tau_2 - k\xi)}$$

(15.10)

Taking into account the "slowness" of the time τ_2, the respective dispersion relation should be written as follows:

$$\omega = 1 + v^2 \frac{1}{2}\alpha^2 k^2 \left(1 + \frac{3}{4}\alpha^2 k^2 A^2\right)$$

(15.11)

One can see that the amplitude-independent part of (15.11) is in accordance with the relation (15.5), while the effective positiveness of nonlinear addition points out the hard type of nonlinearity.

Equation (15.9) is the modified nonlinear Schrodinger equation (NLSE) with the gradient type of the nonlinearity. The standard NLSE admits the localized solution.

However, any localized solutions in Eq. (15.9) is unknown. We will try to examine the possibility of energy localization while dealing with Eq. (15.9). First of all, we replace Eq. (15.9) by its modal representation, taking into account only two resonant NNMs with the wave numbers k_1 and k_2.

$$\chi_0 = \chi_1(\tau_2)\sin(\pi k_1 \xi) + \chi_2(\tau_2)\sin(\pi k_2 \xi) \tag{15.12}$$

After substitution of solution (15.12) into Eq. (15.9), one should use the Galerkin procedure to obtain the equations for complex amplitudes χ_1 and χ_2:

$$
\begin{aligned}
&i\frac{\partial \chi_1}{\partial \tau_2} + \delta\omega_1\chi_1 + \frac{3}{2}\sigma_{11}|\chi_1|^2\chi_1 + \sigma_{12}(2|\chi_2|^2\chi_1 + \chi_2^2\chi_1^*) = 0 \\
&i\frac{\partial \chi_2}{\partial \tau_2} + \delta\omega_2\chi_2 + \frac{3}{2}\sigma_{22}|\chi_2|^2\chi_2 + \sigma_{12}(2|\chi_1|^2\chi_2 + \chi_1^2\chi_2^*) = 0,
\end{aligned}
\tag{15.13}
$$

where $\delta\omega_j$ ($j = 1, 2$) are the frequency shifts (in the "slow" timescale τ_2) from the boundary frequency $\omega_0 = 1$:

$$
\begin{aligned}
\delta\omega_j &= \frac{\alpha^2}{2}\pi^2 k_j^2 \\
\sigma_{ij} &= \frac{3}{16}\alpha^4\pi^4 k_i^2 k_j^2, \quad (j = 1, 2).
\end{aligned}
\tag{15.14}
$$

The Hamiltonian corresponding to Eq. (15.13) can be written as follows:

$$H = \sum_{j=1}^{2}\left(\delta\omega_j|\chi_j|^2 + \frac{3}{4}\sigma_{jj}|\chi_j|^4\right) + \sigma_{12}\left(|\chi_1|^2|\chi_2|^2 + \frac{1}{2}(\chi_1^2\chi_2^{*2} + \chi_1^{*2}\chi_2^2)\right) \tag{15.15}$$

Equation (15.13) possess the "occupation number" integral

$$X = |\chi_1|^2 + |\chi_2|^2, \tag{15.16}$$

which specifies the excitation level of the system.

Because of our interest in the energy localization, we need in the introduction of new weakly interacting variables. They are the linear combination of the resonating modes:

$$\phi_1 = \frac{1}{\sqrt{2}}(\chi_1 + \chi_2); \quad \phi_2 = \frac{1}{\sqrt{2}}(\chi_1 - \chi_2) \tag{15.17}$$

Due to small difference between the modal frequencies, the functions (15.17) are identical to the "coherence domains" coordinates in the discrete nonlinear lattices.

One can notice that the introduction of coherent domains (15.17) makes Eq. (15.13) to be more complicated. However, due to the presence of the integral

(15.16), the dimensionality of the phase space of the system can be reduced. The "occupation number" X parameterizes the total excitation of the system, but the energy distribution is determined by the amplitudes of the coherent domains as well as by the phase shift between them. Actually, taking into account expression (15.16), one can describe the behavior of the coherent domains with two real functions:

$$\phi_1 = \sqrt{X}\cos\theta e^{i\delta_1}; \quad \phi_2 = \sqrt{X}\sin\theta e^{i\delta_2} \tag{15.18}$$

where the variable θ characterizes the relative amplitudes of the coherent domains and the variables δ_j—their phases.

Substituting relationships (15.17) and (15.18) into Eq. (15.13), one can obtain the equations of motion in terms of polar variables (θ, δ_j):

$$
\begin{aligned}
&\sin 2\theta \left(\frac{\partial\theta}{\partial\tau_2} - \frac{1}{2}\left(\delta\omega_1 - \delta\omega_2 - \frac{3}{4}X(\sigma_{11} - \sigma_{22}) \right) \sin\Delta \right. \\
&\left. - \frac{1}{16}X(3\sigma_{11} - 4\sigma_{12} + 3\sigma_{22})\sin 2\Delta \sin 2\theta \right) = 0 \\
&\sin 2\theta \frac{\partial\Delta}{\partial\tau_2} + \left(\left(\delta\omega_1 - \delta\omega_2 - \frac{3}{4}X(\sigma_{11} - \sigma_{22}) \right) \cos\Delta \right. \\
&\left. - \frac{1}{4}X\left((3\sigma_{11} - 2\sigma_{12} + 3\sigma_{22})\cos^2\Delta - 4\sigma_{12} \right)\sin 2\theta \right) \cos 2\theta = 0
\end{aligned}
\tag{15.19}
$$

There are two types of fundamental solutions on the phase plane of the system (see Fig. 15.4). The stationary points corresponding (in the slow time) to NNMs determine the stationary dynamics of the system. Another type of phase trajectories is significant for understanding and description of strongly non-stationary resonant dynamics. They are the LPTs. The motion along the LPT between the states $\theta = 0$ and $\theta = \pi/2$ leads to the redistribution of the energy between the different parts of the system, i.e., between the coherent domains. It was shown that an adequate temporal description of LPT can be obtained in the terms of non-smooth functions which are sawtooth functions and their derivatives in the sense of the distributions theory.

All of these peculiarities of the phase space of system (15.19) are shown in Fig. 15.4 The representative domains of the phase space are bounded by the intervals $0 < \theta < \pi/2$ and $-\pi/2 < \Delta < 3\pi/2$.

The numerical solutions of Eq. (15.19) with the initial conditions corresponding to the immovable point $(\theta = \pi/2, \Delta = \pi/2)$ for two values of the excitation X are shown in Fig. 15.5a, b.

The solutions, which are shown in Fig. 15.5, correspond to a slow redistribution of the energy between the coherent domains. If the initial conditions respect to the LPT, the energy exchange reaches the maximum of the possible amount. The period of such energy exchange may be estimated as the time of the trajectory passing:

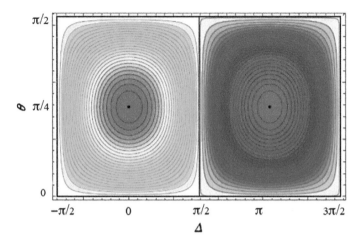

Fig. 15.4 Phase portraits of the system (15.19) for the aspect ratio $L/R = 80$ and the "occupation number" $X = 0.5$. The LPTs are marked out by the *thick black lines*

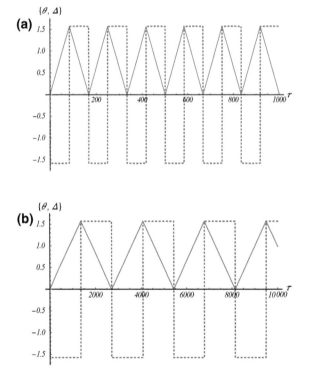

Fig. 15.5 Time evolution of the variables θ (*red solid lines*) and Δ (*blue dashed lines*) for the CNTs with different aspect ratios: **a** $L/R = 20$ and **b** $L/R = 80$. The "occupation number" $X = 0.5$. One should pay the attention that τ is the "slow" time and the real times (in the own period of the RBM) result as the value of τ divided by the square of the small parameter

$$T = \oint d\tau_2 = \oint \frac{d\theta}{d\theta/d\tau_2} \qquad (15.20)$$

The integral in Eq. (15.20) can be estimated from the first of Eq. (15.19) taking into account that $\Delta \approx \pm\pi/2$ on the LPT (see Fig. 15.4) and the transition from $\Delta = -\pi/2$ to $\Delta = \pi/2$ takes no time:

$$T = 2 \int_0^{\pi/2} \frac{d\theta}{d\theta/d\tau_2} \approx \frac{2\pi}{\delta\omega_2 - \delta\omega_1 + \frac{3}{4}(\sigma_{22} - \sigma_{11})} \qquad (15.21)$$

The variation in the period (15.21) with the aspect ratio of the CNT is shown in Fig. 15.6.

To analyze the possibility of the stationary states' instability, one should rewrite Hamiltonian (15.15) in the terms of the polar variables θ and Δ:

$$H(\theta, \Delta) = \frac{X}{2}(\delta\omega_1 + \delta\omega_2 + (\delta\omega_1 - \delta\omega_2)\cos\Delta\sin 2\theta) + 16X^2\Big(3\sigma_{11}(1 + \cos\Delta\sin 2\theta)^2$$
$$+ 3\sigma_{22}(1 - \cos\Delta\sin 2\theta)^2 + 4\sigma_{12}(3\cos^2 2\theta + \sin^2 2\theta\sin^2\Delta)\Big) \qquad (15.22)$$

The conditions of instability may be formulated as follows:

$$\frac{\partial^2 H}{\partial\theta^2} = 0 \qquad (15.23)$$

at $(\theta = \pi/4, \Delta = 0)$ or $(\theta = \pi/4, \Delta = \pi)$. The latter results in the following:

$$X = \frac{2(\delta\omega_2 - \delta\omega_1)}{3(\sigma_{11} - 2\sigma_{12})}, \quad \Delta = 0$$
$$X = \frac{2(\delta\omega_2 - \delta\omega_1)}{3(2\sigma_{12} - \sigma_{22})}, \quad \Delta = \pi \qquad (15.24)$$

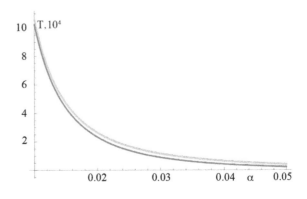

Fig. 15.6 Period of energy exchange versus inverse aspect ratio of the CNT. T is measured in the own period of the RBM. *Blue, orange,* and *green curves* correspond to the excitation $X = 0.01, 0.5,$ and 1, respectively

Taking into account the definition of $\delta\omega_j$ and relationship (15.14), one can see that no bifurcation at the positive values of "occupation number" X exists. Therefore, no localized breather-like excitations can exist in the single-walled CNT. So only the intensive energy exchange is possible in the radial breathing branch.

The analysis performed above is based on the asymptotic expansion of the equations of motion in the framework of the nonlinear Sanders-Koiter elastic thin shell theory. Only two modes in the RB branch were taken into account. Therefore, a verification of our conclusion by the independent numerical methods is needed. Figure 15.3 shows that if the aspect ratio of the CNT is large enough, more than two modes can exist under resonant conditions. So, the influence of the other part of the spectrum is very important for the estimation of the reliability of the obtained results. Our approach includes the direct numerical integration of the modal nonlinear equations of the Sanders-Koiter thin shell theory. The detailed procedure was described for the circumferential flexure modes in Strozzi et al. (2014).

Therefore, we consider the method extremely shortly. In order to carry out the numerical analysis of the CNT dynamics, a two-step procedure was used: (i) The displacement field was expanded by using a double mixed series, and then the Rayleigh–Ritz method was applied to the linearized formulation of the problem, in order to obtain an approximation of the eigenfunctions; (ii) the displacement field was re-expanded by using the linear approximated eigenfunctions, and the Lagrange equations were then considered in conjunction with the nonlinear elastic strain energy to obtain a set of nonlinear ordinary differential equations of motion.

To satisfy the boundary conditions, the displacement field was expanded into series of the Chebyshev orthogonal polynomials.

Such an expansion allows to estimate the natural frequencies (eigenvalues) and modes of vibrations (eigenvectors) under various boundary conditions. The results of performed calculation show that the eigenspectrum values are in good accordance with the estimations made within the framework of reduced Sanders-Koiter theory discussed above (see Fig. 15.3).

In the nonlinear analysis, the full expression of the dimensionless potential energy containing terms up to the fourth order (cubic nonlinearity) is considered. Using the Lagrange equations, a set of nonlinear ordinary differential equations is obtained; these equations must be supplemented with suitable initial conditions on displacements and velocities. This system of nonlinear equations was finally solved with using the implicit Runge–Kutta numerical method with suitable accuracy, precision, and number of steps. The solution of nonlinear equations with initial conditions close to the LPT shows the energy exchange process, the period of which coincides with Eq. (15.21) for the wide interval of aspect ratios and excitation amplitudes (see Fig. 15.7).

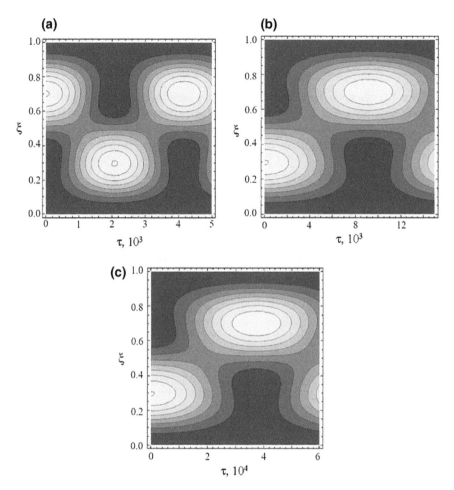

Fig. 15.7 The energy exchange in the CNTs with different aspect ratios: **a** $L/R = 20$, **b** $L/R = 40$, **c** $L/R = 80$. The initial "occupation number" $X = 0.5$. The *dark blue* and *light beige areas* correspond to low and high density of the energy, respectively

15.1.3   Circumferential Flexure Mode

As it was mentioned above, the circumferential flexure mode (CFM) is the most low-frequency optical mode with the gap frequency ~ 20 cm^{-1}. It is a typical shell mode. The main deformations of the CNT at the CF vibrations correspond to the variation in the CNT's cross section without any bending the CNT axis. In the contrast with the RB oscillations, the specific feature of the CF ones is the constancy of the contour length. This circumstance appears in the smallness of the circumferential and shear deformations.

The hypotheses of smallness of circumferential and shear deformations lead to the relations:

$$\varepsilon_\varphi = 0; \quad \varepsilon_{\xi\varphi} = 0 \tag{15.25}$$

(These hypotheses in the relationship of widely used theories of the thin shell were discussed in Kaplunov et al. 2016 in detail.) However, these assumptions do not mean that the displacements included to circumferential and shear deformations are small.

The components of displacement filed should be written as follows:

$$u(\xi, \varphi, \tau) = U_0(\xi, \tau) + U(\xi, \tau)\cos(n\varphi)$$
$$v(\xi, \varphi, \tau) = V(\xi, \tau)\sin(n\varphi) \tag{15.26}$$
$$w(\xi, \varphi, \tau) = W_0(\xi, \tau) + W(\xi, \tau)\cos(n\varphi)$$

On the contrary to the linear theory, one has to take into account axisymmetric constituent of displacement accompanying the oscillations with wave number n. Using relations (15.25), we can express:

$$V(\xi, \tau) = -\frac{1}{n}W(\xi, \tau)$$

$$U(\xi, \tau) = -\frac{\alpha}{n^2}\frac{\partial W(\xi, \tau)}{\partial \xi}$$

$$W_0(\xi, \tau) = -\frac{n^2 + 1}{4n^2}\left((n^2 - 1)^2 W(\xi, \tau) + \alpha^2\left(\frac{\partial W(\xi, \tau)}{\partial \xi}\right)^2\right) \tag{15.27}$$

$$\frac{\partial U_0(\xi, \tau)}{\partial \xi} = -\frac{n^2 + 1}{4n^2}\alpha\left(\frac{\partial W(\xi, \tau)}{\partial \xi}\right)^2$$

Omitting the calculation details, one can write the final equation of motion in terms of radial displacement W (ξ, t):

$$\frac{\partial^2 W}{\partial \tau^2} + \omega_0^2 W - \mu\frac{\partial^2 W}{\partial \xi^2} - \gamma\frac{\partial^4 W}{\partial \xi^2 \partial \tau^2} + \kappa\frac{\partial^4 W}{\partial \xi^4} + aW\frac{\partial}{\partial \tau}\left(W\frac{\partial W}{\partial \tau}\right) = 0 \tag{15.28}$$

where

$$\omega_0^2 = \beta^2\frac{n^2(n^2 - 1)^2}{12(n^2 + 1)}; \mu = \alpha^2\beta^2\frac{(n^2 - 1)(n^2 - 1 + v)}{6(n^2 + 1)}; \gamma = \frac{\alpha^2}{n^2(n^2 + 1)};$$

$$\kappa = \alpha^4\frac{12 + n^4\beta^2}{n^2(n^2 + 1)} \cong \alpha^4\frac{12}{n^2(n^2 + 1)}; a = \frac{(n^2 - 1)^4}{2n^2(n^2 + 1)} \tag{15.29}$$

and only the main order terms are taken into account (see Smirnov et al. 2016a, b for details). The frequency gap ω_0 is small because of the effective thickness smallness.

We introduce the dimensionless time, which is scaled by the gap frequency ω_0: $\tau_0 = \omega_0 \tau$. Also, we introduce a small parameter $\varepsilon = \beta$ and assume that the inverse aspect ratio α has the same order of smallness. Then, taking into account expression (15.29), one can evaluate the orders of smallness for the parameters of Eq. (15.28).

$$\frac{\mu}{\omega_0^2} \sim \varepsilon^2; \gamma \sim \varepsilon^2; \frac{\kappa}{\omega_0^2} \sim \varepsilon^2 \tag{15.30}$$

Equation (15.28) can be used to analyze the effect of various boundary conditions on the spectrum of natural oscillations of CNT. However, the solution for periodic boundary conditions can be considered as a basic one for construction of the solutions under other conditions, as clamped and free edges. The boundary conditions may be taken into account in the frameworks of the dynamic boundary layer concept (Strozzi et al. 2014, Andrianov et al. 2004). The frequency spectrum in the case of simply supported edges can be written as follows:

$$\omega^2 = \frac{\omega_0^2 + \mu \pi^2 k^2 + \kappa \pi^4 k^4}{1 + \gamma \pi^2 k^2} \tag{15.31}$$

where k is a longitudinal wave number corresponding to the number of half-waves along the CNT axis (the conditions corresponding to simply supported CNT are a particular case of periodic boundary conditions).

It is convenient to rewrite Eq. (15.28) in the terms of the complex variables:

$$\Psi = \frac{1}{\sqrt{2}} \left(\frac{\partial W}{\partial \tau} + iW \right) \tag{15.32}$$

with the inverse transformation:

$$W = -\frac{i}{\sqrt{2}} (\Psi - \Psi^*); \frac{\partial}{\partial \tau} W = \frac{1}{\sqrt{2}} (\Psi + \Psi)$$

(the asterisk denotes the complex conjugation).

Taking into account the presence of the small parameter ε, one can extract the "fast" oscillations with the frequency ω_0 (=1 in the current timescale τ_0):

$$\Psi = \varepsilon \chi_0 e^{i \tau_0}. \tag{15.33}$$

The variable χ_0 is a slowly changing function. Performing the multi-scale expansion, the equation for the amplitude of the main order in the "slow" time $\tau_2 = \varepsilon^2 \tau_0$ can be represented as follows:

$$i \frac{\partial \chi_0}{\partial \tau_2} - \frac{\mu - \omega_0^2 \gamma}{2 \omega_0^2} \frac{\partial^2 \chi_0}{\partial \xi^2} + \frac{\partial^4 \chi_0}{\partial \xi^4} - \frac{a}{2} |\chi_0|^2 \chi_0 = 0 \tag{15.34}$$

Equation (15.34) admits the plane wave solution:

$$\chi_0 = A \exp(-i(\omega\tau_2 - k\xi)), \tag{15.35}$$

where A is the amplitude. Solution (15.35) corresponds to the nonlinear normal mode with the dispersion ratio:

$$\omega = \frac{(\mu - \omega_0^2\gamma)k^2 + \kappa k^4}{2\omega_0^2} - \frac{a}{2}A^2 \tag{15.36}$$

As it can be seen, this dispersion relation is in accordance with relation (15.31).

Equation (15.34) differs from the standard Nonlinear SchrodingerEquation (NLSE) because it contains the fourth derivative and, in accordance withprevious procedures, is defined on the finite interval $\xi \ \varepsilon \ [0,1]$. In order toconsider the soliton-like solution for the CNT of the infinite length, one shouldredefine the independent variable as follows: $\xi = x/R$. In such a case $\xi \ \varepsilon \ [-\infty, \infty]$ and the structure of equation (15.34) is preserved, but the parameter α, inthe coefficients μ, γ, k should be equaled to unit. One can show that the function

$$\chi_0 = \sqrt{X_0}e^{-i\epsilon\tau_2}\sec h^2(\lambda\xi), \tag{15.37}$$

is the solution for equation (15.34) with zero conditions at $\xi \to \pm \infty$, if its parameters are determined by the expressions

$$X_0 = \frac{15\left(25\kappa - 8\gamma\mu + 8\gamma^2\Omega_0^2 - 5\sqrt{\kappa}\sqrt{25\kappa - 16\gamma\mu + 16\gamma^2\Omega_0^2}\right)}{64\gamma^2 a_1\Omega_0}$$

$$\lambda = \frac{\sqrt{5\kappa - \sqrt{\kappa}\sqrt{25\kappa - 16\gamma\mu + 16\gamma\mu^2\Omega_0^2}}}{4\sqrt{2}\sqrt{\gamma}\kappa} \tag{15.38}$$

$$\epsilon = \frac{-25\kappa + 8\gamma\mu - 8\gamma^2\Omega_0^2 + 5\sqrt{\kappa\left(25\kappa - 16\gamma\mu + 16\gamma^2\Omega_0^2\right)}}{16\gamma^2\Omega_0}$$

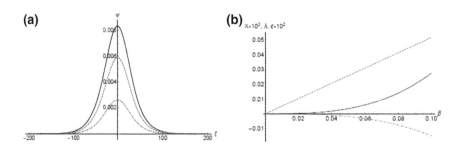

Fig. 15.8 a The breather solution of Eq. (15.34) at the various values of the CNT thickness β (0.1, 0.05, 0.01); **b** The amplitude X (solid curve), the inverse width λ(*dashed curve*) and the "frequency" ϵ (dot-dashed curve) of the solution (15.38) versus CNT thickness β

Solution (15.37) describes the breather-like excitation, which is specified by the "frequency" parameter ε. Its value should be negative, because the localizedsolutions can exist in the gap of the vibration spectrum.

Figure 15.9 show the solution (15.37) and its parameters ε, λ and X_0 at the various values of the effective thickness β.

Nonlinear Eq. (15.34) can be used for the analysis of NNMs interaction. To perform this, one should take into account that the vibration spectrum for any CNT with finite length is discrete, i.e., the longitudinal wave numbers are integers. Let us consider the sum of the resonant NNMs with the wave numbers k_1 and k_2.

$$\chi_0 = \chi_{01}(\tau_2)\sin(\pi k_1 \xi) + \chi_{02}(\tau_2)\sin(\pi k_2 \xi) \tag{15.39}$$

Using the Galerkin procedure, one can obtain the equations for complex amplitudes χ_{01} and χ_{02}:

$$
\begin{aligned}
i\frac{\partial \chi_{01}}{\partial \tau_2} + \delta\omega_1 \chi_{01} - \frac{3a}{8}\left(|\chi_{01}|^2\chi_{01} + 4|\chi_{02}|^2\chi_{01} + 2\chi_{02}^2\chi_{01}^*\right) = 0 \\
i\frac{\partial \chi_{02}}{\partial \tau_2} + \delta\omega_1 \chi_{02} - \frac{3a}{8}\left(|\chi_{02}|^2\chi_{02} + 4|\chi_{01}|^2\chi_{02} + 2\chi_{01}^2\chi_{02}^*\right) = 0
\end{aligned}
\tag{15.40}
$$

where

$$\delta\omega_j = \frac{\mu - \gamma\omega_0^2}{2\omega_0^2}\pi^2 k^2 + \frac{\kappa}{2\omega_0^2}\pi^4 k^4, \quad j = 1, 2$$

are the intervals between the modal frequencies (The frequency shift between the lowest modes ($k_1 = 1$, $k_2 = 2$) is approximately twice smaller than that for the next pair of modes ($k_2 = 2$, $k_3 = 3$)).

The Hamiltonian corresponding to Eqs. (15.40) can be written as follows:

$$
\begin{aligned}
H = \delta\omega_1|\chi_{01}|^2 + \delta\omega_2|\chi_{02}|^2 - \frac{3a}{16}\left(|\chi_{01}|^2\chi_{01} + |\chi_{02}|^2\chi_{02}\right) \\
- \frac{a}{8}\left(4|\chi_{01}|^2|\chi_{02}|^2 + \chi_{01}^2\chi_{02}^{*2} + \chi_{01}^{*2}\chi_{02}^2\right)
\end{aligned}
\tag{15.41}
$$

Equations (15.40), besides the obvious energy integral (15.41), possess another integral:

$$X = |\chi_{01}|^2 + |\chi_{02}|^2, \tag{15.42}$$

which specifies the excitation level of the system.

We introduce new "cluster" variables as the linear combinations of resonating modes with preservation the integral X:

$$\varphi_{01} = \frac{1}{\sqrt{2}}(\chi_{01} + \chi_{02}); \quad \varphi_{02} = \frac{1}{\sqrt{2}}(\chi_{01} - \chi_{02}) \tag{15.43}$$

The cluster variables describe the dynamics of some "coherent" domains of the CNT (Smirnov et al. 2014) [similar to some groups of the particles in the effective discrete one-dimensional chain (Manevitch and Smirnov 2010a, b, c: 2; Smirnov and Manevitch 2011)]. In terms of cluster variables, we can study the processes of intensive energy exchange between the coherent domains and transition to the energy localization on one of the coherent domains. Considering the energy distribution along the nanotube, one can see that the combination $\varphi_1 = 0$ and $\varphi_2 = 0$ corresponds to a predominant energy concentration in certain domain of the CNT, while the rest of which has a lower energy density. The inverse combination $(\varphi_1 = 0, \varphi_2 = 0)$ leads to the opposite energy distribution. In the rest of φ_1 and φ_2 values, the energy distributes more uniformly along the CNT axis.

Because of small difference between modal frequencies, the mentioned domains of CNT demonstrate a coherent behavior similar to beating in the system of two weakly coupled oscillators. Therefore, we can consider these regions as new large-scale elementary blocks, which can be identified as specific elements of the system. These blocks are named the "coherent domains," and they were introduced for the case of the nonlinear chain as the "effective particles" in (Manevitch and Smirnov 2010a, b, c).

The existence of integral of motion (15.42) allows to reduce the dimension of the phase space up to two variables, θ and Δ, which characterize the relationship between the amplitudes and the phase shift between the coherence domains, respectively:

$$\varphi_1 = \sqrt{X} \cos \theta e^{i\delta_1}, \varphi_1 = \sqrt{X} \sin \theta e^{i\delta_2} \qquad (15.44)$$

Substituting these expressions into Eq. (15.40), the equations of motion in the terms of "angular" variables (θ, Δ) can be obtained:

$$\sin 2\theta \left(\frac{\partial \theta}{\partial \tau_2} - \frac{1}{2}(\delta\omega_1 - \delta\omega_2) \sin \Delta + \frac{aX}{8} \sin 2\Delta \sin 2\theta \right) = 0$$

$$\sin 2\theta \frac{\partial \theta}{\partial \tau_2} - \frac{1}{2}(\delta\omega_1 - \delta\omega_2) \cos 2\theta \cos \Delta + \frac{aX}{8} \left(\cos^2 \Delta - 4 \right) \sin 4\theta = 0 \qquad (15.45)$$

Two stationary points with coordinates $(\theta = \pi/4, \Delta = 0)$ and $(\theta = \pi/4, \Delta = \pi)$ correspond to the lowest NNMs χ_{01} and χ_{02}. The trajectory, which separates the NNMs attraction domains, is the LPT.

On the LPT, Eq. (15.45) may be approximately solved in terms of non-smooth functions (Manevitch and Smirnov 2010a, b, c: 3) or integrated by the numerical methods. The numerical solutions of Eq. (15.45) with the initial conditions corresponding to the point $(\theta = 0, \Delta = \pi/2)$ for the various values of the excitation X are shown in Fig. 15.9a–f.

The evolution of $\theta(\tau_2)$ and $\Delta(\tau_2)$ for small value of X, when the system is close to the linear one, is shown in Fig. 15.9a, b. In this case, one can see the non-smooth behavior of the relative amplitudes as well as of the phase shift of the coherent

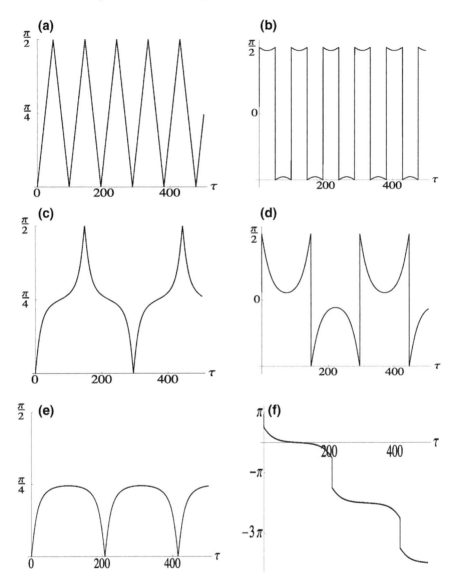

Fig. 15.9 Solutions of Eq. (15.44)—θ (*left panel*) and Δ (*right panel*) at the different occupation numbers X for the CNT (20, 0) ($\alpha = 0.08$ and $\beta = 0.009$): **a, b** $X = 0.1X_{loc}$; **c, d** $X = 0.995X_{loc}$; **e, f** $X = 1.05X_{loc}$. The initial conditions correspond to nearest vicinity of the LPT

domains φ_1 and φ_2. Such a behavior correlates with that any states belonging to the lines $\theta = 0$ or $\theta = \pi/2$ with $\Delta \neq (\pi/2 \pm m\pi)$, in fact, are some "virtual" ones, and they must be passed in the infinitesimal time. The respective solutions should be described in the terms of "non-smooth" functions (Manevitch and Smirnov 2010a, b, c: 3).

Figure 15.9c, d demonstrates that the behavior of the system in the case of large enough X does not qualitatively differ from that in Fig. 15.9a, b.

However, Fig. 15.9e, f exhibit the drastic changes under very small ($\sim 0.5\%$) changes in X. First of all, the variation range of the function θ becomes twice less. It means that the state with $\theta = \pi/2$ is inaccessible, if the initial conditions correspond to $\theta = 0$ and vice versa. The second feature in Fig. 15.9f is the unlimited growth of the function Δ. Such a behavior corresponds to the transit-time trajectories.

Hamilton function (15.41) in terms of angular variables θ and Δ can be written as follows:

$$H = \frac{X}{2}\left(\delta\omega_1 + \delta\omega_2 + (\delta\omega_1 - \delta\omega_2)\cos\Delta\sin 2\theta - \frac{aX}{16}\left(9 + \left(\cos^2\Delta - 4\right)\sin^2 2\theta\right)\right)$$

$$(15.46)$$

Using this expression, one can plot the phase portraits for various values of the parameter X. The representative domains of the phase space are bounded by the intervals $0 \leq \theta \leq \pi/2$ and $-\pi/2 \leq \Delta \leq 3\pi/2$. Two stable stationary points correspond to the normal modes χ_{01} and χ_{02}. The LPT contains two lines, $\theta = 0$ and $\theta = \pi/2$, and two fragments, which connect the pairs of points: $((\theta = 0, \Delta = \pi/2);$ $(\theta = \pi/2, \Delta = \pi/2))$ and $((\theta = 0, \Delta = \pi/2);$ $(\theta = \pi/2, \Delta = \pi/2))$. The analogous trajectory rounds the stationary state χ_{02}. The motion along the LPTs leads to the non-smooth behavior as it is shown in Fig. 15.9.

The stationary state χ_{01} becomes unstable if the parameter X exceeds some threshold, in which value X_{ins} can be calculated from the instability condition:

$$\frac{\partial^2 H}{\partial\theta^2} = 0, \left(\theta = \frac{\pi}{4}, \Delta = 0\right)$$

$$X_{ins} = \frac{8(\delta\omega_2 - \delta\omega_1)}{3a} \tag{15.47}$$

After this threshold, two new NNMs arise nearby the mode χ_{01}. These NNMs correspond to some non-uniform distribution of the energy along the CNT; however, this non-uniformity is weak. The distance between them grows, while the parameter X increases. The main feature of these states consists in that any trajectory surrounding them cannot attain the separatrix, which passes through the unstable stationary state χ_{01}. Any trajectories, which are situated in the gap between the separatrix and the LPT, preserve the possibility to pass from the vicinity of φ_1 state ($\theta = 0$) into the vicinity of φ_2 state ($\theta = \pi/2$) (see Fig. 15.9c, d). This process is accompanied with the slow intensive energy transfer from one part of the CNT to another one and vice versa.

However, the behavior of the solution of Eq. (15.45) is changed drastically if the value of X overcomes next threshold X_{loc}. Its existence results from that the new stationary points move away from the unstable state, while the LPT moves to the unstable state in the vicinity of $\theta = \pi/4$. The principal changes happen when the

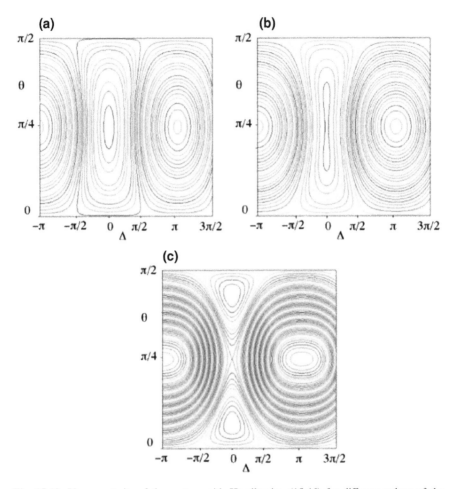

Fig. 15.10 Phase portraits of the system with Hamiltonian (15.46) for different values of the excitation: **a** $X = 0.1X_{loc}$, **b** $X = 0.995X_{loc}$, **c** $X = 1.05X_{loc}$ (see text)

LPT reaches the point ($\theta = \pi/4$, $\Delta = 0$). Then, the gap between the LPT and the separatrix disappears and the only trajectory passing from $\theta = 0$ to $\theta = \pi/2$ is the LPT. With further increasing of the parameter X, the separatrix passing through the unstable stationary points ($\theta = \pi/4$, $\Delta = 0$) and ($\theta = \pi/4$, $\Delta = 2\pi$) arises (see Fig. 15.10c). It separates the phase plane of the system into uncoupled parts, and any trajectories, which start near the $\theta = 0$, cannot attain the value $\theta = \pi/2$ (and vice versa). It means that the energy originally given in a part of CNT is kept in this part. The new LPTs enclose the stationary points, which correspond to the stable NNMs. Figure 15.10c shows the transit-time trajectories, which are in the domain between LPTs and separatrix.

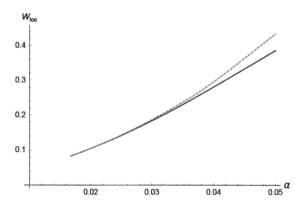

Fig. 15.11 Radial displacement W at the localization threshold for the (20, 0) zigzag CNT versus inverse aspect ratio α: *Solid black* and *dashed red curves* are the analytical predictions based on Eq. (15.49) and the threshold value estimated by the numerical method, respectively. Periodic boundary conditions are considered

The condition of the bifurcation is the degeneration of the energy of the states χ_{01}, φ_1, and φ_2, i.e.,

$$H(\theta = \pi/4, \Delta = 0) = H(\theta = 0, \Delta = \pm\pi/2) = H(\theta = \pi/2, \Delta = \pm\pi/2). \quad (15.48)$$

So, the value of the localization threshold turns out to be:

$$X_{\text{loc}} = \frac{16(\delta\omega_2 - \delta\omega_1)}{3a} \quad (15.49)$$

Figure 15.11 demonstrates the dependence of the localization threshold in the terms of the radial displacement w from the inverse aspect ratio of the CNT. It is clearly seen that the results of the asymptotic analysis are well enough for the long CNTs that agrees with the assumptions, which have been made for the derivation of Eq. (15.49).

Figure 15.12 shows the evolution of the energy distribution along the CNT axis during the MD simulation for several values of the excitation parameter X (Smirnov et al. 2016a, b). The small value X (before instability threshold) corresponds to Fig. 15.12a; the beating phenomenon is well seen. The intermediate X (before localization threshold) relates to Fig. 15.12b; the redistribution of the energy likes a drift along the CNT. Finally, Fig. 15.12c shows the capture of the energy in the initially excited area.

The phenomenon of the partial or full energy exchange is the specific one for the resonating nonlinear normal modes. The result of this interaction is determined by the relations between the parameters of nonlinearity (σ_{ij}) and the frequency differences ($\delta\omega_j$). The most important notion is the limiting phase trajectory, which corresponds to the "elementary process" associated with the energy exchange or capture.

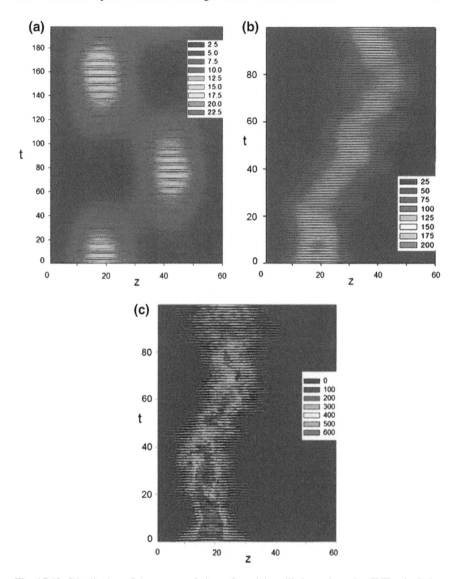

Fig. 15.12 Distribution of the energy of circumferential oscillations along the CNT axis during the MD simulation of (20, 0) CNT with aspect ratio $1/\alpha = 20$. **a** $X = 0.1X_{loc}$, **b** $X = 0.995X_{loc}$, **c** $X = 1.25X_{loc}$. The energy is measured in K, and the time, in the periods of the gap mode

Fig. 15.13 The low-frequency dispersion relations for CNT vibrations corresponding to various circumferential wave number n. *Black*, *red*, and *blue curves* correspond to $n = 0$, $n = 1$, and $n = 2$, respectively

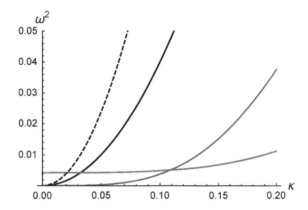

15.2   Coupling Shell- and Beam-Type Oscillations of Single-Walled Carbon Nanotubes

The resonant nonlinear interactions of the CNT vibrational modes with the identical circumferential (azimuthal) wave number have been considered in the previous section. However, the resonant interaction is possible not only for the NNMs, which belong to the same oscillation branch. In Fig. 15.13, the low-frequency part of the spectrum is shown for the CNT with relative length $L/R = 30$ (L and R are the CNT's length and its radius, respectively). One can see that the frequencies of the beam-like and circumferential oscillations are extremely close for the longitudinal wave number $\kappa = 3\pi/L$.

In the framework of the linear theory, these NNMs are independent, and therefore, no interactions between them can exist. In order to reveal the mode interaction and the effects, which can arise as its results, we need in the transition to the nonlinear vibration theory. We consider the CNT oscillations in the framework of the nonlinear Sanders-Koiter theory (Amabili 2008). We demonstrate that the effective reduction in the equations of motion in the combination with the asymptotic analysis allows to study the nonlinear mode coupling and to reveal new stationary oscillations, which are absent in the framework of the linear approach, as well as to describe the non-stationary dynamics under condition of the 1:1 resonance.

15.2.1   The Model

The dimensionless energy of the elastic CNT deformation can be written in the form (15.1).

Using expressions (15.2) for the deformations and curvatures one can vary elastic energy functional (15.1) with respect to the displacements and obtain corresponding equations of motion. When considering the linearized problem for simple-supported CNT, one can represent the displacements as follows:

$$u(\xi,\phi,t) = u(\xi,t)\cos n\phi$$
$$v(\xi,\phi,t) = v(\xi,t)\sin n\phi, \qquad (15.50)$$
$$w(\xi,\phi,t) = w(\xi,t)\cos n\phi.$$

where n is the azimuthal wave number, which can take the integer values $n = 0, 1, 2...$ Taking into account relation (15.50), one can estimate the dispersion relations for different values of n using the linearized approximation of the equations of motion mentioned above. The dispersion relations, which are shown in Fig. 15.13, have been obtained for the values of azimuthal wave number $n = 1$ (beam-like oscillations or BLO) and $n = 2$ (circumferential flexure oscillations or CFO).

However, it is obvious that the full nonlinear set of the equations of motion leads, generally speaking, to unsolvable problem. Therefore, in order to obtain any meaningful results, one should make the physically sensible hypotheses concerning to the shell displacement fields. The specific features of the considered oscillations —BLO and CFO—are the smallness of the circumferential (ε_φ) and the shear (ε_ξ) deformations, in spite of that the displacements, which are included in them may be not small. These assumptions can be written as follows:

$$\varepsilon_{\varphi\xi} = 0; \varepsilon_\varphi = 0. \qquad (15.51)$$

In order to consider the interaction of the oscillations with the different azimuthal wave number, one should rewrite displacements (15.1) as a combination of the partial components:

$$u(\xi,\phi,t) = U_0(\xi,t) + U_1(\xi,t)\cos n_1\phi + U_2(\xi,t)\cos n_2\phi,$$
$$v(\xi,\phi,t) = V_1(\xi,t)\sin n_1\phi + V_2(\xi,t)\sin n_2\phi, \qquad (15.52)$$
$$w(\xi,\phi,t) = W_0(\xi,t) + W_1(\xi,t)\cos n_1\phi + W_2(\xi,t)\cos n_2\phi,$$

where functions $(U_1(\xi, t), V_1(\xi, t), W_1(\xi, t))$, and $(U_2(\xi, t), V_2(\xi, t), W_2(\xi, t))$ describe the BLO and CFO, respectively. In spite of that we do not consider the axisymmetrical oscillations with $n = 0$, the respective adding arises in the nonlinear elastic problem.

Taking into account hypotheses (15.51) and expressions (15.2), we have a possibility to define the relationships between longitudinal, tangential, and radial components of the displacements:

$$V_i(\xi,t) = -\frac{W_i(\xi,t)}{n_i}; U_i(\xi,t) = -\frac{\alpha}{n_i^2}\frac{\partial W_i(\xi,t)}{\partial\xi}, \quad i = 1,2$$

$$W_0(\xi,t) = -\frac{1}{4}\sum_{i=1,2}\frac{1}{n_i^2}\left((n_i^2 - 1)^2 W_i^2(\xi,t) + \alpha^2\left(\frac{\partial W_i(\xi,t)}{\partial\xi}\right)^2\right)$$

$$\frac{\partial U_0(\xi,t)}{\partial\xi} = -\frac{\alpha}{4}\left(\sum_{i=1,2}\frac{n_i^2+1}{n_i^2}\left(\frac{\partial W_i(\xi,t)}{\partial\xi}\right)^2 + \frac{1}{2}\sum_{i=1,2}\left(\frac{n_i^2-1}{n_i^2}\right)^2\left(W_i(\xi,t)\frac{\partial W_i(\xi,t)}{\partial\xi}\right)^2\right)$$

$$\qquad (15.53)$$

Putting expressions (15.52) with accounting (15.53) into (15.1), one can get the equations of motion in the next form:

$$\frac{\partial^2 W_1}{\partial t^2} - \frac{\alpha^2}{2}\frac{\partial^4 W_1}{\partial \xi^2 \partial t^2} + \alpha^4 \frac{12+\beta^2}{24}\frac{\partial^4 W_1}{\partial \xi^4} - \frac{9\alpha^2}{16}\frac{\partial}{\partial \xi}\left(\frac{\partial W_1}{\partial \xi}\frac{\partial}{\partial t}\left(W_2\frac{\partial W_2}{\partial t}\right)\right) = 0$$

$$\frac{\partial^2 W_2}{\partial t^2} + \frac{3\beta^2}{5}W_2 - \frac{\alpha^2\beta^2(3+\nu)}{10}\frac{\partial^2 W_2}{\partial \xi^2} - \frac{\alpha^4}{20}\frac{\partial^4 W_2}{\partial \xi^2 \partial t^2} + \alpha^4\frac{(3+4\beta^2)}{60}\frac{\partial^2 W_2}{\partial \xi^4}$$

$$+ \frac{81}{40}W_2\frac{\partial}{\partial t}\left(W_2\frac{\partial W_2}{\partial t}\right) + \frac{9\alpha^2}{40}\left[W_2\left(\left(\frac{\partial^2 W_2}{\partial \xi \partial t}\right)^2 + 2\frac{\partial^2}{\partial t^2}\left(\frac{\partial W_1}{\partial t}\right)^2\right)\right.$$

$$\left. - \frac{\partial^2 W_2}{\partial t^2}\frac{\partial}{\partial \xi}\left(W_2\frac{\partial W_2}{\partial \xi}\right) - \frac{\partial}{\partial \xi}\left(\frac{\partial W_2}{\partial \xi}\left(\frac{\partial W_2}{\partial t}\right)^2\right)\right] = 0$$

$$(15.54)$$

The azimuthal wave numbers ($n_1 = 1$, $n_2 = 2$) have been taken in order to simplify Eq. (15.54). It is easy to see that the linearization of Eq. (15.54) leads to the uncoupled equations of the BLOs and CFOs. In such a case, we can estimate the dispersion relations for boundary conditions, corresponding to the simple-supported edges:

$$\omega_1^2 = \frac{12+\beta^2}{12(2+\alpha^2\kappa^2)}\alpha^4\kappa^4$$

$$\omega_2^2 = \frac{1}{20+\alpha^2\kappa^2}\left(12\beta^2 + 2\beta^2(3+\nu) + \frac{3+4\beta^2}{3}\alpha^4\kappa^4\right)$$

$$(15.55)$$

where the longitudinal wave number $\kappa = \pi k$ with integer value k specifies the number of half-wavelengths along the CNT axis ($k = 0, 1,...$).

Figure 15.14 shows the dispersion curves (15.55) in comparison with the exact ones, which were estimated by the solution of the full linearized system.

One can see that the correspondence is good enough in the resonant frequency range.

Fig. 15.14 Dispersion relations for BLOs and CFOs. *Blue* and *red curves* correspond to the azimuthal wave numbers $n_1 = 1$ and $n_2 = 2$, respectively. *Solid* and *dashed curves* show the exact values and the values estimated by Eq. (15.55)

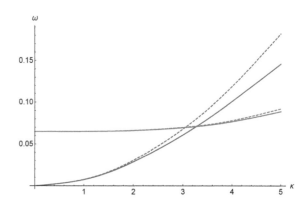

The goal of our study is to reveal the effects of the nonlinear coupling between resonantly interacting BLOs and CFOs. It is obvious that Eq. (15.54) cannot be analyzed directly. The semi-inverse method, which have been developed in (Manevitch and Smirnov 2016; Smirnov and Manevitch 2017), will be discussed in the next section.

15.2.2 Stationary Solutions

The semi-inverse method for the analysis of the complex nonlinear systems was used for the investigation of the dynamics of the discrete nonlinear lattices (Manevitch and Smirnov 2016; Smirnov and Manevitch 2017; Manevitch et al. 2016a, b), the forced oscillations of pendulum (Manevitch and Smirnov 2016; Manevitch et al. 2016a, b), and, in slightly simplified reduction, to the study of the CNT oscillations (Manevitch et al. 2016a, b: 2; Smirnov et al. 2016a, b, 2014). The basis of this method consists in the presentation of the problem in terms of the complex variables with further analysis by the multi-scale expansion method. The feature of some dynamical problems is that the small parameter, which is required for the separation of the timescales, does not present in the initial formulation of the equations. In such a case, this parameter is defined in the processes of the solution. The stationary problem in the framework of the semi-inverse method is somewhat similar to the harmonic balance method (Mikens 2010), but due to the presentation in terms of the complex variables, it turns out more simple and clear. There is an additional advantage of the developed procedure. The complex variables are the classical analogues of the creation and annihilation operators in the formalism of the secondary quantization. In several cases, it may be useful for the comparison of the classical and quantum mechanical problems.

Let us define new variables for the description of the CNT dynamics as follows:

$$\Psi_j(\xi, t) = \frac{1}{\sqrt{2}} \left(\frac{1}{\sqrt{\omega}} \frac{\partial W_j(\xi, t)}{\partial t} + i\sqrt{\omega} W_j(\xi, t) \right), \quad j = 1, 2 \qquad (15.56)$$

The inverse transformation to the initial variables W is written as follows:

$$W_j(\xi, t) = -\frac{i}{\sqrt{2\omega}} \left(\Psi_j(\xi, t) - \Psi_j^*(\xi, t) \right);$$

$$\frac{\partial W_j(\xi, t)}{\partial t} = \sqrt{\frac{\omega}{2}} \left(\Psi_j(\xi, t) + \Psi_j^*(\xi, t) \right), \quad j = 1, 2 \qquad (15.57)$$

where ω is an (yet) undefined frequency and the asterisk means the complex conjugation. First of all, we try to find the stationary solutions for Eq. (15.54), which correspond to the nonlinear normal modes (NNMs) for the coupled BLOs and CFOs. Substituting relations (15.57) into Eq. (15.54), one can find the stationary one-frequency solution in next form:

$$\Psi_j(\xi,t) = \psi_j(\xi)e^{i\omega t}, \quad j = 1,2 \tag{15.58}$$

where functions ψ_j do not depend on the time.

As a result, we obtain the equations which have to determine the profiles of the NNMs:

$$-\frac{\omega}{2}\psi_1 + \alpha^2 \frac{\omega}{4}\frac{\partial^2\psi_1}{\partial\xi^2} + \alpha^4 \frac{12+\beta^2}{48\omega}\frac{\partial^4\psi_1}{\partial\xi^4} + \frac{9\alpha^2}{32}\frac{\partial}{\partial\xi}\left(\psi_2^2\frac{\partial\psi_1^*}{\partial\xi}\right) = 0$$

$$-\frac{5\omega^2-3\beta^2}{10\omega}\psi_2 + \alpha^2 \frac{\omega^2-2\beta^2(3+v)}{40\omega}\frac{\partial^2\psi_2}{\partial\xi^2} + \alpha^4\frac{3+4\beta^2}{120\omega}\frac{\partial^4\psi_2}{\partial\xi^4} - \frac{81}{80}|\psi_2|^2\psi_2$$

$$+\frac{9\alpha^2}{80}\left(\frac{\partial}{\partial\xi}\left(\psi_2^2\frac{\partial\psi_2^*}{\partial\xi}\right) - \psi_2^*\left(\left(\frac{\partial\psi_2}{\partial\xi}\right)^2 - 4\left(\frac{\partial\psi_1}{\partial\xi}\right)^2\right)\right) = 0$$

$$\tag{15.59}$$

It is easy to see that functions

$$\psi_j = \sqrt{X_j}e^{i\kappa_j\xi} \tag{15.60}$$

are the solutions of Eq. (15.59), if the relations

$$\left(-\frac{\omega}{2} - \alpha^2\kappa^2\frac{\omega}{4} + \alpha^4\kappa^4\frac{12+\beta^2}{48\omega}\right) + \frac{9\alpha^2\kappa^2}{32}X_2 = 0$$

$$\frac{1}{6\omega}\left(36\beta^2 - 60\omega^2 + 3\alpha^2\left(2\beta^2(3+v) - \omega^2\right)\kappa^2 + \alpha^4\left(3+4\beta^2\right)\kappa^4\right) \tag{15.61}$$

$$-\frac{9}{4}\left(9 - 2\alpha^2\kappa^2\right)X_2 + 9\alpha^2\kappa^2X_1 = 0$$

are satisfied.

The relationships between the frequency and amplitudes of the BLOs and CFOs can be obtained due to the simple coupling between the modules of functions $|\psi_j|^2 = X_j$ and partial amplitudes A_j:

$$X_j = \frac{\omega}{2}A_j^2 \tag{15.62}$$

which can be seen from the definition (15.58) of the functions ψ_j.

Using relations (15.63), the following expressions for the frequencies of the stationary nonlinear coupling oscillations can be obtained:

$$\omega_1^2 = \frac{4\alpha^4\left(12+\beta^2\right)\kappa^4}{96+48\alpha^2\kappa^2-27\alpha^2\kappa^2 A_2^2}$$

$$\omega_2^2 = \frac{4\left(36\beta^2+6\alpha^2\beta^2(v+3)\kappa^2+\alpha^4\left(4\beta^2+3\right)\kappa^4\right)}{3\left(4(20+\alpha^2\kappa^2)-36\alpha^2\kappa^2 A_1^2+9(9-2\alpha^2\kappa^2)A_2^2\right)}.$$

(15.63)

One can see that expressions (15.63) coincide with dispersion relations (15.55) in the small-amplitude limit $A_1 \to 0$ and $A_2 \to 0$. One should notice that Eqs. (15.63) are valid only for resonant interaction of the modes; i.e., at $\omega_1 \sim \omega_2$, due to that we find the solution for the single-frequency motion (15.59). Figure 15.14 shows the dependences of the frequencies ω_1 and ω_2 on the CFOs amplitude with the various BLOs amplitude at the "resonant" wave number $\kappa = 3\pi$.

15.2.3 Multi-scale Expansion

To analize of the non-stationary solutions of Eq. (15.54) one needs in the method, which allows to describe the time evolution of the BLOs and CFOs in the vicinity of the resonance. This objective may be realized by the multi-scale expansion method, which consists in the separation of the timescales. Such a procedure requires a small parameter. It is obvious that the time separation is better with a smaller value of this parameter.

We consider the "carrier" time t in Eq. (15.59) as the basic "fast" time. Let us define the value "ε" as a small parameter. Its magnitude will be estimated further. We introduce the time hierarchy as follows:

$$\tau_0 = t, \tau_1 = \varepsilon\tau_0, \tau_2 = \varepsilon^2\tau_0, \ldots \tag{15.64}$$

and

$$\frac{\partial}{\partial t} = \frac{\partial}{\partial\tau_0} + \varepsilon\frac{\partial}{\partial\tau_1} + \varepsilon^2\frac{\partial}{\partial\tau_2} + \ldots \tag{15.65}$$

Keeping in mind that the oscillations of the shell should be small enough, we represent the functions ψ_j as follows:

$$\Psi_j(\xi, t) = \varepsilon\left(\psi_{j,0}(\xi, \tau_1, \tau_2, \ldots) + \varepsilon\psi_{j,1}(\xi, \tau_1, \tau_2, \ldots) + \varepsilon^2\psi_{j,2}(\xi, \tau_1, \tau_2, \ldots) + \ldots\right)e^{i\omega\tau_0} \tag{15.66}$$

Substituting expansions (15.66) into Eq. (15.54) and performing the averaging with respect to the "fast" time τ_0, we obtain the equations for the functions $\psi_{j,n}$

$(j = 1, 2, n = 0, 1,...)$ in the different orders of small parameter. Because of the respective procedure has been described above [see, in particular, (Smirnov et al. 2016a, b; Manevitch and Smirnov 2010a, b, c)], we omit the details.

ε^1:

$$-\frac{\omega}{2}\psi_{1,0} + \frac{\alpha^2\omega}{4}\frac{\partial^2\psi_{1,0}}{\partial\xi^2} + \frac{\alpha^4(12+\beta^2)}{48\omega}\frac{\partial^4\psi_{1,0}}{\partial\xi^4} = 0$$

$$\left(-\frac{\omega}{2} + \frac{3\beta^2}{10\omega}\right)\psi_{2,0} + \frac{\alpha^2}{40}\left(\omega - \frac{2\beta^2(3+\nu)}{\omega}\right)\frac{\partial^2\psi_{2,0}}{\partial\xi^2} + \frac{\alpha^4(3+\nu)}{120\omega}\frac{\partial^4\psi_{2,0}}{\partial\xi^4} = 0$$

$$(15.67)$$

One can verify easily that the first (second) of Eq. (15.67) is satisfied exactly for the functions $\psi_{j,0} \sim \exp(i\kappa\xi)$, if the frequency ω corresponds to one of dispersion relations (15.55). However, ω_1 is equal to ω_2 only approximately (see Fig. 15.15). Therefore, when the values of ω_1 and ω_2 are close, Eqs. (15.67) are satisfied with some accuracy, which is determined by the frequencies detuning. This detuning is the required small parameter, which is needed for the separation of the timescales. In such a case, the expressions in Eqs. (15.67) have to be moved into equations of another order of ε. As it will be shown further, the frequency detuning turns out to be of the second order with respect to the parameter ε.

ε^2

$$i\frac{\partial\psi_{1,0}}{\partial\tau_1} - i\frac{\alpha^2}{2}\frac{\partial}{\partial\tau_1}\frac{\partial^2\psi_{1,0}}{\partial\xi^2} - \frac{\omega}{2}\psi_{1,1} + \frac{\alpha^2\omega}{4}\frac{\partial^2\psi_{1,1}}{\partial\xi^2} + \frac{\alpha^4(12+\beta^2)}{48\omega}\frac{\partial^4\psi_{1,1}}{\partial\xi^4} = 0$$

$$i\frac{\partial\psi_{2,0}}{\partial\tau_1} - i\frac{\alpha^2}{20}\frac{\partial}{\partial\tau_1}\frac{\partial^2\psi_{2,0}}{\partial\xi^2} + \left(-\frac{\omega}{2} + \frac{3\beta^2}{10\omega}\right)\psi_{2,1} + \frac{\alpha^2}{40}\left(\omega - \frac{2\beta^2(3+\nu)}{\omega}\right)\frac{\partial^2\psi_{2,1}}{\partial\xi^2}$$

$$+ \frac{\alpha^4(3+\nu)}{120\omega}\frac{\partial^4\psi_{2,1}}{\partial\xi^4} = 0$$

$$(15.68)$$

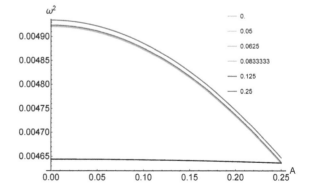

Fig. 15.15 Dependence of the frequencies ω_1 (*black*) and ω_2 (*color curves*) accordingly (15.63) on the amplitude of CFOs at different BLOs' amplitudes for the CNT with $\alpha = 1/30$. The values of BLOs' amplitudes are shown on the *right* of figure. The wave number $\kappa = 3\pi$

Spoken above about the oscillation frequencies and their detuning in Eq. (15.67) is absolutely correct with respect to the expressions in Eq. (15.68), which contain the functions of the first approximation $\psi_{j,1}$. Therefore, one should consider that these terms have to be moved from equations of this order of the small parameter. So, Eq. (15.68) can be written as follows:

$$i\frac{\partial \psi_{1,0}}{\partial \tau_1} - i\frac{\alpha^2}{2}\frac{\partial}{\partial \tau_1}\frac{\partial^2 \psi_{1,0}}{\partial \xi^2} = 0$$
$$i\frac{\partial \psi_{2,0}}{\partial \tau_1} - i\frac{\alpha^2}{20}\frac{\partial}{\partial \tau_1}\frac{\partial^2 \psi_{2,0}}{\partial \xi^2} = 0$$

(15.69)

These equations show that the functions $\psi_{1,0}$ and $\psi_{2,0}$ do not depend on the time τ_1.

The arguments which are similar to mentioned above should be applied to the equations of the next order by the small parameter. According to the series (15.66), the nonlinear terms should be included into equations of the third order of the small parameter. The linear terms from Eq. (15.67) have the same order due to our assumptions about detuning. Omitting some tedious calculations, one can write the equations of the third order as follows:

ε^3

$$i\frac{\partial \psi_{1,0}}{\partial \tau_2} - i\frac{\alpha^2}{2}\frac{\partial}{\partial \tau_2}\frac{\partial^2 \psi_{1,0}}{\partial \xi^2} - \frac{\omega}{2}\psi_{1,0} + \frac{\alpha^2 \omega}{4}\frac{\partial^2 \psi_{1,0}}{\partial \xi^2}$$
$$+ \frac{\alpha^4(12+\beta^2)}{48\omega}\frac{\partial^4 \psi_{1,0}}{\partial \xi^4} + \frac{9\alpha^2}{32}\frac{\partial}{\partial \xi}\left(\psi_{2,0}^2\frac{\partial \psi_{1,0}^*}{\partial \xi}\right) = 0$$
$$i\frac{\partial \psi_{2,0}}{\partial \tau_2} - i\frac{\alpha^2}{20}\frac{\partial}{\partial \tau_2}\frac{\partial^2 \psi_{2,0}}{\partial \xi^2} + \left(-\frac{\omega}{2}+\frac{3\beta^2}{10\omega}\right)\psi_{2,0} + \frac{\alpha^2}{40}\left(\omega - \frac{2\beta^2(3+\nu)}{\omega}\right)\frac{\partial^2 \psi_{2,0}}{\partial \xi^2}$$
$$+ \frac{\alpha^4(3+\nu)}{120\omega}\frac{\partial^4 \psi_{2,0}}{\partial \xi^4} - \frac{81}{80}|\psi_{2,0}|^2\psi_{2,0} + \frac{9\alpha^2}{80}\left(\frac{\partial}{\partial \xi}\left(\psi_{2,0}^2\frac{\partial \psi_{2,0}^*}{\partial \xi}\right)\right.$$
$$\left.- \psi_{2,0}^*\left(\left(\frac{\partial \psi_{2,0}}{\partial \xi}\right)^2 - 4\left(\frac{\partial \psi_{1,0}}{\partial \xi}\right)^2\right)\right) = 0$$

(15.70)

So, we have obtained the evolution equations for the main approximation complex functions $\psi_{j,0}$ under conditions of their resonant interaction with frequency detuning $\sim \varepsilon^2$.

One can notice that the second of Eqs. (15.70), which describes the evolution of circumferential flexure vibrations of the CNT, slightly differs from the respective equation in Smirnov et al. (2016a, b). The main reason is that the previous consideration was based on the another small parameter, which correlates with the gap between the modes' frequencies in the only branch.

One can easily see that the solutions for Eq. (15.70) are the plane wave functions $\exp(i\kappa\,\xi)$ with slowly varying amplitudes, and we will consider their behavior at various amplitudes in the next section.

15.2.4   Analysis of the Steady States Solutions and Non-stationary Dynamics

Let us introduce the new variables, which describe the evolution of the normal modes in the slow time τ_2:

$$\psi_{j,0}(\xi,\tau_2) = \varphi_{j,0}(\tau_2)e^{i\kappa\xi}, \quad j = 1,2 \tag{15.71}$$

Substituting function (15.71) into Eq. (15.70), we obtain the set of ODEs for the functions φ_j:

$$i\left(1 + \frac{\alpha^2\kappa^2}{2}\right)\frac{\partial\varphi_1}{\partial\tau_2} - \left(\frac{\omega}{2} + \frac{\omega}{4}\alpha^2\kappa^2 - \alpha^4\frac{12+\beta^2}{48\omega}\kappa^4\right)\varphi_1 + \frac{9\alpha^2\kappa^2}{32}\varphi_1^*\varphi_2^2 = 0$$

$$i\left(1 + \frac{\alpha^2\kappa^2}{20}\right)\frac{\partial\varphi_2}{\partial\tau_2} - \left(\frac{5\omega^2 - 3\beta^2}{10\omega} + \alpha^2\kappa^2\frac{\omega^2 - 2\beta^2(3+v)}{40\omega} - \alpha^4\kappa^4\frac{3+4\beta^2}{120\omega}\right)\varphi_2$$

$$- \frac{9}{80}\left(9 - 2\alpha^2\kappa^2\right)|\varphi_2|^2\varphi_2 + \frac{9\alpha^2\kappa^2}{20}\varphi_1^2\varphi_2^* = 0$$

$$\tag{15.72}$$

In order to obtain the Hamilton system, we need in the renormalization of the variables. One can show that the functions

$$\chi_1(\tau_2) = \varphi_1(\tau_2); \chi_2(\tau_2) = 4\frac{\sqrt{2+\alpha^2\kappa^2}}{\sqrt{20+\alpha^2\kappa^2}}\varphi_2(\tau_2) \tag{15.73}$$

form the set of the canonical variables for the system with the Hamilton function

$$H = a_1|\chi_1|^2 + a_2|\chi_2|^2 + b_1|\chi_2|^4 + b_2\left(\chi_1^2\chi_2^{*2} + \chi_1^{*2}\chi_2^2\right) \tag{15.74}$$

where

$$
\begin{aligned}
a_1 &= \frac{-12\omega^2(2+\alpha^2\kappa^2)+\alpha^4\kappa^4\left(12+\beta^2\right)}{24\omega(2+\alpha^2\kappa^2)} \\
a_2 &= \frac{36\beta^2 - 60\omega^2 + 3\alpha^2\kappa^2\omega^2 + \alpha^4\kappa^4\left(3+4\beta^2\right)}{6\omega(20+\alpha^2\kappa^2)} \\
b_1 &= -\frac{18(18+5\alpha^2\kappa^2 - 2\alpha^4\kappa^4)}{(20+\alpha^2\kappa^2)^2} \\
b_2 &= \frac{9\alpha^2\kappa^2}{2(20+\alpha^2\kappa^2)}
\end{aligned}
\tag{15.75}
$$

In such a case, the equations of motion can be written as:

$$
i\frac{\partial\chi_j}{\partial\tau_2} = -\frac{\partial H}{\partial\chi_j^*}
\tag{15.76}
$$

One should notice that the respective equations of motion have an additional integral, besides the integral of the energy. It is the integral, which is termed as the "occupation number". In our case, it is expressed as follows:

$$
X = |\chi_1|^2 + |\chi_2|^2
\tag{15.77}
$$

Before starting the analysis of the system with the Hamilton function (15.77), one should discuss the question: what effects do result from the interaction of the BLOs and CFOs? In order to answer this question, let us consider the elastic energy distribution corresponding to considered oscillations. Figure 15.16 shows the "surface" distributions of the energy for the BLOs (Fig. 15.16a), the CFOs (Fig. 15.16b), and their combinations (Fig. 15.16c, d).

The energy distributions along the azimuthal coordinate are shown in Fig. 15.17. These curves have been obtained by integrating the distribution shown in Fig. 15.16 along the longitudinal coordinate.

In the case of non-interacting normal modes, due to a small difference between BLOs' and CFOs' frequencies, the transitions between the energy distributions depicted in Fig. 15.16c, d are similar to the beating in the system of the weakly coupled linear oscillators. However, the nonlinear coupling of the BLOs and CFOs may result in other scenarios. As it was shown above, the nonlinear normal modes do not represent the adequate notions under conditions of 1:1 resonance. It happens due to that the considered system (it may be a nonlinear lattice or a CNT) is separated into some domains with coordinate motion of its components, while the motion in the different domains differs essentially. In such a case, the description of

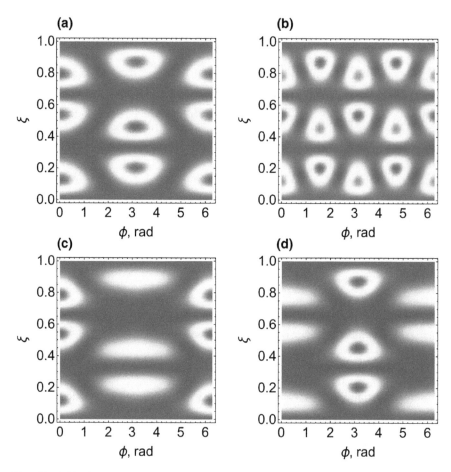

Fig. 15.16 Elastic energy distribution on the surface of the CNT. **a** Beam-like mode; **b** circumferential flexure mode; **c** "sum" of BLO and CFO; **d** "difference" BLO and CFO. $\alpha = 1/30$, $\kappa = 3\pi$; amplitudes $w = 0.05$

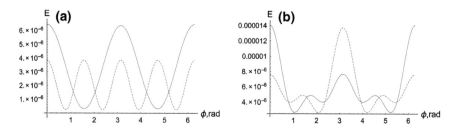

Fig. 15.17 The energy distributions along the azimuthal coordinate. **a** *Solid red* and *dashed blue curves* correspond to BLO and CFO, respectively; **b** *solid red* and *dashed blue curves* correspond to "sum" and "difference" of BLO and CFO, respectively

the system's nonstationary dynamics in terms of the domain coordinates is more appropriate (Manevitch et al. 2016a, b: 2). For the system under consideration, the domain coordinates correspond to the linear combination of the functions χ_1 and χ_2:

$$\sigma_1 = \frac{1}{\sqrt{2}}(\chi_1 + \chi_2); \sigma_2 = \frac{1}{\sqrt{2}}(\chi_1 - \chi_2) \tag{15.78}$$

One can show that the relations $\sigma_1 \ll \sigma_2$ and $\sigma_2 \ll \sigma_1$ correspond to the energy distributions, which are depicted in Fig. 15.16c, d, respectively. Transformation (15.78) preserves integrals (15.74, 15.76).

It is convenient to introduce the polar representation of the variables σ_1 and σ_2. Due to the presence of integral (15.76), one can reduce the phase space of the system for the fixed value of X:

$$\sigma_1 = \sqrt{X} \cos \theta e^{i\delta_1}; \sigma_2 = \sqrt{X} \sin \theta e^{i\delta_2} \tag{15.79}$$

It can be shown that the energy of the system turns out to be dependent on the difference of the phases $\Delta = \delta_1 - \delta_2$ only.

$$H = \frac{1}{4}X(2(a_1 + a_2 - (a_1 - a_2)\sin 2\theta \cos \Delta) \\ + X\left(b_1(1 - \cos \Delta \sin 2\theta)^2 + b_2\left(4 \cos^2 \Delta + (2 - \cos^2 \Delta)\sin 4\theta\right)\right) \tag{15.80}$$

In such a case, the phase space is two-dimensional one and its structure can be studied by the phase portrait method.

The equations of motion in the terms θ and Δ is resulted from the relations:

$$\sin 2\theta \frac{\partial \theta}{\partial \tau_2} = -\frac{\partial H}{\partial \Delta}; \sin 2\theta \frac{\partial \Delta}{\partial \tau_2} = \frac{\partial H}{\partial \theta} \tag{15.81}$$

$$\sin 2\theta \frac{\partial \theta}{\partial \tau_2} = \frac{X}{2}(Xb_1 - a_1 + a_2 - X(b_1 + b_2)\cos \Delta \sin 2\theta) \sin \Delta \sin 2\theta$$

$$\sin 2\theta \frac{\partial \Delta}{\partial \tau_2} = X\left[(Xb_1 - a_1 + a_2) - X(b_1 \cos^2 \Delta - 2b_2(2 - \cos^2 \Delta) \sin 2\theta)\right] \cos 2\theta \tag{15.82}$$

These equations also may be solved in terms of non-smooth functions (Pilipchuk 1996). We will use Eqs. (15.81 and 15.82) for the numerical verification of the trajectories, which will be found on the phase portraits at the different levels of excitations X.

Before starting the study of the phase portraits, one should notice that the excitation level $X \sim 10^{-3}$ corresponds to the amplitude of CNT oscillations $W \sim 4 \times 10^{-3}$ that is appropriate for using of the elastic thin shell theory.

The analysis of the system can be performed by the phase portrait method. Figure 15.18a shows the phase portrait in terms of variable (θ, Δ) for the small excitation level, which corresponds to the occupation number $X = 0.001$.

The topology of the phase portrait is defined by the presence of two stationary points ($\Delta = 0$, $\theta = \pi/4$ and $\Delta = \pi$, $\theta = \pi/4$), which correspond to the normal BLOs and CFOs (Eq. 15.61), respectively. Any trajectories surrounding these stationary states associate with a combination of the BLOs and CFOs. In particular, the trajectory passing through the states with $\theta = 0$ and $\theta = \pi/2$ corresponds to the "domain" variable (15.79) and separates the attraction areas of the NNMs. This trajectory is the limiting phase trajectory (LPT). One should note also that the motion along the LPT is accompanied with the transformation of the energy distribution as it is shown in Fig. 15.16c, d. However, because expression (15.80) is the nonlinear function of the occupation number X, the topology of the phase portrait can be changed while the value of X is varied.

The analysis shows that several qualitative transformations of the phase portrait occur while the occupation number X grows. Figure 15.19 shows the values of the stationary points determining topology of the phase portrait in dependence of the occupation number X.

There are two stable stationary states, which correspond to the NNMs at the small values of X. The first bifurcation happens when the stationary state ($\Delta = \pi$, $\theta = \pi/4$) losses its stability along the θ-direction:

$$\frac{\partial^2 H}{\partial \theta^2}\bigg|_{\Delta=\pi,\theta=\pi/4} = 0 \qquad (15.83)$$

This bifurcation occurs at

$$X = \frac{a_1 - a_2}{2(b_1 + b_2)}. \qquad (15.84)$$

The value of parameter X is equal to 0.001197 at the current parameters of the CNT. Simultaneously, two additional stable states are generated on the line $\Delta = \pi$ with $\theta = \pi/4$.

However, further growth of X leads to the change of the curvature along Δ-direction.

$$\frac{\partial^2 H}{\partial \Delta^2}\bigg|_{\Delta=\pi,\theta=\pi/4} = 0 \qquad (15.85)$$

In such a case, the steady state ($\Delta = \pi$, $\theta = \pi/4$) becomes stable, but two new unstable stationary points on the line $\theta = \pi/4$ arise with $\Delta = \pi$. The respective value of X can be estimated by the relation:

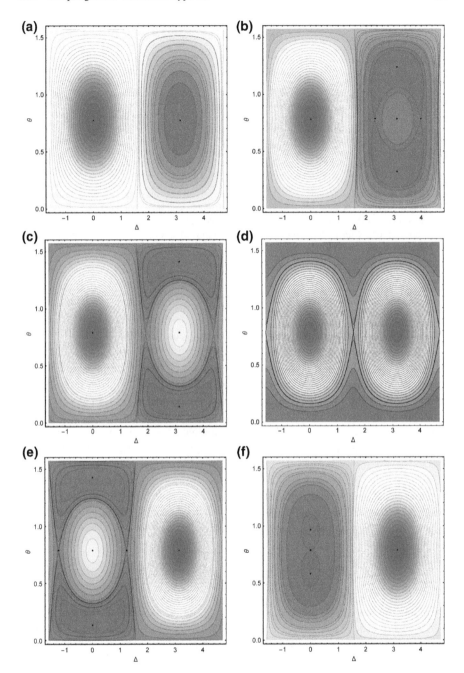

Fig. 15.18 Phase portraits in the variables (Δ, θ) at different values of occupation number X: **a** $X = 0.001$, **b** $X = 0.0015$, **c** $X = 0.00186815$, **d** $X = 0.00246$, **e** $X = 0.00361645$, **f** $X = 0.05$

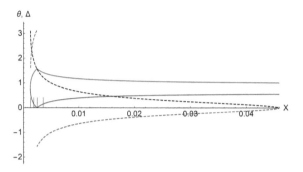

Fig. 15.19 The stationary points' coordinates (θ and Δ) versus occupation number X. *Black and blue dashed curves* show the Δ-coordinates for the unstable stationary points with $\theta = \pi/4$; *solid blue and red curves* show the θ-coordinates for the stable stationary points with $\theta = \pi/4$ and $\Delta = 0$ and π, respectively. *Thin vertical lines* represent the bifurcation values of X. CNT's parameters: $\alpha = 1/30$, $\beta = 0.08$, $v = 0.19$

$$X = \frac{a_1 - a_2}{2(b_1 - b_2)} \tag{15.86}$$

The respective value of the parameter X is equal to 0.00126.

Figure 15.18b shows the phase portrait after these bifurcations ($X = 0.0015$). One can observe that four additional stationary points, which correspond to new NNMs, appear in the quadrant ($\pi/2 < \Delta < 3\pi/2$, $0 < \theta < \pi/2$). Two separatrices passing through the unstable stationary points bound the areas with the limiting variations of the amplitudes. The stationary points with $\Delta = \pi$ correspond to the stable localization of the energy along the azimuthal angle, while the motion along the separatrix leads to fast change of the energy distribution. At the time, the trajectories, which are close to the LPT, preserve passing between the states σ_1 and σ_2. However, the areas which are surrounded by the separatrices are enlarged, while the parameter X grows. The separatrices coincide with the states (σ_1, σ_2) when their energies become to be equal to the energy at the unstable stationary points. It happens when the occupation number X satisfies the relation:

$$X = \frac{(a_1 - a_2)\left(b_1 \pm \sqrt{-b_1(b_1 + 2b_2)}\right)}{(b_1 + 2b_2)^2 + 4b_2^2} \tag{15.87}$$

Figure 15.18c shows the phase portrait when X reaches the lower value (sign + in Eq. (15.87), which is equal to 0.00187. At this moment, the LPT coincides with the separatrix and no trajectory, which couples the σ_1 and σ_2 states, occurs.

Two classes of the trajectories, which lead to the approximately equivalent energy distribution on the CNT surface, arise. The first of them contains the trajectories, which surround the stable stationary points with $\Delta = \pi$ and $\theta = \pi/4$. The motion along such trajectories is confined within some area, which is bounded by

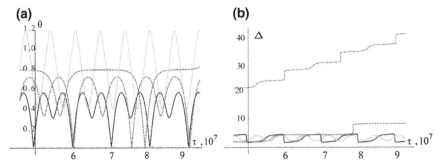

Fig. 15.20 Time evolution of the specific trajectories $\theta(\tau_2)$, $\Delta(\tau_2)$ corresponding to the different initial conditions on the phase portrait. The occupation number $X = 0.001868$ (the bifurcation value). *Black* and *red dot-dashed curves* show the non- stationary "localized" solutions corresponding to LPT and transit-time trajectory, respectively. *Green dotted curves* show some trajectory, surrounding the normal mode, and *blue dashed curves* correspond to the trajectory passing the unstable stationary point. θ and Δ are measured in rad and the time τ in units of the oscillation period $2\pi/\omega$

the LPT passing through the "domain" states σ_1 (or σ_2). The second class is represented as a set of the transit-time trajectories, the "amplitudes" θ of which can change up to $\pi/4$, but the phase Δ grows indefinitely. At the same time, the stationary states with $\Delta = \pi$ and $\theta = \pi/4$ also occur (see Fig. 15.18d). These states correspond to some stationary energy distribution on the CNT surface. Figure 15.20 show the time evolution of the trajectories on Fig. 15.18c, which correspond to the different points on the phase portrait. The black and red dashed curves describe the variables θ and Δ corresponding to the non-stationary solutions.

Next, transformation happens when the "amplitude" θ of the steady states reaches the "domain" value ($\theta = 0$ or $\theta = \pi/2$):

$$X = \frac{a_1 - a_2}{b_1} \qquad (15.88)$$

($X = 0.00246$ at the current parameters of the CNT). At this moment, the stable stationary states with $\Delta = \pi$ disappear and their analogues appear with the phase shift $\Delta = 0$.

After that, while the parameter X grows, the topology of the phase portrait in the quadrant $\pi/2 \ \Delta \ \pi/2$, $0 \ \theta \ \pi/2$ transforms similarly to the previous changes, but it develops in inverse direction. The transit-time trajectories disappear at the value X which is determined by Eq. (15.87) with the sign "−" ($X = 0.003617$). After that the possibility of the full exchange between domains σ_1 and σ_2 arises again (Fig. 15.18e).

Finally, the unstable stationary points with $\theta = \pi/4$ disappear at $X = 0.0459$, but simultaneously the stationary point ($\Delta = 0$, $\theta = \pi/4$) becomes unstable. New stationary states with $\Delta = 0$ and ($\theta < \pi/4$, $\theta > \pi/4$) appear, but their evolution is not interesting from our point of view (Fig. 15.18f).

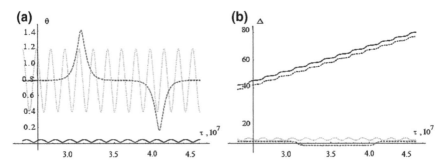

Fig. 15.21 The same as in Fig. 15.20 for the occupation number $X = 0.00246$

The character of the CNT oscillations is defined by the value of the occupation number X and the initial conditions $(\Delta(0), \theta(0))$.

Taking these values, one can integrate Eq. (15.86) numerically and then reinstate the displacement field (u, v, w).

Figures 15.20 and 15.21 present the examples of the behaviors of the angle coordinates θ and Δ corresponding to the different trajectories on the phase portrait for two values of the parameter X. Figures 15.20a, b correspond to the value of X (15.87) when the energy exchange between "domains" σ_1 and σ_2 disappears. The non-stationary behavior corresponding to the passing along the LPT and transit-time trajectories (black and red curves) couples with the variation of the angles θ, Δ with the large period $T \sim 1.5 \times 10^7$. Such a large period is explained by that these trajectories are in the vicinity of the separatrix. One should pay the attention that the phase Δ grows indefinitely for the transit-time trajectories, while it is bounded for the LPT. The green dotted curves on Fig. 15.20 correspond to the evolution of the angles θ, Δ on the trajectory, which is inside the separatrix and surrounds the circumferential normal modes $\theta = \pi/4$, $\Delta = \pi$. Finally, the blue dashed curves show the evolution of the angles variables on the separatrix. Therefore, their behavior is distinguished for others essentially. The presence of two non-identical variations of the angles corresponds to motion along the different branches of the separatrix (see Fig. 15.20d).

The essential distinction of Fig. 15.21 from Fig. 15.20 is that the variations in the angles during passing LPT as well as transit-time trajectory have the extremely small amplitudes because the value of the parameter $X = 0.00246$ is approximately equal to the bifurcation value, when the stable stationary points disappears at $\Delta = \pi$. In such a case, these stationary points are extremely close to the "domain" states σ_1 and σ_2. Therefore, the non-stationary solutions do not practically distinguished from the stationary ones. At the same time, other curves are similar to their analogues on Fig. 15.20.

The curves on Figs. 15.20 and 15.21 have been calculated by the numerical integration of Eqs. (15.81 and 15.82) for two values of the parameter X. Also, we have estimated the behavior of the variables $\theta(\tau_2)$ and $\Delta(\tau_2)$ for other values of X under various initial conditions. In all cases, the data obtained show the excellent agreement with the structure of the phase portrait.

References

Amabili, M.: Nonlinear Vibrations and Stability of Shells and Plates. Cambridge University Press, Cambridge (2008)

Andrianov, I., Awrejcewicz, J., Manevitch, L.: Asymptotical Mechanics of Thin-Walled Structures: A Handbook. Springer, Berlin (2004)

Baker, J., Graves-Morris, P.: Pade approximants. volume 13, 14 of Encyclopedia of Mathematics and Its Application. Addison-Wesley Publishing Co, Reading, Massachusetts (1981)

Chang, T.: Explicit solution of the radial breathing mode frequency of single-walled carbon nanotubes. Acta Mechanica Sinica 23, 159162 (2007)

Chen, L., Kumar, S.: Thermal transport in double-wall carbon nanotubes using heat pulse. J. App. Phys. 110, 074305 (2011)

Chico, L., Perez-Alvarez, R., Cabrillo, C.: Low-frequency phonons in carbon nanotubes: a continuum approach. Phys. Rev. B 73, 075425 (2006)

De Martino, A., Egger, R., Gogolin, A.O.: Phonon-phonon interactions and phonon damping in carbon nanotubes. Phys. Rev. B 79, 205408, 14 (2009)

Dresselhaus, M.S., Eklund, P.C.: Phonos in carbon nanotubes. Adv. Phys. 49, 705 (2000)

Eichler, A., del Álamo Ruiz, M., Plaza, J.A., Bachtold, A.: Strong coupling between mechanical modes in a nanotube resonator, Phys. Rev. Lett. 109, 025503 (2012)

Gambetta, A., Maanzoni, C., Menna, E., Meneghetti, M., Cerullo, G., Lanzani, G., Tretiak, S., Piryatinski, A., Saxena, A., Martin, R.L., Bishop, A.R.: Real-time observation of nonlinear coherent phonon dynamics in single-walled carbon nanotubes. Nat. Phys. 2, 616–620 (2006)

Ghavanloo, E., Fazelzadeh, S.: Vibration characteristics of single-walled carbon nanotubes based on an anisotropic elastic shell model including chirality effect. Appl. Math. Model. 36, 4988 (2012)

Gibson, R.F., Ayorinde, E.O., Wen, Y.-F.: Vibrations of carbon nanotubes and their composites: a review. Comp. Sci. Tech. 67, 1–28 (2007)

Greaney, P.A., Grossman, J.C.: Nanomechanical energy transfer and resonance effects in single-walled carbon nanotubes. Phys. Rev. Lett. 98, 125503 (2007)

Greaney, P.A., Lani, G., Cicero, G., Grossman, J.C.: Anomalous dissipation in single-walled carbon nanotube resonators. Nano Lett. 9, 3699–3703 (2009)

Harik, V.M.: Mechanics of carbon nanotubes: applicability of the continuum-beam models. Comp. Mater. Sci. 24, 328–342 (2002)

Harik, V.M., Gates, T.S., Nemeth, M.P., Applicability of the continuum-shell theories to the mechanics of carbon nanotubes, NASA/CR-2002-211460. ICASE Rep. No 2002-7, ICASE, NASA Langley Research Center, Hampton, Virginia (2002)

Hu, Y.G., Liew, K.M., Wang, Q., Yakobson, B.I.: Nonlocal shell model for elastic wave propogation in single- and double-walled carbon nanotubes. J. Mech. Phys. Solid. 56, 3475–3485 (2008)

Hu, Y.-G., Liew, K., Wang, Q.: Modeling of vibrations of carbon nanotubes. Procedia Eng. 31, 343–347 (2012)

Jiang, J.-W., Wang, J.-S., Li, B.: Elastic and nonlinear stiffness of graphene: a simple approach. Phys. Rev. B 81, 073405 (2010)

Kahn, D., Kim, K.W., Stroscio, M.A.: Quantized vibrational modes of nanospheres and nanotubes in the elastic continuum model, J. Appl. Phys. **89**, 5107 (2001)

Kaplunov, J., Manevitch, L.I., Smirnov, V.V.: Vibrations of an elastic cylindrical shell near the lowest cut-off frequency. Proc. Roy. Soc. Lond. A Math. Phys. Eng. Sci. **472** (2016)

Kürti, J., Żólyomi, V., Kertesz, M., Sun, G.: The geometry and the radial breathing mode of carbon nanotubes: beyond the ideal behaviour, New J. Phys. **5**(1–21), 125 (2003)

Lawler, H.M., Areshkin, D., Mintmire, J.W., White, C.T.: Radial-breathing mode frequencies for single-walled carbon nanotubes of arbitrary chirality: first-principles calculations. Phys. Rev. B **72**, 233403 (2005)

Lebedkin, S., Arnold, K., Kiowski, O., Hennrich, F., Kappes, M.M.: Raman study of individually dispersed single-walled carbon nanotubes under pressure. Phys. Rev. B **73**, 094109 (2006)

Li, B., Wang, J., Wang, L., Zhang, G.: Anomalous heat conduction and anomalous diffusion in nonlinear lattices, single walled nanotubes, and billiard gas channels. Chaos **15**, 015121 (2005)

Liew, K., Wang, Q.: Analysis of wave propagation in carbon nanotubes via elastic shell theories. J. Eng. Sci. **45**, 227 (2007)

Mahan, G.D.: Oscillations of a thin hollow cylinder: carbon nanotubes. Phys. Rev. B **65**, 235402 (2002)

Mahdavi, M., Jiang, L.Y., Sun, X.: Nonlinear vibration of a double-walled carbon nanotube embedded in a polymer matrix. Physica E Low-dimensional Syst. Nanostruct. **43**, 1813–1819 (2011)

Manevitch, L.I., Smirnov, V.V.: Resonant energy exchange in nonlinear oscillatory chains and limiting phase trajectories: from small to large system. volume 518 of CISM Courses and Lectures, Springer, New York, pp. 207–258 (2010a)

Manevitch, L.I., Smirnov, V.V.: Limiting phase trajectories and the origin of energy localization in nonlinear oscillatory chains, Phys. Rev. E **82**, 036602 (2010b)

Manevitch, L.I., Smirnov, V.V.: Limiting phase trajectories and thermodynamics of molecular chains. Dokl. Phys. **55**, 324 (2010c)

Manevitch, L.I., Smirnov, V.V.: Semi-inverse method in the nonlinear dynamics. In: Mikhlin, Y. (ed.) Proceedings of the 5th International Conference on Nonlinear Dynamics, 27–30 Sept 2016, Kharkov, Ukraine, pp. 28–37 (2016)

Manevitch, L.I., Smirnov, V.V., Romeo, F.: Non-stationary resonance dynamics of the harmonically forced pendulum. Cybern. Phys. **5**(3), 91–95 (2016a)

Manevitch, L.I., Smirnov, V.V., Romeo, F.: Stationary and non-stationary resonance dynamics of the finite chain of weaky coupled pendula. Cybern. Phys. **5**(4), 130–135 (2016b)

Maruyama, S.: A molecular dynamics simulation of heat conduction in finite length SWNTs. Phys. B **323**, 193–195 (2002)

Mickens, R.E.: Truly Nonlinear Oscillators: An Introduction to Harmonic Balance, Parameter Expansion, Iteration, and Averaging Methods. World Scientific Publishing Co. Pte. Ltd, Singapore (2010).

Pilipchuk, V.: Analytical study of vibrating systems with strong non-linearities by employing saw-tooth time transformations. J. Sound Vib. **192**(1), 43–64 (1996). doi:10.1006/jsvi.1996. 0175. URL http://www.sciencedirect.com/science/article/pii/S0022460X96901753

Pine, P., Yaish, Y.E., Adler, J.: The effect of boundary conditions on the vibrations of armchair, zigzag, and chiral single-walled carbon nanotubes. J. Appl. Phys. **110**, 124311 (2011)

Rafiee, R., Moghadam, R.M.: On the modeling of carbon nanotubes: A critical review. Compos. Part B **56**, 435–449 (2014)

Rao, A.M., Richter, E., Bandow, S., Chase, B., Eklund, P.C., Williams, K.A., Fang, S., Subbaswamy, K.R., Menon, M., Thess, A., Smalley, R.E., Dresselhaus, G., Dresselhaus, M.S.: Diameter-selective raman scattering from vibrational modes in carbon nanotubes. Science **275**, 187 (1997)

Saito, R., Takeya, T., Kimura, T., Dresselhaus, G., Dresselhaus, M.S.: Raman intensity of single-wall carbon nanotubes. Phys. Rev. B **57**, 4145–4153 (1998)

Savin, A.V., Hu, B., Kivshar, YuS: Thermal conductivity of single-walled carbon nanotubes. Phys. Rev. B **80**, 195423 (2009)

Shi, M.X., Li, Q.M., Huang, Y.: Intermittent transformation between radial breathing and flexural vibration modes in a single-walled carbon nanotube, Proc. R. Soc. A **464**, 1941 (2009a)

Shi, M.X., Li, Q.M., Huang, Y.: Internal resonance of vibrational modes in single-walled carbon nanotubes. Proc. R. Soc. A **465**, 03069 (2009b)

Silvestre, N.: On the accuracy of shell models for torsional buckling of carbon nanotubes. Eur. J. Mech. A **32**, 103 (2012)

Silvestre, N., Wang, C., Zhang, Y., Xiang, Y.: Sanders shell model for buckling of single-walled carbon nanotubes with small aspect ratio. Compos. Struct. **93**, 1683 (2011)

Smirnov, V.V., Manevitch, L.I.: Limiting phase trajectories and dynamic transitions in nonlinear periodic systems. Acoust. Phys. **57**, 271 (2011)

Smirnov, V.V., Manevitch, L.I.: Large-amplitude nonlinear normal modes of the discrete sine lattices. Phys. Rev. E **95**, 022212 (2017)

Smirnov, V.V., Shepelev, D.S., Manevitch, L.I.: Localization of low- frequency oscillations in single-walled carbon nanotubes. Phys. Rev. Lett. **113**, 135502 (2014)

Smirnov, V., Manevitch, L., Strozzi, M., Pellicano, F.: Nonlinear optical vibrations of single-walled carbon nanotubes. 1. Energy exchange and localization of low-frequency oscillations. Phys. D Nonlinear Phenom. **325**, 113–125 (2016a)

Smirnov, V., Manevitch, L., Strozzi, M., Pellicano, F.: Nonlinear optical vibrations of single-walled carbon nanotubes. 1. energy exchange and localization of low-frequency oscillations. Phys. D Nonlinear Phenom. **325**, 113–125 (2016b). doi:http://dx.doi.org/10.1016/j.physd. 2016.03.015. URL http://www.sciencedirect.com/science/article/pii/S0167278915300786

Soltani, P., Saberian, J., Bahramian, R., Farshidianfar, A.: Nonlinear free and forced vibration analysis of a single-walled carbon nanotube using shell model. IJFPS **1**, 47 (2011)

Soltani, P., Ganji, D.D., Mehdipour, I., Farshidianfar, A.: Nonlinear vibration and rippling instability for embedded carbon nanotubes. J. Mech. Sci. Tech. **26**, 985–992 (2012)

Srivastava, D., Makeev, M.A., Menon, M., Osman, M.: Computational nanomechanics and thermal transport in nanotubes and nanowires. J. Nanosci. Nanotechnol. **8**, 1–23 (2007)

Strozzi, M., Manevitch, L.I., Pellicano, F., Smirnov, V.V., Shepelev, D.S.: Low-frequency linear vibrations of single-walled carbon nanotubes: Analytical and numerical models. J. Sound Vib. **333**, 2936–2957 (2014)

Wang, C.Y., Ru, C.Q., Mioduchowski, A.: Applicability and limitations of simplified elastic shell equations for carbon nanotubes. J. Appl. Mech. **71**, 622 (2004)

Yang, W., Wang, R.Z., Song, X.M., Wang, B., Yan, H.: Pressure-induced raman-active radial breathing mode transition in single-wall carbon nanotubes. Phys. Rev. B **75**, 045425 (2007)

Ye, L.-H., Liu, B.-G., Wang, D.-S., Han, R.: Ab initio phonon dispersions of single-wall carbon nanotubes. Phys. Rev. B **69**, 235409 (2004)

Zhang, X., Hu, M., Poulikakos, D.: A low-frequency wave motion mechanism enables efficient energy transport in carbon nanotubes at high heat fluxes. Nano Lett. **12**, 3410 (2012)

Conclusions

The accepted classification of the problems of mathematical physics (in the oscillation and wave theory), first of all, draws a sharp distinction between linear and nonlinear models. Such a distinction is caused by understandable mathematical reasons including the inapplicability of the superposition principle in the nonlinear case. However, in-depth physical analysis allows us to introduce another basis for the classification of ordered oscillation problems, focusing on the difference between the stationary (or nonstationary, but non-resonance) and resonance nonstationary processes. In the latter case, the difference between the linear and nonlinear problems is not fundamental, and a specific technique, equally efficient for description in the same degree for description of both linear and nonlinear resonance nonstationary processes, has been developed. The existence of an alternative approach in the framework of linear theory seems unexpected. Really, the superposition principle allows us to find a solution describing arbitrary nonstationary oscillations as a combination of linear normal modes, which correspond to stationary processes. However, in the systems of weakly coupled oscillators, where resonant nonstationary oscillations can occur, another type of fundamental solution exists. It describes strongly modulated nonstationary oscillations characterized by the maximum possible energy exchange between the oscillators or the clusters of the oscillators (effective particles). Such solutions are referred to as limiting phase trajectories (LPTs). This book demonstrates that the LPT concept suggests a unified approach to the study of such physically different processes as strongly nonstationary energy transfer in a wide range of classical oscillatory systems and quantum dynamical systems with both constant and time-varying parameters. Furthermore, this analogy paves the way for simple mechanical simulation of complicated quantum effects. The role of the LPTs in a deeper understanding and the description of resonance highly nonstationary processes is similar to the role of NNMs in the analysis of the stationary and nonstationary non-resonance, regimes. Moreover, the presented technique can be extended to the models with many degrees of freedom. This technique is based on the statement that every periodic process, independently on the class of its smoothness, can be uniquely expressed as a smooth function of non-smooth variables τ and e or as an element of the algebra of hyperbolic numbers with the basis $(1,e)$ $(e^2=1$, but e does not equal to unit). It is very important that in

© Springer Nature Singapore Pte Ltd. 2018
L.I. Manevitch et al., *Nonstationary Resonant Dynamics of Oscillatory Chains and Nanostructures*, Foundations of Engineering Mechanics,
DOI 10.1007/978-981-10-4666-7

this case the algebraic operations and (under special conditions) differentiation and integration preserve the structure of a hyperbolic number. This property provides applicability and convenience of the corresponding transformations during the process of solving the differential equations.

Interestingly, the hyperbolic numbers, which are frequently used for a simplest illustration of the Clifford algebra, were known since the middle of the nineteenth century as abstract mathematical objects without any connection with vibration *processes. On the other hand, the elliptic complex numbers with the basis $\{1, i\}$ ($i^2 = -1$) and corresponding trigonometric functions turned out, in essence, the main tool for the description of such processes.*

Finally, we highlight differences between the NNMs and the LPTs that have motivated the introduction and the development of the LPT concept:

NNM	LPT
Represents a stationary process independent of initial conditions	Represents a nonstationary process dependent of initial conditions
Is not involved in energy exchange	Corresponds to maximum possible energy exchange between different parts of the system
Can undergo local bifurcation	Can undergo global bifurcation
Can be localized (stationary localization)	Can be localized (nonstationary localization)
Can become an attractor in an active system (synchronization of a traditional type)	Can become an attractor in an active system (synchronization of a new type)
Corresponds to a steady solution in a system subjected to external periodic excitation	Corresponds to maximum energy transfer from a source of external periodic excitation
Is described by smooth sinusoidal functions	Is described by non-smooth functions
Can be extended to systems with infinite numbers of particles	Cannot be extended to systems with infinite numbers of particles but can be considered as a prototype of a breather

CPSIA information can be obtained
at www.ICGtesting.com
Printed in the USA
LVHW01*0004080817
544132LV00001B/36/P